Basic
Mathematics

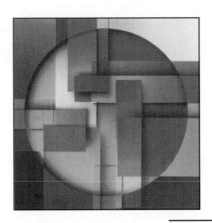

Basic Mathematics

Fourth Edition

Rosanne Proga

PWS PUBLISHING COMPANY

 An International Thomson Publishing Company

Boston • Albany • Belmont • Bonn • Cincinnati • Detroit • London
Madrid • Melbourne • Mexico City • New York • Paris • Singapore
Tokyo • Toronto • Washington

PWS PUBLISHING COMPANY

20 Park Plaza, Boston, MA 02116–4324

 I(T)P ™   International Thomson Publishing
The trademark ITP is used under license.

For more information, contact:

PWS Publishing Co.
20 Park Plaza
Boston, MA 02116

International Thomson Publishing Europe
Berkshire House 168-173
High Holborn
London WC1V 7AA
England

Thomas Nelson Australia
102 Dodds Street
South Melbourne, 3205
Victoria, Australia

Nelson Canada
1120 Birchmount Road
Scarborough, Ontario
Canada M1K 5G4

International Thomson Editores
Campos Eliseos 385, Piso 7
Col. Polanco
11560 Mexico D.F., Mexico

International Thomson Publishing GmbH
Königswinterer Strasse 418
53227 Bonn, Germany

International Thomson Publishing Asia
221 Henderson Road
#05-10 Henderson Building
Singapore 0315

International Thomson Publishing Japan
Hirakawacho Kyowa Building, 31
2-2-1 Hirakawacho
Chiyoda-ku, Tokyo 102
Japan

Library of Congress Cataloging-in-Publication Data
Proga, Rosanne.
    Basic mathematics / Rosanne Proga. — 4th ed.
        p.   cm.
    Includes index.
    ISBN 0-534-94548-1 (acid free)
    1. Mathematics.   I. Title.
QA39.2.P716   1995   94-25641
513′.14—dc20   CIP

 This book is printed on recycled, acid-free paper.

Sponsoring Editor: Susan McCulley Gay
Developmental Editor: Elizabeth A. Rogerson
Production Coordinator: Abigail M. Heim
Production Service: Editorial Services of New England, Inc.
Marketing Manager: Marianne C. P. Rutter
Manufacturing Coordinator: Lisa Flanagan
Editorial Assistant: Judith A. Mustacchia
Interior Designer: David Kelly
Interior Illustrator: G & S Typesetters, Inc.
Cover Designer: Julia Gecha
Typesetter: G & S Typesetters, Inc.
Cover Printer: New England Book Components, Inc.
Text Printer and Binder: Arcata Graphics/Hawkins

Printed and bound in the United States of America.
        95 96 97 98—9 8 7 6 5 4 3 2

▶ *To My Parents*

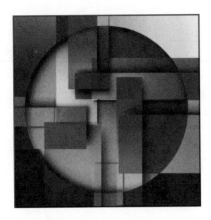

Contents

Preface

Basic Mathematics, Fourth Edition is a practical guide to the fundamentals of arithmetic for students who generally prefer to avoid mathematics. The approach to the subject matter is mature, yet easily understandable, and assumes no prior knowledge of mathematics. The text explains various mathematical techniques thoroughly, because knowing *why* a technique works is an aid in remembering *how* to use it. *Basic Mathematics* provides a wide variety of word problems to reinforce each newly learned skill and to emphasize practical applications of mathematics.

Features

▲ *Explain It in Words* is a new feature included at the end of every chapter review. The exercises require students to express answers in writing to questions involving the application of the concepts presented in each chapter. They can be used as an individual writing exercise or as a group activity providing topics for discussion. By having students explain their use of mathematical concepts verbally or in writing, their understanding of the material will be significantly enhanced.

▲ *Word Problems* reinforce each newly learned skill and emphasize practical applications of mathematics. Word problems involving real-world applications have been added pertaining to whole numbers (Chapter 1), averages (Chapter 2), fractions (Chapter 3), decimals (Chapter 4), proportions (Chapter 5), percents (Chapter 6), temperature (Chapter 9), geometry (Chapter 10), and solving linear equations (Chapter 11).

▲ A *Quick Quiz* appears at the end of each section, enabling students to check their mastery as they progress through the text.

▲ *Calculator Problems* are designated by the symbol 🖩. The appendix describes how to use a pocket calculator. This book may be used with or without a calculator.

▲ *Chapter Summaries* highlight important definitions and calculations.

▲ *Review Exercises* reinforce the material learned.

▲ A *Chapter Test* of 20 to 25 problems has been added at the end of every chapter to provide students with an additional tool to test their understanding of the material. All answers to the chapter tests are provided at the back of the text.

▲ *Examples* have been added to further clarify the discussion of order of operations (Chapter 2), translating words into mathematical expressions (Chapter 2), operations with fractions (Chapter 3), rounding off decimals (Chapter 4), percent increase and decrease (Chapter 6), converting units of measurement (Chapter 9), calculating area (Chapter 10), and solving linear inequalities (Chapter 12).

▲ *Flexibility* is a major consideration in the organization of this text. Instructors who desire an earlier treatment of algebra, for example, may cover the chapters "Introduction to Algebra" (Chapter 11) and "Graphing" (Chapter 12) immediately after "Positive and Negative Numbers" (Chapter 7).

Highlights

▲ *Estimation* is presented as an important problem-solving tool in working with whole numbers (Chapter 1) and decimals (Chapter 4).

▲ *Mixed numbers* appear throughout the discussion of fractions to present an integrated aproach to the basic operations (Chapter 3).

▲ *Algebraic concepts* are introduced early in the text to provide a thorough explanation of the method for solving proportions (Chapter 5). The techniques for setting up and solving equations are also used to solve problems involving percents (Chapter 6).

▲ *Consumer applications* are emphasized throughout the text, especially with problems involving percents and with reading graphs (Chapter 6).

▲ *Geometric concepts* are presented intuitively, using diagrams to enhance students' understanding of the ideas developed (Chapter 10).

▲ *Number systems* and their interrelationships are presented after the discussion of signed numbers (Chapter 7) and roots of numbers (Chapter 8) so students will be better prepared for further work in mathematics.

▲ *Measurement* in the English and metric systems (Chapter 9) is thoroughly discussed, as well as methods for approximating conversions between the two systems, to develop the ability to "think metric."

The Fourth Edition includes the following revisions:

▲ A procedure for determining whether a number is prime or composite has been added (Chapter 2).

▲ The discussion of least common multiple has been expanded (Chapter 2).

▲ Procedures for translating between words and decimal notation have been added (Chapter 4).

▲ The income tax problems have been updated (Chapter 6).

▲ The discussion of raising a signed number to a power has been expanded (Chapter 7).

▲ The concept of sets has been delayed until the discussion of number systems (Chapter 8).

▲ The geometric concepts of point, line, and plane have been introduced (Chapter 10).

▲ The distinction between a constant and a variable has been further clarified (Chapter 11).

Supplements

▲ **Instructor's Manual.**   In the Instructor's Manual, the answer book for even-numbered exercises has been combined with the test bank for added convenience. (The answers to odd-numbered exercises are included in the back of the text.) The test bank includes three test forms for each chapter with problems similar to those in the book, which can be used to supplement exercises in the text or as testing tools. Answers to all tests are provided.

▲ **MathQuest™ Tutorial Software** is based on the interactive AUTHORWARE℗ program. This tutorial runs on Microsoft WINDOWS and Macintosh platforms. Students are given questions to answer. If they give an incorrect answer, the program will respond with hints; then the student has another opportunity to answer the question. If the student answers incorrectly a second time, the program responds with the correct answer and a step-by-step solution. An on-screen, button-operated calculator can be pulled up for additional help. Password-protected teacher utilities are provided to allow access to their students' records. Individual student scorekeeping capabilities allow students to monitor their progress and score in each section of the exercises.

▲ The **EVR Videotape Series** teaches key topics in the text and features a professional math instructor. The videotapes are free to qualified adoptors. These tutorial tapes, produced by Educational Video Resources, can significantly improve students' comprehension and performance in algebra.

▲ **EXP Test for IBM PCs and Compatibles.**   In this computerized test bank, the instructor can select or edit existing multiple-choice, true/false, and open-ended test items; and create and add new items for a more customized test.

▲ **ExamBuilder,** a computerized test bank for the Macintosh, is a simple testing program that allows instructors to view and edit existing tests, as well as create new test items. Questions can be stored by objective and tests can be created using multiple-choice, true/false, fill-in-the-blank, essay, and/or matching formats. Questions can be scrambled to avoid duplicate testing and graphs can be generated and printed. Demonstration disks are available.

Acknowledgments

I would like to express my appreciation to the many people who contributed to the development of this book. I am especially grateful to the following reviewers for their helpful suggestions:

John E. Beris
  Stratton College

Denise Brown
  Collin County Community College

Robert R. Coombs
  Santa Rosa Junior College

Joan S. Gary
  Parkland College

John T. Gordon
Georgia State University

Robert B. Hall
Jefferson State Community College

Richard L. Harms
Pikes Peak Community College

Henry Hartshorn
Kaskaskia College

Laurence C. Huddy, Jr.
Horry-Georgetown Technical College

Mary A. Jackson
Saint Louis Community College

Larry T. Jones
Fayetteville Technical Community College

Laura Coffin Koch
University of Minnesota

Abby D. MacLean
Allegheny College

Pamela E. Matthews
Mount Hood Community College

Loretta Monk
Fayetteville Technical Community College

Barbara J. Peck
Charles County Community College

Leah C. Pierce
Crafton Hills College

Dianne Phelps
Sullivan County Community College

Alban Roques
Louisiana State University at Eunice

Dorothy Schwellenbach
Hartnell College

Terri Seiver
San Jacinto College

Cynthia Shank
Hagerstown Business College

Dorsey Templeton
Pikes Peak Community College

Jean Vincenzi
Saddleback College

Rowena M. Walker
Humphreys College

Ray M. Watson
Westark Community College

The students and colleagues I have encountered throughout my teaching career have provided me with continuous feedback that has helped make each edition of this book better than the last. I would like to acknowledge the assistance of Laurel Technical Services of Redwood City, California, for their error checking. The editors and staff at PWS Publishing have each made invaluable contributions at the various stages of the project. I would also like to express my appreciation for the encouragement provided by my family and friends and to Peter Marzuk for his support and understanding.

Rosanne Proga

Basic
Mathematics

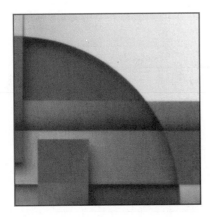

Chapter *1*

Whole Numbers

1.1
The Base Ten System

When we begin to study any new field of knowledge, we first need to learn the vocabulary that is used to communicate the fundamental ideas of the subject. Thus, in order to study mathematics, we must become familiar with the language of mathematics, whose alphabet consists not only of letters, but also of numbers.

In this chapter, we will consider the numbers most commonly encountered—the **whole numbers**. These include 0, 1, 2, 3, 4, 5, 6, 7, 8, 9, 10, 11, 12, . . . The symbol . . . indicates that it is impossible to list all the whole numbers. We express this idea by saying that the set of whole numbers is *infinite*.

The whole numbers can be represented by a diagram called a number line, shown in Figure 1.1. Notice that larger numbers are located to the right of smaller numbers. The arrow on the number line indicates that there is an infinite number of whole numbers.

Figure 1.1

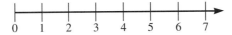

The number system we most frequently use is called the **base ten system**. In this system, ten different digits (0, 1, 2, 3, 4, 5, 6, 7, 8, 9) are used to express all numbers. Since computers basically consist of "off" and "on" switches, machine language is based on the number system of base two, which uses only the digits 0 and 1. The location of each digit, or its **place value**, determines what number that digit expresses.

▶ **DEFINITION** The *place value* of a digit is the name of the location of that digit.

The place value of each digit increases as we move from right to left. Each place value is ten times greater than the one to the right of it. The rightmost location is the ones place, the position immediately to the left of it is the tens place, the next place is the hundreds place, and so on, as shown in Table 1.1.

1

TABLE 1.1
Place Values

	Periods				
	Trillions	Billions	Millions	Thousands	Ones
	Hundred Trillions 100,000,000,000,000 / Ten Trillions 10,000,000,000,000 / Trillions 1,000,000,000,000	Hundred Billions 100,000,000,000 / Ten Billions 10,000,000,000 / Billions 1,000,000,000	Hundred Millions 100,000,000 / Ten Millions 10,000,000 / Millions 1,000,000	Hundred Thousands 100,000 / Ten Thousands 10,000 / Thousands 1,000	Hundreds 100 / Tens 10 / Ones 1

We name numbers according to the location of their digits. For example, the number one thousand has a 1 in the thousands place followed by three zeros: 1,000. After every group of three digits, moving from right to left, we place a comma to make it easier for us to read the number. Each of these three-digit groups is called a **period**. A chart indicating the place values of the first 15 locations included in the first five periods is shown in Table 1.1. The place value names the location of the digit 1 in each numerical representation given. Notice that we place a comma after the trillions, billions, millions, and thousands place. The comma is sometimes omitted after the thousands place in numbers such as 5937 that have only four digits.

The place value of each digit in the number

8,416,923

can be determined using the chart as illustrated, beginning with the ones place and moving from right to left.

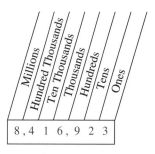

Once we have identified the place value of each digit, we can write the number in words as follows.

8,416,923 in words becomes
eight million, four hundred sixteen thousand, nine hundred twenty-three

The word "and" is not used in the names of whole numbers.

Now let us look at some examples that illustrate how place value is used to assign names to numbers.

EXAMPLE 1 For each of the following numbers, give the place value of each digit and rewrite the number in words.

(a) 325

3: hundreds place
2: tens place
5: ones place

three hundred twenty-five

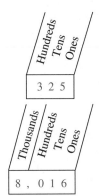

(b) 8,016

8: thousands place
0: hundreds place
1: tens place
6: ones place

eight thousand, sixteen

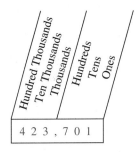

(c) 423,701

4: hundred thousands place
2: ten thousands place
3: thousands place
7: hundreds place
0: tens place
1: ones place

four hundred twenty-three thousand, seven hundred one

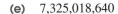

(d) 61,004,072

6: ten millions place
1: millions place
0: hundred thousands place
0: ten thousands place
4: thousands place
0: hundreds place
7: tens place
2: ones place

sixty-one million, four thousand, seventy-two

(e) 7,325,018,640

7: billions place
3: hundred millions place
2: ten millions place
5: millions place
0: hundred thousands place
1: ten thousands place
8: thousands place
6: hundreds place
4: tens place
0: ones place

seven billion, three hundred twenty-five million, eighteen thousand, six hundred forty

▲

EXAMPLE 2   Write each of the following numbers using digits.

(a)   three thousand, eight
3,008

(b)   two hundred fifty-seven thousand, seven hundred twenty-two
257,722

(c)   six million, three thousand, one
6,003,001   ▲

We can also use the concept of place value to rewrite any number in **expanded notation**. This notation will be particularly useful to us when learning how to perform the various arithmetic operations on whole numbers.

To Write a Number in Expanded Notation
1. Identify the place value of each digit.
2. Multiply each digit by its place value.
3. Add the results found in step 2 together.   ▲

We put parentheses around each quantity obtained in step 3 to indicate that multiplication is to be performed before addition.

For example, the number 736 can be written in expanded notation as follows:

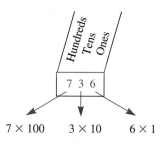

$7 \times 100 \quad 3 \times 10 \quad 6 \times 1$

$736 = 7$ hundreds $+ 3$ tens $+ 6$ ones
$= (7 \times 100) + (3 \times 10) + (6 \times 1)$

Remember that the parentheses are used to indicate that multiplication is performed before addition.

In some forms of expanded notation, the results in parentheses are simplified further as follows:

$736 = (7 \times 100) + (3 \times 10) + (6 \times 1)$
$= 700 + 30 + 6$

We will not perform this extra step when we write numbers in expanded notation.

EXAMPLE 3   Rewrite each of the following numbers in expanded notation.

(a)   85
$85 = 8$ tens $+ 5$ ones $= (8 \times 10) + (5 \times 1)$

(b) 409

409 = 4 hundreds + 0 tens + 9 ones
= (4 × 100) + (0 × 10) + (9 × 1)

(c) 350,081

350,081 = 3 hundred thousands + 5 ten thousands + 0 thousands
+ 0 hundreds + 8 tens + 1 one
= (3 × 100,000) + (5 × 10,000) + (8 × 10) + (1 × 1) ▲

Certain symbols are used in mathematics to represent the various mathematical operations and to compare numbers. The four basic arithmetic operations are indicated by the symbols in the following. The letters a and b are used to represent any two whole numbers.

Four Basic Arithmetic Operations and Their Symbols

Addition: $a + b$ means to add a and b.
Subtraction: $a - b$ means to subtract b from a.
Multiplication: $a \times b$ means to multiply a and b.
Division: $a \div b$ means to divide a by b. ▲

EXAMPLE 4 State what arithmetic operation is indicated by each of the following expressions.

(a) 2×100
Multiply 2 and 100.

(b) $8 - 3$
Subtract 3 from 8.

(c) $4 + b$
Add 4 and b.

(d) $x \div y$
Divide x by y. ▲

A list of symbols used to compare two numbers is shown below. The letters a and b are again used to represent any two whole numbers.

Symbols Used to Compare Numbers

Equality: $a = b$ means a is equal to b.
Inequality: $a \neq b$ means a is not equal to b.
Less than: $a < b$ means a is less than b.
Greater than: $a > b$ means a is greater than b.
Approximate equality: $a \cong b$ means a is approximately equal to b. ▲

Notice that in using the $>$ or $<$ symbol, the wide end of the symbol faces the larger number. Also note the respective locations of a and b on the following number lines.

$a < b$

a is **less than** b indicates that a is to the left of b on a number line.

$a > b$

a is **greater than** b indicates that a is to the right of b on a number line.

EXAMPLE 5 Compare each of the following pairs of numbers represented on this number line. Use the $=, <$, or $>$ sign to indicate the relationship between them.

(a) 3 ? 6

Since 3 is to the left of 6 on the number line, 3 is less than 6.

$3 < 6$

(b) b ? 5

Since b is at the same location as 5 on the number line, b is equal to 5.

$b = 5$

(c) 4 ? a

Since 4 is to the right of a, 4 is greater than a.

$4 > a$

(d) a ? b

Since a is to the left of b, a is less than b.

$a < b$ ▲

EXAMPLE 6 For each of the following expressions, use the $=$ or $\neq$ sign to express the appropriate relationship between the two quantities. For quantities that are unequal, use the greater than ($>$) or less than ($<$) sign to further define the relationship.

(a) 6 ? 16

$6 \neq 16$ 6 is not equal to 16.

$6 < 16$ 6 is less than 16.

(b) $5{,}028 \; ? \; (5 \times 1{,}000) + (2 \times 10) + (8 \times 1)$

$5{,}028 = (5 \times 1{,}000) + (2 \times 10) + (8 \times 1)$

Both sides are equal since the right-hand side is simply 5,028 written in expanded notation.

(c)   82 ? 28
  82 ≠ 28   82 is not equal to 28.
  82 > 28   82 is greater than 28.

(d) 9×10 ? 10×9
  $9 \times 10 = 10 \times 9$
  Since both sides express the number 90, they are equal.

(e) $(6 \times 100) + (2 \times 10)$? $(6 \times 100) + (2 \times 1)$
  6 hundreds + 2 tens ≠ 6 hundreds + 2 ones
  600 + 20 > 600 + 2
  620 > 602
  By rewriting in words, it is apparent that both sides are unequal. Simplifying, we can see that 620 > 602.

(f)   If $a > b$, b ? a.
  $b \neq a$. If a is greater than b, they cannot be equal.
  Draw a number line to show the relationship between a and b. a is greater than b means that a is to the right of b on a number line. Therefore, b is to the left of a, meaning that b is less than a.

$b < a$

$\begin{array}{ccc} + & & + \\ b & & a \end{array} \longrightarrow$

▲

Quick Quiz

		Answers
1.	Write 3,061 in words.	**1.** three thousand, sixty-one
2.	Write 405,790 in expanded notation.	**2.** $(4 \times 100{,}000) + (5 \times 1{,}000)$ $+ (7 \times 100) + (9 \times 10)$
3.	Write sixty thousand, five hundred twelve using digits.	**3.** 60,512

1.1 Exercises

Indicate which of the following statements are true and which are false. For those that are false, change the italicized or underlined expression to make the statement true.

1. The set of whole numbers *does not* include the number 0.

2. The set of whole numbers is *finite*.

3. The number seven thousand, eight is written <u>7,080</u>.

4. Our number system, which uses ten digits, is called the *number ten system*.

For each of the following numbers, give the place value of the digit in parentheses and rewrite the number in words.

5. 58 (5)
6. 29 (9)
7. 806 (8)
8. 321 (2)
9. 4,030 (4)
10. 7,514 (5)
11. 21,009 (2)
12. 39,456 (9)
13. 500,001 (5)
14. 407,200 (7)
15. 1,496,628 (4)
16. 7,820,413 (2)
17. 38,002,540 (8)
18. 943,627,501 (4)
19. 715,151,515 (7)
20. 1,601,070,402 (7)
21. 89,641,372,055 (9)
22. 526,828,424,121 (5)
23. 7,000,000,486,532 (7)
24. 603,030,222,111,300 (6)

Write each of the following numbers using digits.

25. forty-seven
26. ninety-nine
27. six hundred two
28. five hundred seventy-eight
29. one thousand, sixty-six
30. three thousand, four hundred seventeen
31. eight thousand, eight
32. seventy-two thousand, four hundred fifty-nine
33. fifty thousand, six hundred
34. two hundred thousand, two
35. nine hundred forty-four thousand, twenty-six
36. three hundred seven thousand, six hundred sixty-seven
37. seven million, eight
38. three million, nine hundred two thousand, seven hundred fifty-three
39. eleven million, sixty-four thousand, two hundred eighteen
40. sixty-seven million, four hundred seventy thousand, seven
41. two billion, fifty-three million, one hundred three thousand, four hundred eighty-one
42. seven billion, five hundred five million, twenty thousand, four
43. fifty billion, nine million, three hundred thousand, five hundred
44. six hundred eighty billion, two hundred sixty-six million, forty thousand, twenty-eight
45. five trillion, seven hundred five billion, six million, four hundred seventy-four thousand, three hundred two
46. nine hundred trillion, nine hundred thousand, ninety

Write each of the following numbers using expanded notation.

47. 33
48. 51
49. 605
50. 777
51. 4,002
52. 8,115
53. 10,040
54. 63,874
55. 638,518
56. 500,005
57. 30,041,003
58. 48,673,851
59. 999,999,999
60. 348,627,841
61. 5,038,602,124
62. 88,777,666,555
63. 300,002,404,100
64. 5,937,628,416,572
65. 44,290,053,777,119
66. 600,000,007,000,000

State what is meant by each of the following mathematical expressions.

67. 7×3
68. $4 \neq 9$
69. $18 \div 2$
70. $2 < 7$
71. $a = 10$
72. $9 - x$
73. $w > t$
74. $c + d$

Compare each of the following pairs of numbers represented on this number line. Use the
=, >, or < sign to indicate the relationship between them.

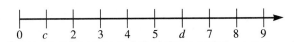

75.	0 ? 3	76.	8 ? 2	77.	d ? 7	78.	4 ? c
79.	8 ? c	80.	c ? 1	81.	6 ? d	82.	d ? c

For each of the following expressions, use the = or ≠ sign to express the appropriate
relationship between the two quantities. For quantities that are unequal, use the greater
than (>) or less than (<) sign to further define the relationship.

83.	45 ? 54	84.	860 ? 8,060
85.	$(3 \times 10) + (2 \times 1)$? 32	86.	637 ? $(6 \times 100) + (3 \times 10) + (7 \times 1)$
87.	4×100 ? 100×4	88.	0×10 ? 0×100
89.	1×10 ? 1×100	90.	$(5 \times 1,000) + (6 \times 100)$? $(5 \times 1,000) + (6 \times 10) + (9 \times 1)$
91.	If $a = b$, then b ? a.	92.	If $a < b < c$, then a ? c.

Answer the following questions.

93. The Pacific Ocean covers sixty-four million square miles. How is this number expressed using digits?

94. The area of the Caspian Sea is 152,239 square miles. What is the place value of 5 in this number?

95. A light-year is the distance light travels in one year. It is equivalent to 5,880,000,000,000 miles. How is this number written in words?

96. The Grand Canyon National Park covers an area of 1,218,375 acres. How is this number written in words?

97. The planet Pluto is 3,675,270,000 miles from the sun. What is the place value of the digit 6 in this number?

98. Twenty years ago, the U.S. gross national product was approximately four trillion, fourteen billion dollars. How is this number expressed using digits?

99. The population of the New York metropolitan area is projected to be 14,648,000 at the turn of the century. How is this number written in words?

100. During a 10-year period, the national debt increased by $1,949,730,000,000. How is this number written in words?

1.2
Addition of Whole Numbers

We will now consider how to perform the four basic arithmetic operations—addition, subtraction, multiplication, and division—on whole numbers. Let us begin our discussion by defining some of the terminology and properties that apply to addition.

▶ **DEFINITION** An *equation* is a mathematical expression that states that two quantities are equal.

Consider the following example. If there are five apples on one table and three apples on another, the total number of apples on both tables is eight. This can be represented by the equation

$$3 + 5 = 8$$

▶ **DEFINITION** To add two numbers *a* and *b*, we can write the equation $a + b = c$, where the numbers *a* and *b* are called *addends* and the result, *c*, is called the *sum*.

For example, in the equation $1 + 2 = 3$, the numbers 1 and 2 are the addends, and the number 3 is the sum.

$$1 + 2 = 3$$
$$\nearrow \quad \uparrow \quad \nwarrow$$
addend addend sum

We can also write this same problem vertically.

$$
\begin{array}{r}
1 \leftarrow \text{addend} \\
+\ 2 \leftarrow \text{addend} \\
\hline
3 \leftarrow \text{sum}
\end{array}
$$

For small numbers, we can determine the result of an addition operation by simply counting the total number of items, using such devices as our fingers or tally marks. The Chinese invented an instrument called the abacus, which consists of sliding beads on parallel wires that enable them to add even very large numbers very rapidly by counting. However, since we use addition so frequently, certain addition facts should be memorized, thus allowing us to build speed and accuracy in performing our calculations.

Table 1.2 gives important addition facts that must be memorized. The numbers to be added, or addends, are chosen from the numbers in the top row and from those in the

TABLE 1.2
Addition Table

+	0	1	2	3	4	⑤	6	7	8	9
0	0	1	2	3	4	5	6	7	8	9
1	1	2	3	4	5	6	7	8	9	10
2	2	3	4	5	6	7	8	9	10	11
3	3	4	5	6	7	8	9	10	11	12
4	4	5	6	7	8	9	10	11	12	13
5	5	6	7	8	9	10	11	12	13	14
6	6	7	8	9	10	11	12	13	14	15
⑦	7	8	9	10	11	⑫	13	14	15	16
8	8	9	10	11	12	13	14	15	16	17
9	9	10	11	12	13	14	15	16	17	18

leftmost column. The sum is obtained by moving your right index finger down the column containing the addend in the top row and moving your left index finger across the row containing the other addend in the leftmost column. The block where your fingers meet, which is the intersection of this row and column, contains the sum.

For example, to add $5 + 7$ we find the number 5 in the top row and the number 7 in the leftmost column. Move one finger down the column containing 5 and another across the row containing 7. Your two fingers should meet at the block containing 12. Thus,

$$5 + 7 = 12$$

Some sums that may be calculated using Table 1.2 are:

$$3 + 4 = 7 \qquad 7 + 9 = 16 \qquad 6 + 6 = 12 \qquad 0 + 5 = 5$$

Recall from Table 1.2 that whenever any number is added to 0, the resulting sum is the same as that number.

Additive Identity Property

The number 0 is said to be the *identity element* of addition, since for any number *a*, $a + 0 = a$ and $0 + a = a$. ▲

For example, $0 + 5 = 5$.

It is also important to note that when adding any two numbers, the same sum is obtained regardless of the *order* of the addends. For example, since

$$3 + 8 = 11 \qquad \text{and} \qquad 8 + 3 = 11$$
$$\text{then} \qquad 3 + 8 = 8 + 3$$

This property is called the **commutative property** of addition and may be stated as follows:

Commutative Property of Addition

For any numbers *a* and *b*, $a + b = b + a$. ▲

According to the *commutative property*, we can change the *order* of the numbers being added without affecting the result.

For example, $1 + 5 = 5 + 1$.

In order to add more than two numbers, we first add any two together, and then we add that result to the remaining numbers to obtain the total sum. For example, to add $5 + 3 + 2$, we first add $5 + 3$ to obtain 8, and then we add $8 + 2$ to obtain a total sum of 10. We can indicate this process by using parentheses, which group together the numbers to be added together first:

$$5 + 3 + 2 = (5 + 3) + 2 = 8 + 2 = 10$$

Notice that we obtain the same result if we first add $3 + 2$ to obtain 5 and then add

5 + 5 to obtain 10. By using parentheses to group the numbers that are added first, we obtain:

$$5 + 3 + 2 = 5 + (3 + 2) = 5 + 5 = 10$$

This example illustrates the **associative property** of addition, which may be stated as follows:

Associative Property of Addition

For any numbers a, b, and c, $(a + b) + c = a + (b + c)$.   ▲

According to the *associative property*, we can change the way in which we *group* the numbers being added without affecting the result.

For example, $(4 + 2) + 7 = 4 + (2 + 7)$.

EXAMPLE 1   State whether the identity, commutative, or associative property of addition is illustrated by each of the following equations.

(a) $8 + 7 = 7 + 8$

Commutative property of addition, since the order is changed.

(b) $2 + (6 + 4) = (2 + 6) + 4$

Associative property of addition, since 6 and 4 are grouped together on the left, and 2 and 6 are grouped together on the right.

(c) $7 + 0 = 7$

Additive identity property.

(d) $5 + (1 + 3) = 5 + (3 + 1)$

Commutative property of addition, since the order of 1 and 3 is reversed. Note that this is not the associative property since the same numbers are grouped together on both sides of the equation.

(e) $(4 + 1) + 3 = 4 + (3 + 1)$

Both the commutative and associative properties of addition, since the order and the grouping are changed.   ▲

In order to add numbers greater than nine, it is easiest to write the problem vertically, lining up the digits with the same place values. We then add the digits in the individual columns separately, moving from right to left and keeping in mind the place value of the digits in each column.

EXAMPLE 2   Add 82 + 56.

Solution

addend	$82 =$	8 tens + 2 ones
addend	$+ 56 =$	5 tens + 6 ones
sum		13 tens + 8 ones

Write the problem vertically and indicate the place value of each digit.
Add the numbers in the ones column:
$2 + 6 = 8$.
Add the numbers in the tens column:
$8 + 5 = 13$.

When the sum of the numbers in any place is greater than nine, the result can be rewritten using the next-higher place value. Thus, in Example 2, 13 tens = 1 hundred + 3 tens, so the final answer is:

1 hundred + 3 tens + 8 ones = 138.   ▲

EXAMPLE 3   Add 385 + 267.

Solution

addend	$385 =$	3 hundreds + 8 tens + 5 ones
addend	$+ 267 =$	2 hundreds + 6 tens + 7 ones
sum		5 hundreds + 14 tens + 12 ones

Since 12 ones = 1 ten + 2 ones, and 14 tens = 1 hundred + 4 tens, the final sum is:

5 hundreds + 1 hundred + 4 tens + 1 ten + 2 ones
= 6 hundreds + 5 tens + 2 ones = 652.   ▲

A shortcut for adding numbers in this fashion is called *carrying*. The procedure involves the three steps listed below.

To Add Whole Numbers

1. Write the problem vertically and line up the numbers with the same place value.

2. Add the numbers in each column separately, moving from right to left.

3. If the sum of any column is greater than nine, put down the digit that is in the ones place and add, or **carry**, the other digit to the next column immediately to the left.   ▲

If we use carrying to add the numbers in Example 3, we obtain the following result.

$$
\begin{array}{r}
1 \\
38\,|5 \\
+\,26\,|7 \\
\hline
2
\end{array}
$$

$5 + 7 = 12$. Put down 2. Carry 1 to the tens place.

$$
\begin{array}{r}
1\,|1 \\
3\,|8\,|5 \\
+\,2\,|6\,|7 \\
\hline
5\,|2
\end{array}
$$

$1 + 8 + 6 = 15$. Put down 5. Carry 1 to the hundreds place.

$$\begin{array}{r} \overset{1}{} \\ 3\,|\,85 \\ +\ 2\,|\,67 \\ \hline 6\,|\,52 \end{array}$$

1 + 3 + 2 = 6. Put down 6.

The sum is 652.

Let us now look at some more examples of addition that illustrate carrying.

EXAMPLE 4 Add 39 + 28 + 83.

Solution

$$\begin{array}{r} 12 \\ 39 \\ 28 \\ +\ 83 \\ \hline 150 \end{array}$$

9 + 8 + 3 = 20. Put down 0. Carry 2 to the tens place.

2 + 3 + 2 + 8 = 15. Put down 5. Carry 1 to the hundreds place. The sum is 150. ▲

EXAMPLE 5 Add 507 + 1,036.

Solution

$$\begin{array}{r} 1 \\ 507 \\ +\ 1{,}036 \\ \hline 1{,}543 \end{array}$$

7 + 6 = 13. Put down 3. Carry 1.
1 + 0 + 3 = 4.
5 + 0 = 5.
The sum is 1,543. ▲

Notice that because of the commutative property, we can change the order of the numbers added together to check our answer.

$$\begin{array}{r} 1 \\ 1{,}036 \\ +\ 507 \\ \hline 1{,}543 \end{array}$$

EXAMPLE 6 Add 3,694,862 + 8,581,077. Check your answer by reversing the addends.

Solution

$$\begin{array}{r} 1\ 1\ \ \ 1 \\ 3{,}694{,}862 \\ +\ 8{,}581{,}077 \\ \hline 12{,}275{,}939 \end{array}$$

Check:

$$\begin{array}{r} 1\ 1\ \ \ 1 \\ 8{,}581{,}077 \\ +\ 3{,}694{,}862 \\ \hline 12{,}275{,}939 \end{array}$$

▲

Estimating Answers in Addition

When adding a column of numbers, it is important to be able to predict roughly the answer so that you can recognize an unreasonable answer caused by a careless mistake. You can do this by approximating each of the addends and mentally adding them up. This process is called *estimation*.

It is very important to estimate answers whenever you use a pocket calculator to do a calculation. Since most mistakes in using a calculator are caused by pressing the wrong buttons, you should always estimate the answer before using a calculator, and then compare the calculator's answer to your estimate.

EXAMPLE 7 Estimate 293 + 512. Then determine the exact answer.

Solution 293 is between 200 and 300 on the number line, but much closer to 300. Therefore 293 ≅ 300. (The symbol ≅ means "is approximately equal to.") Similarly, 512 ≅ 500. Adding 300 + 500 provides us with an estimate of 800. Now we will determine the exact answer:

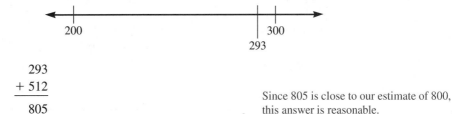

$$
\begin{array}{r}
293 \\
+\ 512 \\
\hline
805
\end{array}
$$

Since 805 is close to our estimate of 800, this answer is reasonable. ▲

EXAMPLE 8 Estimate 597 + 83 + 681 + 3,469. Then determine the exact answer.

Solution Approximate each addend:

$$
\begin{array}{r}
597 \cong 600 \\
83 \cong 100 \\
681 \cong 700 \\
3,469 \cong 3,500 \\
\hline
\end{array}
$$

4,900 is our estimate.

Adding these approximations provides us with an estimate of 4,900. Now we will determine the exact answer:

$$
\begin{array}{r}
1\ 32 \\
597 \\
83 \\
681 \\
+\ 3,469 \\
\hline
4,830
\end{array}
$$

Since 4,830 is close to our estimate of 4,900, this answer is reasonable. ▲

Solving Word Problems Using Addition

Addition is probably the most frequently used arithmetic operation. It is used to solve a wide variety of word problems encountered in daily life. Certain words appearing in word problems such as *add*, *sum*, *plus*, *total*, *increased by*, and *more than*, are a clue that addition may be used to solve the problem. Table 1.3 illustrates how words can be translated into arithmetic expressions using addition.

TABLE 1.3
Expressing Addition

Using Words	Using Symbols
The sum of 5 and 6 is 11.	$5 + 6 = 11$
The total of 6, 9, 3, and 5.	$6 + 9 + 3 + 5$
The number 8 increased by 7.	$8 + 7$
3 more than x is 14.	$x + 3 = 14$
a plus b equals c.	$a + b = c$

EXAMPLE 9 On a certain day the following transactions were processed at a Boston bank: 106 cash deposits, 392 check deposits, 281 cash withdrawals, 434 checks cashed, 27 money order purchases, 15 currency exchanges, and 213 other miscellaneous transactions. What was the total number of transactions conducted on that day?

Solution Add $106 + 392 + 281 + 434 + 27 + 15 + 213$. Let us first estimate the answer by approximating each addend and adding the approximations. Since each number is between 0 and 500, we will look at intervals of 100 to determine each estimate.

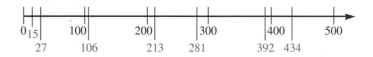

$106 \cong 100$		106 is closer to 100 than to 200.
$392 \cong 400$		392 is closer to 400 than to 300.
$281 \cong 300$		281 is closer to 300 than to 200.
$434 \cong 400$		434 is closer to 400 than to 500.
$27 \cong 0$		27 is closer to 0 than to 100.
$15 \cong 0$		15 is closer to 0 than to 100.
$213 \cong 200$		213 is closer to 200 than to 300.

1,400 is our estimate.

Now let us determine the exact answer.

$$
\begin{array}{r}
106 \\
392 \\
281 \\
434 \\
27 \\
15 \\
+ \quad 213 \\
\hline
1,468
\end{array}
$$

Since 1,468 is close to our estimate of 1,400, this answer is reasonable.

Thus, 1,468 transactions were processed. ▲

Quick Quiz

Add each of the following groups of numbers.

1. $37 + 89$
2. $456 + 797$
3. $3,682 + 56 + 876$

Answers

1. 126
2. 1,253
3. 4,614

1.2 Exercises

The problems marked with the symbol ▦ are calculator problems. The procedure for using a calculator is discussed in the Appendix.

Indicate which of the following statements are true and which are false. For those that are false, change the italicized expression to make the statement true.

1. In the equation $5 + 7 = 12$, the number 12 is called the *sum*.

2. According to the *associative* property of addition, the order of the addends can be changed without changing the sum.

3. When adding numbers that have more than one digit, we add the numbers in the *leftmost* column first.

4. We perform a carry when the sum of the numbers in any column is *greater* than 9.

5. When adding two whole numbers, each addend is always *greater* than or equal to the sum.

6. The number 0 is called the *auxiliary* element of addition.

Perform each of the following additions from memory without referring to the addition table.

7. $0 + 4$	8. $7 + 9$	9. $5 + 9$	10. $4 + 7$	11. $6 + 0$
12. $6 + 9$	13. $9 + 4$	14. $4 + 8$	15. $8 + 9$	16. $7 + 6$
17. $8 + 7$	18. $9 + 9$	19. $6 + 3 + 8$	20. $7 + 0 + 2$	21. $5 + 2 + 9$
22. $9 + 1 + 4$	23. $8 + 7 + 3$	24. $9 + 4 + 7$	25. $4 + 8 + 7$	26. $3 + 9 + 6$

State which property (or properties)—the identity, commutative, or associative of addition—is illustrated by each of the following equations.

27. $7 + 5 = 5 + 7$	28. $3 + (5 + 2) = 3 + (2 + 5)$	29. $8 + (1 + 7) = (8 + 1) + 7$
30. $2 + 9 = 9 + 2$	31. $0 + 8 = 8$	32. $2 + (7 + 4) = (2 + 7) + 4$
33. $8 + (3 + 2) = 8 + (2 + 3)$	34. $(9 + 2) + 5 = 5 + (9 + 2)$	35. $6 + (4 + 2) = (4 + 6) + 2$
36. $8 + (1 + 6) = (8 + 6) + 1$		

Estimate the following sums and then determine the exact answers.

37. $18 + 54$	38. $24 + 81$	39. $127 + 521$	40. $97 + 212$
41. $62 + 13 + 59$	42. $8 + 32 + 77$	43. $323 + 410 + 168$	44. $125 + 684 + 537$
45. $1,021 + 317 + 88$	46. $112 + 3,107 + 498$		

Add each of the following groups of numbers.

47. $18 + 31$	**48.** $53 + 26$	**49.** $87 + 46$	**50.** $75 + 39$
51. $862 + 509$	**52.** $380 + 477$	**53.** $537 + 248$	**54.** $199 + 801$
55. $37 + 86 + 23$	**56.** $52 + 94 + 18$	**57.** $687 + 23 + 15$	**58.** $97 + 324 + 62$
59. $5,029 + 8,653$	**60.** $8,967 + 4,036$	**61.** $39 + 62 + 384 + 7$	**62.** $12 + 284 + 6 + 39$

63. $57,656 + 40,932$ **64.** $61,486 + 29,513$ **65.** $567 + 84 + 67 + 8,248 + 76 + 319$

66. $52 + 928 + 3,516 + 88 + 547 + 43$ **67.** $738,927 + 39,724 + 6,421,507$

68. $5,459,638 + 80,167 + 279,156$ **69.** $3,648 + 294 + 6,527 + 14,687 + 388 + 18,493$

70. $52,468 + 3,284 + 957 + 62,748 + 54,682 + 728$

Translate each of the following statements into an arithmetic expression using addition.

71. The sum of 4 and 9 is 13. **72.** The number 2 increased by 6. **73.** 8 more than 3.

74. The sum of 7 and x. **75.** 5 increased by x is 9. **76.** x more than 8 is 17.

77. The sum of x and y. **78.** The total of a, b, and c is 20.

Solve each of the following word problems.

79. A bookstore sold 512 books in January, 658 books in February, and 486 books in March. What was the total number of books sold for those 3 months?

80. A company employs 12 people in human resources, 43 people in client services, 246 people in product development, and 118 people in sales. What is the total number of people employed in those four departments?

81. Last March, a salesman drove 213 miles from New York to Boston, 457 miles from Boston to Buffalo, 192 miles from Buffalo to Cleveland, and 175 miles from Cleveland to Detroit, calling on a number of customers along the way. What is the total number of miles he put on his car while traveling this route?

82. During the past academic year in the state of Ohio, there were 1,424,000 students enrolled in public elementary schools and 717,000 students enrolled in public secondary schools. How many students were enrolled in Ohio public schools during the past school year?

83. A student budgeted the following amounts to cover her expenses during her freshman year in college: tuition, $5,500; room & board, $3,500; books, $250; transportation, $500; miscellaneous expenses, $400. What did she expect the total cost of her freshman year to be?

84. The area of Greenland is 839,999 square miles, and the area of Iceland is 39,768 square miles. What is the combined area of Greenland and Iceland?

85. A movie theater sold 82 tickets to the 2:00 P.M. show, 53 to the 5:30 P.M. show, 158 to the 8:00 P.M. show, and 144 to the 10:30 P.M. show. What was the total number of tickets sold to all shows?

86. It is estimated that the number of eligible voters in a New England town can be broken down into the following age brackets: 18–24 years, 2,918; 25–44 years, 5,856; 45–64 years, 4,400; 65 years and older, 2,119. What is the total number of eligible voters in that town?

87. In 1 year, the number of passenger cars made by the leading American automobile manufacturers was as follows:

American Motors	164,351
Chrysler Corporation	1,082,274
Ford Motor Company	2,511,888
General Motors	5,316,616

What was the total number of cars manufactured by these four companies that year?

88. The populations of five Midwestern states are as follows:

Ohio	9,706,397
Indiana	4,662,498
Illinois	10,081,158
Michigan	7,823,194
Wisconsin	3,951,777

What is the total population of these five states?

1.3
Subtraction of Whole Numbers

The operation **subtraction** may be defined as the *inverse* of addition. For example, consider the following problem. If ten people are in a room, and three of them leave, how many remain? Using your knowledge of addition, you might solve this problem by asking yourself what number must be added to three to obtain a total of ten.

$$? + 3 = 10$$

It is often easier, however, to pose the problem as ten reduced by three yields what result? Written as an equation this becomes:

$$10 - 3 = ?$$

In both equations the unknown number is 7:

$$7 + 3 = 10$$
$$10 - 3 = 7$$

The second equation illustrates the operation subtraction, and it is the reverse process of the first equation, which illustrates addition.

To *subtract b* from *a*, we can write the equation $a - b = c$, where *a* is the *minuend*, *b* is the *subtrahend*, and *c* is the *difference*. ▲

For example, in the equation $7 - 2 = 5$, 7 is the minuend, 2 is the subtrahend, and 5 is the difference.

$$7 - 2 = 5$$

minuend   subtrahend   difference

We can also write this same problem vertically.

$$7 \leftarrow \quad \text{minuend}$$
$$\underline{-\ 2} \leftarrow \text{subtrahend}$$
$$5 \leftarrow \quad \text{difference}$$

Whenever we do a subtraction problem, we can check our answer by adding the difference to the subtrahend to obtain the minuend: $c + b = a$.

▶ **DEFINITION** Addition and subtraction are said to be *inverse* operations, since if $a - b = c$, then $c + b = a$.

For example, if $9 - 3 = 6$, then $6 + 3 = 9$.

EXAMPLE 1   Subtract each of the following pairs of numbers and check your answers using addition.

(a) $13 - 6$
$\qquad 13 - 6 = 7$ Since $7 + 6 = 13$.

(b) $9 - 4$
$\qquad 9 - 4 = 5$ Since $5 + 4 = 9$.   ▲

Notice that whenever we perform a subtraction operation, the minuend and the subtrahend cannot be reversed without changing the answer to the problem. Thus,

$$11 - 5 \neq 5 - 11$$
$$6 \neq -6$$

The right-hand side yields a negative number that does not belong to the set of whole numbers. On a number line, numbers located to the *left of zero* are called *negative numbers*, and numbers located to the *right of zero* are called *positive numbers*, as shown in Figure 1.2. A negative number is a number less than zero. For example, if the temperature is 3 degrees below zero, we express that temperature as -3 degrees. Whenever the subtrahend is greater than the minuend, the result is a negative number. (Negative numbers will be discussed in detail in Chapter 7.) Since the order of the minuend and subtrahend cannot be changed without creating an inequality, the *commutative* property does *not* hold true for subtraction.

Figure 1.2

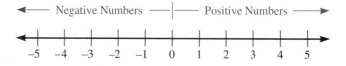

Let us now determine whether the associative property holds true for subtraction, by looking at the following example.

$$\text{Does } (8 - 4) - 2 = 8 - (4 - 2)?$$
$$4 - 2 \ ? \ 8 - 2$$
$$2 \neq 6$$

Thus, the *associative* property does *not* hold true for subtraction.

When more than one subtraction operation appears in a problem, we perform each subtraction in order, moving from left to right, unless otherwise indicated by parentheses.

EXAMPLE 2 Perform the following subtraction operations.

(a) $7 - 3 - 2$
Do the leftmost subtraction first.
$(7 - 3) - 2 = 4 - 2 = 2$

(b) $12 - 6 - 3$
Do the leftmost subtraction first.
$(12 - 6) - 3 = 6 - 3 = 3$ ▲

In order to subtract numbers greater than nine, we write the problem vertically, lining up digits with the same place value. We then subtract the numbers in each column, moving from right to left. If a digit in the ones column of the subtrahend is larger than a digit in the ones column of the minuend, we must rewrite the minuend. Since 1 ten = 10 ones, we can do this by subtracting 1 from the digit in the tens column and adding 10 to the digit in the ones column.

EXAMPLE 3 Subtract 27 from 52.

Solution

minuend $52 = 5 \text{ tens} + 2 \text{ ones} = 4 \text{ tens} + 12 \text{ ones}$
subtrahend $-27 = 2 \text{ tens} + 7 \text{ ones} = 2 \text{ tens} + 7 \text{ ones}$

difference $ 2 \text{ tens} + 5 \text{ ones} = 25$

(1) Write the problem vertically and indicate the place value of each digit.
(2) Since in the ones column 2 < 7, we must rewrite the minuend. Since 1 ten = 10 ones, subtract 1 from the digit in the tens column and add 10 to the digit in the ones column to obtain 4 tens + 12 ones.
(3) Subtract the numbers in the ones column: $12 - 7 = 5$.
(4) Subtract the numbers in the tens column: $4 - 2 = 2$. The result is 25. ▲

A shortcut for subtracting numbers in this fashion is called *borrowing*. The borrowing procedure is outlined here.

To Subtract Whole Numbers

1. Write the number vertically and line up numbers with the same place value.

2. Subtract the numbers in each column separately, moving from right to left.

3. If a digit in the top number is less than the digit that has the same place value in the bottom number, rewrite the top number by subtracting, or borrowing, 1 from the digit immediately to the left of the smaller digit, and adding 10 to the smaller digit. ▲

If we use borrowing to subtract the numbers in Example 3 we obtain the following:

$$\begin{array}{r} 4 \\ \cancel{5}\,^1 2 \\ -2\ 7 \\ \hline \end{array}$$

Since in the ones column $2 < 7$, we must borrow. Since 1 ten = 10 ones, borrow 1 from the 5 in the tens column to get 4 tens, and add 10 ones to the 2 in the ones column to get 12 ones.

$$\begin{array}{r} 4 \\ \cancel{5}\,^1 2 \\ -2\ 7 \\ \hline 5 \end{array}$$

Subtract $12 - 7 = 5$.

$$\begin{array}{r} 4 \\ \cancel{5}\,^1 2 \\ -2\ 7 \\ \hline 2\ 5 \end{array}$$

Subtract $4 - 2 = 2$. The result is 25.

Let us now look at some more examples of subtraction that illustrate borrowing.

EXAMPLE 4 Perform each of the following subtractions and check your answers by using addition.

(a) $516 - 42$

$$\begin{array}{r} 4 \\ \cancel{5}\,^1 1\ 6 \\ -\ \ \ 4\ 2 \\ \hline 4\ 7\ 4 \end{array}$$

Subtract $6 - 2 = 4$.

Since in the tens column $1 < 4$, we must borrow. Since 1 hundred = 10 tens, borrow 1 from the 5 in the hundreds column to get 4 hundreds, and add 10 tens to the 1 in the tens column to get 11 tens.

Subtract $11 - 4 = 7$.

Subtract $4 - 0 = 4$. The result is 474.

Check:
$$\begin{array}{r} 1 \\ 474 \\ +\ 42 \\ \hline 516 \quad \sqrt{} \end{array}$$

(b) $602 - 89$

$$\begin{array}{r} 5 \\ \cancel{6}\,^1 0\ 2 \\ -\ \ \ 8\ 9 \\ \hline \end{array}$$

Since $2 < 9$, we try to borrow from the tens column, but there are no tens. So we borrow 1 from the hundreds column to get 5 hundreds, and add 10 tens to the tens column to get 10 tens.

$$\begin{array}{r} 5\ \ 9 \\ \cancel{6}\,^1 \cancel{0}\,^1 2 \\ -\ \ \ 8\ 9 \\ \hline 5\ 1\ 3 \end{array}$$

Now borrow 1 from the tens column to get 9 tens, and add 10 ones to the ones column to get 12 ones.

$12 - 9 = 3$. $9 - 8 = 1$. $5 - 0 = 5$. The result is 513.

Check:
$$\begin{array}{r} 11 \\ 513 \\ +\ 89 \\ \hline 602 \quad \sqrt{} \end{array}$$

▲

Estimation

It is important to be able to estimate answers when performing subtraction, just as it is when doing addition. Example 5 illustrates how estimation is used to approximate answers to subtraction problems.

EXAMPLE 5   Estimate the following differences. Then determine the exact answers. Check your answers using addition.

(a) $5,182 - 2,328$

Approximate the minuend and subtrahend and subtract:

$$
\begin{array}{r}
5182 \cong 5000 \\
-2328 \cong \underline{2000} \\
\end{array}
$$

3,000 is our estimate.

Now determine the exact answer:

$$
\begin{array}{r}
4 \quad 7 \\
5\,{}^1 1\ 8\,{}^1 2 \\
-2\ 3\ 2\ 8 \\
\hline
2,8\ 5\ 4 \\
\end{array}
$$

Since $2 < 8$, borrow 1 from 8 to get 7 tens, and add 10 to 2 to get 12 ones.

$12 - 8 = 4$. $7 - 2 = 5$.

Since $1 < 3$, borrow 1 from 5 to get 4 thousands, and add 10 to 1 to get 11 hundreds.

$11 - 3 = 8$. $4 - 2 = 2$. The result is 2,854.

Check:
$$
\begin{array}{r}
1 \quad\ 1 \\
2\ 8\ 5\ 4 \\
+2\ 3\ 2\ 8 \\
\hline
5,1\ 8\ 2 \quad \checkmark \\
\end{array}
$$

Since 2,854 is close to our estimate of 3,000, this answer is reasonable.

(b) $1,000 - 327$

Approximate the minuend and subtrahend and subtract:

$$
\begin{array}{r}
1000 \cong 1000 \\
-327 \cong \underline{\ \ 300} \\
\end{array}
$$

700 is our estimate.

Now determine the exact answer:

$$
\begin{array}{r}
0 \\
1\,{}^1 0\ 0\ 0 \\
-\ \ \ 3\ 2\ 7 \\
\hline
\end{array}
$$

Since $0 < 7$, we try to borrow from the tens column, but nothing is there. Since nothing is in the hundreds column either, we borrow 1 from the thousands column to get 0 thousands, and add 10 to the hundreds column to get 10 hundreds.

$$
\begin{array}{r}
0\ 9 \\
1\,{}^1 0\,{}^1 0\ 0 \\
-\ \ \ 3\ 2\ 7 \\
\hline
\end{array}
$$

Borrow 1 from the hundreds column to get 9 hundreds, and add 10 to the tens column to get 10 tens.

$$\begin{array}{r} 0\ 9\ 9 \\ 1\ '{\not0}\ '{\not0}\ '{\not0} \\ -\ \ 3\ 2\ 7 \\ \end{array}$$

Borrow 1 from the tens column to get 9 tens, and add 10 to the ones column to get 10 ones.

$$\begin{array}{r} 0\ 9\ 9 \\ 1\!,\!'{\not0}\ '{\not0}\ '{\not0} \\ -\ \ 3\ 2\ 7 \\ \hline 6\ 7\ 3 \\ \end{array}$$

$10 - 7 = 3.\ \ 9 - 2 = 7.\ \ 9 - 3 = 6.$
The result is 673.

Since 673 is close to our estimate of 700, this answer is reasonable.

Check:
$$\begin{array}{r} 1\,1 \\ 673 \\ +\ \ 327 \\ \hline 1{,}000\ \ \checkmark \\ \end{array}$$

▲

Solving Word Problems Using Subtraction

Word problems that contain words such as *subtract, difference, minus, from, decreased by,* and *less than* can usually be solved using subtraction. Table 1.4 illustrates how words can be translated into arithmetic expressions using subtraction.

TABLE 1.4
Expressing Subtraction

Using Words	Using Symbols
The difference between 8 and 3 is 5.	$8 - 3 = 5$
11 decreased by 4.	$11 - 4$
5 from 12 is 7.	$12 - 5 = 7$
2 less than x is 9.	$x - 2 = 9$
Subtract a from b.	$b - a$

EXAMPLE 6 Fifty years ago, there were approximately 186,500 bachelor's degrees awarded in the United States, and last year there were approximately 944,000 bachelor's degrees awarded. How many more people earned bachelor's degrees last year than did fifty years ago?

Solution Let us first estimate the answer by approximating the minuend and subtrahend and subtracting:

$$\begin{array}{r} 944{,}000 \cong 900{,}000 \\ -\,186{,}500 \cong 200{,}000 \\ \hline 700{,}000 \ \ \text{is our estimate.} \end{array}$$

Now let us find the exact answer:

$$\begin{array}{r} 8\,13\,13 \\ 9\,'4\,'4\!,\!0\ 0\ 0 \\ -1\ 8\ 6{,}5\ 0\ 0 \\ \hline 7\ 5\ 7{,}5\ 0\ 0 \\ \end{array}$$

Since 757,500 is close to our estimate of 700,000, this answer is reasonable.

Check:
$$\begin{array}{r} \overset{111}{757,500} \\ +186,500 \\ \hline 944,000 \quad \checkmark \end{array}$$

Thus 757,500 is close to our estimate of 700,000; this answer is reasonable. ▲

EXAMPLE 7 On the first day of class, 272 students were registered to take Biology I. Since then, 79 students have dropped the course, and 28 students have added the course. How many students are now registered to take Biology I?

Solution Subtract 79 from 272. Then add 28 to the result.

$$\begin{array}{r} \overset{1\,16}{2\,{}^{1}7\,{}^{1}2} \\ -\quad 7\,9 \\ \hline 1\,9\,3 \end{array} \qquad \begin{array}{r} \overset{1\,1}{193} \\ +\quad 28 \\ \hline 221 \end{array}$$

Thus, 221 students are now registered for Biology I. ▲

Quick Quiz

		Answers	
1.	$72 - 49$	**1.**	23
2.	$827 - 658$	**2.**	169
3.	$5,000 - 1,436$	**3.**	3,564

1.3 Exercises

Indicate which of the following statements are true and which are false. For those that are false, change the italicized expression to make the statement true.

1. The *associative* property does not hold true for subtraction.
2. Whenever the subtrahend exceeds the minuend, the difference is *positive*.
3. You can check the result of a subtraction problem by adding the difference to the *subtrahend*.
4. When subtracting numbers that have more than one digit, we begin at the *rightmost* column.

Perform each of the following subtractions from memory.

5.	$15 - 6$	6.	$12 - 7$	7.	$11 - 3$	8.	$16 - 9$	9.	$13 - 6$
10.	$17 - 8$	11.	$14 - 6$	12.	$11 - 7$	13.	$14 - 9$	14.	$16 - 7$

15. $17 - 8$ 16. $15 - 9$ 17. $12 - 3 - 6$ 18. $15 - 7 - 0$ 19. $18 - 5 - 4$

20. $13 - 9 - 2$ 21. $19 - 0 - 9$ 22. $14 - 8 - 6$

Estimate the following differences. Then determine the exact answers.

23. $821 - 285$ 24. $679 - 312$ 25. $1,029 - 506$ 26. $8,172 - 987$

27. $7,178 - 2,954$ 28. $4,938 - 2,068$ 29. $32,197 - 13,465$ 30. $59,238 - 41,069$

Perform each of the following subtractions. Check your answers using addition.

31. $58 - 13$ 32. $85 - 42$ 33. $31 - 18$

34. $83 - 27$ 35. $138 - 92$ 36. $415 - 73$

37. $812 - 167$ 38. $652 - 283$ 39. $206 - 89$

40. $520 - 122$ 41. $713 - 487$ 42. $824 - 398$

43. $5,203 - 584$ 44. $6,046 - 967$ 45. $4,731 - 2,847$

46. $8,485 - 7,942$ 47. $5,013 - 847$ 48. $6,403 - 586$

49. $36,004 - 8,276$ 50. $40,027 - 5,473$ 51. $8,326,412 - 2,718,637$

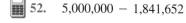 52. $5,000,000 - 1,841,652$ 53. $67,362,107 - 35,684,793$ 54. $62,158,235 - 48,473,829$

Translate each of the following statements into an arithmetic expression using subtraction.

55. 5 less than 12.

56. Subtract 2 from 9.

57. 15 decreased by 7 is 8.

58. 4 less than x.

59. The difference between 10 and 3 is 7.

60. 6 decreased by x.

61. The difference between x and y.

62. a less than b is 9.

Solve each of the following word problems.

63. A stereo receiver that retails for $420 is marked down $59. What is the sale price of this receiver?

64. The Nile River is 4,180 miles long, and the Mississippi River is 2,348 miles long. How much longer is the Nile than the Mississippi?

65. If the population of the United States was 23,261,000 in 1850 and 204,879,000 in 1970, by how much did the population increase during these years?

66. The distance from the sun to Venus is 67,270,000 miles, and the distance from the sun to Saturn is 887,140,000 miles. How much farther from the sun is Saturn than Venus?

67. If a couple bought a house for $62,500 and spent $19,000 on home improvements before selling it for $85,200, how much money did they make on the home?

68. A woman's checking account contained $903 at the beginning of the month. During the month she wrote checks for $37, $129, $425, $62, and $26, and she also deposited $150. What is her balance at the end of the month?

69. This year, 99,529,000 individual income tax returns were filed with the IRS. Last year, 94,496,900 individual returns were filed. How many more individual income tax returns were filed this year than last year?

70. During the past 150 years, 36,200,005 persons immigrated from Europe to the United

States, and 2,859,940 persons immigrated from Asia to the United States. How many more European immigrants were there than Asian?

1.4
Addition and Subtraction of Units of Measurement

Many applications of addition and subtraction involve working with units of measurement. The list of equivalent units of measurement printed on the inside cover of this book will be useful in solving these kinds of problems. Abbreviations for each unit are given in parentheses. You should memorize these relationships, if you do not already know them.

Adding Units of Measurement

All that you have learned about addition can be applied directly to the operation of adding units of measurement. The following examples illustrate how the notion of carrying can be extended to adding units of measurement.

EXAMPLE 1 Add 1 hr 45 min + 2 hr 50 min.

Solution

$$\begin{array}{r} 1 \text{ hr } 45 \text{ min} \\ + 2 \text{ hr } 50 \text{ min} \\ \hline 3 \text{ hr } 95 \text{ min} \end{array}$$

$= 3 \text{ hr} + \overbrace{60 \text{ min} + 35 \text{ min}}$
$= \underline{3 \text{ hr} + 1 \text{ hr}} + 35 \text{ min}$
$= \quad 4 \text{ hr} + 35 \text{ min}$
$= \quad 4 \text{ hr } 35 \text{ min}$

Write the problem vertically and line up like units of measurement.

Add the numbers in the minutes and hours columns separately.

Since 60 min = 1 hr, rewrite 95 min as 1 hr + 35 min.

Carry 1 to the hours column to obtain a final answer of 4 hr 35 min. ▲

EXAMPLE 2 Add 4 yd 1 ft 7 in. + 3 yd 2 ft 9 in.

Solution

$$\begin{array}{r} 4 \text{ yd } 1 \text{ ft } \;7 \text{ in.} \\ +3 \text{ yd } 2 \text{ ft } \;9 \text{ in.} \\ \hline 7 \text{ yd } 3 \text{ ft } 16 \text{ in.} \end{array}$$

$= 7 \text{ yd} + 3 \text{ ft} + \overbrace{12 \text{ in.} + 4 \text{ in.}}$
$= 7 \text{ yd} + \underline{3 \text{ ft} + 1 \text{ ft}} + 4 \text{ in.}$
$= \quad 7 \text{ yd} + 4 \text{ ft} + 4 \text{ in.}$
$= \underline{7 \text{ yd} + 1 \text{ yd}} + 1 \text{ ft} + 4 \text{ in.}$
$= \quad 8 \text{ yd} + 1 \text{ ft} + 4 \text{ in.}$
$= \quad 8 \text{ yd } 1 \text{ ft } 4 \text{ in.}$

Write the problem vertically and line up like units of measurement.

Add the numbers in the inches, feet, and yards columns separately.

Since 12 in. = 1 ft, rewrite 16 in. as 1 ft 4 in.

Carry 1 to the feet column to get 7 yd + 4 ft + 4 in.

Since 3 ft = 1 yd, rewrite 4 ft as 1 yd + 1 ft.

Carry 1 to the yards column to obtain a final answer of 8 yd 1 ft 4 in. ▲

EXAMPLE 3 While on vacation, a tourist bought 2 pounds 7 ounces of chocolate fudge, 10 ounces of vanilla fudge, 1 pound 4 ounces of butterscotch fudge, and 1 pound 5 ounces of strawberry fudge. What was the total amount of fudge purchased?

Solution   Add 2 lb 7 oz + 10 oz + 1 lb 4 oz + 1 lb 5 oz.

$$
\begin{array}{r}
2 \text{ lb } 7 \text{ oz} \\
10 \text{ oz} \\
1 \text{ lb } 4 \text{ oz} \\
+ 1 \text{ lb } 5 \text{ oz} \\
\hline
4 \text{ lb } 26 \text{ oz}
\end{array}
$$

$= 4 \text{ lb} + \overbrace{16 \text{ oz} + 10 \text{ oz}}$ Since 16 oz = 1 lb, rewrite 26 oz as
$= \underbrace{4 \text{ lb} + 1 \text{ lb}} + 10 \text{ oz}$ 1 lb 10 oz.
$= \qquad 5 \text{ lb} + 10 \text{ oz}$
$= \qquad 5 \text{ lb } 10 \text{ oz}$

The tourist purchased 5 pounds 10 ounces of fudge.   ▲

Subtracting Units of Measurement

All that you have learned about subtraction can be applied directly to the operation of subtracting units of measurement. The following examples illustrate how the notion of borrowing can be extended to subtracting units of measurement.

EXAMPLE 4   Subtract 2 ft 9 in. from 5 ft 3 in.

Solution
$$
\begin{array}{r}
5 \text{ ft } 3 \text{ in.} = \quad 4 \text{ ft } 15 \text{ in.} \\
-2 \text{ ft } 9 \text{ in.} = -2 \text{ ft } \ 9 \text{ in.} \\
\hline
2 \text{ ft } \ 6 \text{ in.}
\end{array}
$$

Write the problem vertically and line up like units of measurement.

(1) Since 3 in. < 9 in., we must borrow. Since 12 in. = 1 ft, we borrow 1 ft from the feet column to get 4 ft, and add 12 in. to the inches column to get 15 in.
(2) Subtract the numbers in the inches column: 15 in. − 9 in. = 6 in.
(3) Subtract the numbers in the feet column: 4 ft − 2 ft = 2 ft. The result is 2 ft 6 in.   ▲

EXAMPLE 5   A swimmer began a race at 3:55:44 P.M. and finished at 4:03:18 P.M. How long did it take her to finish the race?

Solution   3:55:44 means 3 hr 55 min 44 sec.
4:03:18 means 4 hr 3 min 18 sec.
Subtract 3 hr 55 min 44 sec from 4 hr 3 min 18 sec.

$$
\begin{array}{r}
4 \text{ hr } \ 3 \text{ min } 18 \text{ sec} = \quad 4 \text{ hr } \ 2 \text{ min } 78 \text{ sec} = \quad 3 \text{ hr } 62 \text{ min } 78 \text{ sec} \\
-3 \text{ hr } 55 \text{ min } 44 \text{ sec} = -3 \text{ hr } 55 \text{ min } 44 \text{ sec} = -3 \text{ hr } 55 \text{ min } 44 \text{ sec} \\
\hline
34 \text{ sec} \qquad\qquad 7 \text{ min } 34 \text{ sec}
\end{array}
$$

(1) Since 18 sec < 44 sec, we must borrow. Since 1 min = 60 sec, borrow 1 min from the minutes column to get 2 min, and add 60 sec to the seconds column to get 78 sec.
(2) Subtract the numbers in the second column:
78 sec − 44 sec = 34 sec.
(3) Since 2 min < 55 min, we must borrow. Since 1 hr = 60 min, borrow 1 hr from the hours column to get 3 hr, and add 60 min to the minutes column to get 62 min.
(4) Subtract the numbers in the minutes column:

(5) Subtract the numbers in the hours column:
$3 \text{ hr} - 3 \text{ hr} = 0 \text{ hr}$.

(6) The result is 7 min 34 sec.

▲

62 min − 55 min = 7 min.

Quick Quiz

Perform the operations indicated.	Answers
1. 18 min 40 sec + 25 min 28 sec	**1.** 44 min 8 sec
2. 5 lb 2 oz − 3 lb 9 oz	**2.** 1 lb 9 oz

1.4 Exercises

Solve each of the following addition problems.

1. 7 ft 7 in.
   +8 ft 6 in.

2. 3 hr 20 min
   +5 hr 45 min

3. 4 lb 12 oz
   +1 lb 8 oz

4. 6 gal 1 qt
   +9 gal 3 qt

5. 7 wk 5 da
   +4 wk 3 da

6. 8 qt 1 pt
   +1 qt 1 pt

7. 5 hr 20 min 45 sec
   +1 hr 40 min 15 sec

8. 2 yd 1 ft 8 in.
   +3 yd 2 ft 6 in.

Solve each of the following subtraction problems.

9. 8 ft 3 in.
   −4 ft 7 in.

10. 6 hr 15 min
    −2 hr 45 min

11. 9 lb 5 oz
    −3 lb 10 oz

12. 5 gal 1 qt
    −1 gal 3 qt

13. 4 wk 2 da
    −3 wk 5 da

14. 5 qt
    −2 qt 1 pt

15. 9 hr 18 min 21 sec
    −3 hr 25 min 48 sec

16. 5 yd 1 ft 2 in.
    −1 yd 1 ft 10 in.

Solve each of the following word problems.

17. If a student worked on his math homework 50 minutes on Saturday and 1 hour 35 minutes on Sunday, how much time did he spend on his math homework that weekend?

18. If the electric cord on a lamp is 2 feet 8 inches long and is attached to an extension cord 3 feet 6 inches long, how far from an electrical outlet can the lamp be placed?

19. A woman buys the following amounts of nuts: 1 pound 3 ounces of pistachios, 2 pounds of peanuts, 10 ounces of cashews, and 1 pound 5 ounces of walnuts. What is the total weight of the nuts she has purchased?

20. At a company picnic, employees brought the following amounts of beverages: 2 gallons 1 quart of lemonade, 3 quarts of cola, and 1 gallon 2 quarts of iced tea. What is the total quantity of beverages brought to the picnic?

21. If a man is 6 feet 2 inches tall, and his son is 5 feet 9 inches tall, how much taller is the father than his son?

22. If a student began studying at 2:45 P.M. and finished at 5:10 P.M., how long did she study?

23. A bag of apples weighs 8 pounds 5 ounces, and a bag of oranges weighs 5 pounds 12 ounces. How much more do the apples weigh than the oranges?

24. An athlete made 4 gallons of cranberry-apple juice by mixing apple juice with 2 gallons 3 quarts of cranberry juice. How much apple juice did he use?

1.5
Multiplication of Whole Numbers

The operation **multiplication** may be defined as repeated addition. For example, if we add five twos together, we obtain:

$$2 + 2 + 2 + 2 + 2 = 10$$

This could also be expressed by saying that the sum of five twos is 10, or 5 times 2 equals 10:

$$5 \times 2 = 10$$

The "times" symbol, $\times$, indicates the operation multiplication. The number 2 in our example, which is the number being multiplied, is called the **multiplicand**. The number 5, which is the number of times the multiplicand is being counted, is called the **multiplier**. The number 10, which is the result, is called the **product**.

When a multiplication problem is written horizontally, the multiplier appears first. When it is written vertically, the multiplicand appears on top:

$$5 \times 2 = 10$$

multiplier   multiplicand   product

$$\begin{array}{r} 2 \leftarrow \text{multiplicand} \\ \times\ 5 \leftarrow \quad \text{multiplier} \\ \hline 10 \leftarrow \qquad \text{product} \end{array}$$

The multiplicand and multiplier are also called **factors**.

▶ **DEFINITION** To multiply two numbers *a* and *b*, we can write the equation $a \times b = c$, where *a* is called the *multiplier*, *b* is called the *multiplicand*, and *c* is called the *product*. *a* and *b* are also referred to as *factors* of the product *c*.

For example, in the expression $3 \times 2 = 6$, 3 is the multiplier, 2 is the multiplicand, and 6 is the product. The numbers 2 and 3 are also called factors of the number 6.

In addition to using a $\times$ to express multiplication, we can use a raised dot, $\cdot$, or put parentheses around some or all of the factors. Thus,

$$3 \times 2 = 3 \cdot 2 = (3)2 = 3(2) = (3)(2)$$

For small numbers, we can determine the results of a multiplication operation by rewriting the problem as repeated addition and calculating the sum. For example,

$$3 \times 8 = 8 + 8 + 8 = 24$$

However, this process can become very cumbersome and time consuming. It is essential, therefore, to memorize the products of all single-digit numbers. The products of all numbers from 0 to 12 are shown in Table 1.5. Some products that may be calculated using this table are:

$$3 \times 0 = 0 \qquad 8 \times 1 = 8 \qquad 4 \times 5 = 20 \qquad 7 \times 5 = 35$$

TABLE 1.5
Multiplication Table

x	0	1	2	3	4	5	6	7	8	9	10	11	12
0	0	0	0	0	0	0	0	0	0	0	0	0	0
1	0	1	2	3	4	5	6	7	8	9	10	11	12
2	0	2	4	6	8	10	12	14	16	18	20	22	24
3	0	3	6	9	12	15	18	21	24	27	30	33	36
4	0	4	8	12	16	20	24	28	32	36	40	44	48
5	0	5	10	15	20	25	30	35	40	45	50	55	60
6	0	6	12	18	24	30	36	42	48	54	60	66	72
7	0	7	14	21	28	35	42	49	56	63	70	77	84
8	0	8	16	24	32	40	48	56	64	72	80	88	96
9	0	9	18	27	36	45	54	63	72	81	90	99	108
10	0	10	20	30	40	50	60	70	80	90	100	110	120
11	0	11	22	33	44	55	66	77	88	99	110	121	132
12	0	12	24	36	48	60	72	84	96	107	120	132	144

Notice that whenever 0 is multiplied by any number, the resulting product is 0. Also recognize that products with 5 as a factor end in 0 or 5. Finally, note that whenever a number is multiplied by 1, the resulting product is the same as that number.

Multiplicative Identity Property

The number 1 is said to be the *identity element* of multiplication, since for any number a, $a \cdot 1 = a$ and $1 \cdot a = a$. ▲

For example, $1 \cdot 4 = 4$.

Let us now examine some of the other properties that apply to multiplication. To illustrate the commutative property, let us compare both sides of this equation:

Does $3 \times 6 = 6 \times 3$?
Using repeated addition we obtain:

$$6 + 6 + 6 = 3 + 3 + 3 + 3 + 3 + 3$$
$$18 = 18$$

Thus, $3 \times 6 = 6 \times 3$

Notice that the *order* of the factors may be changed without changing the result.

Commutative Property of Multiplication

For any numbers a and b, $a \cdot b = b \cdot a$.   ▲

For example, $7 \cdot 4 = 4 \cdot 7$.

To multiply more than two numbers together, we can multiply any two together, and then multiply that result by the remaining factor(s) to obtain the final product. For example, let us compare both sides of this equation:

$$\text{Does } (4 \times 2) \times 3 = 4 \times (2 \times 3)?$$
$$8 \times 3 = 4 \times 6$$
$$24 = 24$$
$$\text{Thus, } (4 \times 2) \times 3 = 4 \times (2 \times 3).$$

According to the associative property, we can change the *grouping* of the factors without affecting the resulting product.

Associative Property of Multiplication

For any numbers, a, b, and c, $a \cdot (b \cdot c) = (a \cdot b) \cdot c$.   ▲

For example, $5 \cdot (6 \cdot 2) = (5 \cdot 6) \cdot 2$.

The commutative and associative properties may sometimes be used to simplify a multiplication problem that contains three or more factors. For example, to multiply

$$5 \times 7 \times 6$$

we might think

$$(5 \times 6) \times 7 = 30 \times 7 = 210$$

The next property to be learned applies to problems that involve both addition and multiplication, as illustrated by this example.

$$\text{Does } 3(7 + 2) = 3(7) + 3(2)?$$
$$3(9) = 21 + 6$$
$$27 = 27$$
$$\text{Thus, } 3(7 + 2) = 3(7) + 3(2).$$

This property is called the **distributive property** of multiplication over addition. It is used whenever the sum of two or more numbers is to be multiplied by another number that appears outside the parentheses. In the example just cited, the factor 3 is *distributed* over each of the addends in parentheses:

$$3(7 + 2) = 3(7) + 3(2)$$

Distributive Property of Multiplication over Addition

For any numbers a, b, and c, $a(b + c) = a(b) + a(c)$.   ▲

For example, $2(4 + 3) = 2(4) + 2(3)$.

The distributive property may also be used to multiply the difference of two numbers by another number. For example, let us compare both sides of this equation:

$$\text{Does } 4(9 - 2) = 4(9) - 4(2)?$$
$$4(7) = 36 - 8$$
$$28 = 28$$
Thus, $4(9 - 2) = 4(9) - 4(2)$.

Distributive Property of Multiplication over Subtraction

For any numbers *a, b,* and *c, a(b − c) = a(b) − a(c)*. ▲

For example, $3(5 - 2) = 3(5) - 3(2)$.

If two sums are multiplied together, the distributive property can be applied twice:

$$(a + b)(c + d) = a(c + d) + b(c + d)$$
$$= a(c) + a(d) + b(c) + b(d)$$

For example,

$$(3 + 5)(2 + 7) = 3(2 + 7) + 5(2 + 7) = 3(2) + 3(7) + 5(2) + 5(7)$$

EXAMPLE 1 State which property (or properties)—the identity, commutative, or associative property of multiplication or the distributive property—is illustrated by each of the following equations.

(a) $3 \cdot (7 \cdot 2) = (3 \cdot 7) \cdot 2$

Associative property of multiplication, since 7 and 2 are grouped together on the left and 3 and 7 are grouped together on the right.

(b) $5 \cdot 1 = 5$

Multiplicative identity property.

(c) $4(5 + 3) = 4(5) + 4(3)$

Distributive property of multiplication over addition, since the factor 4 is distributed over each of the addends in parentheses.

(d) $8 \cdot (3 \cdot 7) = 8 \cdot (7 \cdot 3)$

Commutative property of multiplication, since the 3 and 7 are reversed.

(e) $(2 + 5)(3 + 4) = 2(3 + 4) + 5(3 + 4) = 2(3) + 2(4) + 5(3) + 5(4)$

Distributive property, applied twice to the product of two sums.

(f) $(3 \cdot 9) \cdot 2 = 3 \cdot (2 \cdot 9)$

Commutative and associative properties of multiplication, since both the order and the grouping of the factors are changed. ▲

EXAMPLE 2   Use the distributive property to rewrite each of the following multiplications and then simplify.

(a) $5(2 + 7)$
$$5(2 + 7) = 5(2) + 5(7)$$
$$= 10 + 35 = 45$$

(b) $6(13) + 6(7)$
$$6(13) + 6(7) = 6(13 + 7)$$
$$= 6(20) = 120$$

(c) $8(62) - 8(2)$
$$8(62) - 8(2) = 8(62 - 2)$$
$$= 8(60) = 480$$ ▲

In order to multiply numbers that have more than one digit, we can write the numbers to be multiplied in expanded notation and apply the distributive property.

EXAMPLE 3   Multiply 4×36.

Solution $4 \times 36 = 4(30 + 6)$ Write 36 in expanded notation.
$\qquad\quad = 4(30) + 4(6)$ Apply the distributive property.
$\qquad\quad = 120 + 24 = 144$ Add to obtain the result.   ▲

We can write the same problem vertically and obtain:

$$
\begin{array}{r}
36 \\
\times\ \ 4 \\
\hline
24 \\
120 \\
\hline
144
\end{array}
$$

$4 \times 6 = 24$

$4 \times 30 = 120$

Add to obtain the result of 144.

A shortcut for the same process is:

$$
\begin{array}{rr}
 & 2 \\
\text{multiplicand} & 36 \\
\text{multiplier} & \times\ \ 4 \\
\hline
\text{product} & 144
\end{array}
$$

Write the problem vertically.

Multiply $4 \times 6 = 24$. Put down 4, carry 2 (two tens) to the tens column.

Multiply $4 \times 3 = 12$. Add the 2 that was carried to 12. $12 + 2 = 14$. Put down 14.

If both numbers to be multiplied have more than one digit, we apply the distributive property twice, as shown in Example 4.

EXAMPLE 4   Multiply 32×87.

Solution $32 \times 87 = (30 + 2)(80 + 7)$
$\qquad\qquad = 30(80 + 7) + 2(80 + 7)$
$\qquad\qquad = (30)(80) + (30)(7) + (2)(80) + (2)(7)$
$\qquad\qquad = 2,400 + 210 + 160 + 14 = 2,784$ ▲

Now let us do Example 4 using the shortcut.

$$
\begin{array}{r}
2 \\
1 \\
87 \\
\times\ 32 \\
\hline
174 \\
2\,61 \\
\hline
2{,}784
\end{array}
$$

First multiply 87 by 2. $2 \times 7 = 14$.
Put down 4, carry 1. $2 \times 8 = 16$.
$16 + 1 = 17$. Put down 17.

Now multiply 87 by 30. Think of multiplying 87 by 3 and place the result under the first product, beginning in the tens column. $3 \times 7 = 21$. Put down 1, carry 2.
$3 \times 8 = 24$. $24 + 2 = 26$. Put down 26.

Add the two products to obtain 2,784.

Check:

$$
\begin{array}{r}
1 \\
1 \\
32 \\
\times\ 87 \\
\hline
224 \\
2\,56 \\
\hline
2{,}784\ \ \sqrt{}
\end{array}
$$

Notice that because of the commutative property we can reverse the top number and the bottom number to check our answer.

The steps for multiplying numbers that have more than one digit are outlined in the following.

To Multiply Whole Numbers

1. Write the problem vertically and place the number with the larger number of digits on top and the smaller number below it.

2. Multiply each digit of the top number by the ones digit in the bottom number, moving from right to left.

3. For a product that exceeds nine, write down the rightmost digit of the product and carry the other digit to the next column on the left and write it above the top number. Calculate the next product and be sure to add to that product the digit that was carried.

4. Multiply each digit in the top number by the next digit to the left in the bottom number. Place each product under the previous one calculated, but displaced one column to the left.

5. Repeat step 4 for all remaining digits in the bottom number.

6. Add the products to obtain the final answer. ▲

EXAMPLE 5 Multiply 65×328.

Solution

```
    1 4
    1 4
    3 2 8
  ×   6 5
  ─────────
    1 1 1
    1,6 4 0
    1 9 6 8
  ─────────
    2 1,3 2 0
```

Write the problem vertically with 328 on top.

Multiply 328 by 5. $5 \times 8 = 40$. Put down 0, carry 4. $5 \times 2 = 10$. $10 + 4 = 14$. Put down 4, carry 1. $5 \times 3 = 15$. $15 + 1 = 16$. Put down 16.

Multiply 328 by 6. Place result under first product, but displaced one column to the left. $6 \times 8 = 48$. Put down 8, carry 4. $6 \times 2 = 12$. $12 + 4 = 16$. Put down 6, carry 1. $6 \times 3 = 18$. $18 + 1 = 19$. Put down 19.

Add the products to obtain 21,320. ▲

Quick Quiz

Multiply each of the following pairs of numbers.

1. 4×89
2. 27×58
3. 396×712

Answers

1. 356
2. 1,566
3. 281,952

1.5 Exercises

Indicate which of the following statements are true and which are false. For those that are false, change the italicized expression to make the statement true.

1. Multiplication is repeated *subtraction.*
2. The identity element for multiplication is *0.*
3. When multiplying numbers that have more than one digit, we use the *distributive* property.
4. When multiplying three or more numbers, the way in which we group factors *does not* affect the product.

Perform each of the following multiplications from memory without referring to the multiplication table.

5. 5×5	6. 8×5	7. 4×0	8. 8×8	9. 7×6
10. 3×8	11. 7×7	12. 6×9	13. 1×5	14. 4×7
15. 6×5	16. 6×8	17. 9×5	18. 4×6	19. 9×7
20. 6×6	21. 9×9	22. 5×4	23. 7×3	24. 4×9
25. 8×7	26. 5×7	27. $2 \times 3 \times 7$	28. $8 \times 1 \times 6$	29. $4 \times 2 \times 9$
30. $2 \times 7 \times 5$	31. $5 \times 4 \times 0$	32. $5 \times 3 \times 8$	33. $3 \times 7 \times 3$	34. $6 \times 8 \times 5$

State which property (or properties), the identity, commutative, or associative property of multiplication or the distributive property, is illustrated by each of the following equations.

35. $6 \cdot (3 \cdot 5) = (6 \cdot 3) \cdot 5$

36. $3 \cdot 9 = 9 \cdot 3$

37. $8(5 + 3) = 8(5) + 8(3)$

38. $(7 \cdot 2) \cdot 4 = (2 \cdot 7) \cdot 4$

39. $2(8 - 3) = 2(8) - 2(3)$

40. $5(3 + 6) = (3 + 6)5$

41. $9 \cdot 1 = 9$

42. $(8 \cdot 9) \cdot 7 = 8 \cdot (9 \cdot 7)$

43. $2 \cdot (6 \cdot 3) = 6 \cdot (2 \cdot 3)$

44. $5(4 - 1) + 3(4 - 1) = (5 + 3)(4 - 1)$

Use the distributive property to rewrite each of the following expressions and then simplify.

45. $3(6 + 2)$

46. $4(7 + 5)$

47. $8(7) + 8(3)$

48. $7(8) + 7(2)$

49. $5(7 - 2)$

50. $4(9 - 3)$

51. $(9 + 8)3$

52. $(6 + 7)5$

53. $6(12) + 6(8)$

54. $8(13) + 8(7)$

55. $7(9) - 7(4)$

56. $6(8) - 6(5)$

57. $8(40 + 5)$

58. $4(70 + 5)$

59. $4(38) - 4(8)$

60. $3(47) - 3(7)$

61. $5(73) + 5(7)$

62. $8(56) + 8(4)$

63. $3(94) - 3(4)$

64. $5(88) - 5(8)$

Multiply each of the following pairs of numbers.

65. 6×53

66. 84×4

67. 63×9

68. 5×27

69. 3×98

70. 77×6

71. 41×32

72. 27×53

73. 66×44

74. 33×55

75. 89×64

76. 92×74

77. 47×213

78. 421×36

79. 854×69

80. 38×722

81. 513×282

82. 467×674

83. 621×986

84. 459×729

85. $3,135 \times 832$

86. $741 \times 2,716$

87. $8,629 \times 3,468$

88. $2,738 \times 4,572$

89. $697 \times 37,549$

90. $54,698 \times 274$

1.6
More on Multiplication

Multiplying Numbers with Zeros

Numbers that end in zeros can be multiplied very quickly. Observe the pattern shown in the following examples.

$$7 \times 10 = 70$$
$$7 \times 100 = 700$$
$$7 \times 1,000 = 7,000$$
$$7 \times 10,000 = 70,000$$
$$7 \times 100,000 = 700,000$$

Notice that in each case, the product contains the same number of ending zeros as in the number that is multiplied by 7. Now look at these examples.

$$3 \times 6 = 18$$
$$3 \times 60 = 3 \times 6 \times 10 = (3 \times 6) \times 10$$
$$= 18 \times 10 = 180$$

$$3 \times 600 = 3 \times 6 \times 100 = (3 \times 6) \times 100$$
$$= 18 \times 100 = 1,800$$

$$3 \times 6,000 = 3 \times 6 = 1,000 = (3 \times 6) \times 1,000$$
$$= 18 \times 1,000 = 18,000$$

$$30 \times 60 = 3 \times 10 \times 6 \times 10 = (3 \times 6) \times (10 \times 10)$$
$$= 18 \times 100 = 1,800$$

$$30 \times 600 = 3 \times 10 \times 6 \times 100 = (3 \times 6) \times (10 \times 100)$$
$$= 18 \times 1,000 = 18,000$$

$$300 \times 600 = 3 \times 100 \times 6 \times 600 = (3 \times 6) \times (100 \times 100)$$
$$= 18 \times 10,000 = 180,000$$

Notice that in each case, the product can be obtained by first multiplying 3×6 to get 18, and then attaching to the right of 18 the appropriate number of zeros, which is equal to the sum of the ending zeros in all numbers being multiplied. We have thus established a procedure for multiplying numbers ending in zeros.

To Multiply Numbers Ending in Zero(s)	**1.** Ignore the string of consecutive zeros at the end of each number being multiplied. Find the product of the remaining digits of the numbers being multiplied.
	2. Count the number of ending zeros in all numbers being multiplied. Attach that number of zeros to the right end of the product calculated in step 1. ▲

EXAMPLE 1   Multiply each of the following pairs of numbers.

(a) 3×200
     $3 \times 2 = 6$ Multiply $3 \times 2 = 6$.
     $3 \times 200 = 600$ Attach two zeros to the right of 6.

(b) 40×60
     $4 \times 6 = 24$ Multiply $4 \times 6 = 24$.
     $40 \times 60 = 2,400$ Attach two zeros to the right of 24.

(c) $700 \times 3,000$
     $7 \times 3 = 21$ Multiply $7 \times 3 = 21$.
     $700 \times 3,000 = 2,100,000$ Attach five zeros to the right of 21.

(d) $90 \times 4,200$

$$\begin{array}{r} 1 \\ 42 \\ \times\ \ 9 \\ \hline 378 \end{array}$$

Multiply $9 \times 42 = 378$

$$90 \times 4,200 = 378,000$$

Attach three zeros to the right of 378.

(e) $5,100 \times 680,000$

$$
\begin{array}{r}
4 \\
6\,8 \\
\times\ 5\,1 \\
\hline
6\,8 \\
3\,4\,0 \\
\hline
3,4\,6\,8
\end{array}
$$

Multiply $51 \times 68 = 3,468$.

$5,100 \times 680,000 = 3,468,000,000$ Attach six zeros to the right of 3,468.   ▲

Let us now establish a technique for multiplying numbers that have zeros between the nonzero digits of the number.

EXAMPLE 2   Multiply 308×132.

Solution

$$
\begin{array}{r}
2\,1 \\
1\,3\,2 \\
\times\ 3\,0\,8 \\
\hline
1\,0\,5\,6 \\
0\,0\,0 \\
3\,9\,6 \\
\hline
4\,0,6\,5\,6
\end{array}
$$

Notice that we can omit the second row of zeros if we begin writing the product of 132×3 under the hundreds place as shown on the right. The result is 40,656.

$$
\begin{array}{r}
2\,1 \\
1\,3\,2 \\
\times\ 3\,0\,8 \\
\hline
1\,0\,5\,6 \\
3\,9\,6 \\
\hline
4\,0,6\,5\,6
\end{array}
$$

▲

EXAMPLE 3   Multiply $4,001 \times 6,127$.

Solution

$$
\begin{array}{r}
1\,2 \\
6,1\,2\,7 \\
\times\ \ 4,0\,0\,1 \\
\hline
6\,1\,2\,7 \\
0\,0\,0\,0 \\
0\,0\,0\,0 \\
2\,4\,5\,0\,8 \\
\hline
2\,4,5\,1\,4,1\,2\,7
\end{array}
$$

We can omit the two rows of zeros if we begin writing the product of $4 \times 6,127$ under the thousands place as shown on the right. The result is 24,514,127.

$$
\begin{array}{r}
1\,2 \\
6,1\,2\,7 \\
\times\ \ 4,0\,0\,1 \\
\hline
6\,1\,2\,7 \\
2\,4\,5\,0\,8 \\
\hline
2\,4,5\,1\,4,1\,2\,7
\end{array}
$$

▲

The two previous examples illustrate that every zero that appears in the multiplier causes the product of the multiplicand and the next nonzero digit in the multiplier to be displaced one additional column to the left.

EXAMPLE **4**   Multiply 50,030 × 2,187.

Solution

$$
\begin{array}{r}
4\,3 \\
2\,2 \\
2{,}1\,8\,7 \\
\times \quad 5\,0{,}0\,3\,0 \\
\hline
6\,5\,6\,1\,0 \\
1\,0\,9\,3\,5 \qquad\quad \\
\hline
1\,0\,9{,}4\,1\,5{,}6\,1\,0
\end{array}
$$

Write the product of 2,187 × 3 displaced one column to the left; put a 0 in the ones place as a placeholder.

Write the product of 2,187 × 5 displaced two additional columns to the left.

Add the two products to obtain a result of 109,415,610.

▲

Estimating Answers in Multiplication

It is important to be able to estimate answers to multiplication problems. If you have a rough idea of the result, then you can recognize an unreasonable answer. The following examples illustrate the use of estimation when performing multiplication operations.

EXAMPLE **5**   Estimate the following products. Then determine the exact answers.

(a)   27 × 52

Approximate each factor and then multiply:

$$
\begin{array}{r}
52 \cong \quad 50 \\
27 \cong \quad 30 \\
\hline
1{,}500 \text{ is our estimate.}
\end{array}
$$

Now determine the exact answer:

$$
\begin{array}{r}
5\,2 \\
\times \; 2\,7 \\
\hline
3\,6\,4 \\
1\,0\,4 \quad\; \\
\hline
1{,}4\,0\,4
\end{array}
$$

Since 1,404 is close to our estimate of 1,500, this answer is reasonable.

Notice that the estimate of a product is not as close to the actual answer as is the estimate of a sum or a difference.

(b)   893 × 285

Approximate each factor and then multiply:

$$
\begin{array}{r}
285 \cong \quad 300 \\
893 \cong \quad 900 \\
\hline
270{,}000 \text{ is our estimate.}
\end{array}
$$

Now determine the exact answer:

$$
\begin{array}{r}
6\;4 \\
7\;4 \\
2\;1 \\
2\;8\;5 \\
\times\quad 8\;9\;3 \\
\hline
8\;5\;5 \\
2\;5\;6\;5 \\
2\;2\;8\;0 \\
\hline
2\;5\;4{,}5\;0\;5
\end{array}
$$

Since 254,505 is close to our estimate of 270,000, this answer is reasonable. ▲

Solving Word Problems Using Multiplication

Word problems containing the words *multiply*, *product*, and *times* can usually be solved using multiplication. Table 1.6 illustrates how words can be translated into arithmetic expressions using multiplication.

TABLE 1.6
Expressing Multiplication

Using Words	Using Symbols
Multiply 2 by 3.	2×3
8 times 7 is 56.	$8 \times 7 = 56$
The product of *a* and *b* equals 18.	$a \cdot b = 18$

Many word problems that are solved using multiplication, however, do not contain any of these "clue" words. Consider the following example.

EXAMPLE 6 An engineer earns a weekly salary of $632. What is his yearly salary?

Solution Since there are 52 weeks in a year, multiply 52×632. Let us first estimate the answer by approximating each factor and then multiplying.

$$
\begin{array}{r}
632 \cong \quad 600 \\
52 \cong \quad\;\; 50 \\
\hline
\end{array}
$$

30,000 is our estimate.

Now let us find the exact answer.

$$
\begin{array}{r}
6\;3\;2 \\
\times\quad 5\;2 \\
\hline
1\;2\;6\;4 \\
3\;1\;6\;0 \\
\hline
3\;2{,}8\;6\;4
\end{array}
$$

Since 32,864 is close to our estimate of 30,000, this answer is reasonable. ▲

His yearly salary is $32,864.

Quick Quiz

Multiply each of the following pairs of numbers.	Answers
1. 50×700	1. 35,000
2. 135×40	2. 5,400
3. 407×726	3. 295,482

1.6 Exercises

Multiply each of the following pairs of numbers.

1. 9×30	2. 6×70	3. 50×30	4. 40×70
5. 60×500	6. 90×200	7. 800×800	8. 400×900
9. $8,000 \times 600$	10. $700 \times 3,000$	11. $2,000 \times 70,000$	12. $50,000 \times 5,000$

Multiply each of the following pairs of numbers.

13. 93×806	14. 78×530	15. 506×859	16. 703×487
17. $8,602 \times 535$	18. $396 \times 7,094$	19. $806 \times 5,714$	20. $4,389 \times 702$
21. $5,396 \times 1,270$	22. $7,248 \times 3,007$	23. $7,026 \times 4,681$	24. $3,792 \times 8,704$
25. $2,073 \times 4,603$	26. $5,609 \times 6,078$		

Estimate each of the following products. Then determine the exact answers.

27. 27×51	28. 12×79	29. 41×283	30. 669×31
31. 703×821	32. 393×715	33. $5,262 \times 291$	34. $913 \times 6,884$

Translate each of the following statements into an arithmetic expression using multiplication.

35. The product of 5 and 4 is 20.　　36. Multiply 6 by 7.　　37. 3 times a.
38. The product of x and y is z.

Solve each of the following word problems.

39. If a car can travel 28 miles on 1 gallon of gas, how far can it travel on 18 gallons of gas?

40. If a student studies 26 hours a week, and there are 14 weeks in a semester, how many hours does the student study during an entire semester?

41. A stockholder buys 115 shares of a certain stock that sells for $34 a share. What is her total investment?

42. Lake Huron is 13 times the size of the Great Salt Lake, which covers 1,800 square miles. How large is Lake Huron?

43. The Honolulu public libraries circulate 512,814 books each year. The Cincinnati public libraries circulate 14 times as many books each year. What is the annual circulation of the Cincinnati public libraries?

44. If the population of California last year was 53 times that of Alaska, and 521,943 people were living in Alaska last year, what was the population of California?

45. How many minutes are there in a year?

46. If 1 acre = 43,560 square feet, how many square feet are in 1,386 acres?

1.7
Division of Whole Numbers

The operation **division** may be defined as repeated subtraction. For example, if we were asked to determine how many fives go into 15, we could proceed as follows.

$$
\begin{array}{r}
15 \\
-\ 5 \\
\hline
10 \\
-\ 5 \\
\hline
5 \\
-\ 5 \\
\hline
0
\end{array}
$$
Subtract 5 from 15.

Subtract 5 from the result.

Subtract 5 from the next result.

Stop when the number remaining is less than 5.

Since we were able to subtract three fives before the remaining number was less than five, we can say that five goes into 15 three times. This can be expressed by the equation $15 \div 5 = 3$, where the symbol $\div$ indicates the operation division. This same equation, which states that 15 divided by 5 equals 3, can be written in three different ways:

$$
15 \div 5 = 3 \quad \text{or} \quad \frac{15}{5} = 3 \quad \text{or} \quad 5\overline{\smash)15}^{\,3}
$$

▶ **DEFINITION** *Division* of a number a by another number b is expressed by the equation

$$
a \div b = c \quad \text{or} \quad \frac{a}{b} = c \quad \text{or} \quad b\overline{\smash)a}^{\,c}
$$

where a is the *dividend*, b is the *divisor*, and c is the *quotient*.

For example, in the equation $15 \div 5 = 3$, 15 is the dividend, 5 is the divisor, and 3 is the quotient.

EXAMPLE 1   Perform each of the following divisions using repeated subtraction.

(a) $18 \div 9$

$$
\begin{array}{r}
18 \\
-\ 9 \\
\hline
9 \\
-\ 9 \\
\hline
0
\end{array}
$$
Subtract 9 from 18.

Subtract 9 from the result.

The number remaining is $0 < 9$.

Since we were able to subtract 9 from 18 twice before the number remaining was less than 9, then $18 \div 9 = 2$. The number that remains after the procedure of repeated subtraction is completed is called the **remainder**. Note that *the remainder must always be a whole number that is less than the divisor*. In this example, the remainder is 0. Dividends are said to be *divisible* by a given divisor if the remainder is 0. Thus, in this example, 18 is divisible by 9.

(b) $17 \div 4$

$$
\begin{array}{r}
17 \\
-\ 4 \\
\hline
13 \\
-\ 4 \\
\hline
9 \\
-\ 4 \\
\hline
5 \\
-\ 4 \\
\hline
1
\end{array}
$$

Subtract 4 from 17.

Subtract 4 from the result.

Subtract 4 from the result.

Subtract 4 from the result.

Since $1 < 4$, the remainder is 1.

Since we were able to subtract 4 from 17 four times before obtaining a remainder of 1, we write the answer as 4 R1, which is read 4 remainder 1.

(c) $2 \div 7$

$$
\begin{array}{r}
2 \\
-7 \\
\hline
-5
\end{array}
$$

If we subtract 7 from 2, the result. Is not a whole number, but a negative number.

The established procedure is to subtract the divisor, 7, from the dividend, 2, and from the ensuing results until the remaining result is a *whole number* that is less than the divisor. In this example the dividend, 2, is less than the divisor, 7, and the result of the subtraction is not a whole number. This means that 7 goes into 2 zero times, with a remainder of 2, which is written as 0 R2. Therefore, when dividing whole numbers, if the dividend is less than the divisor, the quotient is 0 and the remainder is equal to the dividend. ▲

We will now use this process of repeated subtraction to illustrate a few principles regarding division.

Any number divided by 1 is equal to the number itself. That is, for any number a, $a \div 1 = a$. As an example, let us consider $3 \div 1$:

$$
\left.\begin{array}{r}
3 \\
-1 \\
\hline
2 \\
-1 \\
\hline
1 \\
-1 \\
\hline
0
\end{array}\right\} 3 \text{ times}
$$

Notice that the number 1 can be subtracted three times from 3. Thus, $3 \div 1 = 3$.

Any nonzero number divided by itself is equal to 1. That is, for any number $a \neq 0$, $a \div a = 1$. As an example, let us consider $8 \div 8$:

$$\left.\begin{array}{r} 8 \\ -8 \\ \hline 0 \end{array}\right\} 1 \text{ time}$$

Notice that the number 8 can be subtracted one time from 8. Thus, $8 \div 8 = 1$.

Zero divided by any nonzero number is equal to zero. That is, for any number $a \neq 0$, $0 \div a = 0$. As an example, let us consider $0 \div 6$. This problem is similar to Example 1(c) since the dividend, 0, is less than the divisor, 6. The quotient is therefore 0 with a remainder equal to the dividend, which is 0. Thus, $0 \div 6 = 0$.

Division by zero is impossible. We say that for any number a, $a \div 0$ is undefined. As an example, let us consider $5 \div 0$.

$$\begin{array}{r} 5 \\ -0 \\ \hline 5 \\ -0 \\ \hline 5 \\ -0 \\ \hline 5 \end{array}$$

Notice that we could continue this process indefinitely and still be no closer to a solution. No matter how many times we subtract 0 from 5, it is impossible to end up with a remainder other than 5. We express this idea by saying that $5 \div 0$ is **undefined.**

We will now look at some examples that illustrate the different notations that can be used to express division.

EXAMPLE 2   Rewrite each of the following division problems using the other two notations, translate the equation into words, and identify the dividend, divisor, and quotient.

(a) $\dfrac{36}{9} = 4$

$36 \div 9 = 4$ $9 \overline{)\,36}$ with quotient 4

Thirty-six divided by nine equals four.

36, dividend; 9, divisor; 4, quotient   Notice that $4 \cdot 9 = 36$.

(b) $56 \div 7 = 8$

$\dfrac{56}{7} = 8$ $7 \overline{)\,56}$ with quotient 8

Fifty-six divided by seven equals eight.

56, dividend; 7, divisor; 8, quotient   Notice that $8 \cdot 7 = 56$.

(c) $6 \overline{)\,54}$ with quotient 9

$54 \div 6 = 9$ $\dfrac{54}{6} = 9$

Fifty-four divided by six equals nine.

54, dividend; 6, divisor; 9, quotient   Notice that $9 \cdot 6 = 54$.

(d) $11 \div 5 = 2 \text{ R}1$

$$\frac{11}{5} = 2 \text{ R}1 \qquad 5 \overline{\smash{\big)}\, 11}^{\,2\,\text{R}1}$$

Eleven divided by five equals two remainder one.

11, dividend;   5, divisor;

2, quotient;   1, remainder   Notice that $(2 \cdot 5) + 1 = 11$.   ▲

The problems in Example 2 lead us to believe that there is a relationship between multiplication and division. In fact, division is the **inverse** of multiplication. For example,

$$24 \div 3 = 8 \quad \text{since} \quad 8 \cdot 3 = 24$$
$$42 \div 6 = 7 \quad \text{since} \quad 7 \cdot 6 = 42$$
$$63 \div 7 = 9 \quad \text{since} \quad 9 \cdot 7 = 63$$

Notice that the product of the quotient and the divisor is equal to the dividend. Thus, multiplication can be used to check the result of a division problem.

▶ *DEFINITION* Multiplication and division are said to be *inverse* operations, since if $a \div b = c$, then $c \cdot b = a$.

For example, if $36 \div 9 = 4$, then $4 \cdot 9 = 36$. Since multiplication is repeated addition, division is repeated subtraction, and addition and subtraction are inverse operations, it makes sense that multiplication and division are also inverse operations.

Notice that whenever we perform a division operation, the dividend and divisor cannot be reversed without changing the answer to the problem. Thus,

$$20 \div 4 \neq 4 \div 20$$
$$5 \neq 0 \text{ R}4$$

Therefore, the *commutative* property does *not* hold true for division.

Let us now determine whether the associative property holds true for division by looking at the following example:

Does $(18 \div 6) \div 3 = 18 \div (6 \div 3)$?

$$3 \div 3 \; ? \; 18 \div 2$$
$$1 \neq 9$$

Thus, the *associative* property does *not* hold true for division.

When more than one division operation appears in a problem, we perform each division in order, moving from left to right, unless otherwise indicated by parentheses.

EXAMPLE 3   Perform each of the following divisions.

(a) $28 \div 7 \div 2$

Do the leftmost division first:

$(28 \div 7) \div 2 = 4 \div 2 = 2$

(b) $48 \div 8 \div 2$

Do the leftmost division first:

$(48 \div 8) \div 2 = 6 \div 2 = 3$ ▲

Let us now consider a technique that uses multiplication instead of repeated subtraction to solve a division problem. This procedure is outlined on the following page.

To Divide Whole Numbers	1. Write the problem using divisor $\lceil$ dividend notation.
	2. Determine the largest number that can be multiplied by the divisor to obtain a product that is less than or equal to the dividend. Write that number above the dividend.
	3. Multiply that number by the divisor.
	4. Subtract the product obtained in step 3 from the dividend. The result is the remainder. ▲

To check the result of a division problem, we can follow these steps:

1. Multiply the quotient by the divisor.
2. Add the remainder.
3. The result should equal the dividend.

EXAMPLE 4 Divide $29 \div 4$ and check your answer.

Solution

$$
\begin{array}{r}
7\ \text{R1} \\
4\overline{)29} \\
-28 \\
\hline
1
\end{array}
$$

Write the problem as shown on the left.

Ask yourself what number multiplied by 4 produces a product less than or equal to 29.

Write 7 in the space for the quotient.

Multiply $7 \times 4 = 28$.

Subtract $29 - 28 = 1$, the remainder. The quotient is 7 R1.

Check:

$$
\begin{array}{r}
7 \\
\times\ 4 \\
\hline
28 \\
+\ 1 \\
\hline
29 \quad \sqrt{}
\end{array}
$$

Multiply $7 \times 4 = 28$.

Add the remainder: $28 + 1 = 29$.

Since 29 is the dividend, the answer is correct. ▲

EXAMPLE 5 Perform each of the following divisions and check your answers.

(a) $53 \div 7$

$$
\begin{array}{r}
7\ \text{R4} \\
7\overline{)53} \\
49 \\
\hline
4
\end{array}
$$

7 goes into 53 seven times. Write down 7.

Multiply $7 \times 7 = 49$.

Subtract $53 - 49 = 4$.

The quotient is 7 R4.

Check:

$$
\begin{array}{r}
7 \\
\times\ 7 \\
\hline
49 \\
+\ 4 \\
\hline
53 \quad \sqrt{}
\end{array}
$$

Multiply quotient and divisor.

Add the remainder

Always be certain that the remainder is less than the divisor. If it isn't, the number chosen for the quotient is not large enough.

(b) $153 \div 9$

In this problem, 9 goes into 153 more than 10 times. To determine exactly how many times, we will use the same procedure more than once.

$$
\begin{array}{r}
17 \\
9\,\overline{)153} \\
90 \\
\hline
63 \\
63 \\
\hline
0
\end{array}
$$

9 goes into 150 ten times. (It is often easier to mentally approximate the dividend by omitting the ones digit.) Put the 1 over the tens place of the dividend.

$9 \times 10 = 90$. $153 - 90 = 63$.

9 goes into 63 seven times. Put the 7 over the ones place of the dividend.

$9 \times 7 = 63$. $63 - 63 = 0$.
The quotient is 17.

Check:
$$
\begin{array}{r}
17 \\
\times\ \ 9 \\
\hline
153 \quad \checkmark
\end{array}
$$

(c) $486 \div 7$

We can save ourselves work by omitting zeros in the numbers we subtract, as long as we remember to line up numbers that have the same place value.

$$
\begin{array}{r}
69\ \text{R3} \\
7\,\overline{)486} \\
42 \\
\hline
66 \\
63 \\
\hline
3
\end{array}
$$

7 goes into 48 about 6 times. Put the 6 over the tens place of the dividend.

$7 \times 6 = 42$. $48 - 42 = 6$.

Bring down the 6 from the dividend. 7 goes into 66 about 9 times. Put the 9 over the ones place of the dividend.

$7 \times 9 = 63$. $66 - 63 = 3$.
The quotient is 69 R3.

Check:
$$
\begin{array}{r}
69 \\
\times\ \ 7 \\
\hline
483 \\
+\ \ \ 3 \\
\hline
486 \quad \checkmark
\end{array}
$$

(d) $2,695 \div 4$

$$
\begin{array}{r}
6\,7\,3\ \text{R3} \\
4\,\overline{)2,6\,9\,5} \\
2\,4 \\
\hline
2\,9 \\
2\,8 \\
\hline
1\,5 \\
1\,2 \\
\hline
3
\end{array}
$$

4 goes into 26 about 6 times. Put the 6 over the hundreds place of the dividend.

$4 \times 6 = 24$. $26 - 24 = 2$.

Bring down 9 from the dividend. 4 goes into 29 about 7 times. Put the 7 over the tens place of the dividend.

$4 \times 7 = 28$. $29 - 28 = 1$.

Bring down the 5 from the dividend. 4 goes into 15 about 3 times. Put 3 over the ones place of the dividend.

$4 \times 3 = 12$. $15 - 12 = 3$.

The quotient is 673 R3.

Check:
$$
\begin{array}{r}
673 \\
\times \quad 4 \\
\hline
2692 \\
+ \quad 3 \\
\hline
2{,}695 \quad \checkmark
\end{array}
$$

(e) $38{,}527 \div 6$

$$
\begin{array}{r}
6{,}421 \ \text{R1} \\
6\,\overline{)38{,}527} \\
36 \\
\hline
25 \\
24 \\
\hline
12 \\
12 \\
\hline
7 \\
6 \\
\hline
1
\end{array}
$$

6 goes into 38 six times.
Put the 6 over the 8 in the dividend.
$6 \times 6 = 36.$ $38 - 36 = 2.$
Bring down 5, the next number in the dividend.
6 goes into 25 four times.
Put the 4 over the 5 in the dividend.
$6 \times 4 = 24.$ $25 - 24 = 1.$
Bring down the 2 from the dividend.
6 goes into 12 two times.
Place the 2 over the 2 in the dividend.
$6 \times 2 = 12.$ $12 - 12 = 0.$
Bring down 7.
6 goes into 7 one time.
Put 1 over the 7.
$6 \times 1 = 6.$ $7 - 6 = 1.$
The quotient is 6,421 R1.

Check:
$$
\begin{array}{r}
6{,}421 \\
\times \quad 6 \\
\hline
38{,}526 \\
+ \quad 1 \\
\hline
38{,}527 \quad \checkmark
\end{array}
$$

(f) $32{,}564 \div 8$

$$
\begin{array}{r}
4{,}070 \ \text{R4} \\
8\,\overline{)32{,}564} \\
32 \\
\hline
56 \\
56 \\
\hline
4
\end{array}
$$

8 goes into 32 four times.
Put 4 over the thousands place of the dividend.
$8 \times 4 = 32.$ $32 - 32 = 0.$
Bring down 5 from the dividend.
8 goes into 5 zero times.
Put 0 over the hundreds place of the dividend.
Bring down 6 from the dividend.
8 goes into 56 seven times.
Put 7 over the tens place of the dividend.
$8 \times 7 = 56.$ $56 - 56 = 0.$
Bring down 4 from the dividend.
8 goes into 4 zero times.
Put 0 over the ones place of the dividend.
The quotient is 4,070 R4.   ▲

Check:
$$
\begin{array}{r}
4{,}070 \\
\times \quad 8 \\
\hline
32{,}560 \\
+ \quad 4 \\
\hline
32{,}564 \quad \checkmark
\end{array}
$$

When the divisor contains more than one digit, we can mentally approximate it by considering only the first digit. However, the numbers we subtract from the dividend must be the exact products of the quotients and the divisor.

EXAMPLE 6 Perform each of the following divisions and check your answers.

(a) $284 \div 32$

$$
\begin{array}{r}
8 \text{ R28} \\
32 \overline{)284} \\
256 \\
\hline
28
\end{array}
$$

30 goes into 280 about 9 times.

$$
\begin{array}{r}
32 \\
\times \ 9 \\
\hline
288 > 284
\end{array}
$$

This guess is too big.
Let us try 8.

Check:

$$
\begin{array}{r}
32 \\
\times \ 8 \\
\hline
256 \\
+ \ 28 \\
\hline
284 \quad \sqrt{}
\end{array}
$$

$$
\begin{array}{r}
32 \\
\times \ 8 \\
\hline
256 < 284
\end{array}
$$

$$
\begin{array}{r}
284 \\
-256 \\
\hline
28 < 32
\end{array}
$$
Since the remainder
is less than the divisor,
this guess is correct.

(b) $8{,}467 \div 53$

$$
\begin{array}{r}
1 \ 5 \ 9 \text{ R40} \\
53 \overline{)8{,}4 \ 6 \ 7} \\
5 \ 3 \\
\hline
3 \ 1 \ 6 \\
2 \ 6 \ 5 \\
\hline
5 \ 1 \ 7 \\
4 \ 7 \ 7 \\
\hline
4 \ 0
\end{array}
$$

50 goes into 80 about 1 time.

$$
\begin{array}{cc}
53 & 84 \\
\times \ 1 & \times 53 \\
\hline
53 < 84 & 31 < 53
\end{array}
$$
Put 1 over the 4.
Bring down the 6.

50 goes into 300 about 6 times.

$$
\begin{array}{r}
53 \\
\times \ 6 \\
\hline
318 > 316
\end{array}
$$
This guess is too big. Try 5.

$$
\begin{array}{cc}
53 & 316 \\
\times \ 5 & -265 \\
\hline
265 < 316 & 51 < 53
\end{array}
$$
Put 5 over the 6.
Bring down the 7.

53 goes into 517 about 9 times.

$$
\begin{array}{cc}
53 & 517 \\
\times \ 9 & -477 \\
\hline
477 < 517 & 40 < 53
\end{array}
$$
Put 9 over the 7.
The remainder is 40.

Check:

$$
\begin{array}{r}
1 \ 5 \ 9 \\
\times \ \ 5 \ 3 \\
\hline
4 \ 7 \ 7 \\
7 \ 9 \ 5 \\
\hline
8{,}4 \ 2 \ 7 \\
+ \ \ \ \ 4 \ 0 \\
\hline
8{,}4 \ 6 \ 7 \quad \sqrt{}
\end{array}
$$

(c) $68{,}921 \div 213$

$$
\begin{array}{r}
3 \ 2 \ 3 \text{ R122} \\
213 \overline{)6 \ 8{,}9 \ 2 \ 1} \\
6 \ 3 \ 9 \\
\hline
5 \ 0 \ 2 \\
4 \ 2 \ 6 \\
\hline
7 \ 6 \ 1 \\
6 \ 3 \ 9 \\
\hline
1 \ 2 \ 2
\end{array}
$$

200 goes into 600 about 3 times.

$$
\begin{array}{cc}
213 & 689 \\
\times \ 3 & -639 \\
\hline
639 < 689 & 50 < 213
\end{array}
$$
Put 3 over the 9.
Bring down the 2.

200 goes into 500 about 2 times.

$$
\begin{array}{cc}
213 & 502 \\
\times \ 2 & -426 \\
\hline
426 < 502 & 76 < 213
\end{array}
$$
Put 2 over the 2.
Bring down the 1.

200 goes into 700 about 3 times.

$$
\begin{array}{cc}
213 & 761 \\
\times \ 3 & -639 \\
\hline
639 < 761 & 122 \text{ remainder}
\end{array}
$$
Put 3 over the 1.

Check:
$$
\begin{array}{r}
3\,2\,3 \\
\times\quad 2\,1\,3 \\
\hline
9\,6\,9 \\
3\,2\,3 \\
6\,4\,6 \\
\hline
6\,8{,}7\,9\,9 \\
+\quad\ \ 1\,2\,2 \\
\hline
6\,8{,}9\,2\,1 \ \ \sqrt{}
\end{array}
$$

▲

Solving Word Problems Using Division

Word problems containing the words *divide* and *quotient* can usually be solved using division. Table 1.7 illustrates how words can be translated into arithmetic expressions using division.

TABLE 1.7
Expressing Division

Using Words	Using Symbols
The quotient of 15 and 3.	$15 \div 3$
Divide 28 by 4.	$28 \div 4$
6 divided by x is 2.	$6 \div x = 2$

Keep in mind that there are word problems that do not contain any of these "clue" words that are also solved using division.

EXAMPLE 7   A company has a total of $21,350 in bonus money to be divided equally among 14 employees. How much should each person receive?

Solution   Divide $21,350 \div 14$.

$$
\begin{array}{r}
1{,}5\,2\,5 \\
14\,\overline{\smash{)}\,2\,1{,}3\,5\,0} \\
1\,4 \\
\hline
7\,3 \\
7\,0 \\
\hline
3\,5 \\
2\,8 \\
\hline
7\,0 \\
7\,0 \\
\hline
0
\end{array}
$$

Each person should receive $1,525.

Check:
$$
\begin{array}{r}
1{,}5\,2\,5 \\
\times\quad\ \ 1\,4 \\
\hline
6\,1\,0\,0 \\
1\,5\,2\,5 \\
\hline
2\,1{,}3\,5\,0 \ \ \sqrt{}
\end{array}
$$

▲

Quick Quiz

			Answers
1.	$2{,}610 \div 9$	**1.**	290
2.	$400 \div 25$	**2.**	16
3.	$5{,}638 \div 107$	**3.**	52 R74

1.7 Exercises

Indicate which of the following statements are true and which are false. For those that are false, change the italicized expression to make the statement true.

1. In the equation $18 \div 3 = 6$, the number 18 is called the *divisor*.

2. A number is said to be evenly divisible by another number if the remainder is equal to *zero*.

3. Division may be defined as repeated *substitution*.

4. To check the result of a division problem, we multiply the quotient by the divisor and *add* the remainder.

5. In a division problem, the remainder must be *greater* than the divisor.

6. Division by *one* is undefined.

Perform each of the following divisions from memory.

7.	$36 \div 4$	**8.**	$15 \div 5$	**9.**	$0 \div 3$	**10.**	$8 \div 0$	**11.**	$64 \div 8$
12.	$42 \div 7$	**13.**	$21 \div 7$	**14.**	$72 \div 9$	**15.**	$28 \div 4$	**16.**	$54 \div 6$
17.	$56 \div 7$	**18.**	$81 \div 9$	**19.**	$16 \div 4$	**20.**	$30 \div 6$	**21.**	$15 \div 1 \div 3$
22.	$0 \div 8 \div 2$	**23.**	$72 \div 8 \div 3$	**24.**	$32 \div 4 \div 2$	**25.**	$42 \div 7 \div 6$	**26.**	$64 \div 8 \div 4$

Perform each of the following divisions. Check your answers.

27.	$52 \div 7$	**28.**	$83 \div 9$	**29.**	$34 \div 6$	**30.**	$71 \div 8$
31.	$25 \div 4$	**32.**	$43 \div 5$	**33.**	$108 \div 6$	**34.**	$72 \div 3$
35.	$208 \div 4$	**36.**	$155 \div 5$	**37.**	$287 \div 7$	**38.**	$576 \div 8$
39.	$113 \div 3$	**40.**	$309 \div 5$	**41.**	$614 \div 4$	**42.**	$293 \div 9$
43.	$4{,}381 \div 3$	**44.**	$8{,}652 \div 7$	**45.**	$5{,}106 \div 4$	**46.**	$2{,}790 \div 8$
47.	$3{,}618 \div 9$	**48.**	$8{,}414 \div 7$	**49.**	$50{,}813 \div 8$	**50.**	$76{,}042 \div 5$
51.	$49{,}303 \div 3$	**52.**	$38{,}412 \div 7$	**53.**	$8{,}627{,}327 \div 4$	**54.**	$2{,}918{,}327 \div 2$
55.	$4{,}000{,}000 \div 3$	**56.**	$7{,}000{,}000 \div 9$	**57.**	$86 \div 21$	**58.**	$93 \div 15$
59.	$62 \div 38$	**60.**	$76 \div 27$	**61.**	$154 \div 13$	**62.**	$294 \div 81$
63.	$367 \div 62$	**64.**	$843 \div 47$	**65.**	$4{,}263 \div 23$	**66.**	$7{,}916 \div 92$
67.	$15{,}347 \div 26$	**68.**	$13{,}205 \div 43$	**69.**	$9{,}246 \div 356$	**70.**	$5{,}481 \div 127$
71.	$7{,}304 \div 281$	**72.**	$8{,}015 \div 673$	**73.**	$36{,}843 \div 502$	**74.**	$13{,}817 \div 709$
75.	$62{,}302 \div 671$	**76.**	$70{,}149 \div 946$	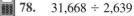 **77.**	$280{,}490 \div 8{,}014$	**78.**	$31{,}668 \div 2{,}639$
79.	$1{,}118{,}042 \div 4{,}678$	**80.**	$4{,}627{,}389 \div 9{,}127$				

Translate each of the following statements into an arithmetic expression using division.

81. The quotient of 36 and 9.

82. 14 divided by 2 equals 7.

83. x divided by y.

84. The quotient of 20 and a is 4.

Solve each of the following word problems.

85. If an inheritance of $7,740 (after taxes) is to be divided equally among nine relatives of the deceased, how much money will each person receive?

86. If 79,143 people voted in a certain election, how many votes would each candidate receive if the votes were split evenly among the three candidates?

87. At the beginning of a card game, a deck of 52 cards is dealt out evenly among three players, and the remaining cards are put in a pile. How many cards does each player receive? How many cards are put in the pile?

88. If 159 marbles are to be divided evenly among 12 children, how many marbles will each child get? How many marbles will be left over?

89. Thirty-two bricks weigh 128 pounds. How much does a single brick weigh?

90. If a station wagon travels 368 miles on 16 gallons of gas, how many miles per gallon does the car get?

91. If a computer programmer earns $33,540 a year, what is her weekly salary?

92. If eight dozen calculus books cost $1,632, what is the price of a single book?

93. An auditorium has a total of 1,176 seats. If there are 84 rows of seats, how many seats are in each row?

94. An investor bought 57 shares of stock for a total cost of $2,166. What was the price of a single share?

95. A machine can fill 2,728 bottles with milk in 1 hour. How long will it take for this machine to fill 35,464 bottles?

96. A wholesale distributor purchased 6,782 radios for a price of $393,356. How much did each radio cost?

1.8
Summary and Review

Key Terms

(1.1) The **whole numbers** include 0, 1, 2, 3, 4, ...

Our number system is called the **base ten system** since it includes ten digits: 0, 1, 2, 3, 4, 5, 6, 7, 8, 9.

The **place value** of a digit is the name of the location of that digit.

A **period** is the name given to every group of three digits in a whole number moving from right to left.

(1.2) An **equation** is a mathematical expression that states that two quantities are equal.

Addition is expressed by the equation $a + b = c$, where a and b are the **addends** and c is the **sum**.

Zero is the **identity element of addition**, since for any number a, $a + 0 = a$ and $0 + a = a$.

According to the **commutative property of addition**, we can change the **order** of the addends without affecting the result.

According to the **associative property of addition**, we can change the **grouping** of the addends without affecting the result.

(1.3) **Subtraction** is expressed by the equation $a - b = c$, where a is the **minuend**, b is the **subtrahend**, and c is the **difference**.

Addition and subtraction are **inverse operations**.

(1.5) **Multiplication** is repeated addition. Multiplication is expressed by the equation $a \cdot b = c$, where a is the **multiplier**, b is the **multiplicand**, and c is the **product**. a and b are also called **factors** of the product c.

One is the **identity element of multiplication** since for any number a, $a \cdot 1 = a$ and $1 \cdot a = a$.

According to the **commutative property of multiplication**, we can change the **order** of the factors without affecting the result.

According to the **associative property of multiplication**, we can change the **grouping** of the factors without affecting the result.

The **distributive property of multiplication over addition** states that for any numbers a, b, and c, $a(b + c) = a(b) + a(c)$.

The **distributive property of multiplication over subtraction** states that for any numbers a, b, and c, $a(b - c) = a(b) - a(c)$.

(1.7) **Division** is repeated subtraction. Division is expressed by the equation $a \div b = c$, where a is the **dividend**, b is the **divisor**, and c is the **quotient**.

The **remainder** is a whole number less than the divisor that is left over after the process of repeated subtraction has been completed.

Division by zero is **undefined** since it is impossible to divide a number by zero.

Multiplication and division are **inverse** operations.

Calculations

(1.1) **To write a whole number in expanded notation**, each digit is multiplied by its corresponding place value, and the results are then added together.

(1.2) **To add whole numbers** when the sum of any column is greater than nine, put down the ones digit of the sum and *carry* the other digit to the next column immediately to the left.

(1.3) **To subtract whole numbers** when a digit in the minuend is less than the digit that has the same place value in the subtrahend, rewrite the minuend by *borrowing* 1 from the digit immediately to the left of the smaller digit, and add 10 to the smaller digit.

(1.5) **To multiply whole numbers** that have more than one digit, multiply each digit in the multiplicand by each digit in the multiplier and add the products.

(1.7) **To divide whole numbers**, determine the largest whole number than can be multiplied by the divisor to obtain a product less than or equal to the dividend. Subtract that product from the dividend. Repeat the process until the result of the subtraction is less than the divisor.

To check the answer to a division problem, multiply the quotient by the divisor and add the remainder. The result should equal the dividend.

Chapter 1
Review Exercises

Indicate which of the following statements are true and which are false. For those that are false, change the italicized expression to make the statement true.

1. Positive numbers are to the *left* of zero on a number line.
2. The associative property *does* hold true for division.
3. Multiplication and *addition* are inverse operations.
4. The base ten system uses only *nine* digits.
5. The number 1 is the identity element for *addition.*
6. The commutative property does not hold true for subtraction and *multiplication.*

(1.1)   Write each of the following numbers in words and in expanded notation.

7. 543
8. 907
9. 4,892
10. 13,069
11. 153,872
12. 890,036
13. 4,079,581
14. 38,500,200
15. 8,491,737,495
16. 3,000,000,000,001

(1.1)   Compare each of the following pairs of numbers represented on the number line. Use the $=$, $>$, or $<$ sign to indicate the relationship between them.

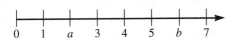

17. $4 \, ? \, 0$
18. $a \, ? \, 2$
19. $6 \, ? \, b$
20. $1 \, ? \, 7$
21. $b \, ? \, a$
22. $5 \, ? \, b$
23. $a \, ? \, 0$
24. $a \, ? \, b$

(1.2, 1.5)   State which property, the additive identity, the commutative or associative property of addition, the multiplicative identity, the commutative or associative property of multiplication, or the distributive property, is illustrated by each of the following expressions.

25. $3 \cdot 9 = 9 \cdot 3$
26. $8 + (3 + 7) = 8 + (7 + 3)$
27. $3(2 + 6) = 3(2) + 3(6)$
28. $(3 \cdot 6) \cdot 2 = 3 \cdot (6 \cdot 2)$
29. $8 \cdot (5 \cdot 4) = 8 \cdot (4 \cdot 5)$
30. $(5 + 2)(8 + 3) = 5(8 + 3) + 2(8 + 3)$
31. $59 \cdot 1 = 59$
32. $(3 + 4) + 7 = 3 + (4 + 7)$
33. $(5 \cdot 3) \cdot 9 = (3 \cdot 5) \cdot 9$
34. $4(13) - 4(5) = 4(13 - 5)$
35. $8 + 4(5) = 8 + 5(4)$
36. $(7 + 3) \cdot 2 = 2 \cdot (7 + 3)$
37. $3(4 + 5) = 3(5 + 4)$
38. $0 + 87 = 87$

(1.2–1.5) Perform each of the operations indicated.

39. $897 + 362$	**40.** $705 - 29$	**41.** $47 \cdot 62$	**42.** $149 \cdot 16$
43. $169 \div 5$	**44.** $1{,}507 + 838$	**45.** $300 - 53$	**46.** $96 \div 7$
47. $18 + 294 + 39$	**48.** $140 + 6{,}073 + 89$	**49.** $1{,}734 \div 11$	**50.** $8{,}369 \div 24$
51. $234 \cdot 407$	**52.** $678 \cdot 44$	**53.** $14{,}003 - 7{,}945$	**54.** $53{,}861 + 92{,}749$
55. $1{,}836 \cdot 493$	**56.** $83{,}492 \div 687$	**57.** $1{,}000{,}000 - 473{,}817$	**58.** $5{,}003 \cdot 7{,}020$

 59. $62{,}859{,}310 - 38{,}672{,}197$

60. $84{,}824 + 62{,}527 + 41{,}689 + 57{,}730$

61. $24{,}862 \cdot 1{,}897$

62. $51{,}228{,}684 \div 6{,}987$

(1.2–1.5) Translate each of the following statements into mathematical expressions.

63. 2 less than 5.

64. w more than 3.

65. The quotient of 33 and 3.

66. The difference between 15 and 6.

67. 8 times a

68. The sum of 5 and 9 is 14.

69. The product of 2 and 8 is 16.

70. The total of x, y, and z.

(1.2–1.5) Solve each of the following word problems.

71. A computer salesman traveled the following numbers of miles during a business week: Monday, 62 miles; Tuesday, 38 miles; Wednesday, 54 miles; Thursday, 107 miles; Friday, 88 miles. What is the total number of miles he traveled that week?

72. A student began her math homework at 7:47 P.M. and finished it 1 hr 25 min later. At what time did she finish the assignment?

73. Last year, the United States admitted 6,739 immigrants from Germany, 12,076 from England, 21,315 from China, and 92,367 from Mexico. What was the total number of immigrants admitted from these countries?

74. At the beginning of the month, a lawyer's checking account had a balance of $418. During the month she made deposits of $250 and $367 and she also wrote checks for $132, $173, $29, $156, $87, $43, and $62. What was her balance at the end of the month?

75. If a long-distance runner began a race at 1:47 P.M. and finished at 4:02 P.M., how long did it take him to run the race?

76. If one bag of groceries weighs 5 pounds 7 ounces and a second bag weighs 4 pounds 10 ounces, what is the combined weight of both bags of groceries?

77. A couple made a down payment of $90 on a refrigerator that cost $495, and they agreed to pay off the rest in 90 days to avoid interest charges. What will be their monthly payment?

78. During the time in which a motorist used 17 gallons of gas, her odometer reading changed from 13,979 miles to 14,438 miles. How many miles per gallon did she get?

79. The distance from New York to San Francisco is 2,571 miles, from San Francisco to Honolulu is 2,393 miles, from Honolulu to Tokyo is 3,853 miles, from Tokyo to Hong Kong is 1,794 miles, from Hong Kong to Calcutta is 1,648 miles, and from Calcutta to Moscow is 3,321 miles. What is the total distance traveled by a plane that flies this route?

80. A teacher's yearly gross salary is $36,000. Every month the following amounts are deducted from his paycheck: $582 for federal taxes, $198 for social security, $141 for state taxes, $124 for retirement fund, and $32 for health insurance. What is his monthly take-home pay?

81. If the population of London is 2,043,000 more than that of Bangkok, and the population of Bangkok is 4,875,000, what is the combined population of Bangkok and London?

82. During her freshman year, a student completed six 4-credit courses, three 2-credit courses, and two 1-credit courses. How many credits did she earn that year?

83. During a football game, a team scored three touchdowns, four field goals, and two extra points. What was its final score? (A touchdown is six points, and a field goal is three points.)

84. During a certain semester at a small college, 258 students took general chemistry, and twice the number of students who took chemistry took calculus. Three times the number of students who took calculus took English composition. What was the total number of students enrolled in these three courses for that semester?

85. A woman is deciding whether to buy a car that costs $6,972 or another that costs $6,144. She plans to make a down payment of $1,500 and pay off the balance over a 3-year period. How much more would her monthly payments be (ignoring interest) if she bought the more expensive car?

86. How many seconds are there in a week?

Chapter 1
Explain It in Words

87. Explain what is meant by the "associative property of addition." Give an example that illustrates this property.

88. Explain what is meant by the "commutative property of multiplication." Give an example that illustrates this property.

89. What is the identity element of addition? Give an example that illustrates this property.

90. What is the identity element of multiplication? Give an example that illustrates this property.

91. Why is our number system called the base ten system?

92. Explain how to check the answer to a division problem using an example.

Chapter 1
Chapter Test

[1.1]   Rewrite each of the following.

1. Write 7,093 in words.

2. Write 6,485 in expanded notation.

[1.2, 1.4]   State which property, the commutative or associative property of addition, the commutative or associative property of multiplication, the additive or multiplicative identity, or the distributive property, is illustrated by each of the following expressions.

3. $6(8 - 3) = 6(8) - 6(3)$

4. $8 \cdot (5 \cdot 4) = (8 \cdot 5) \cdot 4$

5. $12 \cdot 1 = 12$

6. $5 + (3 + 9) = 5 + (9 + 3)$

[1.2]   Add each of the following.

7. $64 + 93$

8. $538 + 284$

9. $35 + 821 + 5{,}673 + 907$

[1.3]   Subtract each of the following.

10. $37 - 19$

11. $708 - 52$

12. $9{,}304 - 763$

[1.4–1.6]   Multiply each of the following.

13. 673×4

14. 76×48

15. $5{,}307 \times 936$

16. $3{,}800 \times 16{,}000$

[1.7]   Divide each of the following.

17. $282 \div 3$

18. $3{,}411 \div 9$

19. $47{,}034 \div 54$

[1.2–1.7]   Translate each of the following into mathematical expressions.

20.   The quotient of seventy-two and nine.

21.   The sum of eight and three is eleven.

22.   Four less than fifteen.

23.   Six times five equals thirty.

[1.2–1.7]   Solve each of the following word problems.

24.   At the beginning of the month, an engineer had a balance of $721 in her checking account. If she wrote checks for $352, $93, $52, and $18, and she made a deposit of $234, what was her balance at end of the month?

25.   A student began his biology homework at 9:38 A.M. and finished it 3 hr 45 min later. At what time did he finish the assignment?

Chapter *2*

More on Whole Numbers

2.1
Exponents

Lengthy arithmetic expressions that occur frequently can be very cumbersome. Therefore, useful shorthand notations have been developed to make writing these expressions less clumsy. For example, we can express the product

$$\underbrace{2 \cdot 2 \cdot 2 \cdot 2 \cdot 2 \cdot 2 \cdot 2}_{7 \text{ twos}} \quad \text{as } 2^7$$

which is read "two to the seventh power." The shorthand notation 2^7 is an exponential expression. Two is called the **base** and 7 is called the **exponent**. The exponent, 7, tells us that the base, 2, appears as a factor seven times. Exponents are used to express repeated multiplication.

▶ *DEFINITION* **The expression b^n means that b appears as a factor n times. b is called the *base* and n is called the *exponent*, or *power*.**

For example, 5^3 means that 5 appears as a factor three times. The base is 5, and the exponent, or power, is 3.

$$\text{base} \rightarrow 5^3 \leftarrow \text{exponent}$$

The expression 5^3 is read as "five to the third power" or "five cubed."

Keep in mind that "five to the third power" simplifies to 125. The exponent, 3, indicates how many times to write the base, 5, as a factor.

$$5^3 = 5 \cdot 5 \cdot 5 = 125$$

Do not confuse this with the expression read as "five times three" which simplifies to 15.

$$5 \cdot 3 = 15$$

EXAMPLE 1 Rewrite each of the following expressions using exponents.

(a) $3 \cdot 3 \cdot 3 \cdot 3$

$$\underbrace{3 \cdot 3 \cdot 3 \cdot 3}_{4 \text{ threes}} = 3^4$$

3 appears as a factor 4 times.

(b) $5 \cdot 5 \cdot 4 \cdot 4 \cdot 4$

$$\underbrace{5 \cdot 5}_{2 \text{ fives}} \cdot \underbrace{4 \cdot 4 \cdot 4}_{3 \text{ fours}} = 5^2 \cdot 4^3$$

5 appears as a factor 2 times, and 4 appears as a factor 3 times.

(c) $9 \cdot 9 \cdot 8 \cdot 7 \cdot 7 \cdot 7$

$$\underbrace{9 \cdot 9}_{2 \text{ nines}} \quad \underbrace{\cdot 8 \cdot}_{1 \text{ eight}} \quad \underbrace{7 \cdot 7 \cdot 7}_{3 \text{ sevens}} = 9^2 \cdot 8^1 \cdot 7^3$$

Notice that since 8 appears as a factor once, it may be rewritten as 8^1, or "8 to the first power." ▲

EXAMPLE 2 Rewrite each of the following numbers as an exponential expression with a base of 2 or 3.

(a) 9

$9 = 3^2$ since $3 \cdot 3 = 9$.

3^2 is read as "3 to the second power" or "3 squared."

(b) 8

$8 = 2^3$ since $2 \cdot 2 \cdot 2 = 8$.

2^3 is read as "2 to the third power" or "2 cubed." ▲

▶ **DEFINITION** For any base b, $b^1 = b$. In other words, any number raised to the first power is the number itself.

For example, $2^1 = 2$, $15^1 = 15$, $304^1 = 304$, and so on.

▶ **DEFINITION** For any base $b \neq 0$, $b^0 = 1$. In other words, any nonzero number raised to the zero power equals one.

For example, $5^0 = 1$, $12^0 = 1$, $153^0 = 1$, and so on.

EXAMPLE 3 Express the following exponential expressions as products, and then simplify to whole numbers.

(a) $2^3 \cdot 3^2$

$$\begin{aligned} 2^3 \cdot 3^2 &= (2 \cdot 2 \cdot 2) \cdot (3 \cdot 3) \\ &= 8 \cdot 9 = 72 \end{aligned}$$

(b) $2^2 \cdot 5^2 \cdot 3^3$

$$\begin{aligned} 2^2 \cdot 5^2 \cdot 3^3 &= (2 \cdot 2) \cdot (5 \cdot 5) \cdot (3 \cdot 3 \cdot 3) \\ &= 4 \cdot 25 \cdot 27 = 100 \cdot 27 = 2,700 \end{aligned}$$

(c) $4^2 \cdot 8^0 \cdot 10^4$

$$\begin{aligned} 4^2 \cdot 8^0 \cdot 10^4 &= (4 \cdot 4) \cdot 1 \cdot (10 \cdot 10 \cdot 10 \cdot 10) \\ &= 16 \cdot 10,000 = 160,000 \end{aligned}$$ ▲

Notice that in Example 3(c), the product $16 \cdot 10^4 = 160,000$, which is the number 16 followed by four zeros.

Any number multiplied by 10^n, where n is a whole number, is that number followed by n zeros.   ▲

EXAMPLE 4   Simplify each of the following exponential expressions.

(a) $6^2 \cdot 10^2$

$6^2 \cdot 10^2 = (6 \cdot 6) \cdot 10^2$

$= 36 \cdot 10^2$

$= 3,600$ 36 followed by two zeros

(b) $3^3 \cdot 10^5$

$3^3 \cdot 10^5 = (3 \cdot 3 \cdot 3) \cdot 10^5$

$= 27 \cdot 10^5$

$= 2,700,000$ 27 followed by five zeros

(c) $2^5 \cdot 10^4$

$2^5 \cdot 10^4 = (2 \cdot 2 \cdot 2 \cdot 2 \cdot 2) \cdot 10^4$

$= 32 \cdot 10^4$

$= 320,000$ 32 followed by four zeros   ▲

Every place value in the base ten system can be rewritten as a **power of ten**, as shown in Table 2.1. Every whole number can thus be rewritten in expanded notation using powers of ten. For example,

$5,823 = (5 \times 1,000) + (8 \times 100) + (2 \times 10) + (3 \times 1)$

$= (5 \times 10^3) + (8 \times 10^2) + (2 \times 10^1) + (3 \times 10^0)$

TABLE 2.1
Place Values as
Powers of Ten

Ones	$10^0 = 1$	Hundred thousands	$10^5 = 100,000$
Tens	$10^1 = 10$	Millions	$10^6 = 1,000,000$
Hundreds	$10^2 = 100$	Ten millions	$10^7 = 10,000,000$
Thousands	$10^3 = 1,000$	Hundred millions	$10^8 = 100,000,000$
Ten thousands	$10^4 = 10,000$	Billions	$10^9 = 1,000,000,000$

EXAMPLE 5   Rewrite each of the following numbers in expanded notation using powers of ten.

(a)   742

$742 = (7 \times 100) + (4 \times 10) + (2 \times 1)$

$= (7 \times 10^2) + (4 \times 10^1) + (2 \times 10^0)$

(b)   30,469

$30,469 = (3 \times 10,000) + (0 \times 1,000) + (4 \times 100) + (6 \times 10) + (9 \times 1)$

$= (3 \times 10^4) + (0 \times 10^3) + (4 \times 10^2) + (6 \times 10^1) + (9 \times 10^0)$

$= (3 \times 10^4) + (4 \times 10^2) + (6 \times 10^1) + (9 \times 10^0)$ ▲

When using exponents, we can detect certain properties that illustrate rules that enable us to simplify our calculations.

Let us first look at the following examples.

$$2^2 \cdot 2^3 = \underbrace{(2 \cdot 2) \cdot (2 \cdot 2 \cdot 2)}_{5 \text{ twos}} = 2^5$$

$$3^3 \cdot 3^4 = \underbrace{(3 \cdot 3 \cdot 3) \cdot (3 \cdot 3 \cdot 3 \cdot 3)}_{7 \text{ threes}} = 3^7$$

$$5^2 \cdot 5^3 \cdot 5^4 = \underbrace{(5 \cdot 5) \cdot (5 \cdot 5 \cdot 5) \cdot (5 \cdot 5 \cdot 5 \cdot 5)}_{9 \text{ fives}} = 5^9$$

In each of these examples, we multiplied exponential expressions of the same base. Notice the relationship between the exponents of the numbers being multiplied and the exponent of the result. In each case, the result could have been found by raising the common base to an exponent that is the sum of all the exponents in the expressions being multiplied. Thus, we have illustrated the following rule.

▶ **RULE 1** To multiply exponential expressions of the same base, keep the base the same and add the exponents.

Using this rule to simplify the previous examples, we obtain:

$$2^2 \cdot 2^3 = 2^{2+3} = 2^5$$
$$3^3 \cdot 3^4 = 3^{3+4} = 3^7$$
$$5^2 \cdot 5^3 \cdot 5^4 = 5^{2+3+4} = 5^9$$

Don't forget that when using this rule the bases must be the same in order for you to add the exponents.

When multiplying very large numbers that end in zeros, it is often convenient to first rewrite each factor using powers of ten, and then use Rule 1 to calculate the product.

EXAMPLE 6 Calculate the following products by first rewriting each factor using powers of ten.

(a) 30×600

$30 \times 600 = (3 \times 10^1) \times (6 \times 10^2)$	Rewrite using powers of ten.
$= (3 \times 6) \times (10^1 \times 10^2)$	Use the commutative and associative properties.
$= 18 \times 10^3$	Multiply powers of the same base.

(b) $8,000 \times 5,000$

$$8,000 \times 5,000 = (8 \times 10^3) \times (5 \times 10^3)$$
$$= (8 \times 5) \times (10^3 \times 10^3)$$
$$= 40 \times 10^6$$
$$= 4 \times 10^1 \times 10^6$$
$$= 4 \times 10^7$$

(c) $70,000 \times 200,000$

$70,000 \times 200,000 = (7 \times 10^4) \times (2 \times 10^5)$	Notice that 14×10^9 is much easier to read than 14,000,000,000. For this reason, very large numbers ending in zeros are almost always expressed using powers of ten.
$= (7 \times 2) \times (10^4 \times 10^5)$	
$= 14 \times 10^9$	

(d) 120,000 × 3,000,000

$$120,000 \times 3,000,000 = (12 \times 10^4) \times (3 \times 10^6)$$
$$= (12 \times 3) \times (10^4 \times 10^6)$$
$$= 36 \times 10^{10}$$

Since 120,000 is 12 followed by four zeros, it can be rewritten as 12×10^4. ▲

Let us now look at the following examples, which illustrate what happens when we raise an exponential expression to a power.

$$(2^2)^3 = (2^2)(2^2)(2^2) = 2^6$$

$$(5^4)^2 = (5^4)(5^4) = 5^8$$

$$(9^5)^4 = (9^5)(9^5)(9^5)(9^5) = 9^{20}$$

Notice that in each case, the base remains unchanged. Also notice that the exponent of each result is the product of the exponents in the original expression. Thus, we have illustrated another rule for simplifying exponential expressions.

▶ **RULE 2** To raise an exponential expression to a power, keep the base the same and multiply the exponents.

Using this rule to simplify the previous examples, we obtain:

$$(2^2)^3 = 2^{2 \cdot 3} = 2^6$$

$$(5^4)^2 = 5^{4 \cdot 2} = 5^8$$

$$(9^5)^4 = 9^{5 \cdot 4} = 9^{20}$$

Be careful of the distinction between multiplying exponential expressions of the same base and raising an exponential expression to a power.

$$5^7 \cdot 5^3 = 5^{7+3} = 5^{10}$$

Here we are multiplying exponential expressions of the same base, so we add the exponents.

$$(5^7)^3 = 5^{7 \cdot 3} = 5^{21}$$

Here we are raising an exponential expression to a power, so we multiply the exponents.

EXAMPLE 7 Simplify each of the following operations using the rules for exponents.

(a) $6^3 \cdot 6^5$

$$6^3 \cdot 6^5 = 6^{3+5} = 6^8$$

Here we are multiplying exponential expressions of the same base, so we add exponents.

(b) $4^3 \cdot 4^5 \cdot 4^7$

$$4^3 \cdot 4^5 \cdot 4^7 = 4^{3+5+7} = 4^{15}$$

Here, again, we are multiplying exponential expressions of the same base, so we add exponents.

(c) $3 \cdot 10^4 \cdot 5 \cdot 10^6$

$$3 \cdot 10^4 \cdot 5 \cdot 10^6 = (3 \cdot 5) \cdot (10^4 \cdot 10^6)$$
$$= 15 \cdot 10^{4+6}$$
$$= 15 \cdot 10^{10}$$

Use the commutative and associative properties.

(d) $(7^4)^2$
$$(7^4)^2 = 7^{4 \cdot 2} = 7^8$$

Here we are raising an exponential expression to a power, so we multiply exponents.

(e) $(8^3)^4 \cdot (5^2)^3$
$$(8^3)^4 \cdot (5^2)^3 = 8^{3 \cdot 4} \cdot 5^{2 \cdot 3}$$
$$= 8^{12} \cdot 5^6$$

Here we are raising two exponential expressions to powers, so we multiply exponents in both cases.

(f) $2^5 \cdot 2^3 \cdot (2^3)^4$
$$2^5 \cdot 2^3 \cdot (2^3)^4 = 2^5 \cdot 2^3 \cdot 2^{3 \cdot 4}$$
$$= 2^5 \cdot 2^3 \cdot 2^{12}$$
$$= 2^{5+3+12}$$
$$= 2^{20}$$

Here we use both rules. First raise the exponential expression to a power by multiplying the exponents. Then multiply the exponential expressions by adding the exponents. ▲

Quick Quiz

		Answers	
1.	Rewrite $7 \cdot 7 \cdot 7 \cdot 2 \cdot 2 \cdot 2 \cdot 2 \cdot 2$ using exponents.	**1.**	$7^3 \cdot 2^5$
2.	Express $4^2 \cdot 5^0 \cdot 3^1$ as a whole number.	**2.**	48
3.	Simplify $6^2 \cdot 6^7$.	**3.**	6^9
4.	Simplify $(3^8)^2$.	**4.**	3^{16}

2.1 Exercises

Indicate which of the following statements are true and which are false. For those that are false, change the italicized word to make the statement true.

1. In the expression b^n, b is called the *exponent*.

2. An exponential expression indicates repeated *addition*.

3. Any nonzero number raised to the zero power equals *one*.

4. When multiplying exponential expressions of the same base, keep the base the same and *multiply* the exponents.

Rewrite each of the following expressions using exponents.

5. $3 \cdot 3 \cdot 3 \cdot 8 \cdot 8$

6. $5 \cdot 5 \cdot 9 \cdot 9 \cdot 9 \cdot 9 \cdot 9$

7. $2 \cdot 2 \cdot 2 \cdot 2 \cdot 2 \cdot 2 \cdot 6 \cdot 7 \cdot 7$

8. $3 \cdot 3 \cdot 4 \cdot 4 \cdot 5 \cdot 5 \cdot 9$

9. $5 \cdot 5 \cdot 5 \cdot 8 \cdot 8 + 5 \cdot 5 \cdot 7 \cdot 7 \cdot 7$

10. $4 \cdot 4 \cdot 4 \cdot 9 \cdot 9 \cdot 9 \cdot 9 + 2 \cdot 2 \cdot 4 \cdot 6 \cdot 6$

Rewrite each of the following numbers as an exponential expression with a base of 2, 3, 5, or 7.

11. 4

12. 27

13. 25

14. 49

15. 32

16. 125

17. 81

18. 16

19. 625

20. 64

Rewrite the following exponential expressions as whole numbers.

21. $2^2 \cdot 5^2$

22. $2^3 \cdot 4^2$

23. $6^0 \cdot 3^2$

24. $7^2 \cdot 2^1$

25. $3^3 \cdot 5^0 \cdot 8^1$

26. $2^2 \cdot 5^3 \cdot 7^0$ 27. $5^1 \cdot 2^2 \cdot 6^2$ 28. $2^5 \cdot 1^3 \cdot 3^2$ 29. $7^2 \cdot 10^3$ 30. $4^3 \cdot 10^2$

31. $2^4 \cdot 10^4$ 32. $3^3 \cdot 10^3$ 📱 33. $6^7 \cdot 11^2$ 📱 34. $3^9 \cdot 4^5$

Write each of the following numbers in expanded notation using powers of ten.

35. 538   36. 914   37. 602   38. 830

39. 7,436   40. 2,951   41. 41,783   42. 20,841

43. 74,062   44. 80,027   45. 340,005   46. 804,070

47. 6,947,089   48. 3,078,614   49. 5,201,840,693   50. 7,003,438,020

Calculate the following products by first rewriting each factor using powers of ten.

51. 70×400 52. 20×900 53. $600 \times 8,000$

54. $9,000 \times 300$ 55. $1,100 \times 7,000$ 56. $6,000 \times 12,000$

57. $40,000 \times 50,000$ 58. $60,000 \times 500,000$ 59. $700,000 \times 300,000$

60. $5,000,000 \times 9,000,000$ 61. $250,000 \times 200,000$ 62. $15,000,000 \times 4,000,000$

Simplify each of the following exponential expressions using the rules for exponents.

63. $3^4 \cdot 3^5$ 64. $8^4 \cdot 8^9$ 65. $4^2 \cdot 4^4$ 66. $5^5 \cdot 5^7$

67. $2^3 \cdot 2^5 \cdot 2^6$ 68. $6^3 \cdot 6^0 \cdot 6^9$ 69. $5^2 \cdot 5^4 \cdot 5^5$ 70. $4^8 \cdot 4^3 \cdot 4^0$

71. $(3^4)^3$ 72. $(8^5)^3$ 73. $(4^2)^5$ 74. $(7^3)^6$

75. $(7^2)^3 \cdot 7^4$ 76. $(9^4)^6 \cdot 9^7$ 77. $5^3 \cdot (5^4)^2$ 78. $4^3 \cdot (4^5)^4$

79. $8^2 \cdot 8^4 \cdot 6^2 \cdot 6^5$ 80. $3^5 \cdot 3^3 \cdot 9^4 \cdot 9^7$ 81. $2^3 \cdot 10^5 \cdot 2^6 \cdot 10^3$ 82. $8^2 \cdot 10^4 \cdot 8^5 \cdot 10^7$

83. $7^8 \cdot 8^6 \cdot 7^5 \cdot 8^2$ 84. $5^4 \cdot 6^3 \cdot 5^9 \cdot 6^7$ 85. $(6^4)^2 \cdot (6^3)^4$ 86. $(8^2)^4 \cdot (8^5)^2$

87. $(4^8)^3 \cdot (4^7)^2$ 88. $(3^5)^3 \cdot (3^4)^2$

2.2
Order of Operations

Now that we have learned how to perform a variety of arithmetic operations, it is important to learn how to solve problems that involve more than one operation. Operations inside parentheses are to be completed before those on the outside. However, problems that contain no parentheses present us with the possibility of different answers depending on the order in which we perform the various operations. For example, if we attempt to simplify $3 + 5 \cdot 4$, we obtain two different answers, depending upon which operation we perform first. Performing addition first we obtain

$$3 + 5 \cdot 4 = (3 + 5) \cdot 4 = 8 \cdot 4 = 32$$

Performing multiplication first we obtain

$$3 + 5 \cdot 4 = 3 + (5 \cdot 4) = 3 + 20 = 23$$

To eliminate this confusion, rules have been developed for the order of operations. They establish a conventional order that must be followed when simplifying arithmetic expressions (see next page). According to the order of operations rules, to solve the previous example correctly, we perform multiplication before addition:

$$3 + 5 \cdot 4 = 3 + (5 \cdot 4) = 3 + 20 = 23$$

The Order of Operations Rules	1. Perform all operations in parentheses first.
	2. Simplify all exponential expressions.
	3. Do multiplication and division as they occur, moving from left to right.
	4. Do addition and subtraction as they occur, moving from left to right. ▲

If there is more than one operation inside parentheses, simplify the expressions inside parentheses using the order of operations rules.

EXAMPLE 1 Simplify each of the following expressions using the order of operations rules.

(a) $8 \cdot (4 + 6)$

$8 \cdot (4 + 6) = 8 \cdot 10 = 80$ We do addition first since it is enclosed in parentheses.

(b) $9 - 2 \cdot 3$

$9 - 2 \cdot 3 = 9 - 6 = 3$ Do multiplication before subtraction.

(c) $6 + 8 \div 2$

$6 + 8 \div 2 = 6 + 4$ Do division before addition.
$\qquad\qquad = 10$

(d) $12 \div 2 \times 6$

$12 \div 2 \times 6 = 6 \times 6$ Do division.
$\qquad\qquad\quad = 36$ Do multiplication.

(e) $48 \div 8 \cdot 5 \div 2$

$48 \div 8 \cdot 5 \div 2 = 6 \cdot 5 \div 2$ Do multiplication and division moving from left to right.
$\qquad\qquad\qquad = 30 \div 2 = 15$

(f) $2^3 \cdot 3^2$

$2^3 \cdot 3^2 = 8 \cdot 9 = 72$ Simplify exponential expressions and then multiply.

(g) $9 \cdot 2^2 - 3 + 5 \cdot 2$

$9 \cdot 2^2 - 3 + 5 \cdot 2 = 9 \cdot 4 - 3 + 5 \cdot 2$ Simplify the exponential expression.
$\qquad\qquad\qquad = 36 - 3 + 10$ Do multiplication.
$\qquad\qquad\qquad = 33 + 10$ Do subtraction.
$\qquad\qquad\qquad = 43$ Do addition. ▲

EXAMPLE 2 Simplify each of the following using the order of operations rules.

(a) $5^2(9 - 3) + 2(8 + 7)$

$5^2(9 - 3) + 2(8 + 7) = 5^2(6) + 2(15)$ Do operations in ( ) first.
$\qquad\qquad\qquad = 25(6) + 2(15)$ Simplify exponential expressions.
$\qquad\qquad\qquad = 150 + 30$ Do multiplication.
$\qquad\qquad\qquad = 180$ Do addition.

(b) $[4(3 + 2) - 3(9 - 7)] \div 7$

$[4(3 + 2) - 3(9 - 7)] \div 7 = [4(5) - 3(2)] \div 7$ Simplify innermost ( ) first.

$\qquad\qquad\qquad = [20 - 6] \div 7$ In [ ] multiply before

$\qquad\qquad\qquad = 14 \div 7$ subtracting.

$\qquad\qquad\qquad = 2$ Do subtraction.

   Do division.

Notice that when parentheses ( ) are nested inside brackets [ ], we simplify the innermost parentheses first, from left to right, and work outward.

(c) $4[3(5 - 3) + (6 - 4)^2]$

$4[3(5 - 3) + (6 - 4)^2] = 4[3(2) + 2^2]$ Simplify innermost ( ) first.

$\qquad\qquad\qquad = 4[6 + 4]$ In [ ] raise to powers, then multiply.

$\qquad\qquad\qquad = 4[10]$ In [ ] do addition.

$\qquad\qquad\qquad = 40$ Do multiplication.

(d) $5^2 \cdot 10^3 + 3^3 \cdot 10^2 + 7^2 \cdot 10^1$

$5^2 \cdot 10^3 + 3^3 \cdot 10^2 + 7^2 \cdot 10^1$

$\qquad\qquad = 25 \cdot 1{,}000 + 27 \cdot 100 + 49 \cdot 10$ First raise to powers.

$\qquad\qquad = 25{,}000 + 2{,}700 + 490$ Next do multiplication.

$\qquad\qquad = 28{,}190$ Finally, do addition.

(e) $3^2 \cdot 3^3 + 3^6$

$3^2 \cdot 3^3 + 3^6 = 3^{2+3} + 3^6$ WARNING: Do not attempt to simplify

$\qquad\qquad = 3^5 + 3^6$ further. We have not established a rule for

   adding powers of the same base. ▲

For problems expressed in words, the wording often indicates which operation should be performed first. When translating words into mathematical expressions, we must remember to use parentheses to enclose those operations that are to be performed first.

EXAMPLE 3   Translate the following words into mathematical expressions and simplify.

(a) 3 less than the product of 7 and 4.

$7 \cdot 4 - 3 = 28 - 3 = 25$ Parentheses are not needed since we perform multiplication before subtraction.

(b) 9 more than the quotient of 6 and 3.

$6 \div 3 + 9 = 2 + 9 = 11$

(c) 5 less than the product of 3 and 4.

$3 \cdot 4 - 5 = 12 - 5 = 7$ Since multiplication is performed before subtraction, no ( ) are needed.

(d) 4 times the sum of 7 and 2.

$4 \cdot (7 + 2) = 4 \cdot 9 = 36$

(e) 6 more than 3 squared.

$3^2 + 6 = 9 + 6 = 15$ Since we simplify powers before multiplying, no ( ) are needed.

(f) The sum of the product of 5 and 4 and the quotient of 18 and 2.

$(5 \cdot 4) + (18 \div 2) = 20 + 9 = 29$ Parentheses are not needed but they do simplify translation.

(g) The difference between the sum of 9 and 8 and the product of 3 and 5.

$(9 + 8) - (3 \cdot 5) = 17 - 15 = 2$ ▲

Quick Quiz

Simplify each of the following expressions.   **Answers**

1. $7 + 4 \cdot 5$ 1. 27
2. $8 \div 2 + 6 \cdot 2 - 9 \div 3$ 2. 13
3. $[(6 - 2)^2 + (9 - 7)^3] \div 4$ 3. 6

2.2 Exercises

Indicate which of the following statements are true and which are false. For those that are false, change the italicized expression to make the statement true.

1. Parentheses enclose the operation to be performed *last*.

2. Unless otherwise indicated by parentheses, we perform addition *before* multiplication.

3. Multiplication and division are performed in *the same* step moving from left to right.

4. If parentheses are nested inside brackets, we evaluate the *innermost* expression first.

Simplify each of the following expressions using the order of operations rules.

5. $5 + 3 \cdot 4$
6. $11 - 2 \cdot 4$
7. $5^2 \cdot 2^2$
8. $3^2 \cdot 2^3$
9. $18 \div 3 + 7$
10. $7 - 21 \div 3$
11. $32 \div 8 + 6 \cdot 2$
12. $5 \cdot 4 - 24 \div 8$
13. $2 \cdot 5^3 + 3^3$
14. $8^0 \cdot 6 - 2^2$
15. $72 \div 8 + 63 \div 9$
16. $56 \div 7 - 24 \div 8$
17. $8(2 + 3)$
18. $5(6 + 2)$
19. $2 \cdot 4^2 + 7^0 \cdot 3$
20. $3^2 \cdot 7 - 2^3 \div 4$
21. $81 \div (13 - 4)$
22. $66 \div (5 + 6)$
23. $56 \div 7 \div 4$
24. $72 \div 8 \div 3$
25. $5(6 - 3)^3$
26. $(5 + 1)^2 \div 4$
27. $(6 \cdot 3 - 5 + 2) \div 3$
28. $(4 \cdot 7 - 6 + 5) \div 9$
29. $6^2 \cdot 10^2 + 5^2 \cdot 10^1$
30. $4^2 \cdot 10^3 + 3^2 \cdot 10^2$
31. $2^3 \cdot 10^3 + 2^3 \cdot 10^2 + 2^3 \cdot 10^1 + 2^3 \cdot 10^0$
32. $3^2 \cdot 10^3 + 3^2 \cdot 10^2 + 3^2 \cdot 10^1 + 3^2 \cdot 10^0$
33. $63 \div 7 + 5 \cdot 2 - 6 \div 3$
34. $8 \cdot 6 - 4 \cdot 3 + 14 \div 2$
35. $3(4 + 5) - 8(6 - 3)$
36. $7(8 - 4) + 3(4 + 3)$
37. $3(8 - 5)^2 - 36 \div (13 - 4)$
38. $48 \div (17 - 9) + 4(7 - 5)^3$
39. $[6(2 + 3) - 4(7 - 2)] \div 5$
40. $[8(13 - 8) - 3(11 - 5)] \div 2$
41. $2[(5 - 3)^2 + 4(10 - 7)]$
42. $5[6(11 - 8) - (12 - 8)^2]$
43. $7^5 + 7^3 \cdot 7^4$
44. $6^4 \cdot 6^5 + 6^2$
45. $5^2 \cdot 5^8 + (5^3)^2$
46. $2^9 \cdot 2^2 + (2^4)^3$

47. $85{,}008 \div 184 + 67 \cdot 23$

48. $347 \cdot 216 \div (4{,}086 - 3{,}978)$

49. $[(13)^5 - (28)^3] \div 3$

50. $(34)(19)^3 + (55)(6)^5$

Find the mistake in each of the following equations and determine the correct answer.

51. $8 + 3 \cdot 2 = 11 \cdot 2 = 22$
52. $4 + 2 \cdot 5 = 6 \cdot 5 = 30$
53. $6 - 2 \div 2 = 4 \div 2 = 2$
54. $9 - 3 \div 3 = 6 \div 3 = 2$
55. $2 \cdot 3^2 = 6^2 = 36$
56. $5 \cdot 2^2 = 10^2 = 100$
57. $8 - 6 + 2 = 8 - 8 = 0$
58. $9 - 5 - 3 = 9 - 2 = 7$
59. $(4 + 5)^2 = 4 + 25 = 29$
60. $(9 - 2)^2 = 9 - 4 = 5$
61. $3(8 - 6) = 24 - 6 = 18$
62. $5(4 + 3) = 20 + 3 = 23$

Translate each of the following word expressions into mathematical expressions and simplify.

63. Twice the sum of 6 and 2.

64. 4 times the difference between 8 and 3.

65. 2 less than the product of 5 and 4.

66. 5 more than the quotient of 15 and 3.

67. 6 more than 2 cubed.

68. 9 less than 5 squared.

69. The sum of the product of 3 and 4 and the quotient of 10 and 2.

70. The difference between the product of 8 and 3 and the product of 6 and 2.

71. The product of the sum of 2 and 5 and the sum of 6 and 3.

72. The product of the difference between 9 and 2 and the sum of 5 and 4.

2.3
Averages

One of the most common applications of division is to calculate an **average**. Our daily living experiences include many references to averages, such as batting averages, average test grades, average income, average speed, and the Dow Jones Industrial Average.

▶ **DEFINITION** The *average* of a group of numbers is a number that is used to represent the entire group. To calculate an average, first add all the numbers in the group, and then divide that sum by the number of numbers that were added together.

In mathematics, the average is often referred to as the *mean*.

EXAMPLE 1 Calculate the average of the numbers 28, 17, 34, 53, 21, 38, 19.

Solution Add the seven numbers. Then divide the total by 7.

$$
\begin{array}{r}
4 \\
28 \\
17 \\
34 \\
53 \\
21 \\
38 \\
19 \\
\hline
210
\end{array}
\qquad
\begin{array}{r}
30 \\
7\,\overline{)210} \\
21 \\
\hline
0
\end{array}
$$

The average is 30.   ▲

Notice that the average of a given group of numbers is always greater than or equal to the smallest number in the group and less than or equal to the largest number in the group. For instance, in Example 1, $17 < 30 < 53$.

EXAMPLE 2 A student received grades of 83, 67, 75, 72, and 88 on his mathematics quizzes. What is his average quiz grade?

Solution Add the five quiz grades. Then divide that sum by 5.

$$
\begin{array}{r}
83 \\
67 \\
75 \\
72 \\
\underline{88} \\
385
\end{array}
\qquad
\begin{array}{r}
77 \\
5\overline{\smash{)}385} \\
\underline{35} \\
35 \\
\underline{35} \\
0
\end{array}
$$

His average quiz grade is 77. ▲

EXAMPLE 3 A stockbroker purchased a variety of stocks for a client at the following prices: six shares at $35 a share, twenty shares at $18 a share, four shares at $40 a share, and ten shares at $27 a share. What was the average price per share of all the stocks purchased?

Solution Calculate the total amount of money spent by multiplying the number of shares purchased at each price by the price per share and then adding those results. Then divide that sum by the total number of shares purchased.

Price Per Share	×	Number of Shares	
$35	×	6	= $ 210
$18	×	20	= $ 360
$40	×	4	= $ 160
$27	×	10	= $ 270
	Total:	40	$1,000

Forty shares were purchased for a total of $1,000. To calculate the average price per share, divide $1,000 by 40:

$$
\begin{array}{r}
25 \\
40\overline{\smash{)}1{,}0\,0\,0} \\
\underline{8\,0} \\
2\,0\,0 \\
\underline{2\,0\,0} \\
0
\end{array}
$$

The average price per share is $25. ▲

When you are asked to calculate an average, it is not always necessary to add up numbers before dividing, because the total is often given in the problem.

EXAMPLE 4 During a particular week at a computer firm, 518 employees worked a total of 19,684 hours. What was the average number of hours worked by each employee that week?

Solution Divide $19,684 \div 518$.

$$
\begin{array}{r}
38 \\
518\overline{)19,684} \\
1554 \\
\hline
4144 \\
4144 \\
\hline
0
\end{array}
$$

On the average, each employee worked 38 hours that week. ▲

Quick Quiz

Find the average of each of the following groups of numbers.

1. 9, 13, 6, 20
2. 216, 462, 381

Answers

1. 12
2. 353

2.3 Exercises

Indicate which of the following statements are true and which are false. For those that are false, change the italicized expressions to make the statement true.

1. When calculating an average, we perform the operation addition *before* division.

2. The *mean* is another name for the average.

3. The average of a group of whole numbers is always *greater than* the largest number in the group.

4. To calculate the average of three numbers, we divide the sum of the numbers by *4*.

Calculate the average of each of the following groups of numbers.

5. 54, 32, 71, 27
6. 89, 65, 38, 41, 72
7. 314, 281, 437
8. 652, 587, 736, 801
9. 52,679, 47,325, 26,839, 39,461
10. 857, 799, 934, 687, 492, 865, 528, 753, 412, 683, 536

Solve each of the following word problems.

11. The high temperature for three days in May was 78°F, 71°F, and 67°F. What was the average high temperature for those three days?

12. The monthly rent for three different two-bedroom apartments is $958, $1,072, and $985. What is the average monthly rent?

13. A nurse took a patient's pulse and got the following readings: 62 beats per minute, 79 beats per minute, 86 beats per minute, and 73 beats per minute. What was the patient's average pulse rate?

14. During eight trials of a science experiment, the amount of time needed for a mouse to complete a maze was recorded to be 14 minutes, 10 minutes, 13 minutes, 12 minutes, 11 minutes, 17 minutes, 16 minutes, and 11 minutes. What was the average time the mouse needed to complete the maze?

15. A ten-year-old practiced the piano for 35 minutes on Monday, 27 minutes on Tuesday, 24 minutes on Wednesday, 42 minutes on Thursday, 18 minutes on Friday, 31 minutes on Saturday, and 26 minutes on Sunday. What was the average length of a daily practice session that week?

16. The heights of the players on the starting lineup of a basketball team are 78 inches, 81 inches, 84 inches, 83 inches, and 79 inches. What is the average height of a player on that team?

17. A secretary types 868 words in 14 minutes. What is his average typing speed in words per minute?

18. A plane flew 4,284 miles in 12 hours. What was its average speed in miles per hour?

19. A salesman drove 235 miles on Monday, 318 miles on Tuesday, 203 miles on Wednesday, 188 miles on Thursday, and 126 miles on Friday. What was the average distance traveled daily during that period?

20. The members of a bowling team bowled the following scores: 198, 160, 214, 231, and 187. What was the average score for that team?

21. On a math quiz worth 10 points, eight students scored 10, twelve students scored 9, five students scored 8, four students scored 5, one student scored 4, and two students scored 2. What was the average score on that quiz?

22. The following bets were made on the outcome of a football game: six people bet $1, three bet $5, four bet $10, one bet $15, and three bet $20. What was the size of the average bet?

23. A contest is advertising that one $7,500 prize, four $2,500 prizes, ten $1,000 prizes, and twenty-five $500 prizes will be awarded to winners. What will be the average amount awarded?

24. The 18 members of a track team ran the following races: five ran the 100-meter race, two ran the 200-meter, three ran the 400-meter, two ran the 1,000-meter, four ran the 5,000-meter, and two ran the 10,000-meter. What was the average distance run by a member of that track team?

25. If 51,100 people died in traffic accidents last year, what was the average number of deaths per day related to motor vehicles last year?

26. If 6,135 students at a certain university bought a total of 79,755 books in one semester, what was the average number of books bought by a student for one semester of study at that university?

27. The area in square miles of five western European countries are: France, 212,973; Spain, 194,884; Italy, 116,304; West Germany, 95,815; and the United Kingdom, 94,249. What is the average area?

28. The number of marriages in the United States for the past eight years was 2,158,802; 2,190,481; 2,282,154; 2,284,108; 2,229,667; 2,152,662; 2,154,807; and 2,178,367. What was the average number of marriages per year for this period?

2.4
Short Division and Tests for Divisibility

Short Division

When dividing by a one-digit number, we can save ourselves some work by using a technique called *short division*. Instead of doing the subtraction to find the remainder for each step as we do in long division, we mentally subtract and write the remainder in front of the next digit in the dividend. Example 1 illustrates the difference between the technique we have been using, long division, and short division.

EXAMPLE 1 Divide $4,524 \div 6$.

Solution

Long Division

$$
\begin{array}{r}
7\,5\,4 \\
6\,\overline{)4{,}5\,2\,4} \\
4\,2 \\
\hline
3\,2 \\
3\,0 \\
\hline
2\,4 \\
2\,4 \\
\hline
0
\end{array}
$$

Short Division

$$6\,\overline{)4\ 5\ {}^{3}2\ {}^{2}4}\quad 7\ 5\ 4$$

6 goes into 45 seven times with a remainder of 3. Put 7 over the dividend. Write 3 in front of the next digit.

6 goes into 32 five times with R2. Put 5 over the dividend. Write 2 in front of the next digit.

6 goes into 24 four times with R0. Put 4 over the dividend.

The answer is 754. ▲

EXAMPLE 2 Divide each of the following numbers using short division.

(a) $31,825 \div 5$

$$5\,\overline{)3\ 1\ {}^{1}8\ {}^{3}2\ {}^{2}5}\quad 6\ 3\ 6\ 5$$

5 goes into 31 six times with R1. Put 6 over the dividend. Write 1 in front of the next digit.

5 goes into 18 three times with R3. Put 3 over the dividend. Write 3 in front of the next digit.

5 goes into 32 six times with R2. Put 6 over the dividend. Write 2 in front of the next digit.

5 goes into 25 five times with R0. Write 5 over the dividend.

The answer is 6,365.

(b) $9,063 \div 7$

$$7\,\overline{)9\ {}^{2}0\ {}^{6}6\ {}^{3}3}\quad 1\ 2\ 9\ 4\ \text{R5}$$

7 goes into 9 one time with R2. Put 1 over the dividend. Write 2 in front of the next digit.

7 goes into 20 two times with R6. Put 2 over the dividend. Write 6 in front of the next digit.

7 goes into 66 nine times with R3. Put 9 over the dividend. Write 3 in front of the next digit.

7 goes into 33 four times with R5. Put 4 over the dividend, and indicate the remainder as in long division. ▲

The answer is 1,294 R5.

Tests for Divisibility

When working with whole numbers, it is often quite useful to be able to determine quickly if a certain number can be evenly divided by another, without actually going through the division process. We will now learn techniques for determining whether or not a number is **divisible** by a given divisor.

▶ **DEFINITION** A number is *divisible* by a given divisor if the division process yields a whole number quotient and a remainder of zero.

Divisibility by 2

A number is divisible by 2 if its ones digit is 0, 2, 4, 6, or 8.   ▲

Numbers divisible by 2 are called *even* numbers. Thus, numbers with a ones digit of 0, 2, 4, 6, or 8 are even. Numbers not divisible by 2 are called *odd* numbers. Thus, numbers with a ones digit of 1, 3, 5, 7, or 9 are odd.

EXAMPLE 3   Determine whether or not each of the following numbers is divisible by 2.

(a)   54
54 is divisible by 2.   Its ones digit is 4.
54 is an even number.

(b)   221
221 is not divisible by 2.   Its ones digit is 1.
221 is an odd number.

(c)   2,000
2,000 is divisible by 2.   Its ones digit is 0.
2,000 is an even number.   ▲

Divisibility by 3

A number is divisible by 3 if the sum of its digits is divisible by 3.   ▲

EXAMPLE 4   Determine whether or not each of the following numbers is divisible by 3.

(a)   265
The sum of the digits is $2 + 6 + 5 = 13$.   13 is not divisible by 3.
Therefore, 265 is not divisible by 3.

(b)   5,187
The sum of the digits is $5 + 1 + 8 + 7 = 21$.   21 is divisible by 3.
Therefore, 5,187 is divisible by 3.

(c) 468,293

The sum of the digits is $4 + 6 + 8 + 2 + 9 + 3 = 32$.   32 is not divisible by 3.

Therefore, 468,293 is not divisible by 3.   ▲

Divisibility by 4

A number is divisible by 4 if the last two digits form a number divisible by 4.   ▲

EXAMPLE 5   Determine whether or not each of the following numbers is divisible by 4.

(a) 536

5<u>36</u>: The last two digits are 36.   36 is divisible by 4.

Therefore, 536 is divisible by 4.

(b) 1,749

1,7<u>49</u>: The last two digits are 49.   49 is not divisible by 4.

Therefore, 1,749 is not divisible by 4.

(c) 69,304

69,3<u>04</u>: The last two digits are 04.   04 is divisible by 4.

Therefore, 69,304 is divisible by 4.   ▲

Divisibility by 5

A number is divisible by 5 if its ones digit is 0 or 5.   ▲

EXAMPLE 6   Determine whether or not each of the following numbers is divisible by 5.

(a) 6,051

6,051 is not divisible by 5 since its ones digit is 1.

(b) 936,275

936,275 is divisible by 5 since its ones digit is 5.   ▲

Divisibility by 6

A number is divisible by 6 if it is divisible by both 2 and 3.   ▲

EXAMPLE 7   Determine whether or not each of the following numbers is divisible by 6.

(a) 582

582 is divisible by 2.   Its ones digit is 2.

The sum of the digits is $5 + 8 + 2 = 15$.   15 is divisible by 3.
582 is divisible by 3.
Therefore, 582 is divisible by 6.

(b) 1,438
1,438 is divisible by 2.   Its ones digit is 8.
The sum of the digits is $1 + 4 + 3 + 8 = 16$.   16 is not divisible by 3.
1,438 is not divisible by 3.
Therefore, 1,438 is not divisible by 6.

(c) 61,275
61,275 is not divisible by 2.   Its ones digit is 5.
Therefore, 61,275 is not divisible by 6. ▲

There is no simple test for divisibility by 7.

Divisibility by 8

A number is divisible by 8 if the last three digits form a number divisible by 8. ▲

EXAMPLE 8   Determine whether or not each of the following numbers is divisible by 8.

(a) 3,265
3,<u>265</u>: The last three digits are 265.

$$\begin{array}{r} 3\ 3\ \text{R}1 \\ 8\overline{)2\ 6\,^25} \end{array}$$ 265 is not divisible by 8.

Therefore, 3,265 is not divisible by 8.

(b) 78,544
78,<u>544</u>: The last three digits are 544.

$$\begin{array}{r} 6\ 8 \\ 8\overline{)5\ 4\,^64} \end{array}$$ 544 is divisible by 8.

Therefore, 78,544 is divisible by 8. ▲

Divisibility by 9

A number is divisible by 9 if the sum of its digits is divisible by 9. ▲

EXAMPLE 9   Determine whether or not each of the following numbers is divisible by 9.

(a)   309
The sum of the digits is
$3 + 0 + 9 = 12.$ 12 is not divisible by 9.
Therefore, 309 is not divisible by 9.

(b)   418,374
The sum of the digits is
$4 + 1 + 8 + 3 + 7 + 4 = 27.$ 27 is divisible by 9.
Therefore, 418,374 is divisible by 9.   ▲

Divisibility by 10

A number is divisible by 10 if its ones digit is 0.   ▲

EXAMPLE 10   Determine whether or not each of the following numbers is divisible by 10.

(a)   4,005
4,005 is not divisible by 10.   Its ones digit is 5.

(b)   821,790
821,790 is divisible by 10.   Its ones digit is 0.   ▲

EXAMPLE 11   The number 4,51_ has a missing ones digit. Fill in the appropriate digit(s) so that the result is divisible by the following numbers.

(a)   Divisible by 2.
The ones digit must be 0, 2, 4, 6, or 8. Therefore, 4,51$\underline{0}$, 4,51$\underline{2}$, 4,51$\underline{4}$, 4,51$\underline{6}$, and 4,51$\underline{8}$ are all divisible by 2.

(b)   Divisible by 3.
The sum of the digits must be divisible by 3.
$451\_ : \underbrace{4 + 5 + 1}_{10} + \_ =$ a number divisible by 3
$10\quad + \_ =$ a number divisible by 3
$10 + \underline{2} = 12$ Divisible by 3.
$10 + \underline{5} = 15$ Divisible by 3.
$10 + \underline{8} = 18$ Divisible by 3.

Therefore, 4,51$\underline{2}$, 4,51$\underline{5}$, and 4,51$\underline{8}$ are all divisible by 3.

(c)   Divisible by 6.
The number must be divisible by both 2 and 3.
The numbers divisible by 3 are 4,51$\underline{2}$, 4,51$\underline{5}$, and 4,51$\underline{8}$.   Example 11(b).
Of these, 4,51$\underline{2}$ and 4,51$\underline{8}$ are also divisible by 2.   Example 11(a).
Therefore, 4,51$\underline{2}$ and 4,51$\underline{8}$ are divisible by 6.   ▲

EXAMPLE 12   The number 81,53_ has a missing ones digit. Fill in the appropriate digit(s) so that the result is divisible by each of the following numbers.

(a)   Divisible by 4.

The last two digits must form a number divisible by 4. 32 and 36 are both divisible by 4. Therefore, 81,532 and 81,536 are both divisible by 4.

(b)   Divisible by 8.

The last three digits must make a number divisible by 8.

$$
\begin{array}{r}
6\ 7 \\
8\overline{\smash{\big)}\,5\ \ 3\,^5 6}
\end{array}
$$

8 goes into 53 six times with R5. Fill in the ones digit so 5_ is divisible by 8. Since 56 is divisible by 8, the ones digit is 6.

Thus, 81,536 is divisible by 8.

(c)   Divisible by 9.

The sum of the digits must be divisible by 9.

$$81{,}53\_ : \underbrace{8 + 1 + 5 + 3}_{17} + \_ = \text{a number divisible by 9}$$

$$17 \qquad + \_ = \text{a number divisible by 9}$$

$$17 \qquad + 1 = 18 \qquad\qquad \text{Divisible by 9.}$$

Therefore, 81,531 is divisible by 9.   ▲

Quick Quiz

Use the divisibility tests to answer each of the following questions.	Answers	
1. Is 458 divisible by 3?	1.	No
2. Is 1,740 divisible by 6?	2.	Yes
3. Is 93,264 divisible by 8?	3.	Yes

2.4 Exercises

Indicate which of the following statements are true and which are false. For those that are false, change the italicized or underlined expression to make the statement true.

1.   A number that is divisible by 10 is *always* divisible by 5.

2.   Numbers divisible by 2 are called *odd*.

3.   A number is divisible by 8 if the last *two* digits make a number divisible by 8.

4.   If a number is divisible by 9, it is also divisible by 2.

Use short division to find each of the following quotients.

5. $436 \div 4$	6. $216 \div 3$	7. $919 \div 7$	8. $813 \div 5$	9. $1{,}344 \div 2$
10. $6{,}284 \div 6$	11. $3{,}719 \div 8$	12. $4{,}602 \div 9$	13. $73{,}841 \div 3$	14. $82{,}036 \div 4$
15. $37{,}245 \div 5$	16. $20{,}493 \div 8$	17. $1{,}003{,}456 \div 7$	18. $3{,}842{,}461 \div 2$	

Use the divisibility tests to determine whether or not each of the following numbers is divisible by 2, 3, 4, 5, 6, 8, 9, and 10. Indicate by which number(s) each is divisible.

19.	182	20.	215	21.	600	22.	416	23.	552
24.	738	25.	1,407	26.	1,233	27.	2,385	28.	1,230
29.	4,356	30.	7,206	31.	8,469	32.	6,210	33.	3,824
34.	5,308	35.	23,186	36.	37,239	37.	86,044	38.	79,600
39.	56,328	40.	63,747	41.	438,816	42.	524,090	43.	360,045
44.	473,284	45.	5,392,074	46.	9,005,635	47.	20,625,080	48.	549,627,088

The number 2,60_ has a missing ones digit. Fill in the appropriate digit(s) so that the result is divisible by the following numbers.

49.	Divisible by 2.	50.	Divisible by 3.	51.	Divisible by 4.	52.	Divisible by 5.
53.	Divisible by 6.	54.	Divisible by 8.	55.	Divisible by 9.	56.	Divisible by 10.

The number 54,89_ has a missing ones digit. Fill in the appropriate digit(s) so that the result is divisible by the following numbers.

57.	Divisible by 2.	58.	Divisible by 3.	59.	Divisible by 4.	60.	Divisible by 5.
61.	Divisible by 6.	62.	Divisible by 8.	63.	Divisible by 9.	64.	Divisible by 10.

A number is divisible by 12 if it is divisible by both 3 and 4. Determine whether or not each of the following numbers is divisible by 12.

65.	288	66.	356	67.	1,536	68.	5,124
69.	2,444	70.	4,698	71.	14,355	72.	92,184

A number is divisible by 15 if it is divisible by both 3 and 5. Determine whether or not each of the following numbers is divisible by 15.

73.	230	74.	555	75.	8,095	76.	6,345
77.	2,961	78.	40,280	79.	10,335	80.	163,550

A number is divisible by 18 if it is divisible by both 2 and 9. Determine whether or not each of the following numbers is divisible by 18.

81.	216	82.	908	83.	4,982	84.	27,846
85.	41,756	86.	328,464	87.	1,029,114	88.	30,478,162

2.5
Prime Numbers and Factoring

When we factor a number, we rewrite it as a product. Sometimes it is possible to factor a number in more than one way. For example, the number 12 may be factored in each of the following ways.

$$12 = 2 \cdot 6$$
$$12 = 3 \cdot 4$$
$$12 = 2 \cdot 2 \cdot 3$$

The last factorization of 12 is the most complete, since 2 cannot be rewritten as a product that does not contain 2, and 3 cannot be rewritten as a product that does not contain 3. The numbers 2 and 3 are therefore called **prime numbers**.

▶ **DEFINITION** A *prime number* is a whole number greater than 1 that is divisible only by 1 and itself.

▶ **DEFINITION** A *composite number* is a whole number greater than 1 that is not prime.

Notice that, by definition, the numbers 0 and 1 are considered neither prime nor composite.

In order to determine which whole numbers are prime, we can use a device called the Sieve of Eratosthenes, developed in ancient times by the Greek mathematician Eratosthenes. This technique requires us to find all the **multiples** of each successive prime number to eliminate whole numbers that are not prime.

▶ **DEFINITION** A *multiple* of a given whole number is the product of that number and another whole number.

To find all the multiples of a given whole number, just multiply that number by every whole number. For example, the multiples of 2 can be determined by multiplying every whole number (0, 1, 2, 3, 4, 5, . . .) by 2 to obtain the following:

Multiples of 2:   0, 2, 4, 6, 8, 10, . . .

Notice that 0 is a multiple of every whole number.

EXAMPLE 1   List the first six multiples of each of the following numbers.

(a) 3
Multiply every whole number by 3.
Multiples of 3: 0, 3, 6, 9, 12, 15

(b) 5
Multiply every whole number by 5.
Multiples of 5: 0, 5, 10, 15, 20, 25

(c) 12
Multiply every whole number by 12.
Multiples of 12: 0, 12, 24, 36, 48, 60   ▲

To find all the prime numbers less than 100, we use the Sieve of Eratosthenes to perform the following steps. The result of this process is shown in Table 2.2.

1. List all the whole numbers starting with 2 that are less than 100.
2. Circle the number 2 and then cross out all multiples of 2.
3. Circle the lowest number that is not crossed out, 3. Cross out all multiples of 3.
4. Circle the lowest number not crossed out, 5. Cross out all multiples of 5.
5. Continue in this manner. Circle the lowest number not crossed out, and then cross out all of its multiples.

6. When all the numbers have been circled or crossed out, you are finished. The circled numbers are prime, and the numbers crossed out are composite. Notice that the number 2 is the only even prime number.

TABLE 2.2
The Sieve of Eratosthenes

		②	③	4̸	⑤	6̸	⑦	8̸	9̸
1̸0̸	⑪	1̸2̸	⑬	1̸4̸	1̸5̸	1̸6̸	⑰	1̸8̸	⑲
2̸0̸	2̸1̸	2̸2̸	㉓	2̸4̸	2̸5̸	2̸6̸	2̸7̸	2̸8̸	㉙
3̸0̸	㉛	3̸2̸	3̸3̸	3̸4̸	3̸5̸	3̸6̸	㊲	3̸8̸	3̸9̸
4̸0̸	㊶	4̸2̸	㊸	4̸4̸	4̸5̸	4̸6̸	㊼	4̸8̸	4̸9̸
5̸0̸	5̸1̸	5̸2̸	㊾	5̸4̸	5̸5̸	5̸6̸	5̸7̸	5̸8̸	㊾
6̸0̸	㉛	6̸2̸	6̸3̸	6̸4̸	6̸5̸	6̸6̸	㊼	6̸8̸	6̸9̸
7̸0̸	㋒	7̸2̸	㋓	7̸4̸	7̸5̸	7̸6̸	7̸7̸	7̸8̸	㋗
8̸0̸	8̸1̸	8̸2̸	㋓	8̸4̸	8̸5̸	8̸6̸	8̸7̸	8̸8̸	㋗
9̸0̸	9̸1̸	9̸2̸	9̸3̸	9̸4̸	9̸5̸	9̸6̸	㊟	9̸8̸	9̸9̸

To determine whether a number is prime or composite, we determine whether or not it is divisible by a prime number. The tests for divisibility are helpful for determining whether a number is divisible by 2, 3, or 5.

EXAMPLE 2 Determine whether each of the following numbers is prime or composite.

(a) 65
65 is divisible by 5.   65 ends in 5.
65 is composite.

(b) 29
29 is not divisible by any prime number.
29 is prime.

(c) 87
87 is divisible by 3.   The sum of the digits is divisible by 3:
87 is composite. $8 + 7 = 15$.   ▲

Now that we are familiar with prime numbers, we can factor any number completely by rewriting it as the product of only prime numbers. **Prime factorization** will be a very useful technique for working with fractions in Chapter 3.

▶ **DEFINITION** The *complete*, or *prime*, factorization of a number is the unique representation of that number as the product of its prime factors.

To completely factor a number, we can use a device called a *factor tree*. First we rewrite the number as the product of any two numbers. Then we rewrite all composite factors as the product of two other numbers. We continue to rewrite all composite factors in this manner, until all factors of the original number are prime. For example, the number 72 can be completely factored as follows:

$72 = 8 \cdot 9$ Rewrite 72 as $8 \cdot 9$.

$= 2 \cdot 4 \cdot 3 \cdot 3$ Factor 8 and 9.

$= 2 \cdot 2 \cdot 2 \cdot 3 \cdot 3$ Factor 4.

The prime factorization of 72 is, therefore,

$$72 = 2 \cdot 2 \cdot 2 \cdot 3 \cdot 3 = 2^3 \cdot 3^2$$

which can be written $2^3 \cdot 3^2$ using exponents.

Let us now consider some further examples of complete factorization.

EXAMPLE 3   Completely factor each of the following numbers.

(a)   81

$81 = 9 \cdot 9$

$= 3 \cdot 3 \cdot 3 \cdot 3$

$81 = 3 \cdot 3 \cdot 3 \cdot 3 = 3^4$

(b)   48

$48 = 6 \cdot 8$

$= 2 \cdot 3 \cdot 2 \cdot 4$

$= 2 \cdot 3 \cdot 2 \cdot 2 \cdot 2$

$48 = 2 \cdot 2 \cdot 2 \cdot 2 \cdot 3 = 2^4 \cdot 3$

(c)   60

$60 = 6 \cdot 10$

$= 2 \cdot 3 \cdot 2 \cdot 5$

$60 = 2 \cdot 2 \cdot 3 \cdot 5 = 2^2 \cdot 3 \cdot 5$

(d)   216

$216 = 3 \cdot 72$

$= 3 \cdot 8 \cdot 9$

$= 3 \cdot 2 \cdot 4 \cdot 3 \cdot 3$

$= 3 \cdot 2 \cdot 2 \cdot 2 \cdot 3 \cdot 3$

$216 = 2 \cdot 2 \cdot 2 \cdot 3 \cdot 3 \cdot 3 = 2^3 \cdot 3^3$ ▲

Quick Quiz

Completely factor each of the following numbers.	Answers
1. 40	1. $2^3 \cdot 5$
2. 66	2. $2 \cdot 3 \cdot 11$
3. 168	3. $2^3 \cdot 3 \cdot 7$

2.5 Exercises

Indicate which of the following statements are true and which are false. For those that are false, change the italicized word to make the statement true.

1. A *composite* number is evenly divisible only by 1 and itself.
2. A nonzero multiple of a number is always *greater* than or equal to that number.
3. The prime factorization of any composite number *is* unique.
4. The number 2 is considered to be *prime*.

State whether each of the following numbers is prime or composite.

5. 9	6. 5	7. 17	8. 27	9. 31	10. 2	11. 15	12. 47
13. 53	14. 67	15. 81	16. 87	17. 93	18. 73	19. 49	20. 59
21. 79	22. 69	23. 23	24. 63	25. 121	26. 97	27. 101	28. 89
29. 105	30. 207	31. 145	32. 113	33. 333	34. 276		

List the first five multiples of each of the following numbers.

35. 4	36. 6	37. 7	38. 9	39. 10
40. 13	41. 15	42. 16	43. 20	44. 25

Completely factor each of the following numbers.

45. 16	46. 36	47. 54	48. 18	49. 42	50. 63
51. 32	52. 80	53. 78	54. 45	55. 56	56. 99
57. 64	58. 24	59. 60	60. 84	61. 108	62. 144
63. 225	64. 169	65. 150	66. 210	67. 178	68. 256
69. 360	70. 425	71. 690	72. 512	73. 700	74. 388
75. 950	76. 1,000	77. 2,500	78. 4,000	79. 6,250	80. 7,500
81. 10,000	82. 35,000				

2.6
Greatest Common Factor (GCF) and Least Common Multiple (LCM)

In this section, we will learn two more techniques that will be helpful for our work with fractions. The first is a procedure to find the **greatest common factor** (**GCF**) of two or more numbers. Recall that **factors** are numbers that are multiplied together to obtain a product. For example, 2 and 3 are factors of 6 because $2 \cdot 3 = 6$.

▶ **DEFINITION** The *greatest common factor*, or *GCF*, of a group of numbers is the largest number by which each number in the group is divisible.

To Find the Greatest Common Factor of a Group of Numbers	1. Completely factor each number in the group.
	2. Find the prime factors common to every number in the group.
	3. The GCF is the product of these common prime factors. ▲

EXAMPLE 1   Find the GCF of 12 and 18.

Solution $12 = 4 \cdot 3 = \boxed{2} \cdot 2 \cdot \boxed{3}$ Factor 12 and 18 completely.
$18 = 6 \cdot 3 = \boxed{2} \cdot 3 \cdot \boxed{3}$ The prime numbers 2 and 3 are factors of both 12 and 18.

$GCF = 2 \cdot 3 = 6$ The GCF is the product of 2 and 3.   ▲

EXAMPLE 2   Find the GCF of each of the following groups of numbers.

(a)   28, 32
$28 = 4 \cdot 7 = \boxed{2} \cdot \boxed{2} \cdot 7$
$32 = 4 \cdot 8 = \boxed{2} \cdot \boxed{2} \cdot 2 \cdot 2 \cdot 2$
$GCF = 2 \cdot 2 = 4$

(b)   64, 96
$64 = 8 \cdot 8 \ = 2 \cdot \boxed{2} \cdot \boxed{2} \cdot \boxed{2} \cdot \boxed{2} \cdot \boxed{2}$
$96 = 3 \cdot 32 = 3 \cdot \boxed{2} \cdot \boxed{2} \cdot \boxed{2} \cdot \boxed{2} \cdot \boxed{2}$
$GCF = 2 \cdot 2 \cdot 2 \cdot 2 \cdot 2 = 32$

(c)   36, 90
$36 = 6 \cdot 6 \ = 2 \cdot 3 \cdot 2 \cdot 3 = 2 \cdot \boxed{2} \cdot \boxed{3} \cdot \boxed{3}$
$90 = 9 \cdot 10 = 3 \cdot 3 \cdot 2 \cdot 5 = \quad \boxed{2} \cdot \boxed{3} \cdot \boxed{3} \cdot 5$
$GCF = 2 \cdot 3 \cdot 3 = 18$

(d)   24, 60, 108
$\quad 24 = 3 \ \cdot 8 \ = 3 \cdot 2 \cdot 2 \cdot 2 = \boxed{2} \cdot \boxed{2} \cdot \boxed{3} \cdot 2$
$\quad 60 = 6 \ \cdot 10 = 2 \cdot 3 \cdot 2 \cdot 5 = \boxed{2} \cdot \boxed{2} \cdot \boxed{3} \cdot 5$
$108 = 12 \cdot 9 \ = 4 \cdot 3 \cdot 3 \cdot 3 = \boxed{2} \cdot \boxed{2} \cdot \boxed{3} \cdot 3 \cdot 3$
$GCF = 2 \cdot 2 \cdot 3 = 12$

The second technique we will learn is a procedure to find the **least common multiple** (**LCM**) of two or more numbers. Recall that a **multiple** of a number is the product of that number and another number. For example, 15 is a multiple of 5 because $5 \cdot 3 = 15$.

▶ **DEFINITION**   The *least common multiple*, or *LCM*, of a group of numbers is the smallest nonzero multiple common to every number in the group.

EXAMPLE 3   Find three common nonzero multiples of 2 and 6, and determine the LCM.

Solution   Multiples of 2: 2, 4, 6, 8, 10, 12, 14, 16, 18, . . .
Multiples of 6: 6, 12, 18, 24, 30, 36, . . .

Three nonzero multiples common to both 2 and 6 are 6, 12, and 18.
Since the LCM is the smallest nonzero common multiple, the LCM of 2 and 6 is 6.

▲

Notice that the LCM of a group of numbers is the smallest positive number that is divisible by every number in the group.

To Find the Least Common Multiple of a Group of Numbers	1. List a few of the nonzero multiples for each number.
	2. If necessary, list additional multiples until you see a multiple that appears in every list.
	3. The smallest multiple common to every list is the LCM. ▲

EXAMPLE 4   Find the LCM of each of the following.

(a)   Find the LCM of 8 and 12.

Solution   Multiples of 8: 8, 16, 24, 32, 48   List multiples of 8.
Multiples of 12: 12, 24, 36, 48   List multiples of 12.
The smallest multiple in both lists is 24.
The LCM is 24.
Notice that 48 is also a common multiple of 8 and 12, but it is not the *least* common multiple.

(b)   Find the LCM of 4, 9, and 18.

Solution   Multiples of 4: 4, 8, 12, 16, 20, 24, 28, 32 36
Multiples of 9: 9, 18, 27, 36, 45
Multiples of 18: 18, 36, 54
The LCM is 36.   ▲

Here is another technique for finding the least common multiple (LCM) that may be easier to use for larger numbers.

To Find the Least Common Multiple of a Group of Numbers	1. Completely factor each number in the group and write each factorization in exponential notation.
	2. List each unique prime factor that appears in any of the factorizations.
	3. Raise each of the prime factors to the highest exponent it is raised to in any of the factorizations.
	4. The LCM is the product of these exponential expressions. ▲

EXAMPLE 5 Find the LCM of 24 and 32.

Solution $24 = 6 \cdot 4 = 2 \cdot 3 \cdot 2 \cdot 2 = 2^3 \cdot 3$ Factor each number completely and write
$32 = 4 \cdot 8 = 2 \cdot 2 \cdot 2 \cdot 2 \cdot 2 = 2^5$ it in exponential notation.

The unique prime factors that appear in the factorization are 2 and 3.
The highest power 2 is raised to is 5: 2^5.
The highest power 3 is raised to is 1: 3^1.

$\text{LCM} = 2^5 \cdot 3^1 = 32 \cdot 3 = 96$ The LCM is the product $2^5 \cdot 3^1$. ▲

EXAMPLE 6 Find the LCM of 12, 42, and 56.

$12 = 3 \cdot 4 = 3 \cdot 2 \cdot 2 = 2^2 \cdot 3$ Factor each number completely and write
$42 = 6 \cdot 7 = 2 \cdot 3 \cdot 7$ it in exponential notation.
$56 = 7 \cdot 8 = 7 \cdot 2 \cdot 2 \cdot 2 = 2^3 \cdot 7$

The unique prime factors appearing in the factorizations are 2, 3, and 7.
The highest exponent 2 is raised to is 3: 2^3.
The highest exponent 3 is raised to is 1: 3^1.
The highest exponent 7 is raised to is 1: 7^1.

$\text{LCM} = 2^3 \cdot 3^1 \cdot 7^1 = 8 \cdot 3 \cdot 7 = 168$ The LCM is the product $2^3 \cdot 3^1 \cdot 7^1$. ▲

Quick Quiz

		Answers
1.	Find the GCF of 18 and 63.	**1.** GCF = 9
2.	Find the LCM of 16 and 36.	**2.** LCM = 144
3.	Find the GCF and LCM of 12, 24, and 66.	**3.** GCF = 6; LCM = 264

2.6 Exercises

Indicate which of the following statements are true and which are false. For those that are false, change the italicized expression to make the statement true.

1. The GCF of a group of numbers is the *sum* of all prime factors common to every number in the set.
2. The LCM of a group of numbers can never equal *zero*.
3. The GCF of a group of numbers is always *greater than* the largest number in the set.
4. The LCM of a group of numbers is always *greater than* the smallest number in the set.

Find the greatest common factor (GCF) of each of the following groups of numbers.

5. 18, 30	**6.** 42, 28	**7.** 64, 40	**8.** 15, 35
9. 60, 36	**10.** 81, 27	**11.** 45, 80	**12.** 54, 66
13. 96, 36	**14.** 48, 72	**15.** 86, 42	**16.** 49, 112

17.	48, 136	18.	108, 81	19.	65, 169	20.	144, 72
21.	36, 24, 48	22.	27, 12, 42	23.	84, 28, 42	24.	96, 32, 48
25.	64, 48, 120	26.	18, 42, 96	27.	15, 45, 90, 75	28.	24, 84, 132, 36

Find the least common multiple (LCM) of each of the following groups of numbers.

29.	4, 6	30.	3, 9	31.	10, 15	32.	12, 16
33.	20, 8	34.	24, 18	35.	36, 27	36.	15, 35
37.	12, 32	38.	48, 21	39.	56, 24	40.	30, 55
41.	8, 4, 12	42.	6, 9, 15	43.	10, 8, 25	44.	24, 16, 12
45.	21, 18, 14	46.	36, 9, 12	47.	20, 36, 48	48.	27, 42, 54
49.	60, 49, 84	50.	24, 42, 96	51.	12, 18, 24, 32	52.	15, 45, 60, 75

Find the GCF and LCM of each of the following groups of numbers.

53.	21, 36	54.	18, 54	55.	6, 30, 54	56.	9, 21, 36
57.	14, 42, 49	58.	18, 45, 63	59.	24, 84, 108	60.	32, 48, 80
61.	16, 28, 36, 48	62.	12, 18, 21, 39				

2.7
Summary and Review

Key Terms

(2.1) **Exponents** express repeated multiplication.

The exponential expression b^n means that b appears as a factor n times. b is the **base**, and n is the **exponent**, or **power**.

For any base b, $b^1 = b$. If $b \neq 0$, $b^0 = 1$.

Powers of ten represent place values in the base ten system.

(2.4) A number is **divisible** by a given divisor if the division process yields a whole number quotient and a remainder of zero.

(2.5) A **prime number** is a whole number greater than 1 that is divisible only by 1 and itself.

A **composite number** is a whole number greater than 1 that is not prime.

A **multiple** of a given whole number is the product of that number and another whole number.

The **complete**, or **prime**, **factorization** of a number is the unique representation of that number as the product of its prime factors.

(2.6) The **greatest common factor**, or **GCF**, of a group of numbers is the largest number by which every number in the group is divisible.

The **least common multiple**, or **LCM**, of a group of numbers is the smallest nonzero multiple common to every number in the group.

Calculations

(2.1) **To multiply a number by 10^n**, where n is a whole number, write the number followed by n zeros.

To multiply exponential expressions of the same base, keep the base the same and add the exponents.

To raise an exponential expression to a power, keep the base the same and multiply the exponents.

(2.3) **To calculate the average** of a group of numbers, first add all the numbers in the group, and then divide that sum by the number of numbers that were added together.

(2.5) **To find all the multiples of a given whole number**, multiply that number by every whole number.

To determine whether a number is prime or composite, we determine whether or not it is divisible by prime numbers.

(2.6) **To find the greatest common factor (GCF)** of a group of numbers, completely factor each number and calculate the product of all prime factors common to every number in the group.

To find the least common multiple (LCM) of a group of numbers, list the nonzero multiples of each number. The smallest number common to every list is the LCM.

To find the least common multiple (LCM) of a group of numbers, completely factor each number and write each factorization in exponential notation. Calculate the product of all unique prime factors, each of which is raised to the highest exponent that appears in any one of the factorizations.

(2.2) **Order of Operations Rules**

1. Perform all operations in parentheses first.
2. Simplify all exponential expressions.
3. Do multiplication and division as they occur, moving from left to right.
4. Do addition and subtraction as they occur, moving from left to right.

(2.4) **Tests for Divisibility**

A number is divisible by 2 if its ones digit is 0, 2, 4, 6, or 8.

A number is divisible by 3 if the sum of its digits is divisible by 3.

A number is divisible by 4 if its last two digits make a number divisible by 4.

A number is divisible by 5 if its ones digit is 0 or 5.

A number is divisible by 6 if it is divisible by both 2 and 3.

A number is divisible by 8 if its last three digits make a number divisible by 8.

A number is divisible by 9 if the sum of its digits is divisible by 9.

A number is divisible by 10 if its ones digit is 0.

Chapter **2**
Review Exercises

Indicate which of the following statements are true and which are false. For those that are false, change the italicized expression to make the statement true.

1. The number 2 *is not* the only prime number that is even.
2. To raise an exponential expression to a power, keep the base the same and *multiply* the exponents.
3. If a number is divisible by *4*, it is also divisible by 2 and 3.
4. For a given set of unique numbers, the GCF is always *greater than* the LCM.
5. Every place value in the base ten system can be written as a power of *ten*.
6. Unless otherwise indicated by parentheses, we do subtraction *before* division.

(2.1)   Simplify each of the following expressions using the rules for exponents.

7. $5^7 \cdot 5^2$ 8. $4^4 \cdot 4^9$ 9. $2^5 \cdot 2^0 \cdot 2^3$ 10. $8^7 \cdot 8^3 \cdot 8^1$ 11. $(7^4)^5$

12. $(9^2)^6$ 13. $(2^3)^4 \cdot (2^5)^3$ 14. $(3^7)^2 \cdot (3^2)^5$ 15. $4^3 \cdot 4^2 + 4^7$ 16. $6^5 + 6^4 \cdot 6^7$

(2.2)   Perform each of the following operations, and express all answers as whole numbers.

17. $6^2 + 2^3$

18. $3^3 - 4^2$

19. $2^3 \cdot 3^4$

20. $4^2 \cdot 7^0$

21. $(9 - 3)^2$

22. $(7 + 3)^5$

23. $4^2 \cdot (3^0)^2$

24. $(2^2)^3 \cdot 5^2$

25. $6 \cdot 3 - 15 \div 5$

26. $12 \div 2 + 4 \cdot 6$

27. $(8 + 1)^2 \div 3^3$

28. $2^5 \div (6 - 2)^2$

29. $5 \cdot 10^3 + 3 \cdot 10^2 + 2 \cdot 10^1 + 6 \cdot 10^0$

30. $9 \cdot 10^3 + 4 \cdot 10^2 + 7 \cdot 10^1 + 5 \cdot 10^0$

31. $72 \div 3^2 + 4$

32. $64 \div 2^3 - 5$

33. $4(6 - 3) + 3(9 - 7)^2$

34. $2(1 + 4)^2 - 5(2 + 6)$

35. $5 + 7 \cdot 3 - 12 \div 4$

36. $2 \cdot 3 + 4 - 9 \div 3$

37. $(8 \div 2^2 \cdot 6 + 7 - 1) \div 6$

38. $48 \div (4 + 6 \cdot 3 - 4^2)$

39. $3[(8 - 3)^2 + 5(11 - 7)]$

40. $[6(9 - 4) + (2 + 3)^3] \div 5$

41. $36,801 \div 423 + (614)(96)$

42. $[(872)(128) + 1,196] \div 476$

43. $(32)^2(74) + (67)(9)^4$

44. $(17)^5 - (36)^3 + (6,839 - 5,216)^2$

(2.4)   Use the divisibility tests to determine whether or not each of the following numbers is divisible by 2, 3, 4, 5, 6, 8, 9, and 10. Indicate by which number(s) each is divisible.

45. 117   46. 204   47. 545   48. 867   49. 2,137

50. 4,698   51. 8,040   52. 9,153   53. 71,265   54. 43,056

(2.5)   State whether each of the following numbers is prime or composite.

55. 13   56. 21   57. 65   58. 29   59. 47

60. 93   61. 51   62. 97   63. 115   64. 201

(2.5)   Completely factor each of the following numbers.

65. 68   66. 96   67. 120   68. 164   69. 245   70. 324   71. 1,500   72. 4,000

(2.6) **Find the greatest common factor (GCF) and least common multiple (LCM) of each of the following groups of numbers.**

73.　32, 56

74.　81, 21

75.　144, 54

76.　72, 104

77.　12, 32, 48

78.　54, 63, 36

79.　69, 18, 72

80.　84, 96, 132

81.　14, 21, 42, 56

82.　24, 32, 36, 48

(2.2) **Translate each of the following expressions into numerical expressions and simplify.**

83.　8 less than the product of 6 and 9.

84.　9 more than 3 cubed.

85.　6 times the sum of 4 and 9.

86.　The sum of the product of 3 and 9 and the quotient of 20 and 4.

87.　The square of the difference between 17 and 3.

88.　The sum of 2 cubed and 5 squared.

89.　3 less than the quotient of 63 and 7.

90.　8 less than the product of 3 and 9.

(2.3) **Solve each of the following word problems.**

91.　A student obtained the scores of 61, 78, 83, 65, and 73 on five examinations. What was his average grade?

92.　A basketball player scored the following points during a six-game series: 18, 29, 24, 20, 31, and 22. What was her scoring average for that series?

93.　The state of New Hampshire has a population of 738,000. If New Hampshire covers an area of roughly 9,000 square miles, what is its population density per square mile?

94.　In a recent survey, the annual salaries of six state governors were reported as follows: Massachusetts, $85,000; New York, $130,000; Ohio, $65,000; California, $85,000; Maine, $70,000; New Hampshire, $66,000. What was the average salary?

95.　If 47,842,375 passengers flew into or out of O'Hare International Airport in Chicago during a one-year period, what was the average number of passengers served daily by the airport last year?

96.　The diameters of the nine planets of our solar system are Mercury, 3,100 miles; Venus, 7,700 miles; Earth, 7,950 miles; Mars, 4,200 miles; Jupiter, 88,700 miles; Saturn, 75,100 miles; Uranus, 32,000 miles; Neptune, 27,700 miles; and Pluto, 1,500 miles. What is the average diameter of the planets in our solar system?

Chapter **2**
Explain It in Words

97.　What is the difference between a prime number and a composite number? Give an example of each.

98.　What is meant by a "multiple" of a number? Give an example.

99.　How do you determine if a number is divisible by 2? Give an example of a number that is divisible by 2 and one that is not.

100.　How do you determine if a number is divisible by 3? Give an example of a number that is divisible by 3 and one that is not.

101.　Explain what is meant by "the greatest common factor" of a group of numbers.

102.　Explain how to find the average of a group of numbers.

Chapter **2**
Chapter Test

[2.1]   Simplify each of the following using the rules for exponents.

1. $7^3 \cdot 7^5$ 2. $(4^2)^4$ 3. $2^7 \cdot 2^3 \cdot 2^6$ 4. $(5^2)^4 \cdot (5^3)^3$

[2.2]   Perform each of the following operations. Express all answers as whole numbers.

5. $6 + 8 \cdot 3$ 6. $48 \div 2^3 - 3$ 7. $5^0 + 36 \div 4 \cdot 3$

8. $5[(4 + 6)^2 - (7 - 4)^4]$

[2.4]   Use the divisibility tests to determine whether or not each of the following numbers is divisible by 2, 3, 4, 5, 6, 8, 9, and 10. Indicate which number(s) each is divisible by.

9. 321   10.   2,340   11.   4,925

[2.5]   Answer each of the following.

12. Is 67 prime or composite?   13.   Completely factor 98.

[2.6]   Find the greatest common factor (GCF) of each of the following groups of numbers.

14. 54, 81   15.   42, 56, 49

[2.6]   Find the least common multiple (LCM) of each of the following groups of numbers.

16. 20, 32   17.   48, 36, 72

[2.2]   Translate each of the following expressions into numerical expressions and simplify.

18. Eight less than four cubed.   19.   Three times the sum of seven and four.

[2.3]   Solve the following word problem.

20. A student obtained quiz grades of 64, 83, 76, 78, and 89. What was her average grade?

Chapter **3**

Fractions

3.1
Definitions

This chapter will deal with numbers called **fractions**. We will start by discussing four different ways in which we use fractions to represent ideas that cannot be adequately conveyed by using only whole numbers.

In the first example, shown in Figure 3.1, a circle is divided into four equal pieces. If we want to focus our attention on only one of those pieces, we can shade one of the four equal pieces of the circle. This one piece can be represented mathematically by the fraction $\frac{1}{4}$, which is read "one-fourth." The fraction $\frac{1}{4}$ means that we are interested in one piece of a whole circle that consists of four equal pieces. Thus, one use of fractions is to represent the parts of a whole.

Figure 3.1

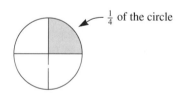

$\frac{1}{4}$ of the circle

As a second example, let us suppose that there are three boxes lined up in front of us, as shown in Figure 3.2, and that a ball is inside one of them. If we are asked to guess which box contains the ball, our chances are one out of three that we will guess correctly. Expressed as a fraction, our chances are represented by $\frac{1}{3}$, or one-third. If two balls are placed inside two different boxes, and we are asked to pick a box containing a ball, our chances of picking correctly are two out of three, or $\frac{2}{3}$ (two-thirds). Thus, a fraction also can mean the number of correct choices out of the total number of choices available.

Figure 3.2
Which box contains the ball? Our chances are one out of three that we guess correctly.

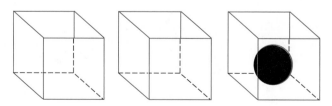

As a third example, let us suppose that the distance between your home and your place of employment is exactly 1 mile. To illustrate this, let us use a number line labeled from 0 (home) to 1 (work), as shown in Figure 3.3. In order to pinpoint any location between work and home, we can use fractions of a mile to represent the distance away from your home. Suppose your car breaks down one morning on your way to work at the place marked with an X in Figure 3.3(a). If we divide the mile into two equal segments, we can estimate your location as approximately $\frac{1}{2}$ mile away from home. If we divide the mile into three equal parts, as in Figure 3.3(b), we obtain a better approximation of $\frac{1}{3}$ mile away. If we divide the mile into four equal parts, as in Figure 3.3(c), we see that the location of your car is between $\frac{1}{4}$ and $\frac{2}{4}$ mile away. Figure 3.3(d) illustrates that the closest approximation is $\frac{2}{5}$ mile away from home. Notice that by dividing the mile into a greater number of parts, we are able to indicate locations with greater accuracy.

Figure 3.3

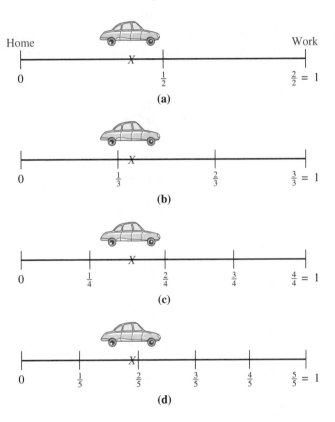

Fractions can also be used simply as another way to represent division. For example, the fraction $\frac{2}{3}$ means $2 \div 3$. We can write $20 \div 5$ as $\frac{20}{5}$, which in simplest form is the number 4.

Now let us apply our understanding of fractions to some problems.

EXAMPLE 1   On a coat that is supposed to have eight buttons, three buttons are missing. What fraction of the buttons is missing?

Solution   Since three out of eight buttons are missing, the fraction that represents the missing buttons is $\frac{3}{8}$.   ▲

EXAMPLE 2   A kennel has nine dogs and five cats. What fraction of the population consists of cats?

Solution   A fraction represents the number of parts of the whole. We must determine the size of the whole population of the kennel, which is $9 + 5 = 14$ animals. Thus, the fraction of cats is 5 out of 14, or $\frac{5}{14}$.   ▲

EXAMPLE 3   If a single playing card is drawn at random from a regular deck of 52 cards, what are the chances that the card is a king?

Solution   There are four kings in a deck of 52 cards. Thus, there are four desirable choices out of a total of 52 possible choices. The chances that the card drawn is a king are, therefore, 4 out of 52, or $\frac{4}{52}$.   ▲

Now that we are familiar with some of the uses of fractions, it is time to learn some formal vocabulary and notation.

▶ **DEFINITION** A *fraction* is any number that can be written in the form $\frac{a}{b}$, where a and b are numbers such that b is not equal to zero. Since $\frac{a}{b}$ means $a \div b$, b cannot be zero since division by zero is undefined. The *terms* of the fraction are a and b. The term a is called the *numerator* and b is called the *denominator*.

For example, in the fraction $\frac{2}{7}$, 2 and 7 are the **terms** of the fraction. The **numerator** is 2, and the **denominator** is 7.

$$\frac{2}{7} \quad \begin{array}{l} \leftarrow \text{ numerator} \\ \leftarrow \text{ denominator} \end{array}$$

▶ **DEFINITION** A *proper fraction* is one in which the numerator is less than the denominator.

Examples of proper fractions are $\frac{1}{5}$, $\frac{3}{41}$, and $\frac{19}{120}$.

▶ **DEFINITION** An *improper fraction* is one in which the numerator is greater than or equal to the denominator.

Examples of improper fractions are $\frac{7}{3}$, $\frac{99}{99}$, and $\frac{287}{103}$.

Characteristics of Fractions

We will now investigate some of the characteristics of fractions by looking at the division problems represented by various fractions. Consider the following examples:

$$\frac{5}{1} = 5 \div 1 = 5$$

$$\frac{33}{1} = 33 \div 1 = 33$$

$$\frac{168}{1} = 168 \div 1 = 168$$

We have illustrated the following point:

Any fraction whose denominator is 1 is equal to its numerator. Likewise, any whole number can be expressed as a fraction whose denominator is 1. ▲

For example, $7 = \frac{7}{1}$, $22 = \frac{22}{1}$, and $300 = \frac{300}{1}$.

Let us now consider these examples:

$$\frac{8}{8} = 8 \div 8 = 1$$

$$\frac{19}{19} = 19 \div 19 = 1$$

$$\frac{153}{153} = 153 \div 153 = 1$$

We have shown the following:

Any fraction whose numerator and denominator are the same (but not zero) is equal to 1. ▲

Observe what happens when the numerator is zero:

$$\frac{0}{6} = 0 \div 6 = 0$$

$$\frac{0}{13} = 0 \div 13 = 0$$

$$\frac{0}{257} = 0 \div 257 = 0$$

These examples show the following:

Any fraction with a numerator of zero and a nonzero denominator is equal to zero. ▲

Recall that division by zero is undefined. It is important to keep that fact in mind when working with fractions since fractions represent division. Therefore, $\frac{9}{0}, \frac{31}{0}, \frac{0}{0}$, and $\frac{405}{0}$ are all examples of fractions (divisions) that are undefined.

Since division by zero is undefined, any fraction with a denominator of zero is undefined. ▲

Mixed Numbers

Improper fractions represent numbers that are greater than or equal to one, since their numerators are greater than or equal to their denominators. Those improper fractions that represent numbers greater than one, but not whole numbers, also can be expressed as **mixed numbers**.

▶ **DEFINITION** A *mixed number* is the sum of a whole number and a proper fraction.

We can gain some understanding of mixed numbers by looking at the names of various locations on the line segment shown in Figure 3.4. In this drawing, a line segment between the values of 0 and 3 is divided into 12 equal pieces. Each mixed number, and its equivalent improper fraction, which is shown below it, appears in color.

Figure 3.4

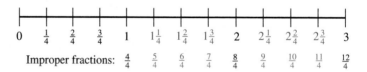

A procedure for changing improper fractions to mixed numbers is shown below.

To Convert an Improper Fraction to a Mixed Number

1. Divide the numerator by the denominator.
2. The quotient is the whole number portion of the mixed number.
3. The remainder is the numerator of the fractional portion of the mixed number. The denominator of the fractional portion is the same as the denominator of the original improper fraction.
4. The mixed number is the sum of the whole number found in step 2 and the fraction found in step 3. ▲

EXAMPLE 4   Convert $\dfrac{83}{4}$ to a mixed number.

Solution $\dfrac{83}{4} = 83 \div 4 = 20\ \text{R}3$

Divide 83 by 4: $4\overline{)83}$

$$\begin{array}{r} 20\ \text{R}3 \\ \hline 4\,\overline{)83} \\ 80 \\ \hline 3 \end{array}$$

The whole number portion is the quotient, 20.
The fractional portion is the remainder over the divisor, $\frac{3}{4}$.
The mixed number is $20 + \frac{3}{4} = 20\frac{3}{4}$

Even though the addition sign between the whole number and the fractional portion is omitted, it is understood to be there.

Therefore, $\dfrac{83}{4} = 20\dfrac{3}{4}$

$20\frac{3}{4}$ is read "twenty and three-fourths."   ▲

EXAMPLE 5   Convert each of the following improper fractions to mixed numbers.

(a) $\dfrac{123}{7}$

$123 \div 7 = 17\ \text{R}4$

Divide 123 by 7: $7\,\overline{)1\ 2^5 3}$ $\begin{array}{r} 1\ 7\ \text{R}4 \end{array}$

The whole number portion is the quotient, 17. The fractional portion is $\frac{4}{7}$.
Thus, $\frac{123}{7} = 17 + \frac{4}{7} = 17\frac{4}{7}$

(b) $\dfrac{1{,}024}{11}$

Divide 1,024 by 11: $11\,\overline{)1{,}024}$

$$\begin{array}{r} 93\ \text{R}1 \\ \hline 11\,\overline{)1{,}024} \\ 99 \\ \hline 34 \\ 33 \\ \hline 1 \end{array}$$

$1{,}024 \div 11 = 93\ \text{R}1$

The whole number portion is 93. The fractional portion is $\frac{1}{11}$.
Thus, $\frac{1{,}024}{11} = 93 + \frac{1}{11} = 93\frac{1}{11}$ ▲

We can think of the problem of changing an improper fraction to a mixed number as a division problem. The numerator is the dividend, and the denominator is the divisor. The mixed number equivalent to the improper fraction is the answer to the division problem. The mixed number is expressed as the sum of the quotient, a whole number, and a fraction whose numerator is the remainder and whose denominator is the divisor.

To convert a mixed number to an improper fraction, we reverse the process that we have just learned. That is, we find the dividend of a division problem, knowing the quotient, remainder, and divisor. The divisor is the denominator of the fractional part of the mixed number, and it is also the denominator of the improper fraction. The numerator of the improper fraction is the dividend, which can be calculated in the same way we check a division problem. Multiply the whole number (quotient) by the denominator (divisor) and add the numerator (remainder). This procedure is outlined on the next page.

To Convert a Mixed Number to an Improper Fraction	1. Multiply the whole number part of the mixed number by the denominator of the fractional part.
	2. Add the numerator of the fractional part to the result of step 1.
	3. The numerator of the improper fraction is the result of step 2. The denominator of the improper fraction is the same as the denominator of the fractional part of the mixed number. ▲

EXAMPLE 6   Convert $8\frac{2}{5}$ to an improper fraction.

Solution $8 \cdot 5 = 40$

Multiply the whole number by the denominator of the fractional part.

$40 + 2 = 42$

Add the numerator of the fractional part to the product of the whole number and the denominator.

$8\frac{2}{5} = \frac{42}{5}$

The numerator of the improper fraction is 42.

The denominator of the improper fraction is 5.   ▲

Notice that the process used to find the numerator of an improper fraction is the same as that used to check the division problem represented by an improper fraction. For example,

$$\frac{42}{5} = 42 \div 5 = 8 \text{ R2}$$

$$\begin{array}{r} 8 \text{ R2} \\ 5\overline{\smash{)}42} \\ \underline{40} \\ 2 \end{array}$$

Check:

$$\begin{array}{r} 8 \quad \text{quotient} \\ \times\ 5 \quad \text{divisor} \\ \hline 40 \\ +\ 2 \quad \text{remainder} \\ \hline 42 \quad \text{dividend} \quad \checkmark \end{array}$$

EXAMPLE 7   Convert each of the following mixed numbers to improper fractions.

(a) $13\frac{7}{10}$

$13 \cdot 10 = 130$

Multiply whole number by denominator.

$130 + 7 = 137$

Add product and numerator.

$13\frac{7}{10} = \frac{137}{10}$

The numerator of the improper fraction is 137.

The denominator of the improper fraction is 10.

We can also solve this problem by setting up our work as follows:

$$13\frac{7}{10} = \frac{(13 \cdot 10) + 7}{10} = \frac{130 + 7}{10} = \frac{137}{10}$$

(b) $53\frac{2}{11}$

$$53\frac{2}{11} = \frac{(53 \cdot 11) + 2}{11} = \frac{583 + 2}{11} = \frac{585}{11}$$ ▲

EXAMPLE 8 A newborn baby boy weighs 8 lb 3 oz. Express his weight in pounds as a mixed number and as an improper fraction.

Solution 8 lb 3 oz = 8 lb + $\frac{3}{16}$ lb   Since 16 oz = 1 lb.

$$= 8\frac{3}{16} \text{ lb}$$

$$8\frac{3}{16} = \frac{(8 \cdot 16) + 3}{16} = \frac{128 + 3}{16} = \frac{131}{16} \text{ lb}$$ Convert $8\frac{3}{16}$ to an improper fraction.

8 lb 3 oz = $8\frac{3}{16}$ lb = $\frac{131}{16}$ lb   ▲

EXAMPLE 9 A train trip from Boston to Philadelphia takes 7 hours 13 minutes. Express this time in hours as a mixed number and as an improper fraction.

Solution 7 hr 13 min = 7 hr + $\frac{13}{60}$ hr   Since 60 min = 1 hr.

$$= 7\frac{13}{60} \text{ hr}$$

$$7\frac{13}{60} = \frac{(7 \cdot 60) + 13}{60} = \frac{420 + 13}{60} = \frac{433}{60}$$ Convert $7\frac{13}{60}$ to an improper fraction.

7 hr 13 min = $7\frac{13}{60}$ hr = $\frac{433}{60}$ hr   ▲

Quick Quiz

		Answers	
1.	Express 3 ÷ 8 as a fraction.	**1.**	$\frac{3}{8}$
2.	Change $\frac{46}{5}$ to a mixed number.	**2.**	$9\frac{1}{5}$
3.	Change $6\frac{3}{8}$ to an improper fraction.	**3.**	$\frac{51}{8}$

3.1 Exercises

Indicate which of the following statements are true and which are false. For those that are false, change the italicized or underlined expression to make the statement true.

1. In the fraction $\frac{7}{8}$, 7 is the *numerator*.
2. The fraction $\frac{12}{12}$ is equal to 12.
3. The fraction $\frac{5}{0}$ is *undefined*.
4. The number $2\frac{3}{4}$ is a *mixed number*.
5. The fraction $\frac{111}{6}$ is called an *invalid* fraction.
6. A proper fraction represents a number *less than* one.
7. The fraction $\frac{5}{20}$ means $20 \div 5$.
8. The *numerator* of a fraction can never be zero.
9. The fraction $\frac{9}{1}$ is equal to 9.
10. In the mixed number $7\frac{3}{5}$, the number 3 represents the *quotient* of a division problem.

Use a fraction to represent the parts of a whole, division, or the chances of a favorable outcome.

11. The shaded portion of this circle.

12. The shaded portion of this box.

13. $39 \div 8$ 14. $3 \div 7$ 15. $1 \div 11$ 16. $28 \div 1$ 17. 6   18. 25

19. The chance of picking a spade at random from a regular deck of 52 cards.

20. The chance of obtaining a head when flipping a fair coin.

Identify each of the following fractions as proper or improper.

21. $\frac{8}{2}$ 22. $\frac{37}{89}$ 23. $\frac{3}{3}$ 24. $\frac{5}{1}$ 25. $\frac{3}{200}$

26. $\frac{0}{50}$ 27. $\frac{52}{7}$ 28. $\frac{63}{63}$ 29. $\frac{300}{299}$ 30. $\frac{645}{700}$

Rewrite each of the following fractions as a division problem and simplify if possible.

31. $\frac{54}{6}$ 32. $\frac{8}{1}$ 33. $\frac{0}{6}$ 34. $\frac{39}{3}$ 35. $\frac{21}{7}$ 36. $\frac{4}{0}$

Use the following diagram, in which a line segment from 0 to 4 is divided into 12 equal pieces, to determine the answers to Problems 37–40.

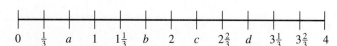

37. Express location a as a fraction.

38. Express location d as a whole number.

39. Express location b as a mixed number and as an improper fraction.

40. Express location c as a mixed number and as an improper fraction.

Convert the following improper fractions to mixed numbers.

41. $\dfrac{7}{4}$

42. $\dfrac{9}{2}$

43. $\dfrac{19}{3}$

44. $\dfrac{23}{6}$

45. $\dfrac{41}{5}$

46. $\dfrac{52}{7}$

47. $\dfrac{82}{3}$

48. $\dfrac{71}{2}$

49. $\dfrac{101}{7}$

50. $\dfrac{181}{9}$

51. $\dfrac{320}{9}$

52. $\dfrac{213}{5}$

53. $\dfrac{543}{7}$

54. $\dfrac{905}{8}$

55. $\dfrac{115}{12}$

56. $\dfrac{200}{11}$

57. $\dfrac{853}{21}$

58. $\dfrac{687}{31}$

Convert the following mixed numbers to improper fractions.

59. $3\dfrac{1}{5}$

60. $6\dfrac{2}{3}$

61. $8\dfrac{2}{9}$

62. $4\dfrac{5}{8}$

63. $4\dfrac{3}{7}$

64. $6\dfrac{1}{4}$

65. $10\dfrac{5}{8}$

66. $12\dfrac{2}{5}$

67. $12\dfrac{3}{8}$

68. $20\dfrac{7}{9}$

69. $21\dfrac{5}{6}$

70. $32\dfrac{4}{5}$

71. $7\dfrac{11}{20}$

72. $9\dfrac{2}{15}$

73. $35\dfrac{9}{10}$

74. $67\dfrac{4}{10}$

75. $52\dfrac{4}{13}$

76. $78\dfrac{5}{11}$

77. $1\dfrac{100}{101}$

78. $2\dfrac{300}{301}$

🖩 79. $69\dfrac{73}{94}$

🖩 80. $82\dfrac{51}{135}$

🖩 81. $137\dfrac{17}{23}$

🖩 82. $389\dfrac{22}{41}$

Answer each of the following questions with an appropriate fraction.

83. In a bus with thirty-two windows, three are broken. What fraction of the windows is broken?

84. In a class of twenty-seven students, five failed their first test. What fraction of the students failed?

85. In a regular deck of fifty-two cards, what fraction of the cards is hearts?

86. If you roll a single die, what is your chance of rolling a 2?

87. In a pack of soda containing five root beers and seven colas, what fraction of the bottles is colas?

88. A bag contains eight hard candies and thirteen chocolate candies. What fraction of the candies is chocolate?

89. If there are eleven adults in a room and seven of them are men, what fraction of the people is men? What fraction is women?

90. A fruit basket contains two bananas, four apples, six oranges, and three pears. What fraction of the fruit consists of apples?

91. A runner finished a race in 7 minutes 21 seconds. Express this time in minutes as a mixed number and as an improper fraction.

92. A woman is 5 feet 7 inches tall. Express her height in feet as a mixed number and as an improper fraction.

93. A bag of peanuts weighs 1 pound 5 ounces. Express this weight in pounds as a mixed number and as an improper fraction.

94. It is estimated that American men between the ages of 25 and 54 watch 28 hours 46 minutes of television on the average per week. Express this amount of time in hours as a mixed number and as an improper fraction.

3.2
Reducing to Lowest Terms

Equivalent Fractions

There is always more than one fraction that can be used to represent the same value. For example,

$$\frac{8}{8} = \frac{3}{3} = \frac{17}{17} = \frac{100}{100} = 1$$

▶ **DEFINITION** Fractions that have the same numerical value are called *equivalent fractions*.

Figure 3.5 illustrates more examples of equivalent fractions. Four line segments of length 1 are divided into two, three, four, and six pieces, respectively. From this diagram we can see that the fractions $\frac{1}{2}, \frac{2}{4}$, and $\frac{3}{6}$ all represent the same distance. These are, therefore, equivalent fractions. We represent this by writing

$$\frac{1}{2} = \frac{2}{4} = \frac{3}{6}$$

Figure 3.5

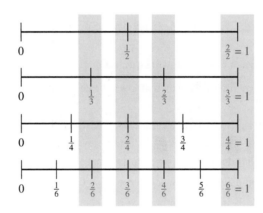

We can also see from Figure 3.5 that

$$\frac{1}{3} = \frac{2}{6}, \qquad \frac{2}{3} = \frac{4}{6}, \qquad \text{and} \qquad \frac{2}{2} = \frac{3}{3} = \frac{4}{4} = \frac{6}{6} = 1$$

We will now look at some examples that show us how to find equivalent fractions.

To find an equivalent fraction of one that is given, multiply or divide the numerator and denominator of the given fraction by the same nonzero number. ▲

EXAMPLE 1 Find an equivalent fraction of $\frac{3}{5}$ with a denominator of 10.

Solution $\dfrac{3}{5} = \dfrac{?}{10}$

To obtain a denominator of 10, we had to multiply 5 by 2.

$$\frac{3 \cdot 2}{5 \cdot 2} = \frac{6}{10}$$

Therefore, we must also multiply the numerator 3 by 2. ▲

Notice that since $\frac{2}{2} = 1$, multiplying both numerator and denominator by 2 in Example 1 is the same as multiplying the original fraction by 1. That is, $\frac{3}{5} \cdot \frac{2}{2} = \frac{3}{5} \cdot 1$. Hence, $\frac{6}{10}$ has the same value as $\frac{3}{5}$.

EXAMPLE 2 Find an equivalent fraction of $\frac{15}{20}$ with a numerator of 3.

Solution $\frac{15}{20} = \frac{3}{?}$

$$\frac{15 \div 5}{20 \div 5} = \frac{3}{4}$$

Divide both numerator and denominator by 5. ▲

EXAMPLE 3 Find two equivalent fractions of $\frac{4}{12}$ with denominators of 3 and 27, respectively.

Solution $\frac{4}{12} = \frac{?}{3}$

$$\frac{4 \div 4}{12 \div 4} = \frac{1}{3}$$

Divide both numerator and denominator by 4.

$$\frac{4}{12} = \frac{?}{27}$$

$$\frac{4}{12} = \frac{1}{3} = \frac{?}{27}$$

There is no whole number we can multiply 12 by to get 27, so we begin with the fraction equivalent to $\frac{4}{12}$ that we just calculated, which is $\frac{1}{3}$.

$$\frac{1 \cdot 9}{3 \cdot 9} = \frac{9}{27}$$

Multiply numerator and denominator of the fraction $\frac{1}{3}$ by 9. ▲

EXAMPLE 4 Find an equivalent fraction for $4\frac{2}{5}$ with a denominator of 15.

Solution $4\frac{2}{5} = \frac{22}{5}$

Change $4\frac{2}{5}$ to an improper fraction.

$$\frac{22}{5} = \frac{?}{15} = \frac{22 \cdot 3}{5 \cdot 3} = \frac{66}{15}$$

Multiply numerator and denominator by 3. ▲

Lowest Terms

Fractions are often easier to understand when the numerator and denominator are as small as possible, or when they are in *lowest terms*.

▶ **DEFINITION** A fraction is in *lowest terms*, or *simplest terms*, if the only common factor of both the numerator and denominator is the **number one.**

If a fraction is not in lowest terms, we can simplify it by **reducing** it to lowest terms.

To *reduce* a fraction to lowest terms, divide both the numerator and denominator by their greatest common factor (GCF). ▲

EXAMPLE 5 Reduce $\frac{15}{60}$ to lowest terms.

Solution
$15 = \boxed{3 \cdot 5}$

$60 = 2 \cdot 2 \cdot \boxed{3 \cdot 5}$

$GCF = 3 \cdot 5 = 15$

$\dfrac{15 \div 15}{60 \div 15} = \dfrac{1}{4}$

Find the greatest common factor of 15 and 60.

Now divide both numerator and denominator by 15. ▲

Another way to reduce a fraction to lowest terms is to completely factor both the numerator and denominator, and then cross out the common factors that appear in both. This technique of crossing out like terms is sometimes called *canceling*.

EXAMPLE 6 Reduce $\frac{12}{18}$.

Solution
$\dfrac{12}{18} = \dfrac{2 \cdot 2 \cdot 3}{2 \cdot 3 \cdot 3} = \dfrac{\cancel{2} \cdot 2 \cdot \cancel{3}}{\cancel{2} \cdot 3 \cdot \cancel{3}} = \dfrac{2}{3}$

$\dfrac{12}{18} = \dfrac{2}{2} \cdot \dfrac{2}{3} \cdot \dfrac{3}{3} = 1 \cdot \dfrac{2}{3} \cdot 1 = \dfrac{2}{3}$

Completely factor the numerator and denominator. Cross out their common factors.
Notice that since $\frac{2}{2}$ and $\frac{3}{3}$ both equal 1, crossing out common factors is the same as eliminating factors of 1. Because of the commutative property of multiplication, it does not matter which two 3s we cancel, as long as one is in the numerator and the other is in the denominator. ▲

Often, it is easiest to divide numerator and denominator by any common factor we can think of and then repeat the process until the fraction is in simplest terms.

EXAMPLE 7 Reduce $\frac{72}{54}$ to lowest terms.

Solution
$\dfrac{72}{54} = \dfrac{72 \div 2}{54 \div 2} = \dfrac{36 \div 3}{27 \div 3} = \dfrac{12 \div 3}{9 \div 3} = \dfrac{4}{3}$

Notice that improper fractions reduced to lowest terms are still improper fractions. ▲

When reducing fractions to lowest terms it is important to remember the following.

Proper fractions reduce to proper fractions and improper fractions reduce to improper fractions. ▲

Sometimes we will show division by a common factor by canceling the original numerator and denominator and writing the results next to them.

EXAMPLE 8   Reduce $\frac{42}{252}$ to lowest terms.

$$\frac{42}{252} = \frac{\overset{21}{\cancel{42}}}{\underset{126}{\cancel{252}}}$$ Divide numerator and denominator by 2.

$$\frac{21}{126} = \frac{\overset{7}{\cancel{21}}}{\underset{42}{\cancel{126}}}$$ Divide numerator and denominator by 3.

$$\frac{7}{42} = \frac{\overset{1}{\cancel{7}}}{\underset{6}{\cancel{42}}} = \frac{1}{6}$$ Divide numerator and denominator by 7. Notice that the numerator of the reduced fraction is 1, not 0, since $7 \div 7 = 1$.   ▲

Instead of showing each individual step, as in Example 8, we usually reduce in the following manner.

$$\frac{42}{252} = \frac{\overset{\overset{\overset{1}{7}}{\cancel{21}}}{\cancel{42}}}{\underset{\underset{\underset{6}{\cancel{42}}}{\cancel{126}}}{\cancel{252}}} \qquad \begin{array}{l} \leftarrow \text{Divide by 7.} \\ \leftarrow \text{Divide by 3.} \\ \leftarrow \text{Divide by 2.} \\ \leftarrow \text{Divide by 2.} \\ \leftarrow \text{Divide by 3.} \\ \leftarrow \text{Divide by 7.} \end{array} \qquad \frac{42}{252} = \frac{1}{6}$$

EXAMPLE 9   There are 5,500 students at a major university, 220 of whom are history majors. Express the fraction of history majors in the student body as a fraction reduced to lowest terms.

Solution $\dfrac{220}{5,500}$ Since 220 out of a total of 5,500 students are history majors.

$$\frac{220}{5,500} = \frac{\overset{\overset{\overset{1}{2}}{\cancel{22}}}{\cancel{220}}}{\underset{\underset{\underset{25}{\cancel{50}}}{\cancel{550}}}{\cancel{5,500}}} \qquad \begin{array}{l} \leftarrow \text{Divide by 2.} \\ \leftarrow \text{Divide by 11.} \\ \leftarrow \text{Divide by 10.} \\ \leftarrow \text{Divide by 10.} \\ \leftarrow \text{Divide by 11.} \\ \leftarrow \text{Divide by 2.} \end{array}$$ Reduce to lowest terms.

$$\frac{220}{5,500} = \frac{1}{25}$$

Therefore, $\frac{1}{25}$ of the student body is history majors.   ▲

Quick Quiz

Reduce each of the following fractions to lowest terms.

Answers

1. $\dfrac{9}{21}$ 1. $\dfrac{3}{7}$

2. $\dfrac{24}{72}$ 2. $\dfrac{1}{3}$

3. $\dfrac{132}{55}$ 3. $\dfrac{12}{5}$

3.2 Exercises

Indicate which of the following statements are true and which are false. For those that are false, change the italicized word to make the statement true.

1. An equivalent fraction is obtained by *multiplying* or *adding* the same number in both numerator and denominator of a given fraction.

2. The fraction $\frac{8}{12}$ *has* been reduced to lowest terms.

3. To reduce a fraction to lowest terms, the operation *subtraction* must be performed on both numerator and denominator.

4. Multiplying both numerator and denominator of a fraction by the same number is the same as multiplying the fraction by *one*.

Change each of the following fractions to equivalent fractions by performing the indicated operation on *both the numerator and the denominator.*

5. $\dfrac{3}{7}$ Multiply by 6.   6. $\dfrac{4}{5}$ Multiply by 2.   7. $\dfrac{21}{49}$ Divide by 7.   8. $\dfrac{32}{48}$ Divide by 8.

9. $\dfrac{8}{13}$ Multiply by 2.   10. $\dfrac{4}{21}$ Multiply by 4.   11. $\dfrac{15}{27}$ Divide by 3.   12. $\dfrac{35}{75}$ Divide by 5.

13. $\dfrac{5}{4}$ Multiply by 7.   14. $\dfrac{6}{5}$ Multiply by 6.   15. $\dfrac{36}{42}$ Divide by 6.   16. $\dfrac{72}{81}$ Divide by 9.

17. $\dfrac{11}{22}$ Divide by 11.   18. $\dfrac{36}{60}$ Divide by 12.   19. $\dfrac{105}{20}$ Divide by 5.   20. $\dfrac{130}{30}$ Divide by 10.

21. $\dfrac{66}{84}$ Divide by 6.   22. $\dfrac{49}{91}$ Divide by 7.   23. $\dfrac{66}{55}$ Divide by 11.   24. $\dfrac{39}{26}$ Divide by 13.

25. $\dfrac{150}{80}$ Divide by 10.   26. $\dfrac{144}{96}$ Divide by 12.

27. $\dfrac{153}{218}$ Multiply by 49.   28. $\dfrac{327}{289}$ Multiply by 34.

29. $\dfrac{702}{4,374}$ Divide by 54.   30. $\dfrac{1,073}{1,769}$ Divide by 29.

Fill in the missing number in each of these equivalent fractions.

31. $\dfrac{3}{6} = \dfrac{?}{2}$ 32. $\dfrac{9}{27} = \dfrac{?}{3}$ 33. $\dfrac{5}{8} = \dfrac{25}{?}$ 34. $\dfrac{4}{16} = \dfrac{1}{?}$

35. $\dfrac{3}{4} = \dfrac{?}{20}$

36. $\dfrac{5}{9} = \dfrac{25}{?}$

37. $\dfrac{9}{36} = \dfrac{3}{?}$

38. $\dfrac{8}{44} = \dfrac{2}{?}$

39. $\dfrac{3}{16} = \dfrac{9}{?}$

40. $\dfrac{5}{21} = \dfrac{10}{?}$

41. $\dfrac{6}{24} = \dfrac{1}{?}$

42. $\dfrac{12}{16} = \dfrac{3}{?}$

43. $\dfrac{12}{48} = \dfrac{?}{12}$

44. $\dfrac{15}{60} = \dfrac{1}{?}$

45. $\dfrac{9}{18} = \dfrac{?}{2}$

46. $\dfrac{3}{27} = \dfrac{?}{9}$

47. $\dfrac{42}{6} = \dfrac{?}{2}$

48. $\dfrac{32}{8} = \dfrac{4}{?}$

49. $\dfrac{72}{45} = \dfrac{8}{?}$

50. $\dfrac{48}{16} = \dfrac{6}{?}$

51. $\dfrac{9}{12} = \dfrac{?}{144}$

52. $\dfrac{3}{16} = \dfrac{24}{?}$

53. $\dfrac{102}{11} = \dfrac{?}{121}$

54. $\dfrac{2}{13} = \dfrac{?}{182}$

55. $\dfrac{30}{225} = \dfrac{?}{15}$

56. $\dfrac{26}{169} = \dfrac{2}{?}$

57. $7 = \dfrac{35}{?}$

58. $9 = \dfrac{63}{?}$

59. $3\dfrac{1}{8} = \dfrac{?}{48}$

60. $5\dfrac{3}{4} = \dfrac{?}{12}$

61. $6\dfrac{2}{3} = \dfrac{60}{?}$

62. $4\dfrac{1}{8} = \dfrac{99}{?}$

63. $\dfrac{5}{9} = \dfrac{?}{6,039}$

64. $\dfrac{7}{8} = \dfrac{2,492}{?}$

65. $\dfrac{2,202}{2,936} = \dfrac{?}{4}$

66. $\dfrac{3,315}{4,641} = \dfrac{5}{?}$

Reduce each of the following fractions to lowest terms.

67. $\dfrac{4}{10}$

68. $\dfrac{6}{8}$

69. $\dfrac{3}{9}$

70. $\dfrac{7}{28}$

71. $\dfrac{9}{12}$

72. $\dfrac{8}{24}$

73. $\dfrac{15}{20}$

74. $\dfrac{21}{56}$

75. $\dfrac{18}{3}$

76. $\dfrac{32}{4}$

77. $\dfrac{16}{28}$

78. $\dfrac{27}{63}$

79. $\dfrac{11}{66}$

80. $\dfrac{15}{75}$

81. $\dfrac{25}{10}$

82. $\dfrac{48}{18}$

83. $\dfrac{18}{36}$

84. $\dfrac{9}{54}$

85. $\dfrac{14}{49}$

86. $\dfrac{63}{70}$

87. $\dfrac{70}{90}$

88. $\dfrac{55}{88}$

89. $\dfrac{42}{88}$

90. $\dfrac{16}{64}$

91. $\dfrac{55}{90}$

92. $\dfrac{28}{40}$

93. $\dfrac{24}{144}$

94. $\dfrac{39}{169}$

95. $\dfrac{120}{270}$

96. $\dfrac{450}{750}$

97. $\dfrac{125}{35}$

98. $\dfrac{180}{150}$

99. $\dfrac{121}{132}$

100. $\dfrac{50}{625}$

101. $\dfrac{48}{256}$

102. $\dfrac{36}{432}$

103. $\dfrac{1,024}{48}$

104. $\dfrac{1,125}{250}$

Answer each of the following questions with a fraction reduced to lowest terms.

105. In a class of 132 students, 22 are over 6 feet tall. What fraction of the total class is more than 6 feet tall?

106. In a recent survey, 63 out of 91 people responded that they regularly use the same brand of toothpaste. What fraction of the people surveyed is loyal to a single brand of toothpaste?

107. A zoologist who makes a monthly salary of $1,250 pays $550 a month for rent. What fraction of her monthly salary does she pay in rent?

108. An administrative assistant who makes a yearly salary of $30,000 pays $8,400 in federal income tax. What fraction of his salary does he pay in federal income tax?

109. An eye clinic was visited by 104 nearsighted people and 13 farsighted people in one day. What fraction of the patients was nearsighted?

110. A photographer took 48 black-and-white pictures and 112 color pictures. What fraction of the photos was black-and-white?

111. Eight hours is equivalent to what fraction of a week?

112. Nine inches is equivalent to what fraction of a yard?

3.3
Addition of Fractions

In Chapter 1 we learned how to perform the four basic arithmetic operations on whole numbers. Now we will establish some rules for arithmetic with fractions, beginning with addition. The following examples will help us understand how these rules were developed.

In the first example, let us imagine that you baked two identical pies for a party and that you divided each pie into five equal pieces, as shown in Figure 3.6. Suppose your guests ate three pieces of the first pie and two pieces of the second pie. Would one pie have been sufficient for the guests?

Looking at Figure 3.6, we can see that the guests ate $\frac{3}{5}$ of the first pie and $\frac{2}{5}$ of the second pie. Adding these two fractions together visually, we can see that the guests consumed a total of $\frac{5}{5}$ of a pie, or one *entire* pie. So one pie would have been enough.

Figure 3.6

$$\frac{3}{5} \quad + \quad \frac{2}{5} \quad = \quad \frac{5}{5} = 1$$

In the second example, imagine that you baked two identical pies, but you did not cut them into the same number of pieces. At the end of the party you noticed that your guests ate $\frac{1}{2}$ of the first pie and $\frac{1}{3}$ of the second pie. How can you determine the total amount of pie that they ate?

Looking at Figure 3.7 we see that this is not an easy problem to do visually, because both pies are not cut into the same number of equal pieces. Therefore, let us slice each pie into an appropriate number of equal pieces—that is, a number that is evenly divisible by both 2 and 3, such as the number 6. As shown in Figure 3.8, $\frac{1}{2}$ of the first pie is equivalent to $\frac{3}{6}$, and $\frac{1}{3}$ of the second pie is equivalent to $\frac{2}{6}$. Adding the two visually, we obtain a total of $\frac{5}{6}$ of a pie.

Figure 3.7

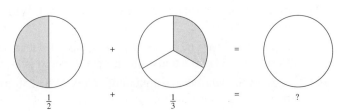

$$\frac{1}{2} \quad + \quad \frac{1}{3} \quad = \quad ?$$

Figure 3.8

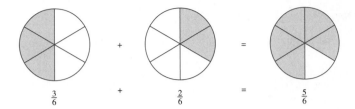

$$\frac{3}{6} \qquad + \qquad \frac{2}{6} \qquad = \qquad \frac{5}{6}$$

Once the two pies are cut into the same number of equal pieces, the denominators of the fractions we wish to add are the same, and the sum can be obtained by adding the numerators. Notice that the common denominator, 6, must be evenly divisible by each of the denominators of the original fractions to be added. Therefore, a common denominator for two or more fractions can be found by multiplying the denominators together. The procedure we have just illustrated is outlined below.

To Add Fractions

1. Find a common denominator.
2. Change each fraction to an equivalent fraction with the common denominator.
3. Add the numerators and write the result over the common denominator.
4. If necessary, reduce the answer to lowest terms. ▲

EXAMPLE 1   Add $\frac{2}{9} + \frac{4}{9}$.

Solution $\dfrac{2}{9} + \dfrac{4}{9} = \dfrac{2 + 4}{9}$ Both fractions already have a common denominator of 9.

$\qquad\qquad = \dfrac{6}{9}$ Add the numerators and place the result over the common denominator.

$\qquad\qquad = \dfrac{2}{3}$ Reduce to lowest terms. ▲

EXAMPLE 2   Add $\frac{2}{5} + \frac{1}{4}$.

$\dfrac{2}{5} + \dfrac{1}{4} = \dfrac{2 \cdot 4}{5 \cdot 4} + \dfrac{1 \cdot 5}{4 \cdot 5}$ A common denominator is $5 \cdot 4 = 20$.

$\qquad\qquad = \dfrac{8}{20} + \dfrac{5}{20}$ Change each fraction to an equivalent fraction with a denominator of 20.

$\qquad\qquad = \dfrac{8 + 5}{20}$ Add the numerators and write the result over the common denominator.

$\qquad\qquad = \dfrac{13}{20}$ ▲

EXAMPLE 3   Add $\frac{3}{8} + \frac{5}{12}$.

$$\frac{3}{8} + \frac{5}{12} = \frac{3 \cdot 9}{8 \cdot 9} + \frac{5 \cdot 6}{12 \cdot 6}$$

One common denominator is 96.
$8 \cdot 12 = 96$.
Another common denominator is 72.
$8 \cdot 9 = 72$ and $12 \cdot 6 = 72$.

$$= \frac{27}{72} + \frac{30}{72}$$

Change each fraction to an equivalent fraction with a denominator of 72.

$$= \frac{27 + 30}{72}$$

Add the numerators and write the result over the common denominator.

$$= \frac{57}{72}$$

$$= \frac{57 \div 3}{72 \div 3} = \frac{19}{24}$$

Reduce to lowest terms.   ▲

Finding the Lowest Common Denominator (LCD)

In order to keep our calculations as simple as possible, we will now introduce a procedure for finding the **lowest common denominator**, or **LCD**, of two or more fractions.

> ▶ **DEFINITION**   The *lowest common denominator*, or *LCD*, is the smallest number that is divisible by the denominators of all fractions being considered.

To find the LCD of two or more fractions, find the least common multiple (LCM) of their denominators.   ▲

The technique for finding the LCM of two or more numbers was discussed in Section 2.6. Example 4 will help us to review the procedure.

EXAMPLE 4   Find the LCD of $\frac{5}{18}$ and $\frac{11}{48}$.

Solution $18 = 3 \cdot 6 = 3 \cdot 2 \cdot 3 = 2 \cdot 3^2$ Factor each denominator completely and
$48 = 6 \cdot 8 = 2 \cdot 3 \cdot 2 \cdot 2 \cdot 2 = 2^4 \cdot 3$ write it in exponential notation.

The highest exponent 2 is raised to is 4: 2^4.   The unique prime factors in the
The highest exponent 3 is raised to is 2: 3^2.   factorizations are 2 and 3.
$LCD = 2^4 \cdot 3^2 = 16 \cdot 9 = 144$ The LCD is the product $2^4 \cdot 3^2$.   ▲

Once the LCD is found, it is important to be able to identify what number each of the original denominators must be multiplied by to obtain the LCD. This number can be found by comparing the prime factorizations of each denominator with the prime factorization of the LCD. Cancel the factors that are common to both the denominator and the LCD to obtain the answer. In Example 4, we can determine what number each denominator must be multiplied by to obtain the LCD as follows:

$$LCD = 2^4 \cdot 3^2 = 2 \cdot 2 \cdot 2 \cdot 2 \cdot 3 \cdot 3$$

$$18 = 2 \cdot 3 \cdot 3$$
$$\text{LCD} = \boxed{2 \cdot 2 \cdot 2} \cdot 2 \cdot 3 \cdot 3 \qquad \text{Multiply by 8.}$$
$$48 = 2 \cdot 2 \cdot 2 \cdot 2 \cdot 3$$
$$\text{LCD} = 2 \cdot 2 \cdot 2 \cdot 2 \cdot 3 \cdot \boxed{3} \qquad \text{Multiply by 3.}$$

We can see, therefore, that we must multiply 18 by 8 and we must multiply 48 by 3 to obtain the LCD.

EXAMPLE 5 Find the LCD of $\frac{7}{90}$ and $\frac{5}{24}$. Then determine what number each denominator must be multiplied by to obtain the LCD.

Solution

$$90 = 10 \cdot 9 = 2 \cdot 5 \cdot 3 \cdot 3 = 2 \cdot 3^2 \cdot 5 \qquad \text{The unique prime factors are 2, 3, and 5.}$$
$$24 = 6 \cdot 4 = 2 \cdot 3 \cdot 2 \cdot 2 = 2^3 \cdot 3 \qquad \text{The highest exponent 2 is raised to is 3: } 2^3.$$
$$\text{LCD} = 2^3 \cdot 3^2 \cdot 5 = 8 \cdot 9 \cdot 5 \qquad \text{The highest exponent 3 is raised to is 2: } 3^2.$$
$$= 72 \cdot 5 = 360 \qquad \text{The highest exponent 5 is raised to is 1: } 5^1.$$
$$\text{LCD} = 2^3 \cdot 3^2 \cdot 5 = 2 \cdot 2 \cdot 2 \cdot 3 \cdot 3 \cdot 5 \qquad \text{To determine what number each}$$
$$90 = 2 \cdot 3 \cdot 3 \cdot 5 \qquad \text{denominator must be multiplied by to}$$
$$\text{LCD} = \boxed{2 \cdot 2} \cdot 2 \cdot 3 \cdot 3 \cdot 5 \qquad \text{obtain the LCD, we compare prime}$$
$$\text{Multiply by 4.} \qquad \text{factorizations.}$$
$$24 = 2 \cdot 2 \cdot 2 \cdot 3$$
$$\text{LCD} = 2 \cdot 2 \cdot 2 \cdot 3 \cdot \boxed{3 \cdot 5}$$
$$\text{Multiply by 15.}$$

Thus, to obtain the LCD we must multiply 90 by 4 and we must multiply 24 by 15. ▲

Let us now add the two fractions in Example 3 by first finding the LCD.

EXAMPLE 6 Add $\frac{3}{8} + \frac{5}{12}$.

Solution

$$8 = 2 \cdot 2 \cdot 2 = 2^3 \qquad \text{Find the LCD. Compare the prime}$$
$$12 = 2 \cdot 2 \cdot 3 = 2^2 \cdot 3 \qquad \text{factorizations. To obtain the LCD we must}$$
$$\text{LCD} = 2^3 \cdot 3 = 8 \cdot 3 = 24 \qquad \text{multiply 8 by 3 and 12 by 2.}$$

$$\frac{3}{8} + \frac{5}{12} = \frac{3 \cdot 3}{8 \cdot 3} + \frac{5 \cdot 2}{12 \cdot 2} \qquad \text{Change each fraction to an equivalent}$$
$$\text{fraction with the same LCD.}$$

$$= \frac{9}{24} + \frac{10}{24}$$

$$= \frac{9 + 10}{24} \qquad \text{Add the numerators and place the result}$$
$$\text{over the LCD.}$$

$$= \frac{19}{24} \qquad \qquad \qquad \qquad ▲$$

Notice that the answer in Example 6 is the same answer as in Example 3. However, since we used 24 instead of 72 as our common denominator, we were able to work with smaller numbers in Example 6, which simplified our calculations.

The following examples show addition of fractions using a vertical format.

EXAMPLE 7 Add $\frac{7}{30} + \frac{5}{42}$.

Solution

$30 = 5 \cdot 6 = 5 \cdot 2 \cdot 3$
$42 = 6 \cdot 7 = 2 \cdot 3 \cdot 7$
$LCD = 2 \cdot 3 \cdot 5 \cdot 7 = 210$

Find the LCD. Compare the prime factorizations. To obtain the LCD we must multiply 30 by 7 and 42 by 5.

$$\frac{7}{30} \cdot \frac{7}{7} = \frac{49}{210}$$
$$+\frac{5}{42} \cdot \frac{5}{5} = \frac{25}{210}$$

Change each fraction to an equivalent fraction with the same LCD.

$$\frac{74}{210}$$

Add the numerators and place the result over the LCD.

$$\frac{74}{210} = \frac{37}{105}$$

Reduce to lowest terms. ▲

EXAMPLE 8 Add $\frac{7}{16} + \frac{11}{12} + \frac{3}{20}$.

Solution

$16 = 2 \cdot 2 \cdot 2 \cdot 2 = 2^4$
$12 = 2 \cdot 2 \cdot 3 = 2^2 \cdot 3$
$20 = 2 \cdot 2 \cdot 5 = 2^2 \cdot 5$
$LCD = 2^4 \cdot 3 \cdot 5 = 16 \cdot 15 = 240$

Find the LCD. To obtain the LCD we must multiply 16 by 15, 12 by 20, and 20 by 12.

$$\frac{7}{16} \cdot \frac{15}{15} = \frac{105}{240}$$
$$\frac{11}{12} \cdot \frac{20}{20} = \frac{220}{240}$$
$$+\frac{3}{20} \cdot \frac{12}{12} = \frac{36}{240}$$

Change each fraction to an equivalent fraction with a denominator of 240.

$$\frac{361}{240}$$

Add the numerators and place the result over the LCD. ▲

Another Look at Mixed Numbers

Now that we have learned how to add fractions, we can examine more closely the meaning of a mixed number. For example, we have already established that $5\frac{2}{3}$ means $5 + \frac{2}{3}$. Since any whole number can be represented by a fraction with a denominator of 1, we can rewrite 5 as $\frac{5}{1}$ and add the two fractions as follows:

$$5\frac{2}{3} = 5 + \frac{2}{3} = \frac{5}{1} + \frac{2}{3}$$

The LCD is 3.

$$= \frac{5 \cdot 3}{1 \cdot 3} + \frac{2}{3}$$

Change $\frac{5}{1}$ to an equivalent fraction with a denominator of 3.

$$= \frac{15}{3} + \frac{2}{3} = \frac{17}{3}$$

Add the numerators.

The result of our addition is the same result we would obtain by converting $5\frac{2}{3}$ to an improper fraction using the technique introduced in Section 3.1.

$$5\frac{2}{3} = \frac{(5 \cdot 3) + 2}{3} = \frac{17}{3}$$

EXAMPLE 9 Use addition of fractions to convert $2\frac{9}{11}$ to an improper fraction.

Solution $2\frac{9}{11} = 2 + \frac{9}{11} = \frac{2}{1} + \frac{9}{11}$ The LCD is 11.

$= \frac{2 \cdot 11}{1 \cdot 11} + \frac{9}{11}$ Change $\frac{2}{1}$ to an equivalent fraction with a denominator of 11.

$= \frac{22}{11} + \frac{9}{11} = \frac{31}{11}$ Add the numerators.   ▲

EXAMPLE 10 At a sales convention, $\frac{1}{6}$ of the participants are wearing glasses and $\frac{2}{9}$ of the participants are wearing contact lenses. What fraction of the people is wearing either glasses or contact lenses?

Solution Add $\frac{1}{6} + \frac{2}{9}$.   Find the total number of people wearing glasses or contacts.

$6 = 2 \cdot 3$ Find the LCD. To obtain the LCD we must
$9 = 3 \cdot 3 = 3^2$ multiply 6 by 3 and 9 by 2.
$\text{LCD} = 2 \cdot 3^2 = 18$

$\dfrac{1}{6} \cdot \dfrac{3}{3} = \dfrac{3}{18}$ Change each fraction to an equivalent
fraction with a denominator of 18.
$+\dfrac{2}{9} \cdot \dfrac{2}{2} = \dfrac{4}{18}$

$\qquad\qquad \dfrac{7}{18}$ Add the numerators and place the result
over the LCD.

$\frac{7}{18}$ of the people is wearing either glasses or contact lenses.   ▲

EXAMPLE 11 On three consecutive days, a jogger ran $\frac{1}{4}$ mile, $\frac{3}{8}$ mile, and $\frac{2}{5}$ mile. What was the total distance he ran on those three days?

Solution Add $\frac{1}{4} + \frac{3}{8} + \frac{2}{5}$.
$4 = 2 \cdot 2 = 2^2$ Find the LCD. To obtain the LCD we must
$8 = 2 \cdot 2 \cdot 2 = 2^3$ multiply 4 by 10, 8 by 5, and 5 by 8.
$5 = 5$
$\text{LCD} = 2^3 \cdot 5 = 40$

$$\frac{1}{4} \cdot \frac{10}{10} = \frac{10}{40}$$

$$\frac{3}{8} \cdot \frac{5}{5} = \frac{15}{40}$$

$$+\frac{2}{5} \cdot \frac{8}{8} = \frac{16}{40}$$

$$\frac{41}{40}$$

Change each fraction to an equivalent fraction with a denominator of 40.

Add the numerators and place the result over the LCD.

$$\frac{41}{40} = 41 \div 40 = 1\frac{1}{40}$$

Convert the answer to a mixed number, which is the way we express distances.

He ran a total of $1\frac{1}{40}$ miles. ▲

Quick Quiz

Add each of the following fractions.

1. $\frac{1}{2} + \frac{1}{3}$

2. $\frac{3}{16} + \frac{5}{12}$

3. $\frac{2}{9} + \frac{7}{24} + \frac{5}{18}$

Answers

1. $\frac{5}{6}$

2. $\frac{29}{48}$

3. $\frac{19}{24}$

3.3 Exercises

Indicate which of the following statements are true and which are false. For those that are false, change the italicized word to make the statement true.

1. In order to add two fractions, you must first find a common *denominator*.

2. The LCD of two fractions is the *largest* number that is evenly divisible by both denominators.

3. When adding two fractions, each greater than zero, the sum is always *greater* than either of the two addends.

4. The sum of two fractions, each greater than zero and less than 1, is *always* less than 1.

Add each of the following fractions. Reduce all answers to lowest terms.

5. $\frac{3}{7} + \frac{1}{7}$

6. $\frac{4}{9} + \frac{2}{9}$

7. $\frac{5}{8} + \frac{3}{8}$

8. $\frac{3}{4} + \frac{1}{4}$

9. $\frac{3}{10} + \frac{5}{10}$

10. $\frac{4}{11} + \frac{5}{11}$

11. $\frac{8}{17} + \frac{5}{17} + \frac{2}{17}$

12. $\frac{4}{21} + \frac{8}{21} + \frac{5}{21}$

13. $\frac{7}{100} + \frac{42}{100} + \frac{66}{100}$

14. $\frac{21}{50} + \frac{16}{50} + \frac{33}{50}$

15. $\frac{1}{3} + \frac{3}{4}$

16. $\frac{3}{7} + \frac{2}{5}$

17. $\frac{7}{12} + \frac{2}{3}$ 18. $\frac{3}{5} + \frac{8}{15}$ 19. $\frac{1}{11} + \frac{2}{3}$ 20. $\frac{3}{4} + \frac{2}{13}$

21. $\frac{7}{10} + \frac{3}{25}$ 22. $\frac{3}{28} + \frac{5}{42}$ 23. $\frac{3}{56} + \frac{1}{8}$ 24. $\frac{1}{6} + \frac{11}{48}$

25. $\frac{12}{7} + \frac{9}{5}$ 26. $\frac{21}{4} + \frac{8}{3}$ 27. $\frac{9}{10} + \frac{37}{100}$ 28. $\frac{81}{100} + \frac{7}{10}$

29. $\frac{5}{48} + \frac{5}{36}$ 30. $\frac{8}{49} + \frac{8}{21}$ 31. $\frac{4}{15} + \frac{3}{50}$ 32. $\frac{7}{40} + \frac{3}{16}$

33. $\frac{5}{18} + \frac{4}{27}$ 34. $\frac{1}{40} + \frac{5}{24}$ 35. $\frac{4}{21} + \frac{5}{63}$ 36. $\frac{7}{52} + \frac{8}{31}$

37. $\frac{3}{11} + \frac{5}{12}$ 38. $\frac{9}{10} + \frac{12}{13}$ 39. $\frac{15}{4} + \frac{27}{5}$ 40. $\frac{24}{7} + \frac{14}{9}$

41. $\frac{37}{5} + \frac{19}{10}$ 42. $\frac{22}{3} + \frac{17}{24}$ 43. $\frac{11}{49} + \frac{9}{14}$ 44. $\frac{7}{52} + \frac{8}{13}$

45. $\frac{1}{2} + \frac{1}{3} + \frac{1}{5}$ 46. $\frac{1}{4} + \frac{1}{6} + \frac{1}{9}$ 47. $\frac{3}{10} + \frac{23}{100} + \frac{561}{1,000}$ 48. $\frac{7}{10} + \frac{69}{100} + \frac{103}{1,000}$

49. $\frac{3}{25} + \frac{2}{15} + \frac{1}{5}$ 50. $\frac{2}{3} + \frac{7}{36} + \frac{5}{27}$ 51. $\frac{3}{8} + \frac{5}{12} + \frac{1}{4}$ 52. $\frac{1}{48} + \frac{3}{16} + \frac{5}{18}$

53. $\frac{5}{16} + \frac{3}{8} + \frac{5}{14}$ 54. $\frac{6}{49} + \frac{3}{56} + \frac{5}{14}$ 55. $\frac{51}{639} + \frac{87}{213}$ 56. $\frac{13}{182} + \frac{47}{728}$

57. $\frac{15}{29} + \frac{25}{49}$ 58. $\frac{11}{37} + \frac{31}{53}$

Use addition of fractions to convert each of the following mixed numbers to an improper fraction.

59. $2\frac{2}{7}$ 60. $6\frac{3}{4}$ 61. $3\frac{1}{17}$ 62. $4\frac{4}{13}$ 63. $13\frac{7}{8}$ 64. $16\frac{3}{4}$

Solve each of the following word problems.

65. On Saturday a student studied $\frac{3}{5}$ hour, and on Sunday she studied an additional $\frac{5}{6}$ hour. What is the total amount of time she spent studying that weekend?

66. A recipe calls for $\frac{2}{3}$ cup of sugar and $\frac{3}{4}$ cup of flour. What is the combined amount of flour and sugar?

67. Of the people attending a conference, $\frac{7}{12}$ have brown eyes and $\frac{2}{5}$ have blue eyes. What fraction of the people has brown eyes or blue eyes?

68. A businessman invested $\frac{3}{10}$ of an inheritance in the stock market and $\frac{5}{8}$ of it in real estate. What fraction of the inheritance did he invest in either the stock market or real estate?

69. A gourmet cook bought $\frac{1}{2}$ pound of swiss cheese, $\frac{3}{8}$ pound of cheddar cheese, and $\frac{1}{4}$ pound of blue cheese. What is the total amount of cheese that he bought?

70. A salesclerk at a fabric store cut a remnant of fabric into three pieces that were of lengths $\frac{3}{4}$ yard, $\frac{5}{8}$ yard, and $\frac{2}{5}$ yard. How long was the original remnant before it was cut?

3.4
Subtraction of Fractions

In this section we will find the difference between two fractions. As we did with whole numbers, we will only work with problems in which the minuend is greater than the subtrahend, so that our results will always be greater than zero. In Chapter 7 we will consider problems in which the results are less than zero.

Comparing Fractions

Before we begin our discussion of subtraction of fractions, let us establish a procedure for determining the larger of two fractions. If the two denominators of the fractions being compared are the same, the procedure is simple. For example, in Figure 3.9 we can see easily that $\frac{5}{8}$ is greater than $\frac{3}{8}$. Using mathematical notation we write $\frac{5}{8} > \frac{3}{8}$. Thus, if two fractions have the same denominator, the larger fraction has the larger numerator.

Figure 3.9

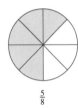 

$$\frac{5}{8} \qquad\qquad \frac{3}{8}$$

Now let us compare two fractions that have different denominators. Look at Figure 3.10 and determine which is larger, $\frac{3}{5}$ or $\frac{2}{3}$. Even a very close inspection of the diagrams leaves us with some doubt, because the shaded areas appear to be roughly the same size. To answer the question with certainty, we need to divide both circles into an equal number of pieces; that is, we need to find a common denominator for the two fractions that we wish to compare. The LCD of 3 and 5 is 15. Converting to equivalent fractions we obtain

$$\frac{3}{5} = \frac{3 \cdot 3}{5 \cdot 3} = \frac{9}{15} \qquad \text{and} \qquad \frac{2}{3} = \frac{2 \cdot 5}{3 \cdot 5} = \frac{10}{15}$$

Since $\frac{9}{15}$ is less than $\frac{10}{15}$, we conclude that $\frac{3}{5}$ is less than $\frac{2}{3}$. This is written as $\frac{3}{5} < \frac{2}{3}$. Thus, $\frac{2}{3}$ is the larger fraction.

Figure 3.10

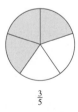 

$$\frac{3}{5} \qquad\qquad \frac{2}{3}$$

We have illustrated a procedure for comparing two fractions, as outlined on the next page.

To Find the Larger of Two Fractions	1. Find the LCD.
	2. Change each fraction to an equivalent fraction with a denominator equal to the LCD.
	3. Compare the two fractions with the same denominator. The one with the larger numerator is the larger fraction.

▲

EXAMPLE 1 Determine which fraction is larger, $\frac{4}{7}$ or $\frac{6}{11}$, and indicate this by placing a $>$ or $<$ sign between them.

Solution $\frac{4}{7}$? $\frac{6}{11}$

$\text{LCD} = 7 \cdot 11 = 77$ Find the LCD.

$\frac{4}{7} \cdot \frac{11}{11} = \frac{44}{77}$ Change each fraction to an equivalent fraction with a denominator of 77.

$\frac{6}{11} \cdot \frac{7}{7} = \frac{42}{77}$

$\frac{44}{77} > \frac{42}{77}$ Compare the equivalent fractions.

Therefore, $\frac{4}{7} > \frac{6}{11}$. ▲

EXAMPLE 2 For each of the following pairs of fractions, indicate which is larger by replacing the ? with a $>$ or $<$ sign.

(a) $\frac{7}{10}$? $\frac{11}{15}$

$10 = 2 \cdot 5$ Find the LCD.
$15 = 3 \cdot 5$
$\text{LCD} = 2 \cdot 3 \cdot 5 = 30$

$\frac{7}{10} \cdot \frac{3}{3} = \frac{21}{30}$ Change each fraction to an equivalent fraction with a denominator of 30.

$\frac{11}{15} \cdot \frac{2}{2} = \frac{22}{30}$

$\frac{21}{30} < \frac{22}{30}$ Compare the equivalent fractions.

Therefore, $\frac{7}{10} < \frac{11}{15}$.

(b) $\dfrac{7}{24} \ ? \ \dfrac{5}{18}$

$24 = 4 \cdot 6 = 2 \cdot 2 \cdot 2 \cdot 3 = 2^3 \cdot 3$ Find the LCD.
$18 = 3 \cdot 6 = 3 \cdot 2 \cdot 3 = 2 \cdot 3^2$
$\text{LCD} = 2^3 \cdot 3^2 = 8 \cdot 9 = 72$

$\dfrac{7}{24} \cdot \dfrac{3}{3} = \dfrac{21}{72}$ Change each fraction to an equivalent
                                            fraction with a denominator of 72.
$\dfrac{5}{18} \cdot \dfrac{4}{4} = \dfrac{20}{72}$

$\dfrac{21}{72} > \dfrac{20}{72}$ Compare the equivalent fractions.

Therefore, $\dfrac{7}{24} > \dfrac{5}{18}$.   ▲

Subtracting Fractions

Let us now begin our discussion of subtraction of fractions with an example. Suppose you have $\frac{1}{2}$ of a pie and you would like to save a piece $\frac{1}{3}$ the size of the whole pie. How much of the pie can you eat and still have $\frac{1}{3}$ of the pie left?

To solve this problem, we need to calculate the difference

$$\frac{1}{2} - \frac{1}{3}$$

Look at Figure 3.11, which illustrates both $\frac{1}{2}$ and $\frac{1}{3}$ of a pie. Both pie pictures must be divided into the same number of equal pieces before they can be compared and subtracted, which is the same as finding the LCD of the two fractions $\frac{1}{2}$ and $\frac{1}{3}$.

$$\text{LCD} = 2 \cdot 3 = 6$$

Figure 3.11

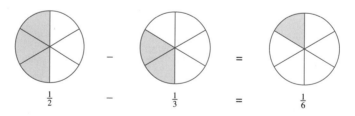

Now change each fraction to an equivalent fraction with a denominator of 6.

$$\frac{1}{2} = \frac{1}{2} \cdot \frac{3}{3} = \frac{3}{6} \qquad \frac{1}{3} = \frac{1}{3} \cdot \frac{2}{2} = \frac{2}{6}$$

Subtract the numerators and write the result over the LCD.

$$\frac{3}{6} - \frac{2}{6} = \frac{3 - 2}{6} = \frac{1}{6}$$

Thus, you can eat $\frac{1}{6}$ of the pie and still have $\frac{1}{3}$ of the pie left.

The procedure that we have illustrated for subtracting two fractions is outlined below.

To Subtract Two Fractions

1. Find the LCD.
2. Change each fraction to an equivalent fraction with a denominator equal to the LCD.
3. Find the difference between the two numerators, and write the result over the common denominator.
4. If necessary, reduce the answer to lowest terms. ▲

EXAMPLE 3   Subtract $\frac{11}{12} - \frac{7}{12}$.

Solution $\dfrac{11}{12} - \dfrac{7}{12} = \dfrac{11 - 7}{12}$ Both fractions have a denominator of 12.

$\qquad\qquad = \dfrac{4}{12}$ Subtract the numerators and put the result over the common denominator.

$\qquad\qquad = \dfrac{1}{3}$ Reduce to lowest terms. ▲

EXAMPLE 4   Subtract $\frac{3}{4} - \frac{5}{14}$.

Solution $4 = 2 \cdot 2 = 2^2$ Find the LCD. To obtain the LCD we must multiply 4 by 7 and 14 by 2.
$14 = 2 \cdot 7$
$\text{LCD} = 2^2 \cdot 7 = 28$

$\dfrac{3}{4} - \dfrac{5}{14} = \dfrac{3 \cdot 7}{4 \cdot 7} - \dfrac{5 \cdot 2}{14 \cdot 2}$ Change each fraction to an equivalent fraction with a denominator of 28.

$\qquad\qquad = \dfrac{21}{28} - \dfrac{10}{28}$

$\qquad\qquad = \dfrac{21 - 10}{28}$ Subtract the numerators and place the result over the LCD.

$\qquad\qquad = \dfrac{11}{28}$ ▲

Now that we know the technique for subtracting fractions, we will use the vertical format for setting up problems.

EXAMPLE 5   Subtract $\frac{9}{28} - \frac{2}{21}$.

Solution $28 = 2 \cdot 2 \cdot 7 = 2^2 \cdot 7$ Find the LCD. To obtain the LCD we must multiply 28 by 3 and 21 by 4.
$21 = 3 \cdot 7$
$\text{LCD} = 2^2 \cdot 3 \cdot 7 = 84$

$$\frac{9}{28} \cdot \frac{3}{3} = \frac{27}{84}$$

$$-\frac{2}{21} \cdot \frac{4}{4} = -\frac{8}{84}$$

$$\frac{19}{84}$$

Change each fraction to an equivalent fraction with a denominator of 84.

Subtract the numerators and place the result over the LCD. ▲

EXAMPLE 6 Subtract $5 - \frac{2}{9}$.

Solution $5 - \frac{2}{9} = \frac{5}{1} - \frac{2}{9}$

$$= \frac{5 \cdot 9}{1 \cdot 9} - \frac{2}{9}$$

$$= \frac{45}{9} - \frac{2}{9} = \frac{43}{9}$$

Rewrite 5 as the improper fraction $\frac{5}{1}$ and proceed as usual. The LCD is 9.
Change $\frac{5}{1}$ to an equivalent fraction with a denominator of 9. Subtract the numerators.

▲

Let us now look at an example that combines both addition and subtraction.

EXAMPLE 7 Calculate $\frac{3}{8} + \frac{1}{3} - \frac{1}{2}$.

Solution $8 = 2 \cdot 2 \cdot 2 = 2^3$
$3 = 3^1$
$2 = 2^1$
$LCD = 2^3 \cdot 3 = 24$

Find the LCD of 8, 3, and 2.

$$\frac{3}{8} \cdot \frac{3}{3} = \frac{9}{24}$$

$$+\frac{1}{3} \cdot \frac{8}{8} = +\frac{8}{24}$$

$$\frac{17}{24}$$

Add $\frac{3}{8} + \frac{1}{3}$.

$$\frac{17}{24} \cdot 1 = \frac{17}{24}$$

$$-\frac{1}{2} \cdot \frac{12}{12} = -\frac{12}{24}$$

$$\frac{5}{24}$$

Then subtract $\frac{17}{24} - \frac{1}{2}$.

The result is $\frac{5}{24}$. ▲

EXAMPLE 8 A banker owns $\frac{7}{8}$ of an acre of land, and his son owns $\frac{1}{6}$ acre less than he does. How much land does his son own?

Solution Subtract $\frac{7}{8} - \frac{1}{6}$.

$8 = 2 \cdot 2 \cdot 2 = 2^3$

$6 = 2 \cdot 3$ Find the LCD.

$\text{LCD} = 2^3 \cdot 3 = 24$

$$\frac{7}{8} \cdot \frac{3}{3} = \frac{21}{24}$$

Change to equivalent fractions and subtract.

$$-\frac{1}{6} \cdot \frac{4}{4} = -\frac{4}{24}$$

$$\overline{\frac{17}{24}}$$

His son owns $\frac{17}{24}$ acre of land.   ▲

Quick Quiz

Perform each of the following operations.   **Answers**

1. $\dfrac{3}{5} - \dfrac{1}{4}$ 1. $\dfrac{7}{20}$

2. $\dfrac{5}{18} - \dfrac{5}{24}$ 2. $\dfrac{5}{72}$

3. $\dfrac{8}{9} - \dfrac{1}{2} + \dfrac{5}{6}$ 3. $\dfrac{11}{9}$

3.4 Exercises

Indicate which of the following statements are true and which are false. For those that are false, change the italicized expression to make the statement true.

1. When two fractions have the same denominator, the larger fraction *always* has the larger numerator.

2. To obtain a result *greater* than zero, we subtract the smaller fraction from the larger fraction.

3. When two fractions have the same numerator, the larger fraction has the *smaller* denominator.

4. In order to subtract a fraction from a whole number, we must first rewrite the whole number as *an improper fraction*.

For each of the following pairs of fractions, indicate which is larger by replacing the ? by a > or < sign.

5. $\dfrac{3}{5}$? $\dfrac{2}{3}$

6. $\dfrac{7}{12}$? $\dfrac{3}{4}$

7. $\dfrac{11}{18}$? $\dfrac{2}{3}$

8. $\dfrac{1}{5}$? $\dfrac{2}{7}$

9. $\dfrac{5}{16}$? $\dfrac{3}{8}$

10. $\dfrac{5}{8} \ ? \ \dfrac{7}{12}$

11. $\dfrac{2}{9} \ ? \ \dfrac{3}{13}$

12. $\dfrac{4}{11} \ ? \ \dfrac{1}{3}$

13. $\dfrac{7}{18} \ ? \ \dfrac{2}{5}$

14. $\dfrac{2}{7} \ ? \ \dfrac{3}{10}$

15. $\dfrac{1}{3} \ ? \ \dfrac{6}{19}$

16. $\dfrac{2}{5} \ ? \ \dfrac{7}{16}$

17. $\dfrac{11}{12} \ ? \ \dfrac{12}{13}$

18. $\dfrac{15}{16} \ ? \ \dfrac{14}{15}$

19. $\dfrac{10}{3} \ ? \ \dfrac{16}{5}$

20. $\dfrac{18}{5} \ ? \ \dfrac{7}{2}$

21. $\dfrac{9}{42} \ ? \ \dfrac{4}{21}$

22. $\dfrac{9}{49} \ ? \ \dfrac{4}{21}$

23. $\dfrac{3}{81} \ ? \ \dfrac{2}{45}$

24. $\dfrac{40}{121} \ ? \ \dfrac{10}{35}$

Perform each of the following subtractions.

25. $\dfrac{6}{7} - \dfrac{2}{7}$

26. $\dfrac{8}{11} - \dfrac{5}{11}$

27. $\dfrac{4}{9} - \dfrac{1}{3}$

28. $\dfrac{3}{4} - \dfrac{7}{12}$

29. $\dfrac{3}{8} - \dfrac{2}{9}$

30. $\dfrac{5}{7} - \dfrac{1}{6}$

31. $\dfrac{5}{6} - \dfrac{1}{9}$

32. $\dfrac{7}{8} - \dfrac{5}{12}$

33. $\dfrac{13}{30} - \dfrac{4}{45}$

34. $\dfrac{11}{24} - \dfrac{7}{36}$

35. $\dfrac{7}{18} - \dfrac{7}{24}$

36. $\dfrac{5}{36} - \dfrac{5}{48}$

37. $\dfrac{8}{13} - \dfrac{2}{39}$

38. $\dfrac{17}{60} - \dfrac{4}{15}$

39. $\dfrac{11}{16} - \dfrac{5}{24}$

40. $\dfrac{9}{16} - \dfrac{5}{64}$

41. $\dfrac{9}{25} - \dfrac{3}{55}$

42. $\dfrac{5}{54} - \dfrac{2}{45}$

43. $\dfrac{5}{24} - \dfrac{5}{84}$

44. $\dfrac{7}{72} - \dfrac{1}{45}$

45. $5 - \dfrac{5}{6}$

46. $8 - \dfrac{2}{3}$

47. $7 - \dfrac{5}{12}$

48. $3 - \dfrac{9}{16}$

49. $\dfrac{38}{294} - \dfrac{13}{588}$

50. $\dfrac{54}{471} - \dfrac{16}{157}$

51. $\dfrac{13}{37} - \dfrac{17}{59}$

52. $\dfrac{52}{67} - \dfrac{23}{41}$

Perform each of the following operations.

53. $\dfrac{1}{6} - \dfrac{1}{24} + \dfrac{7}{18}$

54. $\dfrac{6}{7} - \dfrac{2}{3} + \dfrac{5}{21}$

55. $\dfrac{2}{5} + \dfrac{7}{10} - \dfrac{4}{15}$

56. $\dfrac{8}{9} + \dfrac{2}{3} - \dfrac{5}{6}$

57. $\dfrac{3}{14} - \dfrac{2}{21} + \dfrac{4}{7}$

58. $\dfrac{7}{8} + \dfrac{3}{12} - \dfrac{1}{6}$

59. $\dfrac{4}{5} - \dfrac{1}{4} - \dfrac{1}{3}$

60. $\dfrac{7}{9} - \dfrac{1}{2} - \dfrac{1}{7}$

61. $\dfrac{11}{24} + \dfrac{7}{36} - \dfrac{5}{48}$

62. $\dfrac{9}{56} + \dfrac{11}{16} - \dfrac{3}{28}$

Solve each of the following word problems.

63. A recipe calls for $\frac{3}{4}$ cup brown sugar and $\frac{2}{3}$ cup rolled oats. How much more sugar than oats does the recipe require?

64. A tailor purchased $\frac{5}{8}$ yard of wool fabric and $\frac{3}{4}$ yard of corduroy. How much more corduroy than wool did he buy?

65. The captain of a swim team swam $\frac{3}{8}$ mile, and her teammate swam $\frac{2}{5}$ mile. Who swam farther, the captain or her teammate? By how much?

66. A high school football player completed his math homework in $\frac{4}{5}$ hour, and a basketball player completed his in $\frac{5}{6}$ hour. Who worked faster? By how much?

67. In a suburb of a large midwestern city, $\frac{5}{12}$ of the voters are registered Democrats, and $\frac{7}{18}$ are registered Republicans. Are there more registered Democrats or Republicans?

68. If $\frac{17}{36}$ of the students in a math class received a grade of B or better, and $\frac{11}{24}$ of the students received a grade of D or below, did more students do well or poorly?

3.5
Addition and Subtraction of Mixed Numbers

Since many applications of fractions in daily life involve mixed numbers, we will now investigate some examples that illustrate how to add and subtract mixed numbers.

EXAMPLE 1   Add $3\frac{7}{8} + 5\frac{2}{9}$.

Solution

$$3\frac{7}{8} + 5\frac{2}{9} = \left(3 + \frac{7}{8}\right) + \left(5 + \frac{2}{9}\right)$$ Since a mixed number is the sum of a whole number and a fraction.

$$= (3 + 5) + \left(\frac{7}{8} + \frac{2}{9}\right)$$ By the commutative and associative properties.

$$= 8 + \left(\frac{7}{8} + \frac{2}{9}\right)$$ Add the whole number parts first.

$$= 8 + \left(\frac{7 \cdot 9}{8 \cdot 9} + \frac{2 \cdot 8}{9 \cdot 8}\right)$$ Now add the fractional parts. Convert to equivalent fractions with denominators of 72.

$$= 8 + \left(\frac{63}{72} + \frac{16}{72}\right)$$

$$= 8 + \frac{79}{72}$$ Add the numerators.

$$= 8 + 1 + \frac{7}{72}$$ Convert the result to a mixed number.

$$= 9 + \frac{7}{72}$$ Combine the whole numbers to obtain the final answer.

$$= 9\frac{7}{72}$$ ▲

Example 1 illustrates the procedure for adding mixed numbers, as outlined in the following.

To Add Mixed Numbers

1. Add the whole number parts and fractional parts separately.
2. Change any resulting improper equivalent to a mixed number.
3. Add both sums and express the final answer as a mixed number.   ▲

Let us now look at Example 2, in which we use the vertical format.

EXAMPLE 2   Add $5\frac{3}{4} + 7\frac{5}{12} + 2\frac{1}{8}$.

Solution
$$4 = 2 \cdot 2 = 2^2$$ Find the LCD of the fractional parts.
$$12 = 3 \cdot 2 \cdot 2 = 2^2 \cdot 3$$
$$8 = 2 \cdot 2 \cdot 2 = 2^3$$
$$\text{LCD} = 2^3 \cdot 3 = 8 \cdot 3 = 24$$

$$5\frac{3}{4} = 5 + \frac{3}{4} \cdot \frac{6}{6} = 5 + \frac{18}{24}$$

$$7\frac{5}{12} = 7 + \frac{5}{12} \cdot \frac{2}{2} = 7 + \frac{10}{24}$$

$$+\; 2\frac{1}{8} = 2 + \frac{1}{8} \cdot \frac{3}{3} = 2 + \frac{3}{24}$$

$$\rule{4cm}{0.4pt}$$

$$14 + \frac{31}{24}$$ Add the whole number parts and fractional parts separately.

$$14 + \frac{31}{24} = 14 + 1 + \frac{7}{24}$$ Change $\frac{31}{24}$ to a mixed number.

$$= 15 + \frac{7}{24}$$ Add both sums to simplify.

$$= 15\frac{7}{24}$$ ▲

Example 3 illustrates how to subtract mixed numbers.

EXAMPLE 3   Subtract $5\frac{2}{7} - 3\frac{4}{7}$.

Solution   Minuend $5\frac{2}{7} = 5 + \frac{2}{7}$

Subtrahend $-3\frac{4}{7} = 3 + \frac{4}{7}$

$$\rule{4cm}{0.4pt}$$

After writing the problem vertically and attempting to subtract the whole number parts and fractional parts, we see that the fractional part of the minuend is less than the fractional part of the subtrahend: $\frac{2}{7} < \frac{4}{7}$. We therefore need to rewrite the minuend by borrowing 1 from the whole number part, and adding 1, in the form of $\frac{7}{7}$, to the fractional part:

$$5 + \frac{2}{7} = 4 + 1 + \frac{2}{7} = 4 + \frac{7}{7} + \frac{2}{7} = 4 + \frac{9}{7}$$

$$5 + \frac{2}{7} = 4 + \frac{9}{7}$$ Now subtract the whole number parts and fractional parts.

$$-3 + \frac{4}{7} = 3 + \frac{4}{7}$$

$$\rule{4cm}{0.4pt}$$

$$1 + \frac{5}{7} = 1\frac{5}{7}$$ ▲

We have thus illustrated the procedure for subtracting mixed numbers, as outlined on the following page. We will use this procedure in Example 4 to solve another problem.

To Subtract Mixed Numbers	**1.** Change the fractional parts to equivalent fractions with the same LCD.
	2. If the fractional part of the minuend is less than that of the subtrahend, borrow 1 from the whole number part of the minuend, and add 1 to its fractional part.
	3. Subtract the whole number parts and fractional parts separately. ▲

EXAMPLE 4   Subtract $8\frac{5}{6} - 5\frac{2}{3}$.

$$8\frac{5}{6} = \quad 8 + \frac{5}{6} \quad = \quad 8 + \frac{5}{6}$$

$$-5\frac{2}{3} = -5 + \frac{2}{3}\cdot\frac{2}{2} = -5 + \frac{4}{6}$$

$$\rule{4cm}{0.4pt}$$

$$3 + \frac{1}{6} = 3\frac{1}{6}$$

The LCD of 6 and 3 is 6.

Change each fractional part to an equivalent fraction with a denominator equal to the LCD.

Subtract the whole number and fractional parts separately.   ▲

EXAMPLE 5   Subtract $7\frac{1}{6} - 2\frac{8}{15}$.

Solution $6 = 2 \cdot 3$
$15 = 3 \cdot 5$
$LCD = 2 \cdot 3 \cdot 5 = 30$

Change the fractional parts to equivalent fractions with the same LCD.

$$7\frac{1}{6} = 7 + \frac{1}{6}\cdot\frac{5}{5} = 7 + \frac{5}{30}$$

$$-2\frac{8}{15} = 2 + \frac{8}{15}\cdot\frac{2}{2} = 2 + \frac{16}{30}$$

$$\rule{7cm}{0.4pt}$$

$$7 + \frac{5}{30} = 6 + 1 + \frac{5}{30} = 6 + \frac{30}{30} + \frac{5}{30} = 6 + \frac{35}{30}$$

Since $\frac{5}{30} < \frac{16}{30}$, we must borrow 1 from the whole number part of the minuend and add $1 = \frac{30}{30}$ to its fractional part.

$$7 + \frac{5}{30} = 6 + \frac{35}{30}$$

$$-2 + \frac{16}{30} = 2 + \frac{16}{30}$$

$$\rule{5cm}{0.4pt}$$

$$4 + \frac{19}{30} = 4\frac{19}{30}$$

Subtract the whole number parts and fractional parts separately.

▲

Observe what happens in Example 6 when we subtract a mixed number from a whole number.

EXAMPLE 6   Subtract $7 - 2\frac{5}{8}$.

Solution $7 = 6 + 1 = 6 + \dfrac{8}{8}$

Since there is no fractional part to the minuend, we rewrite 7 as a mixed number.

$$
\begin{aligned}
7 &= 6 + 1 = 6 + \frac{8}{8} \\
-2\frac{5}{8} &= 2 + \frac{5}{8} = 2 + \frac{5}{8} \\
\hline
&\qquad\quad 4 + \frac{3}{8} = 4\frac{3}{8}
\end{aligned}
$$

Now we subtract the whole number parts and fractional parts.

▲

Let us now look at an example that combines both addition and subtraction.

EXAMPLE 7   Calculate $5\frac{1}{8} - 2\frac{3}{4} + 3\frac{5}{6}$.

Solution

$$
\begin{aligned}
5\frac{1}{8} &= 5 + \frac{1}{8} &&= 4 + \frac{9}{8} \\
-2\frac{3}{4} &= 2 + \frac{3}{4} \cdot \frac{2}{2} &&= 2 + \frac{6}{8} \\
\hline
&&& 2 + \frac{3}{8} = 2\frac{3}{8}
\end{aligned}
$$

First subtract $5\frac{1}{8} - 2\frac{3}{4}$.

The LCD of 8 and 4 is 8.

$$
\begin{aligned}
2\frac{3}{8} &= 2 + \frac{3}{8} \cdot \frac{3}{3} = 2 + \frac{9}{24} \\
+3\frac{5}{6} &= 3 + \frac{5}{6} \cdot \frac{4}{4} = 3 + \frac{20}{24} \\
\hline
&\quad 5 + \frac{29}{24} = 5 + 1 + \frac{5}{24} \\
&\qquad\qquad\; = 6 + \frac{5}{24} = 6\frac{5}{24}
\end{aligned}
$$

Now add $2\frac{3}{8} + 3\frac{5}{6}$.

$8 = 2 \cdot 2 \cdot 2 = 2^3$
$6 = 2 \cdot 3$
LCD $= 2^3 \cdot 3 = 24$

▲

We will now solve some words problems that involve addition and subtraction of mixed numbers.

EXAMPLE 8   A flight from Los Angeles to New York that stops in Chicago takes 3 hours 37 minutes to go from Los Angeles to Chicago and 2 hours 43 minutes to go from Chicago to New York. Express the flight time for each part of the trip and the total number of hours of flight time as mixed fractions.

Solution   Los Angeles to Chicago:
3 hr 37 min = 3 hr + $\frac{37}{60}$ hr $3\frac{37}{60}$ hr   Since 1 hr = 60 min.

Chicago to New York:
2 hr 43 min = 2 hr + $\frac{43}{60}$ hr = $2\frac{43}{60}$ hr

$$3\frac{37}{60} = 3 + \frac{37}{60}$$

$$+2\frac{43}{60} = 2 + \frac{43}{60}$$

$$5 + \frac{80}{60}$$

To find the total flight time, add $3\frac{37}{60}$ hr + $2\frac{43}{60}$ hr.

Add the whole number and fractional parts separately.

$$= 5 + 1 + \frac{20}{60}$$

Change $\frac{80}{60}$ to a mixed number.

$$= 6 + \frac{20}{60} = 6\frac{1}{3}$$

Reduce the fractional part to lowest terms.

The total flight time is $6\frac{1}{3}$ hr. ▲

EXAMPLE 9 For a distance of $\frac{1}{4}$ mile, the maximum speed for a greyhound is $39\frac{7}{20}$ mph, and the speed for a human is $27\frac{22}{25}$ mph. How much faster can a greyhound run than a human?

Solution Subtract $39\frac{7}{20} - 27\frac{22}{25}$.

$$39\frac{7}{20} = 39 + \frac{7}{20} \cdot \frac{5}{5} = 39 + \frac{35}{100} = 38 + \frac{135}{100}$$

$$-27\frac{22}{25} = 27 + \frac{22}{25} \cdot \frac{4}{4} = 27 + \frac{88}{100} = 27 + \frac{88}{100}$$

$$11 + \frac{47}{100} = 11\frac{47}{100}$$

$20 = 2^2 \cdot 5$
$25 = 5^2$
LCD $= 2^2 \cdot 5^2 = 100$

A greyhound can run $11\frac{47}{100}$ mph faster than a human. ▲

Quick Quiz

Perform each of the following calculations.

Answers

1. $6\frac{3}{4} + 2\frac{5}{6}$

2. $7\frac{3}{8} - 3\frac{1}{2}$

3. $8\frac{1}{4} - 2\frac{2}{3} + 3\frac{1}{6}$

1. $9\frac{7}{12}$

2. $3\frac{7}{8}$

3. $8\frac{3}{4}$

3.5 Exercises

Perform each of the following additions. Express your answers as mixed numbers reduced to lowest terms.

1. $3\frac{1}{8} + 4\frac{5}{8}$

2. $6\frac{2}{9} + 2\frac{4}{9}$

3. $2\frac{1}{4} + 5\frac{3}{8}$

4. $7\frac{2}{3} + 8\frac{1}{6}$

5. $3\dfrac{3}{4} + 6\dfrac{7}{8}$

6. $1\dfrac{3}{5} + 8\dfrac{9}{10}$

7. $88\dfrac{1}{2} + 47\dfrac{3}{8}$

8. $59\dfrac{2}{9} + 71\dfrac{1}{3}$

9. $27\dfrac{5}{16} + 19\dfrac{3}{4}$

10. $11\dfrac{4}{5} + 24\dfrac{7}{20}$

11. $5\dfrac{1}{3} + 11\dfrac{1}{4}$

12. $9\dfrac{1}{2} + 3\dfrac{2}{7}$

13. $6\dfrac{7}{8} + 10\dfrac{5}{6}$

14. $12\dfrac{8}{9} + 2\dfrac{7}{12}$

15. $7\dfrac{3}{16} + 9\dfrac{9}{20}$

16. $6\dfrac{9}{10} + 3\dfrac{8}{15}$

17. $19\dfrac{5}{12} + 7\dfrac{7}{18}$

18. $8\dfrac{4}{21} + 15\dfrac{9}{14}$

19. $15\dfrac{5}{7} + 3\dfrac{3}{8}$

20. $19\dfrac{3}{4} + 4\dfrac{7}{9}$

21. $93\dfrac{3}{5} + 47\dfrac{4}{25}$

22. $62\dfrac{5}{49} + 88\dfrac{2}{7}$

23. $2\dfrac{22}{45} + 7\dfrac{11}{20}$

24. $5\dfrac{25}{36} + 1\dfrac{13}{16}$

25. $13\dfrac{5}{48} + 4\dfrac{7}{36}$

26. $7\dfrac{4}{63} + 16\dfrac{8}{27}$

27. $53\dfrac{7}{18} + 31\dfrac{5}{54}$

28. $62\dfrac{5}{48} + 27\dfrac{9}{16}$

29. $2\dfrac{3}{5} + 7\dfrac{1}{5} + 4\dfrac{4}{5}$

30. $6\dfrac{7}{8} + 2\dfrac{5}{8} + 9\dfrac{3}{8}$

31. $13\dfrac{3}{4} + 5\dfrac{1}{6} + 27\dfrac{5}{12}$

32. $2\dfrac{1}{2} + 16\dfrac{5}{16} + 33\dfrac{3}{8}$

33. $2\dfrac{5}{6} + 8\dfrac{2}{3} + 4\dfrac{2}{27}$

34. $6\dfrac{5}{12} + 1\dfrac{2}{9} + 3\dfrac{4}{15}$

35. $6\dfrac{2}{3} + \dfrac{5}{12} + 2\dfrac{5}{6}$

36. $4\dfrac{3}{26} + \dfrac{5}{13} + 1\dfrac{8}{39}$

37. $2\dfrac{3}{8} + 1\dfrac{7}{16} + 7\dfrac{5}{24}$

38. $5\dfrac{2}{21} + 2\dfrac{3}{7} + 6\dfrac{8}{35}$

39. $3\dfrac{5}{48} + 8\dfrac{7}{24} + 7\dfrac{5}{12}$

40. $7\dfrac{9}{14} + 4\dfrac{3}{56} + 6\dfrac{5}{28}$

🖩 41. $85\dfrac{35}{817} + 167\dfrac{17}{43}$

🖩 42. $284\dfrac{13}{37} + 58\dfrac{28}{777}$

🖩 43. $473\dfrac{15}{26} + 519\dfrac{22}{35}$

🖩 44. $267\dfrac{38}{41} + 734\dfrac{18}{19}$

Perform each of the following subtractions. Express your answers as mixed numbers reduced to lowest terms.

45. $8\dfrac{3}{4} - 2\dfrac{1}{4}$

46. $7\dfrac{5}{7} - 3\dfrac{2}{7}$

47. $9\dfrac{3}{4} - 4\dfrac{1}{8}$

48. $8\dfrac{4}{5} - 1\dfrac{3}{10}$

49. $4\dfrac{5}{6} - 2\dfrac{5}{12}$

50. $6\dfrac{3}{8} - 3\dfrac{3}{16}$

51. $36\dfrac{2}{3} - 27\dfrac{1}{27}$

52. $25\dfrac{3}{4} - 18\dfrac{2}{5}$

53. $56\dfrac{3}{5} - 29\dfrac{2}{15}$

54. $95\dfrac{5}{24} - 47\dfrac{1}{6}$

55. $5\dfrac{1}{3} - 1\dfrac{2}{3}$

56. $7\dfrac{2}{5} - 3\dfrac{4}{5}$

57. $4\dfrac{2}{9} - 2\dfrac{8}{9}$

58. $9\dfrac{3}{10} - 3\dfrac{7}{10}$

59. $6\dfrac{3}{14} - 2\dfrac{5}{7}$

60. $4\dfrac{2}{9} - 1\dfrac{11}{18}$

61. $32\dfrac{4}{27} - 16\dfrac{5}{9}$

62. $55\dfrac{1}{8} - 29\dfrac{9}{32}$

63. $85\dfrac{5}{9} - 72\dfrac{1}{2}$

64. $55\dfrac{2}{3} - 43\dfrac{7}{8}$

65. $4\dfrac{1}{6} - 3\dfrac{5}{8}$

66. $7\dfrac{3}{8} - 6\dfrac{11}{12}$

67. $8 - 5\dfrac{3}{7}$

68. $9 - 4\dfrac{2}{9}$

69. $15 - 7\dfrac{5}{18}$

70. $23 - 16\dfrac{9}{32}$

71. $54 - 29\dfrac{5}{8}$

72. $61 - 43\frac{3}{4}$

73. $8\frac{5}{42} - 4\frac{5}{14}$

74. $5\frac{5}{54} - 1\frac{5}{18}$

75. $11\frac{3}{56} - 4\frac{4}{7}$

76. $22\frac{6}{45} - 8\frac{5}{9}$

77. $4\frac{7}{36} - 2\frac{3}{48}$

78. $6\frac{5}{42} - 3\frac{2}{63}$

79. $11\frac{5}{6} - 6\frac{3}{8} - 3\frac{1}{24}$

80. $15\frac{11}{12} - 7\frac{1}{4} - 2\frac{1}{6}$

81. $22\frac{9}{20} - 10\frac{3}{5} - 4\frac{2}{15}$

82. $31\frac{5}{6} - 12\frac{7}{18} - 9\frac{11}{36}$

83. $18\frac{7}{20} - 3\frac{3}{32} - 7\frac{9}{16}$

84. $25\frac{8}{9} - 5\frac{7}{18} - 10\frac{4}{27}$

85. $16\frac{8}{15} - 5\frac{4}{25} - 2\frac{2}{75}$

86. $19\frac{3}{40} - 7\frac{7}{24} - 4\frac{9}{16}$

87. $12\frac{71}{288} - 3\frac{5}{16}$

88. $19\frac{7}{24} - 8\frac{59}{360}$

89. $8\frac{9}{14} - 3\frac{5}{27}$

90. $7\frac{11}{36} - 4\frac{8}{19}$

Perform each of the following operations.

91. $6\frac{2}{5} - 2\frac{3}{5} + 4\frac{1}{5}$

92. $7\frac{2}{7} + 3\frac{1}{7} - 5\frac{5}{7}$

93. $8 - 2\frac{3}{4} + 9\frac{4}{5}$

94. $5 - 3\frac{5}{6} + 1\frac{1}{7}$

95. $53\frac{5}{21} - 25\frac{2}{7} + 19\frac{3}{14}$

96. $18\frac{8}{15} + 34\frac{3}{10} - 27\frac{7}{30}$

97. $8\frac{3}{10} + 9\frac{7}{100} - 4\frac{9}{1,000}$

98. $5\frac{7}{1,000} + 3\frac{1}{10} - 2\frac{3}{100}$

99. $8\frac{3}{16} - 6\frac{5}{12} + 3\frac{9}{32}$

100. $5\frac{5}{18} - 2\frac{7}{36} + 7\frac{4}{27}$

Solve each of the following word problems.

101. The Mont Royal Tunnel in Montreal is about $3\frac{1}{5}$ miles long, and the Holland Tunnel between New York and New Jersey is about $1\frac{7}{11}$ miles long. What is the combined length of these two tunnels?

102. A recipe calls for $\frac{1}{4}$ teaspoon ginger, $\frac{7}{8}$ teaspoon salt, $1\frac{1}{2}$ teaspoons cornstarch, and $1\frac{1}{4}$ teaspoons sugar. What is the total amount of these dry ingredients?

103. A swimming pool with a capacity of $25\frac{3}{20}$ gallons contains $8\frac{5}{6}$ gallons of water. How much more water is needed to ll the pool?

104. For city driving, cars average $14\frac{3}{50}$ miles per gallon, and buses average $5\frac{13}{20}$ miles per gallon. How much farther can a car go than a bus on a gallon of gas?

105. A dress that has a $\frac{5}{8}$-inch hem needs to be shortened $1\frac{3}{4}$ inches. What will the hem be after the dress is shortened?

106. A stock opened on the American Stock Exchange at $30\frac{1}{8}$ points and closed at $28\frac{3}{4}$. How much did the stock go down that day?

107. A jogger ran for 2 hours 11 minutes on Saturday and 1 hour 59 minutes on Sunday. Express the number of hours she ran each day and the total number of hours she ran that weekend as mixed numbers.

108. A piece of ham weighs 2 pounds 7 ounces, and a piece of beef weighs 4 pounds 2 ounces. Express the weight of each in pounds as mixed fractions. How many more pounds does the beef weigh than the ham?

109. A dresser is $4\frac{11}{24}$ feet long and a desk is $3\frac{7}{16}$ feet long. If they are placed along a 14-foot wall, how much space along the wall will remain open?

110. A bookshelf $2\frac{5}{8}$ feet long and another $1\frac{7}{12}$ feet long are both filled with books. If the books on both shelves are transferred to a third shelf 6 feet long, how much space will remain on the third bookshelf for additional books?

3.6
Multiplication
of Fractions

To begin our discussion of multiplication of fractions, let us consider a variety of situations in which we have different amounts of pie. For each situation, let us indicate $\frac{1}{2}$ of the total amount of pie available, as shown in Figure 3.12.

Figure 3.12

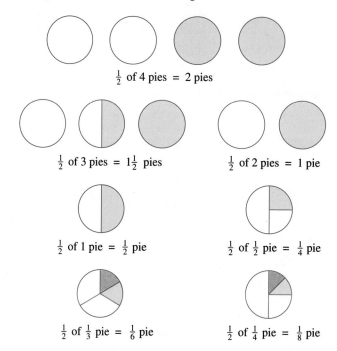

$\frac{1}{2}$ of 4 pies = 2 pies

$\frac{1}{2}$ of 3 pies = $1\frac{1}{2}$ pies $\frac{1}{2}$ of 2 pies = 1 pie

$\frac{1}{2}$ of 1 pie = $\frac{1}{2}$ pie $\frac{1}{2}$ of $\frac{1}{2}$ pie = $\frac{1}{4}$ pie

$\frac{1}{2}$ of $\frac{1}{3}$ pie = $\frac{1}{6}$ pie $\frac{1}{2}$ of $\frac{1}{4}$ pie = $\frac{1}{8}$ pie

Notice that in each situation, the result can be obtained by converting the two numbers separated by the word *of* to fractions, and multiplying the numerators and the denominators. We can conclude, therefore, that the word *of* means to *multiply*.

$$\frac{1}{2} \text{ of } 4 \text{ means } \frac{1}{2} \cdot 4 = \frac{1}{2} \cdot \frac{4}{1} = \frac{1 \cdot 4}{2 \cdot 1} = \frac{4}{2} = 2$$

$$\frac{1}{2} \text{ of } 3 \text{ means } \frac{1}{2} \cdot 3 = \frac{1}{2} \cdot \frac{3}{1} = \frac{1 \cdot 3}{2 \cdot 1} = \frac{3}{2} = 1\frac{1}{2}$$

$$\frac{1}{2} \text{ of } 2 \text{ means } \frac{1}{2} \cdot 2 = \frac{1}{2} \cdot \frac{2}{1} = \frac{1 \cdot 2}{2 \cdot 1} = \frac{2}{2} = 1$$

$$\frac{1}{2} \text{ of } 1 \text{ means } \frac{1}{2} \cdot 1 = \frac{1}{2} \cdot \frac{1}{1} = \frac{1 \cdot 1}{2 \cdot 1} = \frac{1}{2}$$

$$\frac{1}{2} \text{ of } \frac{1}{2} \text{ means } \frac{1}{2} \cdot \frac{1}{2} = \frac{1 \cdot 1}{2 \cdot 2} = \frac{1}{4}$$

$$\frac{1}{2} \text{ of } \frac{1}{3} \text{ means } \frac{1}{2} \cdot \frac{1}{3} = \frac{1 \cdot 1}{2 \cdot 3} = \frac{1}{6}$$

$$\frac{1}{2} \text{ of } \frac{1}{4} \text{ means } \frac{1}{2} \cdot \frac{1}{4} = \frac{1 \cdot 1}{2 \cdot 4} = \frac{1}{8}$$

As another example, let us consider the box shown in Figure 3.13. One-half of the box is lightly shaded. One-third of this shaded area, or $\frac{1}{3}$ of $\frac{1}{2}$, is then shaded again. The area shaded twice is $\frac{1}{6}$ of the total area of the box. Thus,

$$\frac{1}{3} \text{ of } \frac{1}{2} \text{ means } \frac{1}{3} \cdot \frac{1}{2} = \frac{1 \cdot 1}{3 \cdot 2} = \frac{1}{6}$$

Figure 3.13

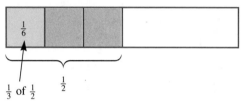

We have therefore illustrated a procedure for multiplying fractions, as outlined below.

**To Multiply
Fractions**

1. Multiply the numerators.

2. Place the result over the product of the denominators.

3. If necessary, reduce the answer to lowest terms.   ▲

EXAMPLE 1   Find each of the following products.

(a) $\dfrac{3}{7} \cdot \dfrac{2}{5}$

$$\frac{3}{7} \cdot \frac{2}{5} = \frac{3 \cdot 2}{7 \cdot 5} = \frac{6}{35}$$

Multiply the numerators. Place the result over the product of the denominators.

(b) $\dfrac{2}{9} \cdot \dfrac{5}{8}$

$$\frac{2}{9} \cdot \frac{5}{8} = \frac{2 \cdot 5}{9 \cdot 8} = \frac{10}{72}$$

Multiply numerators and place result over product of denominators.

$$\frac{10}{72} = \frac{5}{36}$$

Reduce answer to lowest terms.

(c) $\dfrac{3}{4}$ of 5

$$\frac{3}{4} \text{ of } 5 = \frac{3}{4} \cdot 5$$

Remember that *of* means to multiply.

$$= \frac{3}{4} \cdot \frac{5}{1}$$

Change 5 to an improper fraction.

$$= \frac{3 \cdot 5}{4 \cdot 1} = \frac{15}{4} = 3\frac{3}{4}$$

(d) $\dfrac{3}{7} \cdot \dfrac{1}{4} \cdot \dfrac{2}{5}$

$$\dfrac{3}{7} \cdot \dfrac{1}{4} \cdot \dfrac{2}{5} = \dfrac{3 \cdot 1 \cdot 2}{7 \cdot 4 \cdot 5} = \dfrac{6}{140} = \dfrac{3}{70}$$

(e) $\left(\dfrac{2}{3}\right)^3$

$$\left(\dfrac{2}{3}\right)^3 = \dfrac{2}{3} \cdot \dfrac{2}{3} \cdot \dfrac{2}{3}$$

Remember that to cube a number, we must use it as a factor three times.

$$= \dfrac{2 \cdot 2 \cdot 2}{3 \cdot 3 \cdot 3} = \dfrac{8}{27}$$

▲

The procedure for multiplying mixed numbers is outlined below. Example 2 illustrates how to apply the procedure.

To Multiply Mixed Numbers

1. Change each mixed number to an improper fraction.
2. Multiply the numerators.
3. Place the result over the product of the denominators.
4. Express the answer as a mixed number or as a proper fraction reduced to lowest terms. ▲

When performing arithmetic operations on improper fractions, we usually express answers greater than 1 in the form of an improper fraction. When performing arithmetic operations on mixed numbers, we usually express answers greater than 1 in the form of a mixed number.

EXAMPLE 2 Multiply each of the following.

(a) $2\dfrac{1}{3} \cdot 5\dfrac{3}{4}$

$$2\dfrac{1}{3} \cdot 5\dfrac{3}{4} = \dfrac{7}{3} \cdot \dfrac{23}{4}$$

Change each mixed number to an improper fraction.

$$= \dfrac{7 \cdot 23}{3 \cdot 4}$$

$$= \dfrac{161}{12} = 13\dfrac{5}{12}$$

Express the answer as a mixed number.

(b) $\dfrac{1}{8}$ of $7\dfrac{2}{3}$

$$\dfrac{1}{8} \cdot 7\dfrac{2}{3} = \dfrac{1}{8} \cdot \dfrac{23}{3} = \dfrac{23}{24}$$

(c) $4\frac{2}{5} \cdot 1\frac{2}{7}$

$$4\frac{2}{5} \cdot 1\frac{2}{7} = \frac{22 \cdot 9}{5 \cdot 7}$$ Change each mixed number to an improper fraction.

$$= \frac{198}{35}$$

$$= 5\frac{23}{35}$$ Express the answer as a mixed number.

(d) $6\frac{3}{5} \cdot 2\frac{3}{4}$

$$6\frac{3}{5} \cdot 2\frac{3}{4} = \frac{33 \cdot 11}{5 \cdot 4}$$ Change each mixed number to an improper fraction.

$$= \frac{363}{20}$$

$$= 18\frac{3}{20}$$ Express the answer as a mixed number. ▲

To avoid obtaining an answer that needs to be reduced to lowest terms, we divide both numerator and denominator by their common factors before doing any multiplication.

EXAMPLE 3   Perform each of the following multiplications.

(a) $\frac{2}{3} \cdot \frac{3}{5}$

$$\frac{2}{3} \cdot \frac{3}{5} = \frac{2 \cdot \overset{1}{\cancel{3}}}{\underset{1}{\cancel{3}} \cdot 5} = \frac{2}{5}$$ Since $\frac{2}{5} \cdot \frac{3}{3} = \frac{2}{5} \cdot 1 = \frac{2}{5}$.

(b) $\frac{5}{36} \cdot \frac{8}{15}$

$$\frac{5}{36} \cdot \frac{8}{15} = \frac{5 \cdot 8}{36 \cdot 15} = \frac{\overset{1}{\cancel{5}} \cdot \overset{1}{\cancel{2}} \cdot \overset{1}{\cancel{2}} \cdot 2}{\underset{1}{\cancel{2}} \cdot 3 \cdot \underset{1}{\cancel{2}} \cdot 3 \cdot 3 \cdot \underset{1}{\cancel{5}}}$$ Cancel common factors.

$$= \frac{2}{3 \cdot 3 \cdot 3} = \frac{2}{27}$$

(c) $\frac{7}{30} \cdot \frac{12}{21}$

$$\frac{7}{30} \cdot \frac{12}{21} = \frac{\overset{1}{\cancel{7}} \cdot \overset{2}{\cancel{12}}}{\underset{5}{\cancel{30}} \cdot \underset{3}{\cancel{21}}} = \frac{1 \cdot 2}{5 \cdot 3} = \frac{2}{15}$$ Instead of completely factoring numerator and denominator, simply divide both by a series of common factors.

(d) $8\dfrac{1}{3} \cdot 2\dfrac{2}{5}$

$$8\dfrac{1}{3} \cdot 2\dfrac{2}{5} = \dfrac{25}{3} \cdot \dfrac{12}{5}$$

Change each mixed number to an improper fraction.

$$= \dfrac{\overset{5}{\cancel{25}} \cdot \overset{4}{\cancel{12}}}{\underset{1}{\cancel{3}} \cdot \underset{1}{\cancel{5}}}$$

When all the numbers in the denominator are canceled the result is 1.

$$= \dfrac{20}{1}$$

$$= 20$$

(e) $\dfrac{3}{8} \cdot \dfrac{4}{5} \cdot \dfrac{2}{9}$

$$\dfrac{3}{8} \cdot \dfrac{4}{5} \cdot \dfrac{2}{9} = \dfrac{\overset{1}{\cancel{3}} \cdot \overset{1}{\cancel{4}} \cdot \overset{1}{\cancel{2}}}{\underset{4}{\underset{1}{\cancel{8}}} \cdot 5 \cdot \underset{3}{\cancel{9}}} = \dfrac{1}{5 \cdot 3} = \dfrac{1}{15}$$

Notice that when all the numbers in the numerator are canceled, the numerator becomes 1 and not 0.

(f) $1\dfrac{2}{5} \cdot \dfrac{4}{7} \cdot 3\dfrac{1}{8}$

$$1\dfrac{2}{5} \cdot \dfrac{4}{7} \cdot 3\dfrac{1}{8} = \dfrac{7}{5} \cdot \dfrac{4}{7} \cdot \dfrac{25}{3} = \dfrac{\cancel{7} \cdot \cancel{4} \cdot \overset{5}{\cancel{25}}}{\cancel{5} \cdot \cancel{7} \cdot \underset{2}{\cancel{8}}} = \dfrac{5}{2} = 2\dfrac{1}{2}$$

When the result of dividing by a common factor is 1, it is not necessary to write the 1. ▲

Now let us review some of the order of operations rules.

EXAMPLE 4 Perform each of the following operations.

(a) $\dfrac{2}{7}\left(\dfrac{1}{3} + \dfrac{3}{4}\right)$

$$\dfrac{2}{7}\left(\dfrac{1}{3} + \dfrac{3}{4}\right) = \dfrac{2}{7}\left(\dfrac{1 \cdot 4}{3 \cdot 4} + \dfrac{3 \cdot 3}{4 \cdot 3}\right)$$

Remember to do what is in the parentheses first.

$$= \dfrac{2}{7}\left(\dfrac{4}{12} + \dfrac{9}{12}\right)$$

$$= \dfrac{2}{7}\left(\dfrac{13}{12}\right) = \dfrac{\cancel{2} \cdot 13}{7 \cdot \underset{6}{\cancel{12}}}$$

$$= \dfrac{13}{7 \cdot 6} = \dfrac{13}{42}$$

Then do the multiplication.

(b) $\dfrac{3}{4} \cdot \dfrac{2}{9} + \dfrac{1}{2}$

$\dfrac{3}{4} \cdot \dfrac{2}{9} + \dfrac{1}{2} = \dfrac{3 \cdot 2}{4 \cdot 9} + \dfrac{1}{2} = \dfrac{\overset{}{\cancel{3} \cdot \cancel{2}}}{\underset{2 \quad 3}{\cancel{4} \cdot \cancel{9}}} + \dfrac{1}{2}$ Remember to do multiplication before addition.

$= \dfrac{1}{2 \cdot 3} + \dfrac{1}{2} = \dfrac{1}{6} + \dfrac{1}{2}$

$= \dfrac{1}{6} + \dfrac{1 \cdot 3}{2 \cdot 3} = \dfrac{1}{6} + \dfrac{3}{6} = \dfrac{4}{6} = \dfrac{2}{3}$ ▲

The distributive property will often make it easier for us to simplify a problem, as Example 5 illustrates.

EXAMPLE 5   Use the distributive property to simplify the following problems.

(a) $12\left(\dfrac{5}{6} - \dfrac{3}{4}\right)$

$12\left(\dfrac{5}{6} - \dfrac{3}{4}\right) = 12 \cdot \dfrac{5}{6} - 12 \cdot \dfrac{3}{4}$

$= \dfrac{12}{1} \cdot \dfrac{5}{6} - \dfrac{12}{1} \cdot \dfrac{3}{4}$

$= \dfrac{\overset{2}{\cancel{12}} \cdot 5}{1 \cdot \cancel{6}} - \dfrac{\overset{3}{\cancel{12}} \cdot 3}{1 \cdot \cancel{4}} = \dfrac{10}{1} - \dfrac{9}{1}$

$= 10 - 9 = 1$

(b) $\dfrac{2}{3}\left(\dfrac{9}{8} + \dfrac{3}{14}\right)$

$\dfrac{2}{3}\left(\dfrac{9}{8} + \dfrac{3}{14}\right) = \dfrac{2}{3} \cdot \dfrac{9}{8} + \dfrac{2}{3} \cdot \dfrac{3}{14}$

$= \dfrac{\cancel{2} \cdot \overset{3}{\cancel{9}}}{\cancel{3} \cdot \underset{4}{\cancel{8}}} + \dfrac{\cancel{2} \cdot \cancel{3}}{\cancel{3} \cdot \underset{7}{\cancel{14}}} = \dfrac{3}{4} + \dfrac{1}{7}$

$= \dfrac{3 \cdot 7}{4 \cdot 7} + \dfrac{1 \cdot 4}{7 \cdot 4}$

$= \dfrac{21}{28} + \dfrac{4}{28} = \dfrac{25}{28}$ ▲

EXAMPLE 6   A father is 5 feet 8 inches tall, and his daughter is $\frac{3}{4}$ of his height. How tall is his daughter?

Solution   5 ft 8 in. $= 5\frac{8}{12}$ ft $= 5\frac{2}{3}$ ft

$$\frac{3}{4} \text{ of } 5\frac{2}{3} = \frac{3}{4} \cdot 5\frac{2}{3} = \frac{3}{4} \cdot \frac{17}{3} = \frac{3 \cdot 17}{4 \cdot 3} = \frac{17}{4} = 4\frac{1}{4} \text{ ft} \qquad \text{Find } \frac{3}{4} \text{ of } 5\frac{2}{3}.$$

His daughter is $4\frac{1}{4}$ feet tall, which can   Notice that $4\frac{1}{4}$ ft $= 4\frac{3}{12}$ ft $= 4$ ft 3 in.
also be expressed as 4 feet 3 inches.   ▲

EXAMPLE 7   Of the people at a PTA meeting, $\frac{5}{8}$ voted for the Republican candidate for mayor and the rest voted for the Democratic candidate. If ninety-six people attended the PTA meeting, how many people voted for the Democratic candidate?

Solution $\frac{5}{8} \cdot 96 = \frac{5}{8} \cdot \frac{96}{1}$ $\frac{5}{8}$ of 96 voted Republican.

$$= \frac{5 \cdot \overset{12}{96}}{\underset{}{8}}$$

$= 60$ people voted for the Republican candidate.
$96 - 60 = 36$ people voted for the Democratic candidate.   ▲

Quick Quiz

Perform each of the following multiplications.   **Answers**

1. $\frac{3}{8} \cdot \frac{5}{7}$ 1. $\frac{15}{56}$

2. $\frac{4}{9} \cdot \frac{6}{7} \cdot \frac{3}{4}$ 2. $\frac{2}{7}$

3. $5\frac{1}{2} \cdot 2\frac{2}{3}$ 3. $14\frac{2}{3}$

3.6 Exercises

Indicate which of the following statements are true and which are false. For those that are false, change the italicized or underlined expression to make the statement true.

1. The expression $\frac{1}{3}$ of 2 means $\frac{1}{3}$ *divided by* 2.

2. To multiply two fractions, multiply the denominators and *add* the numerators.

3. The product of $5 \cdot \frac{3}{8}$ is equal to $5\frac{3}{8}$.

4. To multiply mixed numbers, first change each mixed number to an *improper fraction*.

Perform each of the following multiplications.

5. $\frac{3}{7} \cdot \frac{2}{9}$ 6. $\frac{1}{5} \cdot \frac{3}{4}$ 7. $\frac{5}{8} \cdot \frac{1}{4}$ 8. $\frac{6}{7} \cdot \frac{2}{5}$ 9. $\frac{6}{7} \cdot \frac{2}{11}$

10. $\frac{1}{2} \cdot \frac{5}{12}$ 11. $\frac{2}{3} \cdot \frac{5}{9}$ 12. $\frac{3}{8} \cdot \frac{7}{10}$ 13. $\frac{8}{9} \cdot \frac{3}{4}$ 14. $\frac{7}{8} \cdot \frac{4}{9}$

15. $\dfrac{7}{16} \cdot \dfrac{2}{49}$ 16. $\dfrac{2}{15} \cdot \dfrac{5}{12}$ 17. $\dfrac{10}{3} \cdot \dfrac{15}{8}$ 18. $\dfrac{12}{5} \cdot \dfrac{20}{3}$ 19. $\dfrac{3}{11} \cdot \dfrac{5}{18}$

20. $\dfrac{7}{24} \cdot \dfrac{3}{16}$ 21. $\dfrac{7}{12} \cdot \dfrac{16}{49}$ 22. $\dfrac{8}{15} \cdot \dfrac{35}{63}$ 23. $\dfrac{5}{64} \cdot \dfrac{8}{3}$ 24. $\dfrac{7}{72} \cdot \dfrac{9}{2}$

25. $\dfrac{3}{4} \cdot \dfrac{1}{7} \cdot \dfrac{5}{8}$ 26. $\dfrac{5}{9} \cdot \dfrac{3}{8} \cdot \dfrac{4}{15}$ 27. $\dfrac{7}{36} \cdot \dfrac{28}{3} \cdot \dfrac{6}{7}$ 28. $\dfrac{5}{6} \cdot \dfrac{54}{7} \cdot \dfrac{3}{25}$ 29. $\dfrac{18}{121} \cdot \dfrac{11}{36} \cdot \dfrac{12}{33}$

30. $\dfrac{21}{32} \cdot \dfrac{3}{56} \cdot \dfrac{64}{15}$ 31. $8 \cdot \dfrac{7}{8}$ 32. $9 \cdot \dfrac{5}{6}$ 33. $\dfrac{2}{7} \cdot 6$ 34. $\dfrac{3}{5} \cdot 4$

35. $4 \cdot 8\dfrac{3}{5}$ 36. $8 \cdot 9\dfrac{1}{4}$ 37. $\dfrac{1}{2} \cdot 2\dfrac{1}{8}$ 38. $\dfrac{1}{9} \cdot 6\dfrac{2}{3}$ 39. $9\dfrac{3}{8} \cdot 5\dfrac{3}{5}$

40. $4\dfrac{2}{3} \cdot 1\dfrac{1}{2}$ 41. $2\dfrac{1}{3} \cdot 3\dfrac{1}{4}$ 42. $4\dfrac{2}{7} \cdot 1\dfrac{3}{4}$ 43. $5\dfrac{1}{5} \cdot 6\dfrac{1}{4}$ 44. $2\dfrac{1}{3} \cdot 4\dfrac{7}{8}$

45. $5\dfrac{1}{3} \cdot 6\dfrac{1}{4}$ 46. $7\dfrac{1}{9} \cdot 3\dfrac{3}{4}$ 47. $6\dfrac{7}{8} \cdot 5\dfrac{3}{5}$ 48. $3\dfrac{3}{4} \cdot 8\dfrac{2}{3}$ 49. $7\dfrac{3}{4} \cdot 2\dfrac{1}{9}$

50. $7\dfrac{1}{7} \cdot 8\dfrac{2}{5}$ 51. $4\dfrac{2}{7} \cdot \dfrac{5}{6} \cdot 3$ 52. $7\dfrac{1}{5} \cdot \dfrac{5}{8} \cdot 4$ 53. $3\dfrac{1}{8} \cdot 4\dfrac{2}{5} \cdot 5\dfrac{1}{3}$ 54. $5\dfrac{2}{5} \cdot 1\dfrac{2}{3} \cdot 6\dfrac{1}{4}$

 55. $\dfrac{37}{52} \cdot \dfrac{65}{81}$ 🖩 56. $\dfrac{63}{98} \cdot \dfrac{27}{48}$ 🖩 57. $82\dfrac{14}{29} \cdot 19\dfrac{18}{31}$ 🖩 58. $58\dfrac{37}{41} \cdot 21\dfrac{73}{85}$

Perform each of the following operations.

59. $\dfrac{3}{4}\left(\dfrac{1}{2} - \dfrac{2}{9}\right)$ 60. $\dfrac{2}{3}\left(\dfrac{5}{8} + \dfrac{7}{12}\right)$ 61. $\dfrac{1}{8} \cdot \dfrac{7}{10} + \dfrac{4}{5}$ 62. $\dfrac{3}{5} \cdot \dfrac{1}{3} + \dfrac{7}{8}$

63. $\dfrac{1}{2} - \dfrac{3}{5} \cdot \dfrac{1}{4}$ 64. $\dfrac{7}{12} + \dfrac{3}{4} \cdot \dfrac{5}{6}$ 65. $\left(\dfrac{1}{3}\right)^3 \cdot \left(\dfrac{1}{2}\right)^2$ 66. $\left(\dfrac{4}{9}\right)^2 \cdot \left(\dfrac{3}{8}\right)^3$

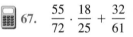 67. $\dfrac{55}{72} \cdot \dfrac{18}{25} + \dfrac{32}{61}$ 🖩 68. $\dfrac{23}{47}\left(\dfrac{18}{29} - \dfrac{31}{84}\right)$

Use the distributive property to simplify each of the following problems.

69. $24\left(\dfrac{5}{8} + \dfrac{2}{3}\right)$ 70. $35\left(\dfrac{5}{7} - \dfrac{2}{5}\right)$ 71. $6\left(\dfrac{2}{15} + \dfrac{1}{6}\right)$ 72. $8\left(\dfrac{3}{4} + \dfrac{7}{12}\right)$

73. $\dfrac{1}{4}\left(\dfrac{4}{7} + \dfrac{8}{21}\right)$ 74. $\dfrac{3}{7}\left(\dfrac{14}{3} - \dfrac{7}{9}\right)$ 75. $\dfrac{3}{11}\left(\dfrac{22}{5} - \dfrac{11}{30}\right)$ 76. $\dfrac{5}{12}\left(\dfrac{24}{7} - \dfrac{9}{14}\right)$

Solve each of the following word problems.

77. A cupcake recipe calls for $\dfrac{2}{3}$ cup of sugar. How much sugar would be required to make $2\dfrac{1}{2}$ times as many cupcakes as this recipe makes?

78. If it takes $\dfrac{2}{9}$ of a gallon of paint to paint a closet, how much paint is required to paint six closets of the same size?

79. In the United States, $\dfrac{4}{5}$ of the people have vision problems that require corrective lenses. Of these people, $\dfrac{2}{7}$ wear contact lenses. What fraction represents the people in the United States who wear contacts?

80. At a PTA meeting $\dfrac{3}{4}$ of the people have brown eyes and $\dfrac{1}{5}$ have blue eyes. If $\dfrac{1}{3}$ of each group

is wearing glasses, what fraction represents the total number of blue-eyed and brown-eyed people wearing glasses?

81. A woman swam four laps of a pool that measures $16\frac{2}{3}$ yards in length. How far did she swim? (A lap is from one end to the other, then back.)

82. A cookie recipe calls for $3\frac{3}{4}$ cups of flour. How much flour would you need to make $\frac{1}{3}$ of this recipe?

83. If a rug measures $8\frac{3}{4}$ feet in length, and a room is $2\frac{1}{2}$ times as long as this rug, how long is the room?

84. A car can travel $21\frac{5}{8}$ miles on 1 gallon of gas. How far can it travel on $\frac{1}{4}$ gallon of gas?

3.7
Division of Fractions

Before we begin our discussion of division of fractions, let us look at some interesting results of multiplication of fractions.

$$\frac{1}{3} \cdot 3 = \frac{1}{3} \cdot \frac{3}{1} = \frac{3}{3} = 1 \qquad 17 \cdot \frac{1}{17} = \frac{17}{1} \cdot \frac{1}{17} = \frac{17}{17} = 1$$

$$\frac{9}{7} \cdot \frac{7}{9} = \frac{63}{63} = 1 \qquad 1\frac{1}{4} \cdot \frac{4}{5} = \frac{5}{4} \cdot \frac{4}{5} = \frac{20}{20} = 1$$

The numbers 3 and $\frac{1}{3}$ are said to be **reciprocals** since their product is one. The fraction $\frac{1}{17}$ is the reciprocal of 17 since $17 \cdot \frac{1}{17} = 1$. Likewise, 17 is the reciprocal of $\frac{1}{17}$.

▶ **DEFINITION** The *reciprocal* of a number is what you must multiply that number by to obtain a product of 1. For any fraction not equal to 0, the reciprocal is obtained by interchanging the numerator and denominator.

EXAMPLE 1   Give the reciprocal of each of the following numbers.

(a) $\frac{2}{3}$

The reciprocal is $\frac{3}{2}$.   Since $\frac{2}{3} \cdot \frac{3}{2} = \frac{6}{6} = 1$.

(b) 100

The reciprocal is $\frac{1}{100}$.   Since $\frac{100}{1} \cdot \frac{1}{100} = \frac{100}{100} = 1$.

(c) $9\frac{1}{5}$

$9\frac{1}{5} = \frac{46}{5}$ First convert $9\frac{1}{5}$ to an improper fraction, and then invert the numerator and denominator.

The reciprocal is $\frac{5}{46}$.   Since $\frac{46}{5} \cdot \frac{5}{46} = 1$.

(d) $\frac{0}{5}$

Since $\frac{0}{5} = 0$, it has no reciprocal. When 0 is multiplied by any number, the product is always 0.   ▲

To illustrate division of fractions, let us consider this example. If we divide $\frac{1}{4}$ of a pie equally among three people, how large of a piece will each person receive? As we can see in Figure 3.14, each person will receive $\frac{1}{12}$ of the whole pie. Mathematically this can be written as

$$\frac{1}{4} \div 3 = \frac{1}{12}$$

Figure 3.14

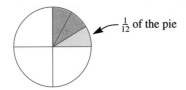

$\frac{1}{12}$ of the pie

As another example, suppose that we want to determine how many sixths go into $\frac{1}{3}$. In other words, what is $\frac{1}{3} \div \frac{1}{6}$? This problem can be illustrated by a box divided into six equal pieces, as shown in Figure 3.15. Each piece represents $\frac{1}{6}$ of the total box. We can see that $\frac{1}{3}$ of the box (shaded area) contains two of these pieces. Thus, $\frac{1}{6}$ goes into $\frac{1}{3}$ exactly 2 times, or

$$\frac{1}{3} \div \frac{1}{6} = 2$$

Figure 3.15

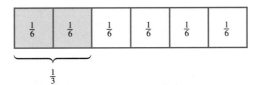

If we closely examine these two examples, we can see that when a fraction is divided by a number, the fraction is multiplied by the reciprocal of the number.

$$\frac{1}{4} \div 3 = \frac{1}{4} \cdot \frac{1}{3} = \frac{1}{12}$$

$$\frac{1}{3} \div \frac{1}{6} = \frac{1}{3} \cdot \frac{6}{1} = \frac{6}{3} = 2$$

We can show *why* this technique works by rewriting each problem as one large fraction.

$$\frac{1}{4} \div 3 \qquad \text{means} \qquad \frac{\frac{1}{4}}{3}$$

This number is called a **complex fraction.**

▶ **DEFINITION** A *complex fraction* is a fraction whose numerator and/or denominator contains fractions.

We can multiply both numerator and denominator of the complex fraction by the reciprocal of the denominator.

$$\frac{\dfrac{1}{4}}{3} = \frac{\dfrac{1}{4} \cdot \dfrac{1}{3}}{3 \cdot \dfrac{1}{3}} = \frac{\dfrac{1}{12}}{1} = \frac{1}{12}$$

Since we get a 1 in the denominator, the answer is the fraction in the numerator.

This technique also works for problems in which fractions are divided by fractions.

EXAMPLE 2   Divide $\frac{3}{7} \div \frac{2}{5}$.

Solution $\dfrac{3}{7} \div \dfrac{2}{5} = \dfrac{\dfrac{3}{7}}{\dfrac{2}{5}} = \dfrac{\dfrac{3}{7} \cdot \dfrac{5}{2}}{\dfrac{2}{5} \cdot \dfrac{5}{2}} = \dfrac{\dfrac{15}{14}}{1} = \dfrac{15}{14}$ ▲

Notice that the answer in Example 2 is the same as the result we would obtain if we simply multiplied the first fraction (dividend) by the reciprocal of the second (divisor).

$$\frac{3}{7} \div \frac{2}{5} = \frac{3}{7} \cdot \frac{5}{2} = \frac{15}{14}$$

We have thus established a procedure for division of fractions, as outlined here.

To Divide Fractions   1. Find the reciprocal of the divisor.

2. Multiply the dividend by the reciprocal of the divisor.

3. If necessary, reduce answer to lowest terms.   ▲

EXAMPLE 3   Perform each of the following divisions.

(a) $\dfrac{7}{9} \div 5$ The reciprocal of 5 is $\frac{1}{5}$.

$\dfrac{7}{9} \div 5 = \dfrac{7}{9} \cdot \dfrac{1}{5} = \dfrac{7}{45}$ Multiply $\frac{7}{9}$ by $\frac{1}{5}$.

(b) $\dfrac{5}{8} \div \dfrac{2}{3}$ The reciprocal of $\frac{2}{3}$ is $\frac{3}{2}$.

$\dfrac{5}{8} \div \dfrac{2}{3} = \dfrac{5}{8} \cdot \dfrac{3}{2} = \dfrac{15}{16}$ Multiply $\frac{5}{8}$ by $\frac{3}{2}$.

(c) $\dfrac{3}{7} \div \dfrac{5}{14}$

$\dfrac{3}{7} \div \dfrac{5}{14} = \dfrac{3}{7} \cdot \dfrac{14}{5}$ The reciprocal of $\frac{5}{14}$ is $\frac{14}{5}$.

$\qquad\qquad = \dfrac{3 \cdot \overset{2}{\cancel{14}}}{\cancel{7} \cdot 5}$ Multiply $\frac{3}{7}$ by $\frac{14}{5}$.

$\qquad\qquad = \dfrac{6}{5}$ Reduce to lowest terms.

(d) $8 \div \dfrac{3}{8}$ The reciprocal of $\frac{3}{8}$ is $\frac{8}{3}$.

$8 \div \dfrac{3}{8} = \dfrac{8}{1} \cdot \dfrac{8}{3} = \dfrac{64}{3}$ Multiply 8 by $\frac{8}{3}$.   ▲

The next examples illustrate how to divide mixed numbers. The procedure used in the examples is outlined below.

To Divide Mixed Numbers	**1.** Change each mixed number to an improper fraction.
	2. Find the reciprocal of the divisor.
	3. Multiply the dividend by the reciprocal of the divisor.
	4. Express the answer as a mixed number or as a proper fraction reduced to lowest terms. ▲

EXAMPLE 4   Perform each of the following divisions.

(a) $3\dfrac{2}{7} \div 4$

$3\dfrac{2}{7} \div 4 = \dfrac{23}{7} \div 4$ Change $3\frac{2}{7}$ to an improper fraction.

$\qquad\qquad = \dfrac{23}{7} \cdot \dfrac{1}{4}$ The reciprocal of 4 is $\frac{1}{4}$.

$\qquad\qquad = \dfrac{23}{28}$

(b) $2\dfrac{5}{8} \div 1\dfrac{2}{5}$

$$2\dfrac{5}{8} \div 1\dfrac{2}{5} = \dfrac{21}{8} \div \dfrac{7}{5}$$ Change mixed numbers to improper fractions.

$$= \dfrac{21}{8} \cdot \dfrac{5}{7}$$ Multiply by the reciprocal.

$$= \dfrac{\overset{3}{\cancel{21}} \cdot 5}{8 \cdot \cancel{7}}$$

$$= \dfrac{15}{8} = 1\dfrac{7}{8}$$ Simplify the result.

(c) $4\dfrac{2}{7} \div 5\dfrac{1}{4}$

$$4\dfrac{2}{7} \div 5\dfrac{1}{4} = \dfrac{30}{7} \div \dfrac{21}{4}$$ Change mixed numbers to improper fractions.

$$= \dfrac{30}{7} \cdot \dfrac{4}{21}$$ Multiply by the reciprocal.

$$= \dfrac{\overset{10}{\cancel{30}} \cdot 4}{7 \cdot \underset{7}{\cancel{21}}}$$

$$= \dfrac{40}{49}$$ Simplify the result.

(d) $6\dfrac{2}{9} \div 5\dfrac{5}{6}$

$$6\dfrac{2}{9} \div 5\dfrac{5}{6} = \dfrac{56}{9} \div \dfrac{35}{6}$$ Change mixed numbers to improper fractions.

$$= \dfrac{56}{9} \cdot \dfrac{6}{35} = \dfrac{\overset{8}{\cancel{56}} \cdot \overset{2}{\cancel{6}}}{\underset{3}{\cancel{9}} \cdot \underset{5}{\cancel{35}}} = \dfrac{16}{15} = 1\dfrac{1}{15}$$ Multiply by the reciprocal and simplify the answer. ▲

EXAMPLE 5   Use the order of operations rules to simplify the following problems.

(a) $\dfrac{3}{5} \div \dfrac{9}{10} \cdot \dfrac{15}{22}$

$$\frac{3}{5} \div \frac{9}{10} \cdot \frac{15}{22} = \left(\frac{3}{5} \div \frac{9}{10}\right) \cdot \frac{15}{22}$$

Multiplication and division are performed in the same step, moving from left to right.

$$= \left(\frac{3}{5} \cdot \frac{10}{9}\right) \cdot \frac{15}{22}$$

$$= \frac{3 \cdot \overset{5}{\cancel{10}} \cdot \overset{5}{\cancel{15}}}{5 \cdot \underset{3}{\cancel{9}} \cdot \underset{11}{\cancel{22}}} = \frac{5}{11}$$

(b) $\dfrac{7}{8} - \dfrac{5}{12} \div \dfrac{10}{3}$

$$\frac{7}{8} - \frac{5}{12} \div \frac{10}{3} = \frac{7}{8} - \left(\frac{5}{12} \div \frac{10}{3}\right)$$

Perform division before subtraction.

$$= \frac{7}{8} - \left(\frac{5}{12} \div \frac{3}{10}\right) = \frac{7}{8} - \frac{\overset{}{\cancel{5}} \cdot \overset{}{\cancel{3}}}{\underset{4}{\cancel{12}} \cdot \underset{2}{\cancel{10}}}$$

$$= \frac{7}{8} - \frac{1}{4 \cdot 2} = \frac{7}{8} - \frac{1}{8}$$

$$= \frac{6}{8} = \frac{3}{4}$$

(c) $\left(\dfrac{1}{2}\right)^3 \div \left(\dfrac{2}{5}\right)^2$

$$\left(\frac{1}{2}\right)^3 \div \left(\frac{2}{5}\right)^2 = \left(\frac{1}{2} \cdot \frac{1}{2} \cdot \frac{1}{2}\right) \div \left(\frac{2}{5} \cdot \frac{2}{5}\right)$$

Raise to powers before dividing.

$$= \frac{1}{8} \div \frac{4}{25}$$

$$= \frac{1}{8} \cdot \frac{25}{4} = \frac{25}{32}$$

▲

EXAMPLE 6   A piece of ribbon that is $7\frac{7}{8}$ yards long is cut into three equal pieces. How long is each piece?

Solution   Divide $7\frac{7}{8} \div 3$.

$$7\frac{7}{8} \div 3 = \frac{63}{8} \div 3 = \frac{63}{8} \cdot \frac{1}{3} = \frac{\overset{21}{\cancel{63}} \cdot 1}{8 \cdot \cancel{3}} = \frac{21}{8} = 2\frac{5}{8} \text{ yd}$$

Each piece is $2\frac{5}{8}$ yards long.   ▲

EXAMPLE 7   A box of chocolate bars weighs $2\frac{2}{3}$ pounds. If there are 24 chocolate bars in the box, how much does each chocolate bar weigh?

Solution Divide $2\frac{2}{3} \div 24$.

$$2\frac{2}{3} \div 24 = \frac{8}{3} \div 24$$

$$= \frac{8}{3} \cdot \frac{1}{24}$$

$$= \frac{8 \cdot 1}{3 \cdot \underset{3}{24}}$$

$$= \frac{1}{9}$$

Each chocolate bar weighs $\frac{1}{9}$ lb.   ▲

Quick Quiz

Perform each of the following divisions.   **Answers**

1. $\dfrac{5}{9} \div 7$ 1. $\dfrac{5}{63}$

2. $\dfrac{3}{8} \div \dfrac{27}{32}$ 2. $\dfrac{4}{9}$

3. $6\dfrac{2}{3} \div 3\dfrac{3}{4}$ 3. $1\dfrac{7}{9}$

3.7 Exercises

Indicate which of the following statements are true and which are false. For those that are false, change the italicized or underlined expression to make the statement true.

1. To divide two fractions, multiply the dividend by the *reciprocal* of the divisor.

2. When you *add* a number and its reciprocal, you obtain a result of one.

3. Three divided by one-third is equal to *one-ninth*.

4. Dividing by $\frac{1}{7}$ is the same as multiplying by $\underline{7}$.

Find the reciprocal of each of the following numbers.

5. $\dfrac{3}{8}$ 6. $\dfrac{7}{12}$ 7. $\dfrac{1}{7}$ 8. $\dfrac{18}{57}$ 9. 11

10. 29 11. 1 12. 0 13. $7\dfrac{3}{4}$ 14. $15\dfrac{2}{3}$

Perform each of the following divisions.

15. $\dfrac{2}{3} \div 9$ 16. $\dfrac{5}{8} \div 7$ 17. $\dfrac{13}{7} \div 4$ 18. $\dfrac{18}{5} \div 5$

19. $\dfrac{3}{4} \div \dfrac{1}{8}$

20. $\dfrac{5}{9} \div \dfrac{1}{4}$

21. $\dfrac{2}{5} \div \dfrac{3}{7}$

22. $\dfrac{3}{8} \div \dfrac{2}{3}$

23. $\dfrac{2}{7} \div \dfrac{3}{5}$

24. $\dfrac{7}{9} \div \dfrac{3}{4}$

25. $\dfrac{14}{15} \div \dfrac{7}{9}$

26. $\dfrac{7}{12} \div \dfrac{5}{8}$

27. $\dfrac{5}{36} \div \dfrac{7}{18}$

28. $\dfrac{3}{16} \div \dfrac{5}{24}$

29. $\dfrac{3}{11} \div \dfrac{5}{22}$

30. $\dfrac{5}{36} \div \dfrac{7}{12}$

31. $\dfrac{14}{3} \div \dfrac{7}{2}$

32. $\dfrac{18}{5} \div \dfrac{9}{4}$

33. $\dfrac{20}{49} \div \dfrac{15}{14}$

34. $\dfrac{5}{12} \div \dfrac{25}{18}$

35. $\dfrac{21}{10} \div \dfrac{6}{25}$

36. $\dfrac{45}{28} \div \dfrac{5}{16}$

37. $8 \div \dfrac{4}{7}$

38. $5 \div \dfrac{3}{8}$

39. $\dfrac{11}{12} \div \dfrac{2}{3} \div \dfrac{33}{2}$

40. $\dfrac{5}{3} \div \dfrac{1}{2} \div \dfrac{5}{6}$

41. $\dfrac{9}{16} \div \dfrac{1}{12} \div \dfrac{3}{4}$

42. $\dfrac{18}{25} \div \dfrac{9}{10} \div \dfrac{1}{3}$

43. $\dfrac{3}{5} \div 1\dfrac{1}{2}$

44. $4\dfrac{7}{8} \div \dfrac{3}{8}$

45. $5\dfrac{1}{3} \div \dfrac{4}{5}$

46. $\dfrac{1}{8} \div 3\dfrac{5}{8}$

47. $1\dfrac{2}{5} \div 3\dfrac{1}{2}$

48. $5\dfrac{1}{3} \div 7\dfrac{5}{7}$

49. $6\dfrac{3}{4} \div 4\dfrac{2}{5}$

50. $5\dfrac{2}{3} \div 2\dfrac{3}{7}$

51. $7\dfrac{1}{3} \div 2\dfrac{1}{2}$

52. $8\dfrac{1}{2} \div 1\dfrac{5}{6}$

53. $2\dfrac{1}{12} \div 3\dfrac{3}{4}$

54. $5\dfrac{1}{3} \div 1\dfrac{5}{9}$

55. $6\dfrac{1}{8} \div 1\dfrac{3}{4}$

56. $3\dfrac{1}{3} \div 2\dfrac{6}{7}$

57. $2\dfrac{2}{9} \div 5\dfrac{7}{11}$

58. $4\dfrac{3}{5} \div 8\dfrac{2}{3}$

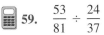 59. $\dfrac{53}{81} \div \dfrac{24}{37}$

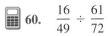 60. $\dfrac{16}{49} \div \dfrac{61}{72}$

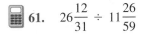 61. $26\dfrac{12}{31} \div 11\dfrac{26}{59}$

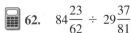 62. $84\dfrac{23}{62} \div 29\dfrac{37}{81}$

Perform each of the following operations.

63. $\dfrac{1}{3} \cdot \dfrac{8}{15} \div \dfrac{4}{5}$

64. $\dfrac{7}{9} \cdot \dfrac{5}{6} \cdot \dfrac{3}{10}$

65. $\dfrac{1}{9} \div \dfrac{5}{9} \cdot \dfrac{15}{49} \div \dfrac{3}{14}$

66. $\dfrac{1}{7} \cdot \dfrac{2}{9} \div \dfrac{4}{27} \cdot \dfrac{7}{2}$

67. $\left(\dfrac{2}{7} + \dfrac{3}{5}\right) \div \dfrac{1}{5}$

68. $\left(\dfrac{3}{4} - \dfrac{5}{9}\right) \div \dfrac{2}{3}$

69. $\dfrac{1}{3} \div \dfrac{1}{4} + \dfrac{5}{6}$

70. $\dfrac{2}{3} + \dfrac{5}{8} \div \dfrac{15}{4}$

🖩 71. $\left(\dfrac{51}{67} + \dfrac{11}{19}\right) \div \dfrac{15}{43}$ 🖩 72. $\dfrac{43}{77} \div \dfrac{18}{53} \cdot \dfrac{27}{31}$

Solve each of the following word problems.

73. One lap of an indoor track is $\dfrac{1}{8}$ mile long. How many laps must a jogger run to go $\dfrac{2}{3}$ mile?

74. A car went $\dfrac{11}{12}$ mile on $\dfrac{1}{30}$ gallon of gas. How many miles per gallon did the car get?

75. A box of cookies weighs $10\dfrac{1}{2}$ oz. If the box contains 30 cookies, how much does each cookie weigh?

76. A pitcher contains $1\dfrac{3}{4}$ quarts of lemonade. If this is divided equally among six people, how much lemonade does each person receive?

77. A total of $3\dfrac{1}{8}$ pounds of chocolate is to be divided evenly among ten children. How much chocolate should each child receive?

78. An eight-story building is $90\dfrac{2}{3}$ feet high. What is the height of each story?

79. How many boxes can be filled with $7\frac{2}{9}$ pounds of crackers if each box has a capacity of $1\frac{2}{3}$ pounds?

80. A drinking glass can hold $1\frac{1}{3}$ cups of liquid. How many glasses of this size can be filled from a pitcher that contains $14\frac{2}{3}$ cups of water?

81. If $11\frac{1}{4}$ pints of root beer are divided equally among 15 cans, how much root beer is in each can?

82. An ant can travel $1\frac{1}{2}$ miles in $2\frac{3}{4}$ hours. What is its average speed in miles per hour?

83. The planet Jupiter makes one rotation on its axis in $9\frac{5}{6}$ hours. How many times will it rotate in $44\frac{1}{4}$ hours?

84. A bus traveled $65\frac{7}{8}$ miles in $1\frac{5}{12}$ hours. What was its average speed in miles per hour?

3.8
Simplifying Complex Fractions

We have already introduced one technique to simplify complex fractions—that is, multiplying numerator and denominator by the reciprocal of the denominator. For example:

$$\frac{\dfrac{2}{9}}{\dfrac{3}{5}} = \frac{\dfrac{2}{9} \cdot \dfrac{5}{3}}{\dfrac{3}{5} \cdot \dfrac{5}{3}} = \frac{\dfrac{10}{27}}{1} = \frac{10}{27}$$

Notice that the same result can be obtained by multiplying both numerator and denominator by the LCD of the fractions $\frac{2}{9}$ and $\frac{3}{5}$, which is 45.

$$\frac{\dfrac{2}{9}}{\dfrac{3}{5}} = \frac{\dfrac{2}{9} \cdot 45}{\dfrac{3}{5} \cdot 45} = \frac{\dfrac{2}{9} \cdot \dfrac{\overset{5}{\cancel{45}}}{1}}{\dfrac{3}{\cancel{5}} \cdot \dfrac{\overset{9}{\cancel{45}}}{1}} = \frac{2 \cdot 5}{3 \cdot 9} = \frac{10}{27}$$

The second technique produces a result in which there is a whole number in both the numerator and denominator. This procedure is very useful when evaluating complicated complex fractions.

We will use this technique in the following example.

EXAMPLE 1

Simplify $\dfrac{\dfrac{3}{5}}{\dfrac{4}{7}}$.

$$\frac{\dfrac{3}{5}}{\dfrac{4}{7}} = \frac{\dfrac{3}{5} \cdot 35}{\dfrac{4}{7} \cdot 35}$$

The LCD of $\frac{3}{5}$ and $\frac{4}{7}$ is 35.
Multiply numerator and denominator by 35.

$$= \frac{\dfrac{3}{\cancel{5}} \cdot \overset{7}{\cancel{35}}}{\dfrac{4}{\cancel{7}} \cdot \overset{5}{\cancel{35}}}$$

$$= \frac{21}{20}$$

▲

EXAMPLE 2 Simplify the following complex fraction.

$$\frac{\frac{1}{3} + \frac{1}{2}}{\frac{5}{6}}$$

Solution $\dfrac{\frac{1}{3} + \frac{1}{2}}{\frac{5}{6}}$

Find the LCD of all fractions that appear in the numerator and denominator. The LCD of $\frac{1}{3}, \frac{1}{2}$, and $\frac{1}{6}$ is 6.

$$\frac{\left(\frac{1}{3} + \frac{1}{2}\right) \cdot 6}{\frac{5}{6} \cdot 6} = \frac{\frac{1}{3} \cdot \frac{6}{1} + \frac{1}{2} \cdot \frac{6}{1}}{\frac{5}{6} \cdot \frac{6}{1}}$$

Multiply both numerator and denominator of the complex fraction by the LCD. Evaluate the numerator using the distributive property.

$$= \frac{2 + 3}{5} = \frac{5}{5} = 1$$ ▲

In general, complex fractions that contain arithmetic operations in the numerator and/or denominator can be simplified in one of two ways. Both methods are outlined here.

To Simplify Complex Fractions

Simplify the numerator and denominator separately. Then multiply the numerator by the reciprocal of the denominator.

OR

Find the LCD of all fractions that appear in the numerator and denominator. Then multiply both numerator and denominator by the LCD, using the distributive property. ▲

The following example illustrates both techniques.

EXAMPLE 3 Simplify the following complex fraction.

$$\frac{\frac{5}{8} - \frac{1}{3}}{\frac{1}{4} + \frac{7}{12}}$$

(a) $$\frac{\frac{5}{8} - \frac{1}{3}}{\frac{1}{4} + \frac{7}{12}} = \frac{\frac{5 \cdot 3}{8 \cdot 3} - \frac{1 \cdot 8}{3 \cdot 8}}{\frac{1 \cdot 3}{4 \cdot 3} + \frac{7}{12}} = \frac{\frac{7}{24}}{\frac{10}{12}}$$

Simplify numerator and denominator separately.

$$= \frac{7}{24} \div \frac{10}{12} = \frac{7}{24} \cdot \frac{12}{10}$$

Now multiply the numerator by the reciprocal of the denominator.

$$= \frac{7 \cdot \cancel{12}}{\cancel{24} \cdot 10} = \frac{7}{2 \cdot 10} = \frac{7}{20}$$

(b) $8 = 2^3$
$3 = 3$
$4 = 2^2$
$12 = 2^2 \cdot 3$
LCD $= 2^3 \cdot 3 = 24$

Find the LCD of $\frac{5}{8}, \frac{1}{3}, \frac{1}{4}$, and $\frac{7}{12}$.

$$\frac{\left(\dfrac{5}{8} - \dfrac{1}{3}\right) \cdot 24}{\left(\dfrac{1}{4} + \dfrac{7}{12}\right) \cdot 24} = \frac{\left(\dfrac{5}{8} \cdot 24\right) - \left(\dfrac{1}{3} \cdot 24\right)}{\left(\dfrac{1}{4} \cdot 24\right) + \left(\dfrac{7}{12} \cdot 24\right)}$$

Multiply numerator and denominator by 24, using the distributive property.

$$= \frac{\left(\dfrac{5}{\cancel{8}} \cdot \dfrac{\overset{3}{\cancel{24}}}{1}\right) - \left(\dfrac{1}{\cancel{3}} \cdot \dfrac{\overset{8}{\cancel{24}}}{1}\right)}{\left(\dfrac{1}{\cancel{4}} \cdot \dfrac{\overset{6}{\cancel{24}}}{1}\right) + \left(\dfrac{7}{\cancel{12}} \cdot \dfrac{\overset{2}{\cancel{24}}}{1}\right)} = \frac{(5 \cdot 3) - 8}{6 + (7 \cdot 2)} = \frac{15 - 8}{6 + 14}$$

$$= \frac{7}{20}$$ ▲

EXAMPLE 4 Simplify the following complex fraction.

$$\frac{5\dfrac{7}{18} + 2\dfrac{1}{2}}{9\dfrac{3}{4} - 4\dfrac{5}{12}}$$

Solution $18 = 2 \cdot 3^2$
$2 = 2$
$4 = 2^2$
$12 = 2^2 \cdot 3$
LCD $= 2^2 \cdot 3^2 = 36$

Find the LCD of all fractions.

$$\frac{5\dfrac{7}{18} + 2\dfrac{1}{2}}{9\dfrac{3}{4} - 4\dfrac{5}{12}} = \frac{\left(\dfrac{97}{18} + \dfrac{5}{2}\right) \cdot 36}{\left(\dfrac{39}{4} - \dfrac{53}{12}\right) \cdot 36}$$

Change each mixed number to an improper fraction and multiply numerator and denominator by 36.

$$= \frac{\left(\dfrac{97}{18} \cdot 36\right) + \left(\dfrac{5}{2} \cdot 36\right)}{\left(\dfrac{39}{4} \cdot 36\right) - \left(\dfrac{53}{12} \cdot 36\right)} = \frac{\left(\dfrac{97}{\cancel{18}} \cdot \dfrac{\overset{2}{\cancel{36}}}{1}\right) + \left(\dfrac{5}{\cancel{2}} \cdot \dfrac{\overset{18}{\cancel{36}}}{1}\right)}{\left(\dfrac{39}{\cancel{4}} \cdot \dfrac{\overset{9}{\cancel{36}}}{1}\right) - \left(\dfrac{53}{\cancel{12}} \cdot \dfrac{\overset{3}{\cancel{36}}}{1}\right)}$$

$$= \frac{(97 \cdot 2) + (5 \cdot 18)}{(39 \cdot 9) - (53 \cdot 3)} = \frac{194 + 90}{351 - 159} = \frac{284}{192} = \frac{71}{48} = 1\frac{23}{48}$$ ▲

Quick Quiz

Simplify each of the following complex fractions.   **Answers**

1. $\dfrac{\dfrac{5}{6} - \dfrac{3}{4}}{\dfrac{2}{3}}$

2. $\dfrac{2\dfrac{1}{3}}{\dfrac{1}{2} + \dfrac{4}{9}}$

1. $\dfrac{1}{8}$

2. $2\dfrac{8}{17}$

3.8 Exercises

Simplify each of the following complex fractions.

1. $\dfrac{\dfrac{3}{8}}{\dfrac{5}{6}}$

2. $\dfrac{\dfrac{2}{5}}{\dfrac{7}{9}}$

3. $\dfrac{\dfrac{8}{11}}{\dfrac{5}{22}}$

4. $\dfrac{\dfrac{7}{18}}{\dfrac{13}{27}}$

5. $\dfrac{\dfrac{2}{5} + \dfrac{4}{3}}{\dfrac{8}{9}}$

6. $\dfrac{\dfrac{3}{4} + \dfrac{5}{7}}{\dfrac{1}{2}}$

7. $\dfrac{\dfrac{3}{4}}{\dfrac{9}{16} - \dfrac{1}{2}}$

8. $\dfrac{\dfrac{4}{5}}{\dfrac{11}{15} - \dfrac{2}{3}}$

9. $\dfrac{\dfrac{2}{9} + \dfrac{5}{6}}{\dfrac{1}{2} + \dfrac{2}{3}}$

10. $\dfrac{\dfrac{5}{12} + \dfrac{1}{4}}{\dfrac{2}{3} + \dfrac{5}{6}}$

11. $\dfrac{\dfrac{5}{24} + \dfrac{3}{8}}{\dfrac{7}{12} - \dfrac{1}{6}}$

12. $\dfrac{\dfrac{4}{5} + \dfrac{6}{15}}{\dfrac{2}{3} - \dfrac{7}{30}}$

13. $\dfrac{\dfrac{5}{7} - \dfrac{1}{4}}{\dfrac{1}{2} + \dfrac{5}{8}}$

14. $\dfrac{\dfrac{1}{2} - \dfrac{2}{5}}{\dfrac{3}{10} + \dfrac{6}{7}}$

15. $\dfrac{\dfrac{11}{15} - \dfrac{7}{12}}{\dfrac{3}{4} - \dfrac{1}{30}}$

16. $\dfrac{\dfrac{11}{21} - \dfrac{2}{9}}{\dfrac{6}{7} - \dfrac{1}{3}}$

17. $\dfrac{\dfrac{5}{8} + \dfrac{1}{2}}{4\dfrac{3}{4}}$

18. $\dfrac{\dfrac{7}{9} + \dfrac{2}{3}}{5\dfrac{1}{2}}$

19. $\dfrac{3\dfrac{5}{8}}{\dfrac{3}{4} - \dfrac{7}{12}}$

20. $\dfrac{1\dfrac{7}{9}}{\dfrac{2}{3} - \dfrac{5}{18}}$

21. $\dfrac{\dfrac{3}{10}}{2\dfrac{1}{5} + 4\dfrac{7}{15}}$

22. $\dfrac{\dfrac{1}{6}}{5\dfrac{7}{18} + 1\dfrac{5}{12}}$

23. $\dfrac{7\dfrac{4}{5} - 3\dfrac{1}{4}}{\dfrac{2}{3}}$

24. $\dfrac{3\dfrac{4}{7} - 1\dfrac{2}{21}}{\dfrac{5}{6}}$

25. $\dfrac{3\frac{5}{12} - 1\frac{2}{15}}{\frac{7}{30} + 1\frac{3}{20}}$

26. $\dfrac{\frac{8}{21} + 3\frac{2}{3}}{4\frac{1}{2} - 2\frac{3}{7}}$

27. $\dfrac{3\frac{2}{5} - 2\frac{1}{6}}{7\frac{1}{2} + \frac{4}{5}}$

28. $\dfrac{3\frac{2}{9} + 2\frac{5}{18}}{5\frac{5}{6} - \frac{7}{27}}$

29. $\dfrac{3\frac{2}{9} + 4\frac{1}{2}}{1\frac{5}{18} + 6\frac{2}{3}}$

30. $\dfrac{3\frac{5}{6} + 5\frac{3}{4}}{2\frac{7}{12} + 6\frac{1}{3}}$

31. $\dfrac{5\frac{3}{10} - 2\frac{5}{6}}{7\frac{4}{5} - 3\frac{1}{2}}$

32. $\dfrac{6\frac{7}{12} - 2\frac{1}{6}}{3\frac{5}{8} - 1\frac{3}{4}}$

3.9
Summary and Review

Key Terms

(3.1) A **fraction** is any number that can be written in the form $\frac{a}{b}$, where a and b are numbers such that b is not equal to zero. The **terms** of the fraction are a and b. The term a is called the **numerator**, and b is called the **denominator**.

A **proper fraction** is one in which the numerator is less than the denominator.

An **improper fraction** is one in which the numerator is greater than or equal to the denominator.

A **mixed number** is the sum of a whole number and a proper fraction.

(3.2) **Equivalent fractions** are fractions that have the same numerical value.

A fraction is in **lowest terms**, or **simplest terms**, if the only common factor of both the numerator and denominator is the number one.

(3.3) The **lowest common denominator**, or **LCD**, is the smallest number that is divisible by the denominators of all fractions being considered.

(3.7) The **reciprocal** of a number is what you must multiply that number by to obtain a product of one.

A **complex fraction** is a number whose numerator and/or denominator contains fractions.

Calculations

(3.1) **To convert an improper fraction to a mixed number**, divide the numerator by the denominator. The quotient is the whole number portion of the mixed number, and the fractional portion of the mixed number is the remainder divided by the divisor.

To convert a mixed number to an improper fraction, multiply the whole number part by the denominator, add the product to the numerator, and put the result over the same denominator.

(3.2) **To find an equivalent fraction of one that is given**, multiply or divide the numerator and denominator of the given fraction by the same number.

To reduce a fraction to lowest terms, divide both numerator and denominator by their greatest common factor. When performing arithmetic operations on fractions, answers should be reduced to lowest terms.

(3.3) **To find the lowest common denominator (LCD) of two or more fractions**, completely factor each denominator, and calculate the product of all unique prime factors, each of which is raised to the highest exponent that appears in the complete factorization of any denominator. In other words, find the LCM of the denominators.

To add (or subtract) fractions, find the LCD and change each fraction to an equivalent fraction with a denominator equal to the LCD. Add (or subtract) the numerators and place the result over the LCD.

(3.4) **To find the larger of two fractions**, change each to an equivalent fraction with a denominator equal to the LCD and compare the numerators. The larger fraction has the larger numerator.

(3.5) **To add mixed numbers**, add the whole number parts and fractional parts separately, change any resulting improper fraction to a mixed number, and add the sums together.

To subtract mixed numbers, change the fractional parts to equivalent fractions with the same LCD. If the fractional part of the minuend is less than that of the subtrahend, borrow one from the whole number part of the minuend, and add one to its fractional part. Then subtract the whole number and fractional parts separately.

(3.6) **To multiply fractions**, multiply the numerators and place the result over the product of the denominators.

To multiply mixed numbers, change each mixed number to an improper fraction. Then multiply the numerators and place the result over the product of the denominators.

(3.7) **To find the reciprocal of a nonzero fraction**, interchange the numerator and the denominator.

To divide two fractions, multiply the first (dividend) by the reciprocal of the second (divisor).

To divide mixed numbers, change each mixed number to an improper fraction, and multiply the dividend by the reciprocal of the divisor.

(3.8) **To simplify a complex fraction**, multiply the numerator and denominator by the LCD of all fractions that appear in the numerator and denominator.

Chapter **3**
Review Exercises

Indicate which of the following statements are true and which are false. For those that are false, change the italicized word to make the statement true.

1. A mixed number is the *product* of a whole number and a fraction.

2. The commutative property holds true for addition and *subtraction* of fractions.

3. Fractions in which the numerator is greater than or equal to the denominator are called *improper* fractions.

4. To convert a mixed number to an improper fraction, you multiply the whole number by the *denominator* and add the numerator.

5. A *reduced* fraction is obtained by multiplying the numerator and denominator by the same whole number.

6. Any fraction that has a *one* as the denominator is equal to the numerator.

State which of the following problems are correct and which are incorrect. For those that are incorrect, replace the underlined answer with the correct one.

7. $3 \cdot \dfrac{5}{8} = 3\underline{\dfrac{5}{8}}$

8. $\dfrac{1}{2} + \dfrac{3}{3} = \underline{\dfrac{4}{5}}$

9. $\dfrac{0}{9}$ is $\underline{\text{undefined}}$.

10. $\dfrac{1}{3} \cdot \dfrac{9}{5} + \dfrac{1}{5} = \underline{\dfrac{2}{3}}$

11. $2 - \dfrac{1}{6} = 1\underline{\dfrac{5}{6}}$

12. $\dfrac{132}{144} = \underline{\dfrac{11}{12}}$

13. $\dfrac{5}{8}$ is $\underline{\text{less}}$ than $\dfrac{3}{5}$.

14. The reciprocal of $1\dfrac{1}{8}$ is $\underline{8}$.

(3.3–3.8) Perform each of the following operations. Be sure to reduce all answers to lowest terms.

15. $\dfrac{3}{8} + \dfrac{1}{4}$

16. $\dfrac{5}{9} + \dfrac{2}{3}$

17. $\dfrac{5}{12} - \dfrac{1}{8}$

18. $\dfrac{5}{6} - \dfrac{2}{15}$

19. $\dfrac{8}{11} \cdot \dfrac{5}{3}$

20. $\dfrac{3}{8} \cdot \dfrac{7}{12}$

21. $\dfrac{7}{16} \div \dfrac{1}{20}$

22. $\dfrac{42}{9} \div \dfrac{10}{6}$

23. $\dfrac{3}{18} + \dfrac{5}{27}$

24. $\dfrac{7}{16} + \dfrac{11}{24}$

25. $6\dfrac{3}{4} + \dfrac{2}{9}$

26. $\dfrac{3}{5} \cdot 2\dfrac{7}{8}$

27. $7 \div \dfrac{1}{3}$

28. $\dfrac{3}{8} \div 9$

29. $\dfrac{9}{28} + \dfrac{1}{21}$

30. $\dfrac{5}{36} + \dfrac{7}{48}$

31. $5\dfrac{3}{4} - \dfrac{5}{12}$

32. $8\dfrac{2}{3} - \dfrac{7}{15}$

33. $5\dfrac{1}{8} + 6\dfrac{2}{3}$

34. $3\dfrac{5}{9} + 2\dfrac{7}{12}$

35. $7\dfrac{5}{6} - 2\dfrac{7}{8}$

36. $9\dfrac{1}{3} - 5\dfrac{8}{15}$

37. $\dfrac{17}{72} \div 3\dfrac{7}{9}$

38. $5\dfrac{2}{7} \div \dfrac{3}{4}$

39. $3\dfrac{2}{25} \cdot 8\dfrac{4}{7}$

40. $6\dfrac{5}{9} \cdot 5\dfrac{3}{4}$

41. $7\dfrac{5}{6} \div 2\dfrac{3}{4}$

42. $5\dfrac{1}{4} \div 2\dfrac{5}{8}$

43. $\dfrac{7}{15} \cdot \dfrac{35}{18} \div \dfrac{49}{12}$

44. $\dfrac{3}{32} \div \dfrac{5}{8} \cdot \dfrac{5}{16}$

45. $8\dfrac{4}{9} + 3\dfrac{5}{18} - 2\dfrac{2}{27}$

46. $7\dfrac{3}{5} - 2\dfrac{7}{20} + 9\dfrac{3}{4}$

47. $\dfrac{5}{3}\left(\dfrac{12}{5} - \dfrac{1}{2}\right)$

48. $\dfrac{2}{5}\left(\dfrac{8}{9} + \dfrac{1}{4}\right)$

49. $\dfrac{6}{7} \cdot \dfrac{5}{9} + \dfrac{2}{3}$

50. $\dfrac{5}{7} - \dfrac{1}{3} \cdot \dfrac{1}{4}$

51. $\left(\dfrac{1}{3}\right)^2 + \dfrac{5}{12}$

52. $\dfrac{7}{9} - \left(\dfrac{1}{2}\right)^3$

53. $\left(\dfrac{3}{4}\right)^3 - \left(\dfrac{1}{8}\right)^2$

54. $\left(\dfrac{2}{3}\right)^2 + \left(\dfrac{5}{6}\right)^2$

55. $\dfrac{\frac{5}{13}}{\frac{7}{39}}$

56. $\dfrac{\frac{7}{55}}{\frac{3}{11}}$

57. $\dfrac{8\frac{1}{6}}{18\frac{2}{3}}$

58. $\dfrac{5\frac{1}{8}}{20\frac{3}{4}}$

59. $\dfrac{\frac{7}{9} + \frac{5}{6}}{\frac{11}{12}}$

60. $\dfrac{\frac{8}{15} - \frac{1}{3}}{\frac{2}{5}}$

61. $\dfrac{\frac{21}{25}}{\frac{4}{5} + \frac{2}{15}}$

62. $\dfrac{\frac{7}{36}}{\frac{8}{9} - \frac{5}{12}}$

63. $\dfrac{\frac{2}{3} + \frac{1}{2}}{\frac{3}{4} + \frac{5}{12}}$

64. $\dfrac{\frac{3}{16} + \frac{1}{2}}{\frac{3}{4} + \frac{5}{12}}$

65. $\dfrac{\frac{5}{9} + 2\frac{3}{4}}{\frac{7}{12} - \frac{1}{8}}$

66. $\dfrac{5\frac{1}{6} - \frac{3}{4}}{\frac{2}{9} + \frac{5}{18}}$

67. $\dfrac{5\frac{2}{7} - 3\frac{1}{3}}{2\frac{5}{6} + 1\frac{9}{42}}$

68. $\dfrac{3\frac{2}{3} - 1\frac{4}{63}}{5\frac{3}{7} - 2\frac{1}{9}}$

69. $\dfrac{4\frac{3}{4} + 5\frac{2}{3}}{9\frac{5}{6} - 6\frac{1}{2}}$

70. $\dfrac{3\frac{3}{4} + 1\frac{1}{2}}{5\frac{4}{5} - 2\frac{7}{10}}$

Solve each of the following word problems.

71. A jogger ran $3\frac{1}{4}$ miles on Saturday and $2\frac{7}{8}$ miles on Sunday. What is the total number of miles she ran that weekend?

72. In a small town in New Hampshire, $\frac{1}{12}$ of the people are between 6 feet and $6\frac{1}{2}$ feet tall. Of the town's population, $\frac{3}{64}$ are over $6\frac{1}{2}$ feet tall. What fraction of the people in this town are over 6 feet tall?

73. A utility stock had an opening price on the New York Stock Exchange on Monday of $34\frac{3}{8}$ points and a closing price on Friday of $36\frac{1}{4}$ points. By how many points did the value of the stock increase that week?

74. The average life span for a 20-year-old male living in the United States is $68\frac{7}{9}$ years, and that for a 20-year-old male living in Denmark is $71\frac{1}{12}$ years. On the average, how much greater is the male life expectancy for the Dane than for the American?

75. A cookie recipe calls for $2\frac{2}{3}$ cups of sugar and $3\frac{1}{2}$ cups of flour. How much flour and how much sugar would you need to make a double recipe?

76. A race car driver drove 200 miles in $1\frac{1}{5}$ hours. What was the car's average speed in miles per hour?

77. If $\frac{5}{12}$ of the people attending a convention are Democrats and $\frac{4}{9}$ are Republicans, are there more Democrats or Republicans at that convention? What fraction of the people is affiliated with neither party?

78. If $\frac{2}{9}$ of the students in a history class received a grade above C, and $\frac{8}{21}$ received a grade below C, what fraction of the students received a grade of C?

79. A company produces soda bottles that have a capacity of $1\frac{1}{8}$ pints each. How many bottles will $20\frac{1}{4}$ pints of soda fill?

80. At a well-known university, there are 2,550 freshmen. Two-thirds of the freshmen take mathematics, and $\frac{1}{5}$ of those taking mathematics also take economics. How many freshmen at this school study both mathematics and economics?

81. On a certain day, $\frac{3}{14}$ of a bank's customers made check deposits, $\frac{1}{6}$ made cash deposits, and $\frac{4}{7}$ made withdrawals. If the bank had 546 customers on that day, how many more people made withdrawals than deposits?

82. $\frac{3}{4}$ of the people who work at a particular company drive 10 miles or less to work, and $\frac{1}{6}$ drive more than 10 miles to work. $\frac{4}{9}$ of those who drive to work use carpools. What fraction of the people who work at this company use carpools?

Chapter 3
Explain It in Words

83. Explain how to convert an improper fraction to a mixed number. Use an example.

84. Explain how to convert a mixed number to an improper fraction. Use an example.

85. Explain how to multiply two fractions. Use an example.

86. What is the reciprocal of a number? Give an example.
87. Explain how to reduce a fraction to lowest terms. Use an example.
88. Explain how to find the LCD of two fractions. Use an example.

Chapter **3**
Chapter Test

[3.1, 3.2, 3.7] **Answer each of the following.**

1. Convert $\frac{41}{4}$ to a mixed number.
2. Convert $2\frac{7}{8}$ to an improper fraction.
3. Reduce $\frac{21}{63}$ to lowest terms.
4. Find the reciprocal of $4\frac{3}{4}$.

[3.3] **Add each of the following.**

5. $\frac{1}{3} + \frac{2}{9}$
6. $\frac{2}{5} + \frac{3}{20}$

[3.4] **Subtract each of the following.**

7. $\frac{9}{10} - \frac{3}{5}$
8. $\frac{5}{6} - \frac{3}{4}$

[3.5] **Perform each of the following operations.**

9. $5\frac{3}{4} + 3\frac{1}{3}$
10. $7 - 2\frac{2}{9}$
11. $8\frac{2}{7} - 2\frac{1}{2}$
12. $5\frac{3}{8} + 1\frac{5}{12} - 3\frac{5}{6}$

[3.6] **Multiply each of the following.**

13. $\frac{2}{5} \cdot \frac{4}{9}$
14. $\frac{7}{9} \cdot \frac{3}{5}$
15. $5\frac{2}{3} \cdot 9$
16. $5\frac{2}{5} \cdot 2\frac{1}{3}$

[3.7] **Divide each of the following.**

17. $\frac{3}{10} \div \frac{2}{5}$
18. $\frac{20}{7} \div \frac{5}{9}$
19. $2\frac{3}{5} \div 5$
20. $4\frac{1}{3} \div 2\frac{3}{4}$

[3.6–3.8] **Perform each of the following operations.**

21. $\frac{7}{8} \div \frac{1}{4} \cdot \frac{2}{7}$
22. $\frac{1}{2}\left(\frac{6}{5} - \frac{2}{3}\right)$
23. $\dfrac{\frac{5}{18}}{\frac{4}{9} - \frac{1}{3}}$

[3.5–3.7] **Solve each of the following word problems.**

24. A train can travel 294 miles in $5\frac{1}{4}$ hours. What is its average speed in miles per hour?
25. A jogger ran $1\frac{5}{8}$ miles on Monday and $2\frac{1}{2}$ miles on Wednesday. What is the total number of miles he ran on those two days?

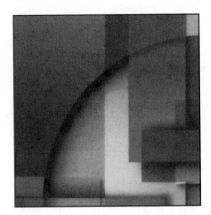

Chapter *4*

Decimals

4.1
Decimals and Place Value

In this chapter, we will increase our knowledge of place value by discussing the values of digits located to the right of the ones place. These are called *decimal places* and are represented by powers of 10 that appear in the denominator.

The decimal places are named from left to right beginning with the tenths place, which is represented by the fraction $\frac{1}{10}$ (one tenth). The value of the next place to the right is obtained by multiplying one-tenth by $\frac{1}{10}$. Thus, to the right of the tenths place is the hundredths place, represented by the fraction $\frac{1}{100}$ (one hundredth). A chart showing the first six decimal places to the right of the ones place, along with the six corresponding place values to the left of the ones place, is given in Table 4.1.

TABLE 4.1
Decimal Places

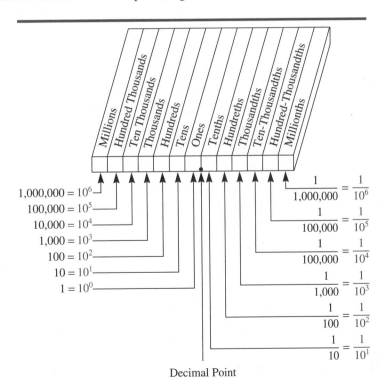

$1,000,000 = 10^6$
$100,000 = 10^5$
$10,000 = 10^4$
$1,000 = 10^3$
$100 = 10^2$
$10 = 10^1$
$1 = 10^0$

$\dfrac{1}{1,000,000} = \dfrac{1}{10^6}$

$\dfrac{1}{100,000} = \dfrac{1}{10^5}$

$\dfrac{1}{100,000} = \dfrac{1}{10^4}$

$\dfrac{1}{1,000} = \dfrac{1}{10^3}$

$\dfrac{1}{100} = \dfrac{1}{10^2}$

$\dfrac{1}{10} = \dfrac{1}{10^1}$

Decimal Point

Notice that the place values located to the left of the ones place all end in *s*: ten*s*, hundred*s*, thousand*s*, and so on. The decimal places, located to the right of the ones place, all end in *ths*: ten*ths*, hundred*ths*, thousand*ths*, and so on. The fractions that represent the decimal places, such as $\frac{1}{10}$, $\frac{1}{100}$, and $\frac{1}{1,000}$, are examples of **decimal fractions**.

▶ **DEFINITION** A *decimal fraction* is a fraction whose denominator can be expressed as a power of 10.

If the numerator of a decimal fraction is a whole number, the name of the fraction will tell us the place value of the rightmost digit in the numerator.

EXAMPLE 1 Write the name of each of the following decimal fractions in words, and give the place value of each digit in the numerator.

(a) $\dfrac{3}{10}$

three tenths
3 is in the tenths place.

(b) $\dfrac{8}{1,000}$

eight thousandths
8 is in the thousandths place.

(c) $\dfrac{16}{100}$

sixteen hundredths
6 is in the hundredths place.
1 is in the tenths place.

One place to the left of the hundredths place is the tenths place.

(d) $\dfrac{4,853}{10,000}$

four thousand, eight hundred fifty-three ten-thousandths
3 is in the ten-thousandths place.
5 is in the thousandths place.
8 is in the hundredths place.
4 is in the tenths place.

(e) $\dfrac{521}{10}$

five hundred twenty-one tenths
1 is in the tenths place.
2 is in the ones place.
5 is in the tens place.

Notice that $\frac{521}{10}$ is an improper fraction that can be rewritten as $52\frac{1}{10}$ (fifty-two and one tenth).

▲

Numbers that can be expressed as decimal fractions are called **decimals**.

▶ **DEFINITION** A *decimal* is a number that can be represented as a fraction whose denominator is a power of 10 greater than or equal to 1.

Writing decimals as decimal fractions can be very awkward, so it is more common to express decimals in *decimal notation*. When a number is written in decimal notation, the digits in the various decimal places are written to the right of the ones place, separated from the ones place by a *decimal point*, a symbol that looks like a period.

As an example, let us look at how the following numbers are written as decimal fractions and in decimal notation.

$$\text{seven tenths} \quad \frac{7}{10} \quad = .7$$

$$\text{seven hundredths} \quad \frac{7}{100} \quad = .07$$

$$\text{seven thousandths} \quad \frac{7}{1,000} \quad = .007$$

$$\text{seven ten-thousandths} \quad \frac{7}{10,000} \quad = .0007$$

The number of zeros in the denominator of a decimal fraction indicates the number of places to the right of the decimal point when it is written in decimal notation. If necessary, zeros are written between the decimal point and the first nonzero digit to act as placeholders. For example,

$$\text{4 zeros} \quad \frac{61}{10,000} = .0061 \quad \text{4 decimal places}$$

This number is read as sixty-one ten-thousandths. The rightmost digit appears in the ten-thousandths place. The two zeros between the decimal point and the number 6 are necessary placeholders.

We can also write .0061 as 0.0061. Even though the first zero is not a necessary placeholder in this case, it is often included to make the number easier to read.

EXAMPLE 2 Write each of the following decimal fractions in decimal notation and translate it into words.

(a) $\dfrac{4}{100}$

$\dfrac{4}{100} = 0.04$

four hundredths

(b) $\dfrac{81}{1,000}$

$\dfrac{81}{1,000} = 0.081$

eighty-one thousandths

(c) $\dfrac{747}{100,000}$

$$\dfrac{747}{100,000} = 0.00747$$

seven hundred forty-seven hundred-thousandths ▲

When a number greater than 1 is written in decimal notation, we read the whole number and decimal portions separately, replacing the decimal point by the word *and*. For example,

548.372

is read as

five hundred forty-eight *and* three hundred seventy-two thousandths

The place value of each digit can be indicated as follows:

In decimal notation, we know that zeros to the right of the decimal point between the decimal point and the first nonzero digit are essential. Other zeros frequently are not. For example, the value of a number is not affected by zeros to the right of the decimal point that appear after the *last* nonzero digit, or by zeros to the left of the decimal point that appear in front of the first nonzero digit. Zeros that appear between the decimal point and the first nonzero digit on *either side* of the decimal point are necessary placeholders. For example, in the number

0400.020

the first and last zeros do not affect the value of the number, but the middle three zeros are necessary placeholders.

0400.020 = 400.02

This number is read as four hundred and two hundredths.

The procedure for translating a decimal into words is outlined below.

To Translate a Decimal into Words

1. Write the number to the left of the decimal point as you would any whole number.

2. Replace the decimal point by the word "and."

3. Write the number to the right of the decimal point as you would a whole number followed by the name of the place value of the last digit. ▲

EXAMPLE 3   Indicate the place value of each nonzero digit in the following decimals, and translate the decimal into words.

(a)   45.8

```
4     5   .   8
↑     ↑       ↖
tens  ones    tenths
```

forty-five and eight tenths

(b)   6.02

```
6   .   0   2
↑           ↑
ones        hundredths
```

six and two hundredths

(c)   89.6305

```
8     9   .   6       3          0   5
↑     ↑       ↑       ↖              ↖
tens  ones    tenths  hundredths     ten-thousandths
```

eighty-nine and six thousand, three hundred five ten-thousandths

(d)   500.007

```
5     0   0   .   0   0   7
↑                         ↑
hundreds                  thousandths
```

five hundred and seven thousandths

(e)   2,002.0202

```
2         0   0   2   .   0   2      0   2
↑                 ↑           ↑          ↑
thousands         ones        hundredths  ten-thousandths
```

two thousand, two and two hundred two ten-thousandths   ▲

A decimal fraction whose numerator is greater than or equal to its denominator has a value that is greater than or equal to 1, and is sometimes called an improper decimal fraction. To write an improper decimal fraction in decimal notation, we determine the place value of the last digit in the numerator and put the decimal point in the appropriate location. For example, in the decimal fraction

$$\frac{527}{10} \qquad \text{five hundred twenty-seven tenths}$$

7 is in the tenths place. Therefore, the decimal point is placed between the 2 and the 7:

$$\frac{527}{10} = 52.7 \qquad \text{fifty-two and seven tenths}$$

To Translate Words into Decimal Notation	**1.** Write the expression to the left of the word "and" as a whole number.
	2. Replace the word "and" by a decimal point. If "and" does not appear in the expression, write the digit 0 followed by a decimal point.
	3. Write the rest of the expression as a whole number without commas. If necessary, insert zeros immediately to the right of the decimal point so that the last digit appears in the place value named at the end of the word expression. ▲

EXAMPLE 4 Write each of the following improper decimal fractions in decimal notation and translate it into words.

(a) $\dfrac{71}{10}$

7.1   1 is in the tenths place.
seven and one tenth

(b) $\dfrac{8,352}{100}$

83.52   2 is in the hundredths place.
eighty-three and fifty-two hundredths

(c) $\dfrac{738,461}{1,000}$

738.461   1 is in the thousandths place.
seven hundred thirty-eight and four hundred sixty-one thousandths   ▲

Now that we are familiar with decimal notation, we can use our knowledge of place value to rewrite decimals in expanded notation. For example, to express the number 49.613 in expanded notation, we write

$$49.613 = (4 \times 10) + (9 \times 1) + \left(6 \times \frac{1}{10}\right) + \left(1 \times \frac{1}{100}\right) + \left(3 \times \frac{1}{1,000}\right)$$

Using powers of ten,

$$49.613 = (4 \times 10^1) + (9 \times 10^0) + \left(6 \times \frac{1}{10^1}\right) + \left(1 \times \frac{1}{10^2}\right) + \left(3 \times \frac{1}{10^3}\right)$$

EXAMPLE 5   Write each of the following numbers in decimal notation.

(a)   thirty-seven hundredths
0.37   The last digit, 7, is in the hundredths place.

(b)   six hundred nine ten-thousandths
0.0609   The last digit, 9, is in the ten-thousandths place.

(c)   eighty-two and fifty-four millionths
82.000054   Replace the word "and" by a decimal point. The last digit, 4, is in the millionths place.   ▲

EXAMPLE 6   Translate each of the following numbers into decimal notation and then rewrite it in expanded notation.

(a)   six and three hundredths

$$6.03 = (6 \times 1) + \left(3 \times \frac{1}{100}\right)$$

$$= (6 \times 10^0) + \left(3 \times \frac{1}{10^2}\right)$$

(b)   twenty-nine and sixty-two thousandths

$$29.062 = (2 \times 10) + (9 \times 1) + \left(6 \times \frac{1}{100}\right) + \left(2 \times \frac{1}{1,000}\right)$$

$$= (2 \times 10^1) + (9 \times 10^0) + \left(6 \times \frac{1}{10^2}\right) + \left(2 \times \frac{1}{10^3}\right)$$

(c)   eight hundred fifteen and seven hundred forty-two thousandths

$$815.742 = (8 \times 100) + (1 \times 10) + (5 \times 1) + \left(7 \times \frac{1}{10}\right)$$

$$+ \left(4 \times \frac{1}{100}\right) + \left(2 \times \frac{1}{1,000}\right)$$

$$= (8 \times 10^2) + (1 \times 10^1) + (5 \times 10^0) + \left(7 \times \frac{1}{10^1}\right)$$

$$+ \left(4 \times \frac{1}{10^2}\right) + \left(2 \times \frac{1}{10^3}\right)$$

(d)   six thousand, two and nine hundred-thousandths

$$6,002.00009 = (6 \times 1,000) + (2 \times 1) + \left(9 \times \frac{1}{100,000}\right)$$

$$= (6 \times 10^3) + (2 \times 10^0) + \left(9 \times \frac{1}{10^5}\right)$$ ▲

One of the most common uses of decimal notation is our monetary system. The smallest unit of money is one cent, and 100 cents = 1 dollar. The dollar is the monetary unit most frequently used, and since 1 cent = $\frac{1}{100}$ dollar, any amount of money can be expressed in dollars as a decimal with two places to the right of the decimal point—the tenths and hundredths place. For example, 450 cents (¢) is equivalent to $\frac{450}{100}$ dollars ($), which, expressed as a decimal, is 4.50 dollars. Using monetary notation this becomes $4.50, which is read as "four dollars and fifty cents." Notice that even though 4.50 = 4.5, we leave the zero in the hundredths place when expressing units of money to make them easier to read.

It is important to be able to write monetary units in words for the purpose of writing checks. All dollar amounts are written out in words on checks, and the cents are written as the numerator of a decimal fraction with a denominator of 100, as shown in Figure 4.1.

Figure 4.1

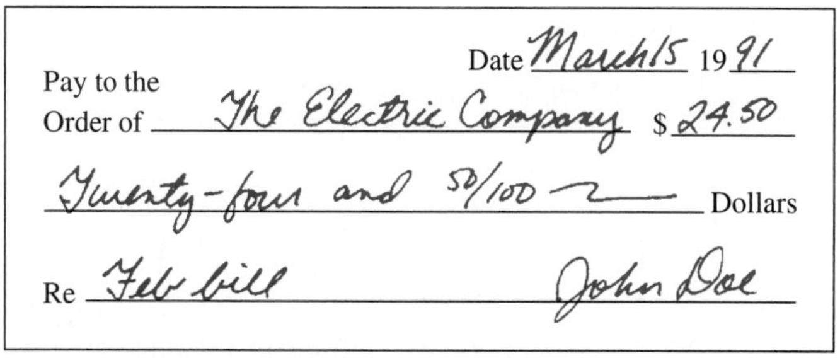

EXAMPLE 7 Write the following amounts of money as decimal fractions and as decimals in monetary notation. Rewrite the amount in words as it would appear on a check.

(a) 33¢

33 cents = $\frac{33}{100}$ dollars = $0.33

Zero and 33/100 dollars

(b) 4,700¢

4,700 cents = $\frac{4,700}{100}$ dollars = $47.00,   If there are zeros in both the tenths place

or $47   and hundredths place, these places are

Forty-seven and 00/100 dollars   sometimes omitted.

(c) 203¢

203 cents = $\frac{203}{100}$ dollars = $2.03

Two and 3/100 dollars

(d) 8 dollars and 630 cents

$8 + \frac{630}{100}$ dollars = $8 + 6 + \frac{30}{100}$ dollars = $14.30

Fourteen and 30/100 dollars

(e) 561,004¢

561,004 cents = $\frac{561,004}{100}$ dollars = $5,610.04

Five thousand, six hundred ten and 4/100 dollars   ▲

Quick Quiz

1. Write 8.27 in words.

2. Write sixty-two and five thousandths in decimal notation.

3. Write 9.031 in expanded notation.

Answers

1. eight and twenty-seven hundredths

2. 62.005

3. (9×10^0)
$+ \left(3 \times \dfrac{1}{10^2}\right)$
$+ \left(1 \times \dfrac{1}{10^3}\right)$

4.1 Exercises

Indicate which of the following statements are true and which are false. For those that are false, change the italicized word to make the statement true.

1. The place value two places to the right of the decimal point is called the *tenths* place.

2. One cent is equivalent to *one-tenth* of a dollar.

3. Decimal fractions have a power of ten in the *numerator*.

4. Zeros that appear between the decimal point and the first nonzero digit are *necessary*.

Write each of the following decimal fractions in decimal notation.

5. $\dfrac{3}{10}$

6. $\dfrac{5}{10}$

7. $\dfrac{8}{100}$

8. $\dfrac{32}{100}$

9. $\dfrac{5}{1,000}$

10. $\dfrac{19}{1,000}$

11. $\dfrac{529}{1,000}$

12. $\dfrac{8}{10,000}$

13. $\dfrac{21}{10,000}$

14. $\dfrac{621}{10,000}$

15. $\dfrac{3}{100,000}$

16. $\dfrac{48}{100,000}$

17. $\dfrac{127}{100,000}$

18. $\dfrac{69}{1,000,000}$

19. $\dfrac{6,589}{1,000,000}$

20. $\dfrac{973,642}{1,000,000}$

21. $\dfrac{73}{10}$

22. $\dfrac{804}{10}$

23. $\dfrac{519}{100}$

24. $\dfrac{1,362}{100}$

25. $\dfrac{2,837}{1,000}$

26. $\dfrac{50,505}{1,000}$

27. $\dfrac{70,803}{10,000}$

28. $\dfrac{6,328,174}{10,000}$

For each of the following decimals, give the place value of the digit in parentheses, and rewrite the number in words.

29. 0.3 (3)

30. 0.04 (4)

31. 0.53 (3)

32. 0.835 (5)

33. 5.9 (9)

34. 7.06 (0)

35. 47.8 (8)

36. 60.02 (2)

37. 38.974 (4)

38. 82.7091 (9)

39. 800.65327 (7)

40. 302.305301 (1)

41. 3,603.46 (4)

42. 4,000.0073 (7)

43. 500.4689 (9)

44. 302.00005 (5)   **45.** 8,067.0004 (4)   **46.** 80,305.0702 (2)

47. 362,006.200008 (8)   **48.** 6,002,003.008007 (7)

Write each of the following numbers in decimal notation.

49. four tenths

50. five hundredths

51. six thousandths

52. eighty-seven thousandths

53. four ten-thousandths

54. seventy-eight ten-thousandths

55. three hundred eighty-six ten-thousandths

56. fifty-three hundred-thousandths

57. six millionths

58. five hundred six millionths

59. two thousand, fifty-nine millionths

60. sixty-two thousand, four hundred thirty-eight millionths

61. two and six tenths

62. seven and fifty-four hundredths

63. three hundred eight and twelve hundredths

64. six thousand and six thousandths

65. eleven thousand and fifty-eight hundred-thousandths

66. five million, six hundred and forty-three millionths

Write each of the following decimals in expanded notation.

67. 5.2   **68.** 8.06   **69.** 36.46   **70.** 89.364

71. 230.005   **72.** 5,627.364   **73.** 30,005.6278   **74.** 200,002.00002

75. 50,083,070.0406   **76.** 4,070,801.050602

Write the following amounts of money as decimals in monetary notation, and rewrite them in words as they would appear on a check.

77. 16¢   **78.** 89¢   **79.** 374¢

80. 620¢   **81.** 4,648¢   **82.** 7,923¢

83. 80,567¢   **84.** 368,902¢   **85.** 5 dollars and 214 cents

86. 27 dollars and 480 cents

Answer each of the following questions.

87. Jupiter is 5.203 astronomical units away from the sun. What is the place value of the digit 3 in this number?

88. The cost of a new refrigerator is $659.00. How would this amount be written out on a check?

89. The moon makes one revolution around the earth in twenty-seven and thirty-two hundredths days. How is this number written in decimal notation?

90. The average annual rainfall in Providence, Rhode Island, is 42.75 inches. How is this number written out in words?

4.2
Rounding Off

It is often very awkward to work with numbers that have a large number of digits. For example, if the computerized scale at a supermarket indicates that a bag of tomatoes weighs 1.51827 pounds, the grocer will probably tell you that it weighs 1.5 pounds. This approximation was calculated by **rounding off** the number 1.51827 to the nearest tenth to obtain 1.5.

▶ **DEFINITION** The procedure used to estimate a number to a given degree of precision is called *rounding off*.

When rounding off a number, we state the place value of the last digit of our result to indicate the precision of the approximation. For example, the number 1.51827 can be rounded off in the following ways.

To the nearest tenth	1.5
To the nearest hundredth	1.52
To the nearest thousandth	1.518
To the nearest ten-thousandth	1.5183

Notice that in rounding off the number 1.51827 to the nearest hundredth, 1.52 is a better approximation than 1.51, so we added 1 to the digit in the hundredths place. This process is called *rounding up*. When rounding off 1.51827 to the nearest thousandth, 5.18 is a better approximation than 5.19, so the digit in the thousandths place stayed the same. This process is called *rounding down*.

To Round Off a Number

1. Examine the digit immediately to the right of the place value to which you are rounding off.

2. If the digit to the right of the place to be rounded is 5 or greater, round up. That is, add 1 to the digit in the place value to which you are rounding.

3. If the digit to the right of the place to be rounded is less than 5, round down. That is, the digit in the place value to which you are rounding stays the same. ▲

EXAMPLE 1 Round off the following numbers to the nearest hundredth.

(a) 3.6248

2 is in the hundredths place. 4 is on its right.

$4 < 5$ Round down: 2 stays the same.

$3.6248 \cong 3.62$ The symbol $\cong$ means "is approximately equal to."

(b) 53.475

7 is in the hundredths place. 5 is on its right.

$5 = 5$ Round up: $7 + 1 = 8$.

$53.475 \cong 53.48$

(c) 0.297

9 is in the hundredths place. 7 is on its right.

$7 > 5$ Round up: $9 + 1 = 10$. Write 0 in the hundredths place and carry 1 to the tenths place to obtain .30.

$0.297 \cong 0.30$ Notice that we leave the zero in the hundredths place to indicate precision to the nearest hundredth. ▲

When rounding off numbers to places left of the ones place, it is necessary to replace the digits to the right of the place to which we are rounding off with zeros. The zeros act as necessary placeholders. This is shown in Example 2.

EXAMPLE 2 Round off each of the following numbers as indicated.

    (a) 2,583 to the nearest hundred

2,583	5 is in the hundreds place. 8 is on its right.
2,583 ≅ 2,600	Since $8 > 5$, round up: $5 + 1 = 6$. Replace the digits to the right of the hundreds place with zeros.

    (b) 53,072 to the nearest ten

53,072	7 is in the tens place. 2 is on its right.
	Since $2 < 5$, round down: 7 stays the same.
53,072 ≅ 53,070	Replace the digit to the right of 7 with a zero.

    (c) 299,534 to the nearest thousand

299,534	9 is in the thousands place. 5 is on its right.
	Since $5 = 5$, round up: $9 + 1 = 10$. Write 0 in the thousands place and carry 1 to the ten thousands place to obtain 10. Write 0 in the ten thousands place and carry 1 to the hundred thousands place to
299,534 ≅ 300,000	obtain 300,000. ▲

EXAMPLE 3 Round off the number 7,439,855 as indicated.

    (a) To the nearest thousand.

7,439.855	
7,000	$4 < 5$, so round down: 7 stays the same. Notice that the zeros are necessary placeholders.

    (b) To the nearest hundred.

7,439.855	
7,400	$3 < 5$, so round down: 4 stays the same.

    (c) To the nearest ten.

7,439.855	
7,440	$9 > 5$, so round up: $3 + 1 = 4$.

    (d) To the nearest whole number.

7,439.855	
7,440	$8 > 5$, so round up: $9 + 1 = 10$. Write 0 in the ones place and carry 1 to the tens place to obtain 40.

    (e) To the nearest tenth.

7,439.855	
7,439.9	$5 = 5$, so round up: $8 + 1 = 9$.

(f) To the nearest hundredth.
7,439.8$\underline{55}$
7,439.86   5 = 5, so round up: 5 + 1 = 6.   ▲

EXAMPLE 4   Round off each of the following numbers as indicated.

(a)   16.78954   nearest thousandth

16.78$\underline{9}$54   9 is in the thousandths place.

16.78954 ≅ 16.790   5 = 5, so round up. Add 1 to the
digit in the thousandths place:
16.789 + .001 = 16.790
Zero is a necessary placeholder.

(b)   350.997   nearest hundredth

350.9$\underline{9}$7   9 is in the hundredths place.

350.997 ≅ 351.00   7 > 5, so round up. Add 1 to the
digit in the hundredths place:
350.99 + .01 = 351.00
The zeros are necessary placeholders.

(c)   0.659951   nearest ten-thousandth

0.659$\underline{9}$51   9 is in the ten-thousandths place.

0.659951 ≅ 0.6600   5 = 5, so round up. Add 1 to the
digit in the ten-thousandths place:
0.6599 + .0001 = 0.6600
The zeros are necessary placeholders.   ▲

You may be asked to round off a number expressed in dollars to a degree of precision
that is expressed in cents. It is helpful, therefore, to know the place values of the locations
to the right of the decimal point in terms of both dollars and cents. As an example, let us
determine the place value of each digit in the number $6.52814. Remember that since
1 dollar = 100 cents, $6.52814 = 652.814¢. Table 4.2 gives the place value of each digit
in terms of both dollars and cents.

TABLE 4.2
Place Values in
Dollars and Cents

| | Place Value $6.52814 = 652.814¢ | |
Digit	Dollars	Cents
6	ones	hundreds
5	tenths	tens
2	hundredths	ones
8	thousandths	tenths
1	ten-thousandths	hundredths
4	hundred-thousandths	thousandths

EXAMPLE 5   Round off $20.59748 to the places indicated.

(a)   To the nearest dollar.

2\underline{0}$.59748

$21   5 = 5, so round up.

(b) To the nearest tenth of a dollar.
$20.59748
$20.6   9 > 5, so round up.

(c) To the nearest cent.
$20.59748
$20.60   7 > 5, so round up.

(d) To the nearest tenth of a cent.
$20.59748
$20.597   4 < 5, so round down.

(e) To the nearest hundredth of a cent.
$20.59748
$20.5975   8 > 5, so round up.   ▲

Quick Quiz

Round off each of the following numbers as indicated.	**Answers**
1. 4.857 (nearest tenth)	**1.** 4.9
2. 267.416 (nearest hundred)	**2.** 300
3. $85.95 (nearest dollar)	**3.** $86

4.2 Exercises

Indicate which of the following statements are true and which are false. For those that are false, change the italicized expression to make the statement true.

1. When rounding off, we examine the digit immediately to the *right* of the place value to which we are rounding off.

2. When the digit examined is *5 or less,* we round down.

3. Rounding off to the nearest *hundredth* of a dollar is the same as rounding off to the nearest tenth of a cent.

4. To round up, the digit in the place value to which we are rounding is *increased by one.*

Round off each of the following decimals to the nearest tenth, the nearest hundredth, and the nearest thousandth.

5. 3.2785	**6.** 0.1063	**7.** 13.9628	**8.** 82.3974	**9.** 7.0058	**10.** 3.2109
11. 0.5020	**12.** 8.0305	**13.** 62.8299	**14.** 27.3906	**15.** 38.0705	**16.** 52.5060
17. 4.7979	**18.** 5.9898	**19.** 67.9999	**20.** 11.9999		

Round off each of the following decimals to the nearest hundred, the nearest ten, and the nearest whole number.

21. 6,846.7	**22.** 9,320.3	**23.** 7,045.8	**24.** 340.69	**25.** 709.38	**26.** 281.97
27. 695.7	**28.** 219.5	**29.** 7,080.4	**30.** 3,020.7	**31.** 499.6	**32.** 529.9
33. 985.9	**34.** 976.5	**35.** 60,709.92	**36.** 34,050.99		

Round off each of the following numbers as indicated.

37. 325,873   (nearest ten thousand)

38. 738,359   (nearest hundred thousand)

39. 6.05038   (nearest ten-thousandth)

40. 8.29731   (nearest hundredth)

41. 2.5874095   (nearest millionth)

42. 95,601,382   (nearest million)

43. 69.9583   (nearest tenth)

44. 499.71   (nearest whole number)

45. 9,999.99   (nearest tenth)

46. 909.999   (nearest hundredth)

Round off each of the following dollar amounts to the values indicated.

47. $367.69   (nearest dollar)

48. $204.27   (nearest dollar)

49. $32.845   (nearest cent)

50. $65.602   (nearest cent)

51. $88.59   (nearest tenth of a dollar)

52. $12.47   (nearest tenth of a dollar)

53. $5.0305   (nearest tenth of a cent)

54. $8.4297   (nearest tenth of a cent)

55. $6.83402   (nearest hundredth of a cent)

56. $3.76995   (nearest hundredth of a cent)

Answer each of the following questions.

57. The Panama Canal is 50.7 miles long. What is its length rounded off to the nearest mile?

58. Saturn is 9.539 astronomical units away from the sun. What is this distance rounded off to the nearest hundredth of an astronomical unit?

59. A runner completed a race in 53.684 seconds. What was the runner's time rounded off to the nearest tenth of a second?

60. Hudson Bay covers 475,800 square miles. What is its area rounded off to the nearest thousand square miles?

61. A color television is on sale for $499.99. What is its price rounded off to the nearest dollar?

62. A gallon of gasoline sells for $1.599. What is the price per gallon rounded off to the nearest cent?

4.3
Addition and Subtraction of Decimals

We will now show how to add and subtract decimals. In order to demonstrate why the procedure for adding decimals works, we will add two decimals by using what we know about adding fractions. As an example, let us consider how to add $3.6 + 9.5$ by changing each decimal to a mixed number as follows:

$$3.6 = 3\frac{6}{10}$$

$$+9.5 = 9\frac{5}{10}$$

$$12\frac{11}{10} \qquad \text{Add the whole number and fractional parts separately.}$$

$$= 12 + 1 + \frac{1}{10} \qquad \text{Convert: } \frac{11}{10} + 1\frac{1}{10}$$

$$= 13\frac{1}{10} \qquad \text{Add the whole numbers.}$$

$$= 13.1 \qquad \text{Convert to decimal notation.}$$

Notice that this same result can be obtained by simply adding the digits with the same place value, just as we do with whole numbers, moving from right to left and carrying when necessary. Using this procedure for the previous example we obtain

$$
\begin{array}{r}
1 \\
3.6 \leftarrow \text{addend} \\
+\ 9.5 \leftarrow \text{addend} \\
\hline
13.1 \leftarrow \text{sum}
\end{array}
$$

Thus, we have illustrated the addition procedure outlined here.

To Add Decimals

1. Write the numbers to be added vertically and line up the decimal points.

2. Add all digits with the same place value, beginning with the rightmost column. Carry to the next column when the sum of the digits in any column is greater than 9.

3. Be sure to place a decimal point in the sum in the correct location. Check that all its digits are in line with the digits in the addends with the same place value. ▲

Example 1 provides some additional illustrations of adding decimals.

EXAMPLE 1 Add each of the following.

(a) 18.36 + 40.962

$$
\begin{array}{r}
1\ 1\ \ \ \\
18.36 \\
+40.962 \\
\hline
59.322
\end{array}
$$

(b) 83.76 + 0.0673 + 9.73

$$
\begin{array}{r}
11\ 1\ \ \ \\
83.7600 \\
0.0673 \\
+\ 9.7300 \\
\hline
93.5573
\end{array}
$$

Sometimes we add zeros to the right of the digits following the decimal point to help us line up the addends correctly.

(c) 3,067 + 0.3678 + 402.071 + 2.14095

$$
\begin{array}{r}
1\ \ 1\ 1\ \ \ \ \ \\
3,067. \\
0.3678 \\
402.0710 \\
+\ \ \ \ 2.14095 \\
\hline
3,471.57975
\end{array}
$$

Notice that even though 3,067 can be written without a decimal point, the decimal point is located to the right of the ones place: 3,067 = 3,067.

(d) $4.03 + $0.67 + $995.84

$$
\begin{array}{r}
\overset{111\ 1}{} \\
\$4.03 \\
0.67 \\
+995.84 \\
\hline
\$1,000.54
\end{array}
$$

When adding monetary amounts, the dollar sign usually appears in the first addend and in the sum.

▲

We will now demonstrate why the procedure for subtracting decimals works, by using what we know about subtracting fractions. As an example, let us consider how to subtract $8.2 - 3.8$ by converting each decimal to a mixed number as follows:

$$8.2 = 8\frac{2}{10} = 7\frac{12}{10}$$

$$-3.8 = 3\frac{8}{10} = 3\frac{8}{10}$$

$$\rule{3cm}{0.4pt}$$

$$4\frac{4}{10} = 4.4$$

Notice that the same result can be obtained by simply subtracting digits with the same place value, moving from right to left and borrowing when necessary. Using this procedure for the previous example we obtain the following.

$$
\begin{array}{r}
7 \\
8\!.2 \leftarrow \text{minuend} \\
-3.8 \leftarrow \text{subtrahend} \\
\hline
4.4 \leftarrow \text{difference}
\end{array}
$$

Thus, we have established the subtraction procedure outlined below.

To Subtract Decimals

1. Write the numbers to be subtracted vertically and line up the decimal points.

2. If the number of decimal places in the subtrahend exceeds that in the minuend, write the necessary number of zeros to the right of the last decimal place in the minuend.

3. Subtract all digits with the same place value, beginning in the rightmost column and borrowing when necessary.

4. Insert a decimal point in the correct location in the difference.

▲

Now we will look at some additional examples of subtracting decimals.

EXAMPLE 2 Subtract each of the following.

(a) $27.4 - 12.9$

$$
\begin{array}{r}
{}^{6} \\
2\ 7\overset{|}{.}4 \\
-1\ 2.9 \\
\hline
1\ 4.5
\end{array}
$$

(b) $102.3 - 8.62$

$$
\begin{array}{r}
9\,1\,1\,1\,2 \\
1\,{}^{|}0\,{}^{|}2\overset{|}{.}3\,{}^{|}0 \\
-\qquad 8.6\ 2 \\
\hline
9\ 3.6\ 8
\end{array}
$$

Add a zero to the right of the last decimal place in the minuend.

(c) $462 - 26.528$

$$
\begin{array}{r}
5\,1\,1\ \ 9\ \ 9 \\
4\ 6\,{}^{|}2.\,{}^{|}0\,{}^{|}0\,{}^{|}0 \\
-\quad 2\ 6.\ 5\ 2\ 8 \\
\hline
4\ 3\ 5.\ 4\ 7\ 2
\end{array}
$$

When adding zeros to the right of a whole number, be certain to insert a decimal point to the right of the ones place.

(d) $\$5,200 - \$92.95$

$$
\begin{array}{r}
1\ \ 9\ \ 9\ \ 9 \\
\$5,2\,{}^{|}0\,{}^{|}0.\,{}^{|}0\,{}^{|}0 \\
-\qquad 9\ 2.\ 9\ 5 \\
\hline
\$5,1\ 0\ 7.\ 0\ 5
\end{array}
$$

▲

We will now use the estimation process introduced in Chapter 1 to approximate the answers to problems involving addition and subtraction of decimals.

EXAMPLE 3 Estimate each of the following sums and differences. Then determine the exact answers.

(a) $58.93 + 129.67 + 70.14$

$$
\begin{array}{r}
58.93 \cong 60 \\
129.67 \cong 130 \\
70.14 \cong 70 \\
\hline
260
\end{array}
$$

Approximate each addend by rounding off to the nearest ten.

Our estimate is 260.

Now determine the exact answer:

$$
\begin{array}{r}
58.93 \\
129.67 \\
70.14 \\
\hline
258.74
\end{array}
$$

Since 258.74 is close to our estimate of 260, this answer is reasonable.

(b) $3{,}797.3 - 412.5$

$$
\begin{array}{r}
3{,}797.3 \cong 3{,}800 \\
412.5 \cong 400 \\
\hline
3{,}400
\end{array}
$$

Approximate the minuend and subtrahend by rounding them off to the nearest hundred.
Our estimate is 3,400.

Now determine the exact answer:

$$
\begin{array}{r}
3{,}797.3 \\
412.5 \\
\hline
3{,}384.8
\end{array}
$$

Since 3,384.8 is close to our estimate of 3,400, this answer is reasonable.   ▲

EXAMPLE 4   At the beginning of the month, an architect had a balance of $482.76 in his checking account. During the month, he wrote checks for $53.62, $137.00, $86.99, $398.12, and $78.26. He also made deposits of $55.89, $125.00, and $738.64. What was his balance at the end of the month?

Solution

$$
\begin{array}{r}
\overset{222\ 1}{} \\
\$482.76 \\
55.89 \\
125.00 \\
738.64 \\
\hline
\$1{,}402.29
\end{array}
$$

Add the deposits to the original balance.

$$
\begin{array}{r}
\overset{331\ 1}{} \\
\$53.62 \\
137.00 \\
86.99 \\
398.12 \\
78.26 \\
\hline
\$753.99
\end{array}
$$

Add the amounts of the checks written.

$$
\begin{array}{r}
\overset{13\ \ 9\ \ 11}{} \\
\$\ 1{,}4\,0\,2.2\,9 \\
-\ \ \ \ 7\,5\,3.9\,9 \\
\hline
\$\ \ \ 6\,4\,8.3\,0
\end{array}
$$

Then subtract the total of the check amounts.

His balance at the end of the month was $648.30.   ▲

Quick Quiz

Perform each of the operations indicated.	Answers
1. $6.389 + 29.47$	1. 35.859
2. $61.43 - 47.85$	2. 13.58
3. $87.13 + 5.62 - 13.7$	3. 79.05

4.3 Exercises

Indicate which of the following statements are true and which are false. For those that are false, change the italicized expression to make the statement true.

1. To add or subtract decimals, write the problem *horizontally* and line up the decimal points.

2. When adding and subtracting decimals, we follow *the same* rules for borrowing and carrying as apply to adding and subtracting whole numbers.

3. When adding decimals, we add the digits with the same place value, moving from *left to right*.

4. When subtracting decimals, the difference is *always* greater than the subtrahend.

Add each of the following decimals.

5. $5.3 + 6.8$
6. $4.6 + 9.7$
7. $14.28 + 32.67$
8. $27.84 + 16.29$
9. $3.864 + 12.7$
10. $35.8 + 2.169$
11. $\$45.83 + \$0.92$
12. $\$0.57 + \$86.75$
13. $0.6783 + 153.6$
14. $289.9 + 0.0481$
15. $5.3 + 20.82 + 0.067$
16. $93 + 4.863 + 0.39$
17. $1.863 + 17 + 1289.6 + 0.4519$
18. $0.6793 + 4382.7 + 12.7 + 2.1869$
19. $\$362.94 + \$487.02 + \$13.62$
20. $\$576.03 + \$0.79 + \$93.67$
21. $37.289 + 0.86897 + 2.9935$
22. $0.05916 + 8.9472 + 67.00847$

Subtract each of the following decimals.

23. $7.4 - 2.8$
24. $9.2 - 1.7$
25. $\$13.07 - \$4.95$
26. $\$81.31 - \$3.86$
27. $53.1 - 7.432$
28. $71.3 - 0.8516$
29. $\$163.84 - \$4.78$
30. $\$781.36 - \$5.07$
31. $6.073 - 0.6354$
32. $4.602 - 0.8307$
33. $3 - 0.2774$
34. $7 - 0.5382$
35. $420 - 9.934$
36. $809 - 5.6214$
37. $701.3 - 69.4795$
38. $2,000 - 0.728$

Perform each of the operations indicated.

39. $8.9 - 6.2 + 3.7$
40. $4.1 + 5.2 - 7.6$
41. $18.47 + 406.049$
42. $502.06 - 289.74 + 21.48$
43. $\$0.37 + \$82.03 - \$5.95$
44. $\$560 - \$92.37 + \$25.62$
45. $8,000 - 2.6384 + 647.29$
46. $6,002.999 + 211.001 - 152.78$
47. $\$3,000 - \$15.95 + \$167.82$
48. $\$55,000 - \$7,999.99 + \$72.88$
49. $28.013 + 7.5861 - 13.426785$
50. $87.41 - 39.217039 + 4.09582$

Estimate each of the following sums and differences. Then determine the exact answers.

51. $48.41 + 121.73$
52. $61.07 + 299.34$
53. $60.2 + 259.3 + 29.7$
54. $28.7 + 819.6 + 41.2$
55. $840.7 - 19.8$
56. $469.1 - 30.4$
57. $5,693.9 - 301.5$
58. $8,802.3 - 598.7$

Solve each of the following word problems.

59. An artist bought a brush for $5.98, paints for $18.24, and paper for $12.57. If she gave the cashier $40, how much change would she receive?

60. A cab driver bought some items at a health food store which cost $3.92, $0.89, $1.29, and $2.79. How much change would he receive from a $10 bill?

61. A jogger ran the following distances in a single week: 3.25 miles, 2.8 miles, 3.5 miles, 2.75 miles, 3.1 miles, 2.9 miles, and 2.5 miles. What was the total distance she ran that week?

62. If a $498 stereo receiver is marked down $29.99, what is its sale price?

63. A man bought items at a grocery store that cost $1.89, $0.43, $3.89, $0.65, $0.39, $1.29, $5.62, $0.59, and $1.19. He had coupons for some of these items worth $0.15, $0.40, and $0.25. What was his grocery bill?

64. If it takes the earth 365.256 days to revolve about the sun, and it takes Mercury 87.969 days to revolve about the sun, how much longer does it take the earth than Mercury to revolve about the sun?

65. If a student bought a notebook for $1.57, a pen for $0.49, typing paper for $1.89, and a chocolate bar for $0.35, how much change would he receive from a $10 bill?

66. If the St. Lawrence Seaway is 2,400 miles long, and the Panama Canal is 50.7 miles long, how much longer is the St. Lawrence Seaway than the Panama Canal?

67. If a secretary's monthly take-home pay is $889.45, and he spends $465 for rent, $15.86 for electricity, $38.17 for gas, $52.43 for phone, and $132 for car payments, how much does he have left for other expenses?

68. A student's checking account showed a balance of $255.39 at the beginning of the month. During the month she wrote checks for $15.23, $128.75, $62.45, $89.36, and $12.71, and she made a deposit of $193.67. What was her balance at the end of the month?

4.4
Multiplication of Decimals

In order to establish a procedure for multiplying decimals, we can convert each decimal to an equivalent fraction, and then apply the rules for multiplying fractions. As an example, let us consider the following products.

$$0.2 \times 3 = \frac{2}{10} \times \frac{3}{1} = \frac{6}{10} = 0.6$$

$$0.02 \times 3 = \frac{2}{100} \times \frac{3}{1} = \frac{6}{100} = 0.06$$

$$0.2 \times 0.3 = \frac{2}{10} \times \frac{3}{10} = \frac{6}{100} = 0.06$$

$$0.002 \times 3 = \frac{2}{1,000} \times \frac{3}{1} = \frac{6}{1,000} = 0.006$$

$$0.02 \times 0.3 = \frac{2}{100} \times \frac{3}{10} = \frac{6}{1,000} = 0.006$$

$$0.002 \times 0.3 = \frac{2}{1,000} \times \frac{3}{10} = \frac{6}{10,000} = 0.0006$$

$$0.02 \times 0.03 = \frac{2}{100} \times \frac{3}{100} = \frac{6}{10,000} = 0.0006$$

Notice that each product can be determined by multiplying 2×3 and inserting the appropriate number of zeros between the decimal point and the number 6. In each example, the number of decimal places in the product is equal to the sum of the number of decimal places in all numbers being multiplied. We have thus illustrated a procedure for multiplying decimals, as outlined below.

To Multiply Decimals

1. Multiply the numbers as usual.
2. Put a decimal point in the answer so that the number of decimal places in the answer equals the **sum** of the number of decimal places in all the numbers multiplied together.   ▲

Let us look at some additional examples of multiplying decimals.

EXAMPLE 1 Multiply each of the following decimals.

(a) 0.36×0.81

$$
\begin{array}{r}
\overset{4}{} \\
0.3\,6 \\
\times\quad 0.8\,1 \\
\hline
3\,6 \\
2\,8\,8 \\
\hline
0.2\,9\,1\,6
\end{array}
$$

Two decimal places.
Two decimal places.

The number of decimal places in the product is equal to $2 + 2 = 4$.

(b) 0.052×8.6

$$
\begin{array}{r}
\overset{1}{} \\
1 \\
0.0\,5\,2 \\
\times\quad 8.6 \\
\hline
3\,1\,2 \\
4\,1\,6 \\
\hline
0.4\,4\,7\,2
\end{array}
$$

Three decimal places.
One decimal place.

The number of decimal places in the product is equal to $3 + 1 = 4$.

(c) 43.7×0.00035

$$
\begin{array}{r}
1\,2 \\
1\,3 \\
4\,3.7 \\
\times\quad 0.0\,0\,0\,3\,5 \\
\hline
2\,1\,8\,5 \\
1\,3\,1\,1 \\
\hline
0.0\,1\,5\,2\,9\,5
\end{array}
$$

The number of decimal places in the product is equal to $1 + 5 = 6$.

(d) $\$59.95 \times 60$

$$
\begin{array}{r}
5\,5\,3 \\
\$\quad 5\,9.9\,5 \\
\times\quad\quad 6\,0 \\
\hline
\$3,5\,9\,7.0\,0
\end{array}
$$

The number of decimal places in the product is equal to $2 + 0 = 2$.   ▲

We will now investigate how to multiply decimals by powers of ten. For example,

$$2.5 \times 10^1 = 2.5 \times 10 = 25$$

$$2.5 \times 10^2 = 2.5 \times 100 = 250$$

$$2.5 \times 10^3 = 2.5 \times 1{,}000 = 2{,}500$$

Notice that in each case, the product can be obtained by moving the decimal point to the right the number of places that is indicated by the power of ten (see box).

To *multiply* a decimal by 10^n, where n is a whole number, move the decimal point n places to the *right*.　　　　　▲

EXAMPLE 2   Find each of the following products.

(a) 0.3×10^4

$0.3 \times 10^4 = 0.3000 = 3{,}000$ Move the decimal point four places to the right.

(b) 82.73×10^3

$82.73 \times 10^3 = 82.730 = 82{,}730$ Move the decimal point three places to the right.

(c) 6.41896×10^5

$6.41896 \times 10^5 = 6.41896 = 641{,}896$ Move the decimal point five places to the right.

(d) 25.3×10^9

$25.3 \times 10^9 = 25.300000000$ Move the decimal point nine places to the right.

$\qquad\qquad = 25{,}300{,}000{,}000$ ▲

Now let us look at some examples of multiplying decimals by whole numbers that are multiples of powers of ten.

EXAMPLE 3   Multiply each of the following numbers.

(a) 1.5×300

$1.5 \times 300 = 1.5 \times 3 \times 100$	Rewrite 300 in expanded notation.
$= (1.5 \times 3) \times 10^2$	Express 100 as a power of 10.
$= 4.5 \times 10^2$	Multiply using associative property.
$= 450$	Multiply result by power of 10.

(b) $7.2 \times 20{,}000$

$7.2 \times 20{,}000 = 7.2 \times 2 \times 10{,}000$	Rewrite 20,000 in expanded notation.
$= (7.2 \times 2) \times 10^4$	Express 10,000 as a power of 10.
$= 14.4 \times 10^4$	Multiply using associative property.
$= 144{,}000$	Multiply result by power of 10.

(c) $0.012 \times 80{,}000$

$$0.012 \times 80{,}000 = \frac{12}{1{,}000} \times 8 \times 10{,}000 \qquad \text{Rewrite factors.}$$

$$= (12 \times 8) \times \left(\frac{1}{1{,}000} \times 10{,}000\right) \quad \text{By the commutative and associative properties.}$$

$$= 96 \times \frac{10{,}000}{1{,}000} \qquad \text{Multiply.}$$

$$= 96 \times 10 \qquad \text{Reduce to lowest terms.}$$

$$= 960 \qquad \text{Multiply.}$$

(d) $150{,}000{,}000 \times 0.0003$

$$150{,}000{,}000 \times 0.0003 = 15 \times 10{,}000{,}000 \times \frac{3}{10{,}000} \qquad \text{Rewrite factors.}$$

$$= (15 \times 3) \times \frac{10{,}000{,}000}{10{,}000} \qquad \text{By the commutative and associative properties.}$$

$$= 45 \times 1{,}000 \qquad \text{Multiply and reduce.}$$

$$= 45{,}000 \qquad \text{Multiply.} \qquad \blacktriangle$$

Decimals are used often as part of a shorthand notation for very large or very small numbers.

EXAMPLE 4 Rewrite the following numbers.

(a) 3.2 million

$$3.2 \text{ million} = 3.2 \times 1{,}000{,}000$$
$$= 3.2 \times 10^6$$
$$= 3{,}200{,}000$$

(b) 1.8 thousandths

$$1.8 \text{ thousandths} = 1.8 \times \frac{1}{1{,}000}$$
$$= \frac{1.8}{1{,}000}$$
$$= \frac{1.8}{1{,}000} \times \frac{10}{10}$$
$$= \frac{18}{10{,}000}$$
$$= 0.0018$$

(c) 7.4 billion

$$7.4 \text{ billion} = 7.4 \times 1{,}000{,}000{,}000$$
$$= 7{,}400{,}000{,}000$$

(d) 4.61 billion

$$4.61 \text{ billion} = 4.61 \times 1{,}000{,}000{,}000$$
$$= 4{,}610{,}000{,}000$$

(e) 2.7 hundredths

$$2.7 \text{ hundredths} = 2.7 \times \frac{1}{100}$$

$$= \frac{2.7}{100} \times \frac{10}{10}$$

$$= \frac{27}{1,000}$$

$$= 0.027$$ ▲

When working with decimals, we sometimes encounter arithmetic expressions that contain more than one operation. To simplify these expressions, we follow the same order of operations rules that apply when working with whole numbers.

EXAMPLE 5 Simplify each of the following expressions.

(a) $3(1.8 + 2.4)$

$$3(1.8 + 2.4) = 3(4.2)$$
$$= 12.6$$

Perform the operation in ( ) first. Then multiply the sum by 3.

$$\begin{array}{r} 1.8 \\ +2.4 \\ \hline 4.2 \end{array} \qquad \begin{array}{r} 4.2 \\ \times \quad 3 \\ \hline 12.6 \end{array}$$

(b) $(0.05)(1.2)^2$

$$(0.05)(1.2)^2 = (0.05)(1.44)$$
$$= 0.0720$$

Square 1.2. Then multiply result by 0.05.

$$\begin{array}{r} 1.2 \\ \times \quad 1.2 \\ \hline 2\,4 \\ 1\,2 \\ \hline 1.4\,4 \end{array} \qquad \begin{array}{r} 1.4\,4 \\ \times \quad 0.0\,5 \\ \hline 0.0\,7\,2\,0 \end{array}$$

(c) $[5.8 - 4(0.7)]^3$

$$[5.8 - 4(0.7)]^3 = [5.8 - 2.8]^3$$
$$= (3.0)^3$$
$$= 27$$

In [ ], do multiplication before subtraction. Then cube result. ▲

EXAMPLE 6 The planet Mercury is 0.387 astronomical units away from the sun. If one astronomical unit is equivalent to 93 million miles, how far away in miles is Mercury from the sun?

Solution $$0.387 \times 93 \text{ million} = 0.387 \times 93 \times 1,000,000$$
$$= (0.387 \times 93) \times 10^6$$
$$= 35.991 \times 10^6$$
$$= 35,991,000 \text{ miles}$$

Multiply.

$$\begin{array}{r} 0.3\,8\,7 \\ \times \quad 9\,3 \\ \hline 1\,1\,6\,1 \\ 3\,4\,8\,3 \\ \hline 3\,5.9\,9\,1 \end{array}$$ ▲

Quick Quiz

Multiply each of the following decimals.

1. 0.8×0.09
2. 7.3×0.21
3. 1.48×60.7

Answers

1. 0.072
2. 1.533
3. 89.836

4.4 Exercises

Indicate which of the following statements are true and which are false. For those that are false, change the italicized expression to make the statement true.

1. When multiplying decimals, the number of decimal places in the product is equal to the *sum* of the number of decimal places in all factors.

2. To multiply a decimal by 10^n, where n is a whole number, move the decimal point n places to the *left*.

3. The number 6.9 million is equal to the *sum* of 6.9 and 1,000,000.

4. If two decimals, each greater than zero and less than one, are multiplied together, the product is *less than* either factor.

Multiply each of the following decimals.

5. 0.07×0.2
6. 0.5×0.06
7. 0.28×0.008
8. 0.04×0.096
9. 3.8×0.81
10. 0.62×9.7
11. 0.534×2.7
12. 6.32×8.09
13. $\$6.28 \times 7.4$
14. $\$18.09 \times 6.2$
15. 0.0071×0.038
16. 0.6704×0.605
17. 15.07×3.012
18. 7.0063×217.14
19. 382.68×0.00079
20. 0.000603×579.28
21. $7,006.08 \times 0.0731$
22. 0.2053×5607.39
23. $\$2,999.99 \times 12$
24. $\$5,085.75 \times 52$
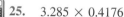 25. 3.285×0.4176
26. 728.8×456.3
27. 562.7×329.4
28. 0.8394×6.196

Find each of the following products.

29. 0.8×10^5
30. 2.93×10^4
31. 0.0031×10^2
32. 2.0084×10^2
33. 0.3602005×10^4
34. 0.0370059×10^3
35. 5.7×10^8
36. 6.2×10^7
37. 0.02803×10^7
38. 0.30705×10^8
39. $4.8 \times 3,000$
40. $7,000 \times 0.25$
41. 0.003×9.4
42. 1.84×400
43. $60,000 \times 0.379$
44. $0.807 \times 50,000$
45. $3,000 \times 0.09$
46. 0.005×800
47. $70,000 \times 0.002$
48. $0.0004 \times 6,000$

Rewrite the following numbers using digits.

49. 4.4 million
50. 7.2 million
51. 8.9 hundredths
52. 5.3 thousandths
53. 64 thousandths
54. 814 hundredths
55. 1.75 thousand
56. 2.5 billion

Simplify each of the following expressions.

57. $7(3.4 + 9.6)$
58. $8(18.4 - 11.4)$
59. $(0.05)(0.02)^2$
60. $(0.01)^3(0.08)$

61. $2[1.5 + (30)(0.03)]$

62. $3[4.2 + (0.25)(60)]$

63. $(1.5 + 3.3)(7.2 - 5.6)$

64. $(9.4 - 8.5)(6.8 + 5.2)$

65. $(7.2 + 9.8)^2$

66. $(0.018 + 0.002)^3$

67. $5[(3.8 + 6.9) - 4.7]$

68. $3[18.8 - (9.3 + 7.5)]$

69. $[(0.5)(0.2)]^3(16.8 - 9.3 + 1.5)^2$

70. $[5.6 + (4.5)(3.2)][(0.6)(0.5)]^3$

Solve each of the following word problems.

71. If one bagel costs $0.65, how much does a dozen bagels cost?

72. If 1 kilogram is equal to 2.2 pounds, how many pounds does a 76-kilogram person weigh?

73. If 1 meter is equal to 39.37 inches, how many inches are there in 27 meters?

74. If salami costs $1.29 per pound, and ham costs $2.35 per pound, what is the total cost of 1.5 pounds of salami and 2.25 pounds of ham? Round off your answer to the nearest cent.

75. The population of Finland is 4,770,000, and the population of Denmark is 1.08 times that of Finland. What is the population of Denmark?

76. If a mechanic makes $442.33 a week, what is his yearly salary?

77. An electrician earns $12.65 an hour and is paid 1.5 times her hourly wage for every hour above 40 hours worked in a single week. What would her pay be for a 47.5-hour work week? Round off to the nearest cent.

78. A welder worked the following hours each day during a single week: 7.8, 8.1, 4.2, 8.0, and 7.9. If he earns $10.45 an hour, what were his gross earnings that week?

79. A psychiatrist bought a new car and made a down payment of $2,500. If he pays off the remainder in 36 monthly payments of $186.25 each, what is the total amount he will pay for the car?

80. A notebook costs 98¢, a biology book costs 12.5 times the price of the notebook, and a mathematics book costs 1.2 times the price of the biology book. What is the price of the mathematics book?

81. The area of the Indian Ocean is approximately equal to 24 times the area of the Caribbean Sea. If the area of the Caribbean Sea is 1.05 million square miles, what is the area of the Indian Ocean?

82. Between 1840 and 1920 approximately 4.2 million people emigrated from Great Britain to the United States. If during that same period one-seventh of that number emigrated from France to the United States, how many people emigrated from France to the United States between 1840 and 1920?

83. To determine state income tax, a person must subtract from his or her gross salary $2,000 for a spouse's exemption (if filing jointly) and $700 per dependent, and then multiply that result by 0.05375. How much state income tax will a couple with three children have to pay if they file jointly and earn a total of $46,530.00? Round off your answer to the nearest cent.

84. A house that costs $80,000.00 is purchased with a down payment of $15,000.00 and a 30-year mortgage that specifies monthly payments of $521.67. How much more than the original selling price do the buyers agree to pay for the house by taking out a mortgage?

4.5
Division of Decimals

In order to establish a procedure for dividing decimals, we can rewrite any division problem as a fraction, and then reduce it to lowest terms. For example,

$$0.8 \div 0.4 = \frac{0.8}{0.4}$$

$$= \frac{0.8}{0.4} \times \frac{10}{10}$$

Multiply the numerator and denominator by 10 to make the denominator a whole number.

$$= \frac{8}{4} = 2$$

We will now investigate some additional examples of dividing decimals in this manner.

EXAMPLE 1   Divide each of the following decimals.

(a) $0.9 \div 0.03$

$$0.9 \div 0.03 = \frac{0.9}{0.03}$$

Rewrite the division as a fraction.

$$= \frac{0.9}{0.03} \times \frac{100}{100}$$

Multiply the numerator and denominator by 100 to make the denominator a whole number.

$$= \frac{90}{3} = 30$$

Reduce to lowest terms.

(b) $0.36 \div 0.004$

$$0.36 \div 0.004 = \frac{0.36}{0.004}$$

Rewrite as a fraction.

$$= \frac{0.36}{0.004} \times \frac{1,000}{1,000}$$

Multiply the numerator and denominator by 1,000 to make the denominator a whole number.

$$= \frac{360}{4} = 90$$

Reduce.

(c) $0.042 \div 0.0007$

$$0.042 \div 0.0007 = \frac{0.042}{0.0007}$$

$$= \frac{0.042}{0.0007} \times \frac{10,000}{10,000}$$

Multiply the numerator and denominator by 10,000 to make the denominator a whole number.

$$= \frac{420}{7} = 60$$

▲

Notice that for each division in Example 1, we first multiplied numerator and denominator by the lowest power of ten necessary to make the denominator a whole number. The same result can be achieved simply by moving the decimal point in the numerator and denominator the same number of places to the right.

For example, to divide

$$0.0072 \div 0.0009 = \frac{0.0072}{0.0009}$$

move the decimal point four places to the right in both the numerator and denominator and then perform the division:

$$\frac{0.\underline{0072}}{0.\underline{0009}} = \frac{72}{9} = 8$$

If we write this same problem using the alternative notation for division,

$$0.\underline{0009}\,\overline{)0.\underline{0072}}$$

we indicate that the decimal point has been moved by placing the caret symbol ($\wedge$) in the new location before proceeding with the division. The decimal point in the quotient must be placed immediately above the caret in the dividend as shown:

$$0.0009_{\wedge}\,\overline{)\,0.0072_{\wedge}}^{\;8.}$$

We can now summarize the procedure for dividing decimals, as outlined here.

To Divide Decimals

1. Move the decimal point in the divisor the necessary number of places to the right to make it a whole number, and place a caret in its new location.

2. Move the decimal point in the dividend the same number of places to the right, and place a caret in its new location.

3. Put a decimal point in the quotient immediately above the caret in the dividend.

4. Proceed with the technique for dividing whole numbers, noting the location of the decimal point in the quotient. ▲

Let us now look at some additional examples of dividing decimals.

EXAMPLE 2 Divide each of the following decimals.

(a) $0.02904 \div 0.004$

$$\begin{array}{r} .726 \\ 0.04_{\wedge}\,\overline{)\,0.02_{\wedge}904} \\ \underline{2\;8} \\ 10 \\ \underline{8} \\ 24 \\ \underline{24} \\ 0 \end{array}$$

Move the decimal point two places to the right.

Check:
$$\begin{array}{r} 0.726 \\ \times\;\;\;0.04 \\ \hline 0.02904 \quad \sqrt{} \end{array}$$

(b) $0.040678 \div 0.473$

$$
\begin{array}{r}
.086 \\
0.473_\wedge \overline{)0.040_\wedge 678} \\
37\ 84 \\
\hline
2838 \\
2838 \\
\hline
0
\end{array}
$$

Move the decimal point three places to th right. Notice that zero is a necesary placeholder in the quotient.

Check:
$$
\begin{array}{r}
0.473 \\
\times \quad 0.086 \\
\hline
2838 \\
3784 \\
\hline
0.040678 \quad \sqrt{}
\end{array}
$$

(c) $114 \div 0.06$

$$
\begin{array}{r}
19\ 00. \\
0.06_\wedge \overline{)114.00_\wedge} \\
6 \\
\hline
54 \\
54 \\
\hline
0
\end{array}
$$

Recall that the decimal point in the whole number 114 is to the right of the ones place.

Notice that it is necessary to insert two zeros before moving the decimal point in the dividend, and that zeros are necessary placeholders in the quotient.

Check:
$$
\begin{array}{r}
1,900 \\
\times \quad 0.06 \\
\hline
114.00 \quad \sqrt{}
\end{array}
$$

▲

In problems where the division process does not yield a remainder of zero, we are often asked to round off the quotient to a specified place value. In order to do so, it is necessary to carry out the division procedure until the quotient contains a digit that is one place value to the right of that to which we intend to round off. For example, to round off to the nearest hundredth, we must carry out the division to the thousandths place.

EXAMPLE 3 Calculate the following quotients and round off as indicated.

(a) $1.3682 \div 0.017$ (nearest hundredth)

$$
\begin{array}{r}
80.482 \cong 80.48 \\
0.017_\wedge \overline{)1.368_\wedge 200} \\
1\ 36 \\
\hline
82 \\
68 \\
\hline
140 \\
136 \\
\hline
40 \\
34 \\
\hline
6
\end{array}
$$

Carry out division to the thousandths place.

Check:
$$
\begin{array}{r}
80.482 \\
\times \quad 0.017 \\
\hline
563374 \\
80482 \\
\hline
1368194 \\
+ \qquad 6 \\
\hline
1.368200
\end{array}
$$

$\sqrt{}$

(b) $0.436 \div 3.7$ (nearest thousandth)

$$
\begin{array}{r}
.1178 \cong 0.118 \\
3.7\overline{)0.4\,3600} \\
\underline{37} \\
66 \\
\underline{37} \\
290 \\
\underline{259} \\
310 \\
\underline{296} \\
14
\end{array}
$$

Carry out division to the ten-thousandths place.

Check: $\begin{array}{r} 0.1178 \\ \times\ \ \ \ 3.7 \\ \hline 8246 \\ 3534 \\ \hline 43586 \\ +\ \ \ \ \ 14 \\ \hline 0.43600 \ \checkmark \end{array}$

(c) $\$259 \div 12$ (nearest cent)

$$
\begin{array}{r}
\$\ 21.583 \cong \$21.58 \\
12\overline{)\$259.000} \\
\underline{24} \\
19 \\
\underline{12} \\
70 \\
\underline{60} \\
100 \\
\underline{96} \\
40 \\
\underline{36} \\
4
\end{array}
$$

Carry out division to the nearest tenth of a cent or nearest thousandth of a dollar.

Check: $\begin{array}{r} 21.583 \\ \times\ \ \ \ \ 12 \\ \hline 43166 \\ 21583 \\ \hline 258996 \\ +\ \ \ \ \ \ \ 4 \\ \hline 259.000 \ \checkmark \end{array}$

▲

Now we will learn how to divide decimals by powers of ten.

$$3.2 \div 10^1 = \frac{3.2}{10} = \frac{3.2}{10.0} = \frac{.32}{1.0} = 0.32$$

$$3.2 \div 10^2 = \frac{3.2}{100} = \frac{03.2}{100.0} = \frac{.032}{1.0} = 0.032$$

$$3.2 \div 10^3 = \frac{3.2}{1,000} = \frac{003.2}{1000.0} = \frac{.0032}{1.0} = 0.0032$$

Notice that in each case, the quotient can be obtained by moving the decimal point to the left the number of places that is indicated by the power of ten.

To *divide* a decimal by 10^n, where n is a whole number, move the decimal point n places to the *left*. ▲

EXAMPLE 4   Find each of the following quotients.

(a) $0.83 \div 10^2$

$0.83 \div 10^2 = 00.83 = 0.0083$ Move the decimal point two places to the left.

(b) $5.29 \div 100,000$

$100,000 = 10^5$

$5.29 \div 10^5 = 00005.29 = 0.0000529$ Move the decimal point five places to the left.

(c) $\dfrac{8,678.432}{10^3}$

$8.678.432 \div 10^3 = 8678.432 = 8.678432$ Move the decimal point three places to the left.   ▲

When division is expressed by fractional notation, we use the order of operations rules to *simplify the numerator and denominator first*, before doing the division. For example, let us simplify the following expression.

$$\frac{7.3 + 2.7}{(0.5)^2} = \frac{10.0}{0.25} = \frac{10.00}{0.25} = \frac{1,000}{25} = 40$$

Notice that this expression could be rewritten with parentheses to indicate the operation performed first, as follows:

$$\frac{7.3 + 2.7}{(0.5)^2} = (7.3 + 2.7) \div (0.5)^2 = 10 \div 0.25 = 40$$

EXAMPLE 5 Simplify each of the following divisions in fractional notation.

(a) $\dfrac{15.6 - 8.4}{0.07 + 0.05}$

$$\dfrac{15.6 - 8.4}{0.07 + 0.05} = \dfrac{7.2}{0.12} = \dfrac{720}{12} = 60$$

Perform the operations in the numerator and the denominator. Reduce to lowest terms.

(b) $\dfrac{(0.03)^3}{5.81 - 5.801}$

$$\dfrac{(0.03)^3}{5.81 - 5.801} = \dfrac{0.000027}{0.009} = \dfrac{0.027}{9} = 0.003$$

(c) $\dfrac{(0.3)(0.2)^3}{(0.02)(0.5)^2}$

$$\dfrac{(0.3)(0.2)^3}{(0.02)(0.5)^2} = \dfrac{(0.3)(0.008)}{(0.02)(0.25)}$$

$$= \dfrac{0.0024}{0.0050} = \dfrac{24}{50}$$

$$= \dfrac{48}{100} = 0.48$$ ▲

The computation of averages is an important application of division of decimals. Remember that to calculate an average, we first add all the numbers in the group, and then we divide that sum by the number of numbers that were added together.

EXAMPLE 6 During the fall semester, a student obtained the following grades on his math quizzes: 78, 62, 83, 77, and 86. Each quiz was worth 100 points. What was his average quiz grade for that semester?

Solution

```
  78
  62
  83
  77
  86
 ───
 386
```

Add the five scores.

```
      7 7.2
  5 │ 3 8 6.0
      3 5
      ───
        3 6
        3 5
        ───
          1 0
          1 0
          ───
             0
```

Divide the total by 5.

His average grade is 77.2. ▲

EXAMPLE 7 An investor earned $516 in dividends for 80 shares of stock. What was the dividend per share?

Solution The dividend per share is the total amount of dividends divided by the number of shares of stock, or $516 ÷ 80.

$$
\begin{array}{r}
\$\ \ \ 6.45 \\
80\ \overline{\smash{)}\$5\ 1\ 6.0\ 0} \\
4\ 8\ 0 \\
\hline
3\ 6\ 0 \\
3\ 2\ 0 \\
\hline
4\ 0\ 0 \\
4\ 0\ 0 \\
\hline
0
\end{array}
$$

The dividend per share is $6.45. ▲

Quick Quiz

Calculate each of the following quotients.	Answers
1. 0.63 ÷ 0.009	**1.** 70
2. 0.3648 ÷ 0.08	**2.** 4.56
3. 0.4053 ÷ 2.1	**3.** 0.193

4.5 Exercises

Indicate which of the following statements are true and which are false. For those that are false, change the italicized word to make the statement true.

1. To divide a decimal by 10^n, where n is a whole number, move the decimal point n places to the *right*.

2. A *caret* indicates the new location of a decimal point.

3. When simplifying a fractional expression, we perform the arithmetic operations in the numerator and denominator *before* dividing the numerator by the denominator.

4. To round off a quotient, carry out the division until the quotient has a digit in the location immediately to the *left* of the place value to which you intend to round off.

Calculate the following quotients.

5. 0.81 ÷ 0.09
6. 0.56 ÷ 0.8
7. 0.0049 ÷ 0.7
8. 2.4 ÷ 0.003
9. 16.9 ÷ 0.0013
10. 1.08 ÷ 0.0012
11. 0.0221 ÷ 1.7
12. 0.00225 ÷ 1.5
13. 0.00252 ÷ 0.018
14. 3.78 ÷ 0.0021
15. 73.6 ÷ 0.0032
16. 148.5 ÷ 4.5
17. 5.096 ÷ 9.8
18. 38.64 ÷ 8.4
19. 0.4088 ÷ 0.0073
20. 0.1624 ÷ 0.0056
21. 17,860 ÷ 0.038
22. 59,520 ÷ 0.064

Calculate the following quotients and round off as indicated.

23. $4.2 \div 0.9$ (nearest ten-thousandth)

24. $6.1 \div 0.7$ (nearest hundredth)

25. $168 \div 0.9$ (nearest tenth)

26. $771 \div 0.7$ (nearest whole number)

27. $20.8 \div 0.03$ (nearest thousandth)

28. $19.7 \div 0.07$ (nearest hundredth)

29. $\$17.89 \div 6$ (nearest cent)

30. $\$52.37 \div 3$ (nearest tenth of a cent)

31. $689 \div 0.27$ (nearest hundredth)

32. $457 \div 0.65$ (nearest thousandth)

33. $946 \div 0.31$ (nearest whole number)

34. $673 \div 0.84$ (nearest tenth)

35. $\$254.99 \div 18$ (nearest tenth of a dollar)

36. $\$567.45 \div 14$ (nearest tenth of a dollar)

37. $53.485 \div 0.002154$ (nearest whole number)

38. $10.005 \div 24.3$ (nearest ten-thousandth)

39. $0.6284 \div 9.217$ (nearest hundred-thousandth)

40. $82,547 \div 0.00764$ (nearest hundred)

Find each of the following quotients.

41. $7.83 \div 10^4$

42. $92,174 \div 10^7$

43. $0.006005 \div 10$

44. $0.0638 \div 1,000$

45. $90.009 \div 10^2$

46. $3,070 \div 10^5$

47. $\dfrac{281.7}{10^6}$

48. $\dfrac{80.4}{10^3}$

49. $\dfrac{0.5386}{100}$

50. $\dfrac{3,684}{10,000}$

Simplify each of the following expressions.

51. $\dfrac{5.8 + 9.2}{0.003}$

52. $\dfrac{13.6 - 2.6}{0.022}$

53. $\dfrac{(0.2)^3}{(0.05)^2}$

54. $\dfrac{(0.3)^2}{(0.001)^3}$

55. $\dfrac{7.3 + 12.7}{20 - 19.5}$

56. $\dfrac{23.6 - 8.1}{13.8 + 1.7}$

57. $\dfrac{(0.2)^3(0.3)^2}{(0.5)^2}$

58. $\dfrac{(0.06)^2}{(0.1)^2(0.2)^2}$

59. $\dfrac{(0.04)^3}{5.01 - (0.1)^2}$

60. $\dfrac{8.04 - (0.2)^2}{(4)(0.05)^2}$

61. $\dfrac{(0.3)^3}{4.61 - 4.601}$

62. $\dfrac{(0.6)^2}{7.31 - 7.301}$

63. $4.2 \div 0.7 + (0.2)(0.6)$

64. $0.36 \div 0.06 + (0.8)(0.3)$

65. $[3.8 + (0.4)(3)] \div (0.5)^2$

66. $[(6)(0.4) + 15.6] \div (0.3)^2$

67. $(0.2)^4 + 0.048 \div 0.6$

68. $(0.9)^2 - 0.56 \div 0.8$

69. $6.4 \div 0.08 \div (0.2)^2$

70. $(0.2)^5 \div 0.004 \div 0.8$

Solve each of the following word problems.

71. If an engineer's yearly salary is $37,000, what is her weekly salary? Round off to the nearest cent.

72. If a car travels 513 miles on 17.2 gallons of gasoline, how many miles per gallon (correct to the nearest tenth) does the car get?

73. If chicken cost 79¢ per pound, how many pounds of chicken (correct to the nearest tenth) could you buy for $5.00?

74. A batting average is calculated by dividing the number of hits by the number of times at

bat and rounding off that result to the nearest thousandth. If a player made 23 hits in 89 times at bat, what was his batting average?

75. Iceland has an area of 39,702 square miles and a population of 230,000. What is the average population density per square mile? Round off your answer to the nearest thousandth.

76. During a practice session, a swimmer completed a 100-meter course in the following times: 50.45 seconds, 50.53 seconds, 49.89 seconds, 51.06 seconds, and 50.93 seconds. What was his average time during that practice session?

77. During a particular academic year, a student received one A, two B's, four C's, and one D. The numerical equivalents for these grades are A = 4.0, B = 3.0, C = 2.0, and D = 1.0. What was the student's GPA (grade point average) for that year? Round off your answer to the nearest tenth.

78. A jogger can run 4 miles in 38 minutes 6 seconds. What is the jogger's average speed in miles per hour? Round off to the nearest tenth.

79. A race car driver finished a 500-mile race in 3 hours 6 minutes 27 seconds. What was the driver's average speed in miles per hour? Round off your answer to the nearest tenth.

80. A woman bought a condominium and obtained a 15-year mortgage that requires her to pay a total of $92,574. What are her monthly payments?

4.6
Fractions and Decimals

Since any decimal can also be expressed as a fraction, it is often useful to convert decimals to fractions and fractions to decimals. To convert a decimal to a fraction, we simply rewrite the number as a decimal fraction and reduce to lowest terms. For example,

$$0.35 = \frac{35}{100} = \frac{7}{20}$$

EXAMPLE 1 Convert each of the following decimals to fractions.

(a) 0.06

$$0.06 = \frac{6}{100} = \frac{3}{50}$$ Rewrite as a decimal fraction and reduce to lowest terms.

(b) 0.125

$$0.125 = \frac{125}{1,000} = \frac{5}{40} = \frac{1}{8}$$ Rewrite as a decimal fraction and reduce.

(c) 0.0032

$$0.0032 = \frac{32}{10,000} = \frac{8}{2,500} = \frac{4}{1,250} = \frac{2}{625}$$ Rewrite and reduce.

(d) 8.25

$$8.25 = 8\frac{25}{100} = \frac{825}{100} = \frac{165}{20} = \frac{33}{4}$$ ▲

Recall that fractional notation expresses the operation division. Therefore, to convert a fraction to a decimal, simply divide the numerator by the denominator. For example,

$$\frac{3}{4} = 3 \div 4 = 0.75$$

$$\begin{array}{r} .75 \\ 4\overline{)3.00} \\ \underline{28} \\ 20 \\ \underline{20} \\ 0 \end{array}$$

These two procedures are outlined below.

To convert a decimal to a fraction, rewrite the decimal as a decimal fraction and reduce it to lowest terms.

To convert a fraction to a decimal, divide the numerator by the denominator. ▲

Because we obtained a remainder of zero when calculating the decimal representation for $\frac{3}{4}$, 0.75 is called a **terminating decimal**.

▶ **DEFINITION** A *terminating decimal* is the exact representation of a fraction. That is, a terminating decimal is the quotient obtained by dividing the numerator of a fraction by the denominator when the remainder eventually becomes zero.

Sometimes, when converting a fraction to its decimal representation, the long division process will never yield a remainder of zero, regardless of how far we carry out the division. For example, to convert $\frac{1}{3}$ to a decimal we must divide 1 by 3, as illustrated.

$$\frac{1}{3} = 1 \div 3 = 0.333 \ldots$$

$$\begin{array}{r} .333 \\ 3\overline{)1.000} \\ \underline{9} \\ 10 \\ \underline{9} \\ 10 \\ \underline{9} \\ 1 \end{array}$$

The decimal we obtained as our answer is called a **repeating decimal**. The symbol . . . indicates that the pattern of digits repeats itself indefinitely.

▶ **DEFINITION** A *repeating decimal* is the representation of a fraction in which a pattern of digits repeats itself indefinitely.

We can also write the decimal representation for $\frac{1}{3}$ as

$$\frac{1}{3} = 0.\overline{3}$$

The bar over the 3 indicates that the digit 3 is repeated indefinitely.

EXAMPLE 2 Convert each of the following fractions to decimals.

(a) $\dfrac{5}{8}$

$$\frac{5}{8} = 5 \div 8 = 0.625$$

$$
\begin{array}{r}
.625 \\
8\,\overline{)5.000} \\
48 \\
\hline
20 \\
16 \\
\hline
40 \\
40 \\
\hline
0
\end{array}
$$

The result is a terminating decimal.

(b) $\dfrac{4}{11}$

$$\frac{4}{11} = 4 \div 11 = 0.3636\cdots = 0.\overline{36}$$

$$
\begin{array}{r}
.3636 \\
11\,\overline{)4.0000} \\
33 \\
\hline
70 \\
66 \\
\hline
40 \\
33 \\
\hline
70 \\
66 \\
\hline
4
\end{array}
$$

Notice that the digits 36 will repeat themselves indefinitely. We place a bar over both digits that repeat.

The result is a repeating decimal.

(c) $\dfrac{38}{25}$

$$\frac{38}{25} = 1.52$$

$$
\begin{array}{r}
1.52 \\
25\,\overline{)38.00} \\
25 \\
\hline
130 \\
125 \\
\hline
50 \\
50 \\
\hline
0
\end{array}
$$

The result is a terminating decimal. ▲

Any fraction whose numerator is a whole number, and whose denominator is a whole number other than zero, can be represented as a terminating or repeating decimal. Sometimes, however, when a decimal does not terminate, a pattern of repeating digits does not become apparent, even after carrying out the division four or five places. In such cases, instead of carrying out the division further, we can approximate the answer by rounding off.

EXAMPLE 3 Convert $\frac{4}{13}$ to a decimal rounded off to the nearest ten-thousandth.

Solution $\frac{4}{13} = 4 \div 13 \cong 0.30769 \cong 0.3077$

Carry out the division to the hundred-thousandths place and round off.

$$
\begin{array}{r}
.30769 \\
13\overline{)4.00000} \\
\underline{39} \\
100 \\
\underline{91} \\
90 \\
\underline{78} \\
120 \\
\underline{117} \\
3
\end{array}
$$

▲

Quick Quiz

Answers

1. Convert 0.075 to a fraction.   1. $\frac{3}{40}$

2. Convert $\frac{8}{9}$ to a decimal.   2. $0.\overline{8}$

3. Convert $\frac{3}{17}$ to a decimal correct to the nearest   3. 0.176
   thousandth.

4.6 Exercises

Indicate which of the following statements are true and which are false. For those that are false, change the italicized expression to make the statement true.

1. A *continuing* decimal is the representation of a fraction in which a pattern of digits repeats itself indefinitely.

2. To convert a fraction to a decimal, divide the *denominator by the numerator*.

3. If the long division process eventually yields a remainder of zero, the fraction can be represented as a *terminating* decimal.

4. Fractions whose numerators are less than their denominators can be rewritten as decimals *greater* than one.

Convert each of the following decimals to fractions reduced to lowest terms.

5.	0.08	6.	0.46	7.	0.16	8.	0.36
9.	0.024	10.	0.125	11.	0.0075	12.	0.0028
13.	0.3125	14.	0.625	15.	2.5	16.	6.4
17.	12.8	18.	5.75	19.	15.45	20.	70.75
21.	40.15	22.	20.05	23.	4.0375	24.	8.2025

Convert each of the following fractions to decimals. Carry out the division until the remainder is zero or until a pattern of repeating digits becomes apparent.

25. $\dfrac{1}{4}$ 26. $\dfrac{1}{3}$ 27. $\dfrac{2}{5}$ 28. $\dfrac{7}{8}$ 29. $\dfrac{5}{11}$ 30. $\dfrac{3}{8}$

31. $\dfrac{5}{32}$ 32. $\dfrac{5}{6}$ 33. $\dfrac{4}{15}$ 34. $\dfrac{5}{16}$ 35. $\dfrac{4}{9}$ 36. $\dfrac{3}{25}$

37. $\dfrac{7}{18}$ 38. $\dfrac{11}{15}$ 39. $\dfrac{3}{4}$ 40. $\dfrac{1}{8}$ 41. $\dfrac{2}{9}$ 42. $\dfrac{2}{55}$

43. $\dfrac{8}{3}$ 44. $\dfrac{7}{4}$ 45. $\dfrac{13}{9}$ 46. $\dfrac{15}{11}$ 47. $\dfrac{19}{16}$ 48. $\dfrac{26}{15}$

49. $\dfrac{367}{12}$ 50. $\dfrac{461}{15}$ 51. $\dfrac{727}{3}$ 52. $\dfrac{555}{9}$ 53. $\dfrac{2}{21}$ 54. $\dfrac{6}{7}$

 55. $\dfrac{678}{999}$ 56. $\dfrac{281}{999}$

Find a decimal approximation for each of the following fractions correct to the place value indicated.

57. $\dfrac{3}{19}$ (nearest thousandth)   58. $\dfrac{6}{21}$ (nearest thousandth)

59. $\dfrac{7}{31}$ (nearest hundredth)   60. $\dfrac{6}{17}$ (nearest thousandth)

61. $\dfrac{13}{15}$ (nearest ten-thousandth)   62. $\dfrac{5}{9}$ (nearest ten-thousandth)

63. $\dfrac{6}{11}$ (nearest hundred-thousandth)   64. $\dfrac{5}{14}$ (nearest hundredth)

65. $\dfrac{2}{7}$ (nearest thousandth)   66. $\dfrac{9}{14}$ (nearest ten-thousandth)

67. $\dfrac{18}{7}$ (nearest hundredth)   68. $\dfrac{29}{13}$ (nearest thousandth)

69. $\dfrac{368}{9}$ (nearest hundredth)   70. $\dfrac{555}{19}$ (nearest thousandth)

 71. $\dfrac{521}{683}$ (nearest thousandth)   72. $\dfrac{37}{749}$ (nearest ten-thousandth)

4.7
Summary and Review

Key Terms

(4.1)   A **decimal** is a number that can be represented as a fraction whose denominator is a power of 10 greater than or equal to 1.

A **decimal fraction** is a fraction whose denominator can be expressed as a power of 10.

(4.2)   **Rounding off** is the procedure used to estimate a number to a given degree of precision.

(4.6)   A **terminating decimal** is the exact representation of a fraction. It is the quotient obtained by dividing the numerator of the fraction by the denominator when the remainder eventually becomes zero.

A **repeating decimal** is the representation of a fraction in which a pattern of digits repeats itself indefinitely.

Calculations

(4.1) **To translate a decimal into words**, write the number to the left of the decimal point as a whole number, replace the decimal point by the word "and," and write the number to the right of the decimal point as a whole number followed by the name of the place value of the last digit.

To translate words into decimal notation, write the expression to the left of the word "and" as a whole number, replace the word "and" by a decimal point, and write the rest of the expression as a whole number without commas. If necessary, insert zeros immediately to the right of the decimal point so that the last digit appears in the place value named at the end of the written expression.

(4.2) **To round off a number**, examine the digit immediately to the right of the place value to which you are rounding off. If that digit is 5 or greater, add 1 to the digit in the place value to which you are rounding (*round up*). If the digit to the right is less than 5, retain the digit in the place value to which you are rounding (*round down*).

(4.3) **To add (or subtract) decimals**, write the numbers vertically and line up the decimal points. Add (or subtract) the digits in each column separately, moving from right to left, and bring down the decimal point.

(4.4) **To multiply decimals**, calculate the product as usual, and put a decimal point in the appropriate location so that the number of decimal places in the product equals the sum of the number of decimal places in all factors.

To multiply a decimal by 10^n, where n is a whole number, move the decimal point n places to the right.

(4.4) **To divide decimals**, move the decimal point in the divisor and in the dividend the same number of places to the right in order to make the divisor a whole number. Place a caret in the new location of each decimal point. Put a decimal point in the quotient immediately above the caret in the dividend, and apply the technique for dividing whole numbers.

To divide a decimal by 10^n, where n is a whole number, move the decimal point n places to the left.

(4.6) **To convert a decimal to a fraction**, rewrite the number as a decimal fraction and reduce it to lowest terms.

To convert a fraction to a decimal, divide the numerator by the denominator.

Chapter 4
Review Exercises

Indicate which of the following statements are true and which are false. For those that are false, change the italicized expression to make the statement true.

1. The decimal places are located to the *left* of the decimal point.

2. *Rounding off* is a procedure for estimating numbers.

3. To find the average of a set of numbers, we use the operations addition and *subtraction*.

4. *All* fractions whose numerators and denominators are nonzero whole numbers can be represented as either terminating or repeating decimals.

5. When adding and *multiplying* decimals, we write the problem vertically and line up the decimal points.

6. If the numerator of a fraction is evenly divisible by the denominator, the fraction is equivalent to a *repeating* decimal.

(4.1) **Write each of the following decimals in words and in expanded notation.**

7. 8.7   8. 6.2   9. 5.87   10. 12.04

11. 43.628   12. 70.296   13. 806.5614   14. 4,083.0609

(4.2) **Round off each of the following decimals to the places indicated.**

15. 4.537   (nearest tenth)   16. 8.984   (nearest hundredth)

17. 17.86   (nearest whole number)   18. 8.2786   (nearest thousandth)

19. 55.44   (nearest ten)   20. 467.329   (nearest hundredth)

21. 953.628   (nearest hundred)   22. 74.4893   (nearest thousandth)

23. $5.0477   (nearest tenth of a cent)   24. $27.568   (nearest tenth of a dollar)

(4.3–4.5) **Perform each of the following operations.**

25. $623.56 + 4.872$ 26. $6.3784 + 297.69$

27. $741.36 - 3.89$ 28. $83 - 2.743$

29. 6.42×0.03 30. 13.93×0.005

31. $142.8 \div 0.06$ 32. $872.4 \div 0.4$

33. $5.468 + 12.82 - 8.9273$ 34. $48.1 + 362.76 - 7.9857$

35. 8.534×0.38 36. 31.746×0.26

37. $5.6079 \div 0.67$ 38. $3.9312 \div 0.72$

39. $632.50 + $3,028.99$ 40. $1,813.45 - 546.89

41. 89.99×28 42. $220.02 \div 57$

43. $58.473 - 9.5841 + 127.8 - 88.63$ 44. $17.357 - 9.461 + 752.3 - 68.598$

45. $7.13 \times 0.04 \times 5.2$ 46. $8.5 \times 0.26 \times 0.009$

47. $5.67 \div 27 \div 0.03$ 48. $4.82 \div 0.004 \div 0.25$

49. $7.2 \times 400,000$ 50. 0.009×0.0005

51. $(3.2 + 6.3) \div 0.05$ 52. $(2.6 + 1.6) \div 0.06$

53. $(0.6)^2 - (0.4)^3$ 54. $(1.1)^2 - (0.2)(0.8)$

55. $0.63 \div 30,000$ 56. $432 \div 0.0002$

57. $(1.2)^2 \div (3.2 - 2.8)$ 58. $(1.4)^2 \div (0.2)^2$

59. $(4.7 + 3.7) \div (7.3 - 6.6)$ 60. $(9.3 - 1.8) \div (4.2 + 1.8)$

61. $[8.143 + 0.007 - (0.3)(0.5)]^2$ 62. $[5.219 + 0.081 - 3.3]^3$

63. $5.4309 + 621.847 + 0.5649$ 64. $7,004 - 5.3682 - 29.637$

65. 3.218×0.4527 66. $62.2836 \div 0.657$

(4.4–4.5) **Find each of the following products and quotients.**

67. 0.0053×10^3 68. 862×10^5 69. $\dfrac{0.078}{10^6}$ 70. $\dfrac{3.892}{10^7}$

71. 16.894×10^0 72. $\dfrac{4.5831}{10^0}$ 73. $9,209 \times 1,000$ 74. $\dfrac{260,047}{10,000}$

(4.6) **Convert each of the following decimals to fractions reduced to lowest terms.**

75. 0.12   76. 0.08   77. 0.025   78. 0.004   79. 0.00016   80. 0.00068

81. 0.00225   82. 0.00144   83. 4.4   84. 7.5   85. 20.45   86. 247.5

(4.6) **Convert each of the following fractions to decimals. Round off only when indicated.**

87. $\dfrac{2}{3}$ 88. $\dfrac{3}{8}$ 89. $\dfrac{7}{16}$

90. $\dfrac{5}{6}$ 91. $\dfrac{8}{11}$ 92. $\dfrac{9}{25}$

93. $\dfrac{5}{7}$ (nearest thousandth)   94. $\dfrac{9}{21}$ (nearest ten-thousandth)   95. $\dfrac{12}{5}$

96. $\dfrac{23}{8}$ 97. $\dfrac{52}{11}$ 98. $\dfrac{86}{13}$ (nearest thousandth)

(4.3–4.5) **Solve each of the following word problems.**

99. If a baseball player made seventeen hits in sixty-nine times at bat, what was his batting average? Round off to the nearest thousandth.

100. A carpenter earns $13.65 an hour and gets time-and-a-half for overtime (more than 40 hours worked in a single week). What would be the carpenter's salary for a 47.25-hour work week?

101. Before a salesman began a two-day trip, the odometer on his car read 29,086.9 miles. At the end of the first day, the reading was 29,586.7 miles, and at the end of the second day, the reading was 30,070.2 miles. Which day did he travel the longer distance? What was the total distance that he traveled on these two days combined?

102. If seven umbrellas cost $96.95, what is the cost of two umbrellas?

103. During a student's undergraduate years at a state university, she received the following grades: five A's, eleven B's, fourteen C's, two D's, and one F. If all courses are worth the same number of credit hours, what was the student's GPA (grade point average)? (Recall that A = 4.0, B = 3.0, C = 2.0, D = 1.0, and F = 0.0.) Round off your answer to the nearest tenth.

104. The scale on a map indicates that 1 inch represents 12.8 miles. If the distance between two cities on that map is 4.5 inches, what is the actual distance between them?

105. A retailer buys scarves at a cost of $51.96 a dozen. He sells the scarves for $7.98 apiece. If he sold 5 scarves yesterday, how much profit did he make on the 5 scarves?

106. A nurse took a patient's temperature every hour and recorded these temperatures (in °F): 100.2, 99.8, 98.7, 99.2, 99.4, 100.4, and 101.1. What was the patient's average temperature for this 7-hour period? Round off your answer to the nearest tenth.

107. If you bought three identical notebooks and received $4.06 in change from a $10 bill, what was the price of a single notebook?

108. During a 585-mile trip, a car averaged 32.5 miles per gallon and used a full tank of gas. How much would it cost to refill the tank with gas that costs $1.545 a gallon?

Chapter 4
Explain It in Words

109. Explain what is meant by a "terminating decimal." Give an example.
110. Explain what is meant by a "repeating decimal." Give an example.
111. Explain how to convert a decimal to a fraction. Use an example.
112. Explain how to convert a fraction to a decimal. Use an example.
113. Explain how to round off a decimal. Use an example.
114. Explain how to multiply two decimals. Use an example.

Chapter 4
Chapter Test

[4.1] Write each of the following in words and in expanded notation.

1. 3.97   2. 73.475

[4.2] Round off each of the following to the places indicated.

3. 8.479   4. 63.26   5. $4.7825
   (nearest hundredth)   (nearest whole number)   (nearest tenth of a cent)

[4.3] Add each of the following.

6. 284.36 + 91.43   7. 9.0518 + 617.48 + 31.8724   8. 53.67 + 89,014

[4.3] Subtract each of the following.

9. 76.08 − 52.37   10. 562.5 − 8.309   11. 209 − 93.728

[4.4] Multiply each of the following.

12. 9.26 × 0.09   13. 73.532 × 4.7   14. 46,000 × 0.05

[4.5] Divide each of the following.

15. 0.2316 ÷ 0.004   16. 5993.15 ÷ 6.7   17. 378 ÷ 0.07

[4.4–4.5] Find each of the following products and quotients.

18. 0.0093×10^4 19. $\dfrac{8,261.7}{10^5}$

[4.6] Convert each of the following fractions to decimals.

20. $\dfrac{62}{4}$ 21. $\dfrac{19}{50}$ 22. $\dfrac{3}{11}$

[4.3–4.5] Solve each of the following word problems.

23. An architect bought items at a stationery store. The items cost $3.79, $2.29, $1.59, $0.89, and $0.69. How much change would she receive from a $20 bill?

24. A draftsman makes $8.75 an hour and gets time-and-a-half for overtime (more than 40 hours worked in a single week). What would be his wage for a 45.2-hour work week?

25. The total rainfall in inches for a midwestern city during the months of April through September was 6.1, 2.2, 3.3, 1.6, 1.4, and 2.8. What is the average monthly rainfall for those months?

Chapter **5**

Ratio and Proportion

5.1
Ratio   In this chapter we will learn how fractions can be used to express a relationship between two quantities. For example, if a class of students contains 13 men and 27 women, we can say that the ratio of men to women is 13 to 27, or $13:27$. Ratios are usually written as fractions in which the numerator is the first quantity being compared, and the denominator is the second. Thus, the ratio of men to women in this case can also be written as $\frac{13}{27}$.

> ▶ **DEFINITION** A *ratio* is a fraction that expresses a relationship between two quantities. The ratio of a to b can be written in the form $a:b$ or $\frac{a}{b}$. The first quantity being compared is placed in the numerator, and the second quantity is placed in the denominator.

EXAMPLE 1   In a basket of fruit, there are 5 apples, 6 oranges, and 3 bananas.

(a)   What is the ratio of oranges to apples?

Since there are 6 oranges and 5 apples, the ratio of oranges to apples is $6:5$ or $\frac{6}{5}$.

(b)   What is the ratio of bananas to oranges?

Since there are 3 bananas and 6 oranges, the ratio of bananas to oranges is $3:6$ or $\frac{3}{6}$. $\frac{3}{6}$ reduces to $\frac{1}{2}$, so the ratio expressed in lowest terms is $\frac{1}{2}$ or $1:2$.

(c)   What fraction of the fruit is oranges?
First add the number of apples, oranges, and bananas to find the total amount of fruit.

$5 + 6 + 3 = 14$ fruit

Since there are 6 oranges and 14 pieces of fruit, the ratio of oranges to fruit is $6:14$ or $\frac{6}{14}$. $\frac{6}{14}$ reduces to $\frac{3}{7}$, so $\frac{3}{7}$ of the fruit is oranges.   ▲

When using ratios, it is important to keep in mind the units in which the two quantities being compared are expressed. In order to compare these two quantities in a meaningful way, they must be expressed in the same units. A list of equivalent units of measurement is given on the inside cover of this book. To convert from one unit of measurement to another, we can multiply the quantity we wish to convert by a fraction equivalent to 1. For example, since 12 inches = 1 foot, the fractions

$$\frac{1 \text{ ft}}{12 \text{ in.}} \quad \text{and} \quad \frac{12 \text{ in.}}{1 \text{ ft}}$$

both are equal to 1. This can be shown by rewriting the numerators as follows:

$$\frac{1 \text{ ft}}{12 \text{ in.}} = \frac{12 \text{ in.}}{12 \text{ in.}} = 1$$

$$\frac{12 \text{ in.}}{1 \text{ ft}} = \frac{1 \text{ ft}}{1 \text{ ft}} = 1$$

Fractions such as this are sometimes called **unit fractions**.

▶ **DEFINITION** A *unit fraction* is a fraction equivalent to 1.

We can convert 92 inches to feet, for example, by multiplying 92 inches by the unit fraction $\frac{1 \text{ ft}}{12 \text{ in.}}$.

$$92 \text{ in.} = 92 \text{ in.} \times 1$$

$$= \frac{92 \text{ in.}}{1} \times \frac{1 \text{ ft}}{12 \text{ in.}}$$

$$= \frac{92}{12} \text{ ft}$$

$$= \frac{23}{3} \text{ ft} = 7\frac{2}{3} \text{ ft}$$

Notice that we cancel out common units in the numerator and denominator just as we do with common factors.

We determine the appropriate unit fraction to use for a conversion by looking at the units we need to cancel for the conversion to take place. If we had tried to multiply

$$92 \text{ in.} \times \frac{12 \text{ in.}}{1 \text{ ft}}$$

the inches would not have canceled, and our attempt at conversion would not have worked.
We will now summarize the procedure for converting units of measurement.

To Convert from One Unit of Measurement to Another

Multiply by the unit fraction that has the old unit in the denominator and the new unit in the numerator. ▲

For example, to convert 24 ounces to pounds, multiply 24 ounces by the unit fraction $\frac{1 \text{ lb}}{16 \text{ oz}}$:

$$24 \text{ oz} = \frac{24 \text{ oz}}{1} \times \frac{1 \text{ lb}}{16 \text{ oz}} = \frac{24}{16} \text{ lb} = \frac{3}{2} \text{ lb} = 1\frac{1}{2} \text{ lb}$$

EXAMPLE 2 Convert each of the following units of measurement.

(a) Convert 3 gallons to quarts

$$3 \text{ gal} = \frac{3 \text{ gal}}{1} \times \frac{4 \text{ qt}}{1 \text{ gal}}$$

Multiply by the unit fraction that has qt in the numerator and gal in the denominator.

$$= \frac{3 \text{ gal}}{1} \times \frac{4 \text{ qt}}{1 \text{ gal}}$$

$$= 12 \text{ qt}$$

(b) Convert 25 minutes to hours

$$25 \text{ min} = \frac{25 \text{ min}}{1} \times \frac{1 \text{ hr}}{60 \text{ min}}$$

Multiply by the unit fraction that has hours in the numerator and minutes in the denominator.

$$= \frac{25 \text{ min}}{1} \times \frac{1 \text{ hr}}{60 \text{ min}}$$

$$= \frac{25}{60} \text{ hr}$$

$$= \frac{5}{12} \text{ hr}$$

Divide numerator and denominator by 5. ▲

We will now look at some examples that require us to convert units of measurement in order to compare two quantities.

EXAMPLE 3 Express each of the following comparisons as a ratio with the same units in the numerator and denominator. Reduce to lowest terms.

(a) 8 nickels to 3 dimes

$$8 \text{ nickels} = 8 \text{ nickels} \times 1$$

$$= \frac{8 \text{ nickels}}{1} \times \frac{1 \text{ dime}}{2 \text{ nickels}}$$

$$= \frac{8}{2} \text{ dimes}$$

$$= 4 \text{ dimes}$$

We can convert 8 nickels to dimes by multiplying 8 nickels by a unit fraction with nickels in the denominator (so that nickels cancel) and dimes in the numerator. Since 2 nickels = 1 dime, this unit fraction is $\frac{1 \text{ dime}}{2 \text{ nickels}}$.

Thus,

$$8 \text{ nickels} : 3 \text{ dimes} = 4 \text{ dimes} : 3 \text{ dimes}$$

$$= \frac{4 \text{ dimes}}{3 \text{ dimes}}$$

$$= \frac{4}{3}$$

Express each quantity in the same units.

Rewrite as a fraction and simplify.

The ratio of 8 nickels to 3 dimes is $\frac{4}{3}$.

(b) 15 minutes to 5 hours

$$15 \text{ min} = 15 \text{ min} \times 1$$

$$= \frac{15 \cancel{\text{min}}}{1} \times \frac{1 \text{ hr}}{60 \cancel{\text{min}}}$$

$$= \frac{15}{60} \text{ hr}$$

$$= \frac{1}{4} \text{ hr}$$

We can convert 15 min to hours by multiplying 15 min by a unit fraction with minutes in the denominator (so minutes cancel) and hours in the numerator. Since 1 hr = 60 min, this unit fraction is $\frac{1 \text{ hr}}{60 \text{ min}}$.

Thus, $15 \text{ min} : 5 \text{ hr} = \frac{1}{4} \text{ hr} : 5 \text{ hr}$

Express each quantity in the same units.

$$= \frac{\frac{1}{4} \cancel{\text{hr}}}{5 \cancel{\text{hr}}}$$

Rewrite as a fraction.

$$= \frac{\frac{1}{4} \times 4}{5 \times 4}$$

Simplify.

$$= \frac{1}{20}$$

The ratio of 15 minutes to 5 hours is $\frac{1}{20}$.

(c) 8 pints to $5\frac{1}{2}$ quarts

$$5\frac{1}{2} \text{ qt} = 5\frac{1}{2} \text{ qt} \times 1$$

$$= \frac{11}{2} \cancel{\text{qt}} \times \frac{2 \text{ pt}}{1 \cancel{\text{qt}}}$$

$$= \frac{22}{2} \text{ pt}$$

$$= 11 \text{ pt}$$

We can convert $5\frac{1}{2}$ qt to pints by multiplying $5\frac{1}{2}$ qt by a unit fraction with quarts in the denominator (so that quarts cancel) and pints in the numerator. Since 2 pt = 1 qt, this unit fraction is $\frac{2 \text{ pt}}{1 \text{ qt}}$.

Thus, $8 \text{ pt} : 5\frac{1}{2} \text{ qt} = 8 \text{ pt} : 11 \text{ pt}$

Express each quantity in the same units and simplify.

$$= \frac{8 \cancel{\text{pt}}}{11 \cancel{\text{qt}}}$$

$$= \frac{8}{11}$$

The ratio of 8 pints to $5\frac{1}{2}$ quarts is $\frac{8}{11}$.

(d) 4 yards to 200 inches

It is not always necessary to use one of the units given as the common unit to compare the two quantities. In this problem we will convert both units to feet.

$$4 \text{ yd} = 4 \text{ yd} \times 1$$

$$= 4 \text{ yd} \times \frac{3 \text{ ft}}{1 \text{ yd}}$$

Since 3 ft = 1 yd, we can convert 4 yd to feet.

$$= 12 \text{ ft}$$

$$200 \text{ in.} = 200 \text{ in.} \times 1$$

$$= 200 \text{ in.} \times \frac{1 \text{ ft}}{12 \text{ in.}}$$

Since 12 in. = 1 ft, we can convert 200 in. to feet.

$$= \frac{200}{12} \text{ ft}$$

$$= \frac{50}{3} \text{ ft}$$

Thus, $4 \text{ yd} : 200 \text{ in.} = 12 \text{ ft} : \frac{50}{3} \text{ ft}$

Express the ratio of 4 yd to 200 in. in feet.

$$= \frac{12 \text{ ft}}{\frac{50}{3} \text{ ft}}$$

Express the ratio as a fraction.

$$= \frac{12 \times 3}{\frac{50}{3} \times 3}$$

Simplify.

$$= \frac{36}{50}$$

Reduce to lowest terms.

$$= \frac{18}{25}$$

The ratio of 4 yards to 200 inches is $\frac{18}{25}$. ▲

One of the most common examples of a ratio is a **rate**. Rates express the relationship between two unlike quantities.

▶ **DEFINITION** A *rate* is a ratio that compares two different quantities and has a denominator equal to 1.

For example, if you traveled 150 miles in 3 hours, your rate of speed could be calculated by rewriting the ratio of miles to hours as an equivalent ratio with a denominator of one. The ratio of miles to hours is

$$\frac{150 \text{ mi}}{3 \text{ hr}} = \frac{150 \text{ mi} \times \frac{1}{3}}{3 \text{ hr} \times \frac{1}{3}} = \frac{50 \text{ mi}}{1 \text{ hr}}$$

Your rate of speed is $\frac{50 \text{ mi}}{1 \text{ hr}}$, which is more commonly written as 50 mi/hr or 50 miles per hour. Notice that the / is read as "per."

EXAMPLE 4   Before lunch, a secretary typed nine pages in 50 minutes. After lunch, she typed 15 pages in $1\frac{1}{4}$ hours. If each page typed was approximately the same length, did she type faster before or after lunch?

Solution   Determine her rate before lunch:

$$\frac{9 \text{ p}}{50 \text{ min}} = 0.18 \text{ p/min}$$

Before lunch, the ratio of pages to minutes was 9 to 50.

Determine her rate after lunch in the same units as her rate before lunch:

$$1 \text{ hr} = 1\frac{1}{4} \text{ hr} \times 1$$

We must convert $1\frac{1}{4}$ hr to minutes. Since 1 hr = 60 min, we can multiply $1\frac{1}{4}$ hr by the unit fraction $\frac{60 \text{ min}}{1 \text{ hr}}$.

$$= \frac{5}{4} \text{ hr} \times \frac{60 \text{ min}}{1 \text{ hr}}$$

$$= \frac{5(60)}{4} \text{ min}$$

$$= 75 \text{ min}$$

So, she typed 15 p in 75 min.

Her rate after lunch was, therefore,

$$\frac{15 \text{ p}}{75 \text{ min}} = \frac{1 \text{ p}}{5 \text{ min}} = 0.20 \text{ p/min}$$

Thus, she typed faster after lunch. $0.18 \text{ p/min} < 0.20 \text{ p/min}$.   ▲

Ratios are especially useful for determining *unit prices*; that is, for finding the price of a single item given the price of a larger quantity of the items. For example, if soup is priced at 5 cans for a dollar, the price of one can could be determined by examining the ratio of dollars to cans, which is $1:5$ or $\frac{1 \text{ dollar}}{5 \text{ cans}}$. Changing $\frac{1}{5}$ to a decimal by dividing the denominator into the numerator so that the denominator is equal to 1, $\frac{1}{5} = \frac{0.20}{1}$, we obtain the unit price of \$0.20/can. We want the denominator of the ratio to be equal to 1, because unit pricing is a rate.

EXAMPLE 5   Determine the unit price of each of the following items with the given prices.

(a)   Soap—3 bars for \$0.99

$$\frac{0.99 \text{ dollar}}{3 \text{ bars}}$$

Find the ratio of dollars to bars.

$$\frac{0.99}{3} = \frac{0.33}{1}$$

Unit price is \$0.33/bar.

(b) Socks—5 pairs for $7.29

$$\frac{7.29 \text{ dollars}}{5 \text{ pairs}}$$

Find the ratio of dollars to pairs.

$$\frac{7.29}{5} = \frac{1.458}{1}$$

Unit price is $1.458/pair.

(c) Orange juice—16 fluid ounces for 52¢

$$\frac{52 \text{ cents}}{16 \text{ fl oz}}$$

Find the ratio of cents to fluid ounces.

$$\frac{52}{16} = \frac{3.25}{1}$$

Unit price is 3.25¢/fl oz. ▲

EXAMPLE 6 A 16-fluid-ounce bottle of vegetable oil costs $0.94, and a 24-fluid-ounce bottle costs $1.29. Determine which size is the better buy.

Solution $$\frac{0.94 \text{ dollar}}{16 \text{ fl oz}} = \frac{0.94}{16} = \frac{0.05875}{1}$$

Find the ratio of dollars to fluid ounces for the smaller size.

Unit price is $0.05875/fl oz for the 16-fl-oz bottle.

$$\frac{1.29 \text{ dollars}}{24 \text{ fl oz}} = \frac{1.29}{24} = \frac{0.05375}{1}$$

Find the ratio of dollars to fluid ounces for the larger size.

Unit price is $0.05375/fl oz for the 24-fl-oz bottle.
Since the unit price of the 24-fl-oz size is less, it is the better buy. ▲

Quick Quiz

Express each of the following ratios as a fraction without units. Reduce to the lowest terms.

Answers

1. 9 to 36

2. $3\frac{1}{2}$ to $5\frac{1}{4}$

3. 3 yards to 8 feet

1. $\frac{1}{4}$

2. $\frac{2}{3}$

3. $\frac{9}{8}$

5.1 Exercises

Indicate which of the following statements are true and which are false. For those that are false, change the italicized expression to make the statement true.

1. The ratio $x:y$ can be written as a fraction in which x is the *denominator*.
2. A ratio expresses the operation *subtraction*.
3. If the units in the numerator and denominator are *the same*, they cancel.
4. Changing the order of the two quantities being compared *does not* affect the value of the ratio.

Express each of the following ratios as a fraction reduced to lowest terms.

5. $3:5$
6. $7:9$
7. $8:2$
8. $11:1$
9. 5 to 8
10. 12 to 3
11. $88:4$
12. $6:100$
13. $4:56$
14. $63:3$
15. $\frac{1}{4}:3$
16. $6:\frac{2}{3}$
17. $\frac{5}{8}:\frac{2}{3}$
18. $\frac{2}{9}:\frac{3}{5}$
19. $3\frac{1}{2}:8\frac{1}{4}$
20. $6\frac{2}{3}:1\frac{3}{4}$
21. 2.7 to 0.9
22. 0.0032 to 0.04

Convert each of the following units of measurement.

23. Convert 4 yards to feet.
24. Convert 3 hours to minutes.
25. Convert 20 nickels to quarters.
26. Convert 35 days to weeks.
27. Convert 5 quarts to pints.
28. Convert 9 pints to fluid ounces.
29. Convert 18 inches to feet.
30. Convert 36 ounces to pounds.
31. Convert $\frac{5}{8}$ pound to ounces.
32. Convert $1\frac{3}{4}$ gallons to quarts.

Express each of the following comparisons as a ratio with the same units in the numerator and denominator. Reduce to lowest terms.

33. 3 weeks to 18 days
34. 16 months to 2 years
35. 19 nickels to 5 quarters
36. 14 dimes to 29 nickels
37. 64 items to 4 dozen
38. 6 dozen to 99 items
39. 42 inches to 4 feet
40. 10 feet to 100 inches
41. 6 dollars to 15 quarters
42. 81 nickels to 8 dollars
43. 7 pints to 3 quarts
44. 5 gallons to 35 quarts
45. 12 ounces to 2 pounds
46. $\frac{3}{4}$ pound to 8 ounces
47. 12 fluid ounces to 2 pints
48. 3 tablespoons to 8 teaspoons
49. 9 months to $\frac{1}{2}$ year
50. $3\frac{1}{4}$ days to 100 hours
51. $9\frac{1}{3}$ dozen to 400 items
52. 1,000 pounds to $2\frac{1}{2}$ tons
53. 4.25 dollars to 75 nickels
54. 150 dimes to 9.35 dollars
55. 3 gallons to 18 pints
56. 30 cups to 7 quarts
57. 8 yards to 48 inches
58. 2 miles to 176 yards
59. 720 seconds to $\frac{1}{2}$ hour
60. $1\frac{3}{4}$ hours to 6,000 seconds

Answer each of the following questions.

61. In a math class, 4 students have blue eyes, 2 students have green eyes, and 14 students have brown eyes.

a. What is the ratio of blue-eyed to brown-eyed students?
b. What is the ratio of brown-eyed to green-eyed students?
c. What fraction of the students is blue-eyed?
d. What fraction of the students has blue or green eyes?
e. What fraction of the students does not have green eyes?

62. A bag contains 9 red marbles, 12 blue marbles, 8 white marbles, and 6 green marbles.
    a. What is the ratio of red marbles to green marbles?
    b. What is the ratio of red marbles to blue marbles?
    c. What fraction of the marbles in the bag is green?
    d. What fraction of the marbles in the bag is either blue or white?
    e. What fraction of the marbles in the bag is not red or white?

63. A serving of oatmeal has the following nutritional value.

Protein	5 grams
Carbohydrate	18 grams
Fat	2 grams
Potassium	0.045 gram
Sodium	0.005 gram

    a. What is the ratio of fat to carbohydrate?
    b. What is the ratio of potassium to sodium?
    c. What is the ratio of potassium to protein?
    d. What fraction of the total nutritional value contains protein?
    e. What fraction of the total nutritional value contains carbohydrate or fat?

64. A lawyer has the following monthly budget.

Rent	$575
Utilities	$145
Food	$230
Transportation	$120
Clothing	$170
Miscellaneous	$260

    a. What is the ratio of her allowance for transportation to that for clothing?
    b. What is the ratio of her allowance for food to that for rent?
    c. What is the ratio of her rent to the amount budgeted for utilities?
    d. What fraction of her total budget does she allocate for miscellaneous expenses?
    e. What fraction of the total budget does she allocate for food and clothing?

Solve each of the following word problems.

65. A regular deck of 52 cards contains 13 hearts. What ratio represents the fraction of cards in a regular deck that is hearts?

66. An orange has 75 calories and an apple has 115 calories. What is the ratio of calories in an apple to calories in an orange?

67. A station wagon traveled 435 miles on 15 gallons of gas. A sedan traveled 124 miles on 4 gallons of gas. Which car got better mileage?

68. In a single day, an umbrella factory produced 12,000 umbrellas, and 72 of these were defective. What ratio represents the fraction of umbrellas produced that was defective?

69. A 16-ounce box of rice sells for 88¢. What is the unit price of 1 ounce of rice?

70. A 13-ounce can of tuna costs $2.34. What is the unit price of 1 ounce of tuna?

71. An acid solution contains 5 fluid ounces of hydrochloric acid and 95 fluid ounces of water. What is the ratio of acid to water?

72. If a cake recipe requires $2\frac{1}{4}$ cups of flour and $\frac{2}{3}$ cup of sugar, what is the ratio of flour to sugar?

73. The area of Cambodia is 70,000 square miles, and the area of Ireland is 26,600 square miles. What is the ratio of the area of Ireland to that of Cambodia?

74. The population of Brazil is 122,000,000, and the population of Haiti is 5,000,000. What is the ratio of the population of Brazil to that of Haiti?

75. A can of frozen orange juice concentrate that makes 24 fluid ounces sells for 57¢. A quart of ready-made orange juice sells for 88¢. Determine the unit price of each per fluid ounce (1 qt = 32 fl oz). Which is the better buy?

76. An 8-ounce package of cheese that contains 12 slices costs $1.29. What is the unit price per slice? What is the unit price per ounce?

77. A jogger ran for $\frac{3}{8}$ hour on Monday and for 20 minutes on Tuesday. Express the ratio of these two times as a fraction with the same units in the numerator and denominator. On which day was the jogger's running time longer?

78. A football player drank $\frac{2}{3}$ pint of water followed by 12 fluid ounces of orange juice. Express the ratio of water to orange juice as a fraction with the same units in the numerator and denominator. Did he drink more water or more orange juice?

79. A sports car traveled 217 miles in $3\frac{1}{2}$ hours. What was its average rate of speed?

80. A jogger can run $3\frac{1}{5}$ miles in 40 minutes. Another jogger can run 5.4 miles in 1 hour. Find the rate of each jogger and determine who runs faster.

81. The population of Virginia is approximately 5,000,000. If the area of the state is almost 40,000 square miles, what is the average population density per square mile of Virginia?

82. The Yukon River is 1,900 miles long and the Mekong River is 2,500 miles long. What is the ratio of the length of the Yukon to that of the Mekong?

83. A dozen eggs sells for $1.05, and another carton containing 8 eggs sells for 68¢. Determine the unit price for each carton. Which is the better buy?

84. If $\frac{3}{4}$ pound of freshly ground coffee costs $1.95, what is the unit price per pound?

85. A seamstress purchased 20 inches of blue ribbon and $\frac{5}{8}$ yard of red ribbon. Express the ratio of blue ribbon to red ribbon as a fraction with the same units in the numerator and denominator. Which piece of ribbon is longer?

86. A secretary spoke on the phone to a client for 27 minutes, and then the secretary's boss spoke to the client for 0.4 hour. Express the ratio of these two times as a fraction with the same units in the numerator and denominator. Who spoke longer to the client?

87. The population of Canada is 23,850,000. If Canada covers an area of 3,851,809 square miles, what is the average population density per square mile of Canada? Round off your answer to the nearest hundredth.

88. A 13-ounce can of coffee sells for $2.19. What is the unit price per ounce? Round your answer to the nearest tenth of a cent.

89. A runner completed a 6.2-mile race in 1 hour 8 minutes. What was her average rate of speed in miles per hour? Round off your answer to the nearest hundredth.

90. A toaster factory produced 13,457 toasters, of which 83 were defective. Find a decimal that represents the fraction of toasters that was defective. Round off your answer to the nearest hundred-thousandth.

5.2
Proportion

As was illustrated in the previous section, more than one equivalent ratio is often used to express the same relationship between two quantities. For example, if a purse contains 4 nickels and 12 dimes, the ratio of nickels to dimes is 4 to 12, or 1 to 3. This can be written mathematically as

$$4:12::1:3$$

and is read, "4 is to 12 as 1 is to 3."

This statement of equality between two ratios is called a **proportion**, and it most frequently appears in fractional notation. That is,

$$\frac{4}{12} = \frac{1}{3}$$

▶ **DEFINITION** A *proportion* is a statement of equality between two ratios. It can be written in the form

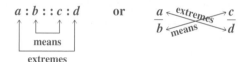

The numbers *a*, *b*, *c*, and *d* are called the *terms* of the proportion. The terms *a* and *d* are called the *extremes*. The terms *b* and *c* are called the *means*.

In our example of the ratio of 4 nickels to 12 dimes, the numbers 4 and 3 are the *extremes*, and the numbers 12 and 1 are the *means*.

4 : 12 : : 1 : 3

means

extremes

EXAMPLE 1 For each of the following proportions, identify the extremes and the means.

(a) $2:9::6:27$

Extremes are 2 and 27.

Means are 6 and 9.

(b) $\dfrac{3}{24} = \dfrac{2}{16}$

Extremes are 3 and 16.

Means are 24 and 2. ▲

An important property of proportions can be illustrated by comparing the product of the means to the product of the extremes. For example, in the proportion $\frac{1}{2} = \frac{3}{6}$, the product of the extremes, 1×6, is equal to the product of the means, 2×3, since both products are equal to 6. This property is called the **fundamental property of proportions**.

The Fundamental Property of Proportions

In any proportion, the product of the extremes is equal to the product of the means.

If $$\frac{a}{b} = \frac{c}{d}$$

then $$a \cdot d = b \cdot c$$ ▲

The products of $a \cdot d$ and $b \cdot c$ are sometimes called **cross products**, because they can be obtained by cross multiplying.

$$\frac{a}{b} \bowtie \frac{c}{d}$$

$$ad = bc$$

Notice that when we use letters to represent numbers, we often omit the symbol $\times$ or $\cdot$ used to express the operation multiplication.

product of extremes = product of means

We can show that the fundamental property of proportions is true for the proportion $\frac{a}{b} = \frac{c}{d}$ by multiplying both sides of the LCD of the two ratios, which is bd.

$$(bd)\frac{a}{b} = \frac{c}{d}(bd)$$

Now, cancel common factors in the numerator and denominator.

$$\frac{\cancel{b}da}{\cancel{b}} = \frac{cb\cancel{d}}{\cancel{d}}$$

$$ad = bc$$

EXAMPLE 2 Use the fundamental property of proportions to determine whether each of the following statements is true or false.

(a) $$\frac{4}{16} = \frac{5}{20}$$

Does $(4)(20) = (16)(5)$?   Cross multiply to determine the products
$\qquad\qquad 80 = 80$ of the extremes and the means.

$\dfrac{4}{16} = \dfrac{5}{20}$ is a true statement.

(b) $$\frac{6}{11} = \frac{7}{12}$$

Does $(6)(12) = (11)(7)$?   Cross multiply.
$\qquad\qquad 72 \neq 77$

$\dfrac{6}{11} = \dfrac{7}{12}$ is a false statement.

(c) $\dfrac{\dfrac{2}{3}}{\dfrac{6}{8}} = \dfrac{\dfrac{6}{9}}{\dfrac{6}{8}}$

Does $(2)\left(\dfrac{9}{8}\right) = \left(\dfrac{3}{8}\right)(6)$? Cross multiply.

$$\dfrac{18}{8} = \dfrac{18}{8}$$

$\dfrac{\dfrac{2}{3}}{\dfrac{6}{8}} = \dfrac{\dfrac{6}{9}}{\dfrac{6}{8}}$ is a true statement. ▲

We can use a similar technique to determine the larger of two ratios whose numerators and denominators are all greater than zero. Suppose that

$$\frac{a}{b} > \frac{c}{d}$$

We can multiply both sides of the inequality by bd to obtain

$$(bd)\frac{a}{b} > \frac{c}{d}(bd)$$

$$\frac{\not b da}{\not b} > \frac{cb\not d}{\not d}$$

$$ad > bc \qquad \text{The extremes are greater than the means.}$$

We have therefore shown that

If the first ratio is greater than the second, the product of the extremes is greater than the product of the means.

Similarly, if

$$\frac{a}{b} < \frac{c}{d}$$

we can multiply both sides of the inequality by bd to obtain

$$(bd)\frac{a}{b} < \frac{c}{d}(bd)$$

$$\frac{\not b da}{\not b} < \frac{cb\not d}{\not d}$$

$$ad < bc \qquad \text{The extremes are less than the means.}$$

We have now also shown that

If the first ratio is less than the second, the product of the extremes is less than the product of the means.

EXAMPLE 3 Determine which of the following ratios is greater, and place the appropriate sign ($>$ or $<$) between them.

(a) $\dfrac{3}{8}$? $\dfrac{1}{3}$

$3(3)$? $8(1)$ Cross multiply.
$9 > 8$ The extremes are greater than the means.

$\dfrac{3}{8} > \dfrac{1}{3}$ The first ratio is greater than the second.

(b) $\dfrac{4}{3}$? $\dfrac{7}{5}$

$4(5)$? $3(7)$ Cross multiply.
$20 < 21$ The extremes are less than the means.

$\dfrac{4}{3} < \dfrac{7}{5}$ The first ratio is less than the second.

(c) $\dfrac{\frac{2}{3}}{\frac{3}{4}}$? $\dfrac{\frac{6}{11}}{\frac{7}{2}}$

$\dfrac{2}{3}\left(\dfrac{7}{2}\right)$? $4\left(\dfrac{6}{11}\right)$ Cross multiply.

$\dfrac{14}{6}$? $\dfrac{24}{11}$ Simplify.

$\dfrac{7}{3}$? $\dfrac{24}{11}$ Reduce to lowest terms.

$7(11)$? $3(24)$ Since it is still not clear which ratio is greater, we cross multiply again.
$77 > 72$ The extremes are greater than the means.

$\dfrac{\frac{2}{3}}{\frac{3}{4}} > \dfrac{\frac{6}{11}}{\frac{7}{2}}$ The first ratio is larger than the second.

Notice that the same result can be obtained by simplifying the complex fractions in the original ratios before cross multiplying.

$\dfrac{\frac{2}{3} \cdot 3}{4 \cdot 3} = \dfrac{2}{12} = \dfrac{1}{6}$ Simplify $\dfrac{\frac{2}{3}}{4}$.

$\dfrac{\frac{6}{11} \cdot 22}{\frac{7}{2} \cdot 22} = \dfrac{12}{77}$ Simplify $\dfrac{\frac{6}{11}}{\frac{7}{2}}$.

$$\frac{1}{6} \ ? \ \frac{12}{77}$$

Restate in simplified terms and cross multiply.

$$77 > 72$$

The extremes are greater than the means.

$$\frac{\frac{2}{3}}{\frac{3}{4}} > \frac{\frac{6}{11}}{\frac{7}{2}}$$

The first ratio is greater than the second. ▲

Example 4 illustrates how a word problem can be solved by using this same technique to compare ratios.

EXAMPLE 4 A Japanese car can travel 20 miles on 0.75 gallon of gas, and an American car can travel 160 miles on 6.2 gallons of gas. Which car gets the better gas mileage?

Solution

Japanese car   American car

$$\frac{20 \text{ miles}}{0.75 \text{ gallon}} \ ? \ \frac{160 \text{ miles}}{6.2 \text{ gallons}}$$

Set up two ratios that express the mileage for each car, and determine which is greater.

$$(20)(6.2) \ ? \ (0.75)(160)$$

Cross multiply.

$$124 > 120$$

The extremes are greater than the means.

Japanese car $>$ American car

The first ratio is larger.

Thus, the Japanese car gets better mileage.   ▲

Note that since multiplication by a two-digit number is usually easier than division by a two-digit number, this method of comparing ratios is easier than determining the actual mileage for each car.

Quick Quiz

Determine which of the following ratios is larger and put the appropriate sign ($>$ or $<$) between them.

Answers

1. $\dfrac{3}{5} \ ? \ \dfrac{5}{8}$

1. $3(8) < 5(5)$

$$\frac{3}{5} < \frac{5}{8}$$

2. $\dfrac{2\frac{2}{5}}{5\frac{1}{4}} \ ? \ \dfrac{1\frac{3}{7}}{3\frac{1}{3}}$

2. $\left(\dfrac{12}{5}\right)\left(\dfrac{10}{3}\right) > \left(\dfrac{21}{4}\right)\left(\dfrac{10}{7}\right)$

$$\frac{2\frac{2}{5}}{5\frac{1}{4}} > \frac{1\frac{3}{7}}{3\frac{1}{3}}$$

3. $\dfrac{1.1}{0.03} \ ? \ \dfrac{0.7}{0.02}$

3. $(1.1)(0.02) > (0.03)(0.7)$

$$\frac{1.1}{0.03} > \frac{0.7}{0.02}$$

5.2 Exercises

Indicate which of the following statements are true and which are false. For those that are false, change the italicized or underlined expression to make the statement true.

1. A *proportion* is a statement of equality between two ratios.

2. In the proportion $a:b::c:d$, the terms _a and b_ are the means.

3. When comparing two ratios, if the product of the means exceeds the product of the extremes, the first ratio is *greater* than the second.

4. If two ratios are equal, the cross products are *equal*.

For each of the following proportions, identify the extremes and the means, and rewrite the expression in words.

5. $6:9::2:3$

6. $1:7::3:21$

7. $\dfrac{5}{2} = \dfrac{20}{8}$

8. $\dfrac{24}{6} = \dfrac{4}{1}$

9. $\dfrac{1}{2}:11::1:22$

10. $4:9::\dfrac{4}{3}:3$

11. $\dfrac{2\frac{1}{3}}{8} = \dfrac{7}{24}$

12. $\dfrac{7}{6\frac{2}{5}} = \dfrac{5\frac{1}{4}}{4\frac{4}{5}}$

Use the fundamental property of proportions to determine whether each of the following statements is true or false.

13. $4:9::2:3$

14. $6:18::2:6$

15. $\dfrac{1}{3} = \dfrac{8}{24}$

16. $\dfrac{9}{2} = \dfrac{27}{6}$

17. $5:3 = 7:4$

18. $2:7 = 6:21$

19. $\dfrac{5}{21} = \dfrac{6}{20}$

20. $\dfrac{9}{36} = \dfrac{2}{8}$

21. $\dfrac{13}{2} = \dfrac{17}{3}$

22. $\dfrac{8}{27} = \dfrac{16}{54}$

23. $\dfrac{\frac{3}{4}}{\frac{5}{8}} = \dfrac{\frac{8}{15}}{\frac{4}{9}}$

24. $\dfrac{\frac{1}{6}}{\frac{3}{11}} = \dfrac{\frac{2}{3}}{\frac{9}{10}}$

25. $\dfrac{\frac{6}{11}}{\frac{5}{22}} = \dfrac{\frac{4}{15}}{\frac{1}{9}}$

26. $\dfrac{\frac{3}{5}}{\frac{7}{15}} = \dfrac{\frac{4}{7}}{\frac{5}{9}}$

27. $\dfrac{3\frac{1}{4}}{\frac{2}{3}} = \dfrac{8}{1\frac{1}{3}}$

28. $\dfrac{2\frac{7}{8}}{3\frac{1}{2}} = \dfrac{4\frac{2}{3}}{6\frac{2}{9}}$

29. $\dfrac{2.1}{0.7} = \dfrac{3.3}{1.1}$

30. $\dfrac{8.4}{0.42} = \dfrac{6.0}{0.3}$

31. $\dfrac{3.2}{1.2} = \dfrac{5.6}{2.1}$

32. $\dfrac{8.6}{4.2} = \dfrac{7.3}{3.6}$

33. $\dfrac{527}{61} = \dfrac{242}{28}$

34. $\dfrac{184}{95} = \dfrac{473}{245}$

35. $\dfrac{0.0417}{8.62} = \dfrac{0.254}{52.6}$

36. $\dfrac{0.32}{1.2} = \dfrac{0.0128}{0.048}$

Determine which of the following ratios is greater, and put the appropriate sign ($>$ or $<$) between them.

37. $\dfrac{4}{9} \; ? \; \dfrac{5}{11}$

38. $\dfrac{2}{3} \; ? \; \dfrac{5}{8}$

39. $\dfrac{7}{16} \; ? \; \dfrac{3}{7}$

40. $\dfrac{5}{18} \; ? \; \dfrac{2}{7}$

41. $\dfrac{\frac{3}{8}}{\frac{5}{6}} \; ? \; \dfrac{\frac{1}{4}}{\frac{4}{9}}$

42. $\dfrac{\frac{1}{3}}{\frac{7}{12}} \; ? \; \dfrac{\frac{3}{8}}{\frac{9}{14}}$

43. $\dfrac{\frac{3}{7}}{\frac{3}{44}} \; ? \; \dfrac{\frac{4}{5}}{\frac{1}{8}}$

44. $\dfrac{\frac{3}{8}}{\frac{2}{9}} \; ? \; \dfrac{\frac{1}{3}}{\frac{4}{19}}$

45. $\dfrac{5\frac{5}{8}}{4\frac{1}{2}} \; ? \; \dfrac{1\frac{2}{3}}{1\frac{1}{5}}$

46. $\dfrac{3\frac{7}{8}}{6\frac{1}{4}} \; ? \; \dfrac{3\frac{1}{5}}{5\frac{5}{7}}$

47. $\dfrac{1.2}{3.5} \; ? \; \dfrac{0.15}{0.5}$

48. $\dfrac{0.43}{0.04} \; ? \; \dfrac{0.21}{0.02}$

49. $\dfrac{5.2}{0.13} \; ? \; \dfrac{0.11}{0.003}$

50. $\dfrac{0.05}{0.72} \; ? \; \dfrac{0.44}{6.2}$

51. $\dfrac{20.7}{0.039} \; ? \; \dfrac{7.93}{0.015}$

52. $\dfrac{0.0067}{4.25} \; ? \; \dfrac{0.501}{318}$

53. $\dfrac{849}{0.616} \; ? \; \dfrac{23.4}{0.017}$

54. $\dfrac{0.0042}{0.119} \; ? \; \dfrac{0.031}{0.879}$

Use the techniques for comparing ratios presented in this section to solve each of the following word problems.

55. A baseball player for the Boston Red Sox made 15 hits in 39 times at bat. A player for the Chicago White Sox made 28 hits in 71 times at bat. Which player has the higher batting average?

56. A sports car can travel 300 miles on 18 gallons of gas, and a luxury car can travel 400 miles on 25 gallons of gas. Which car gets better mileage?

57. Deodorant soap costs 99¢ for four bars, and facial soap costs $1.49 for six bars. If all bars are the same size, which brand is the better buy?

58. If a 7-ounce tube of toothpaste costs $1.29, and a 12-ounce tube costs $2.25, which size is the better buy?

59. A Russian skater finished a 500-meter race in 37 seconds, and a U.S. skater finished a 1,000-meter race in 1 minute 13 seconds. Who skated faster?

60. Chicken wings cost $1.76 for 3.2 pounds, and chicken legs cost $2.22 for 4.1 pounds. Which is more expensive, chicken legs or wings?

61. A 32-ounce bottle of dishwashing detergent sells for $2.19. An 18.5-ounce bottle sells for $1.25. Which size is the better buy?

62. In May, 17.2 gallons of gas sold for $32.49. In August, 12.9 gallons sold for $23.87. During which month was the price of gas higher?

63. Alabama has 148 hospitals and admits 752,371 patients per year. Florida admits 1,608,248 patients yearly and has a total of 245 hospitals. Which state has the higher ratio of hospitals to yearly admissions?

64. Last year, there were 2,425,000 marriages and 1,187,000 divorces. Ten years ago, there were 2,152,662 marriages and 1,036,000 divorces. Was the ratio of divorces to marriages higher last year or ten years ago?

5.3
Solving Proportions

Thus far, we have looked only at proportions whose terms are all numbers. Proportions often contain an unknown term that is represented by a letter. The letter that is used to represent an unknown quantity is called a **variable**.

For example, in the expression $5 \cdot x$, the number 5 is multiplied by the variable x. The number 5 is often called the **coefficient** of x. The **coefficient** of a variable is the number that is multiplied by the variable. When expressing the product of a number and a variable, the multiplication sign is usually omitted.

$$5 \cdot x = 5x$$

EXAMPLE 1 Identify the variables and their coefficients in each of the following mathematical expressions.

(a) $12y$

y is a variable.

12 is the coefficient of y.

(b) $\dfrac{3}{4}w$

w is a variable.

$\frac{3}{4}$ is the coefficient of w.

(c) t

t is a variable.

Since $t = 1 \cdot t$, the coefficient of t is 1.

(d) $(8)(3)x$

$(8)(3)x = 24x$

x is a variable.

24 is the coefficient of x. ▲

The fundamental property of proportions, which states that the product of the extremes is equal to the product of the means, is used to find the value of the unknown term in a proportion. For example, to find the value of x in the proportion

$$\frac{3}{7} = \frac{9}{x}$$

we first cross multiply:

$$3 \cdot x = 7 \cdot 9$$

Since our goal is to solve for x, we need to perform some mathematical operation on both sides of the equation that will isolate the variable x on the left to obtain

$$x = \text{the unknown quantity}$$

Since $x = 1 \cdot x$, we can isolate x by multiplying both sides of the equation by the reciprocal of 3, which is $\frac{1}{3}$.

$$\frac{1}{3} \cdot 3 \cdot x = 7 \cdot 9 \cdot \frac{1}{3}$$

$$1 \cdot x = \frac{63}{3}$$

$$x = 21$$

To check whether or not our solution is correct, we substitute the value obtained for x into the original equation and determine if the resulting proportion is a true statement.

Check:
$$\frac{3}{7} \; ? \; \frac{9}{21}$$
$$(3)(21) = (7)(9)$$
$$63 = 63 \; \checkmark$$

This procedure is summarized here.

To Solve a Proportion for an Unknown Term

1. Cross multiply to set the product of the extremes equal to the product of the means.

2. Multiply both sides of the equation by the reciprocal of the coefficient of the unknown.

3. Check the solution by substituting the value obtained for the variable into the original equation. ▲

EXAMPLE 2 Find the value of the unknown term in each of the following proportions.

(a) $\dfrac{5}{9} = \dfrac{x}{4}$ Cross multiply.

$5(4) = 9x$ Switch the left side and right side of the
$9x = 5(4)$ equation so the variable will be on the left.
$9x = 20$ The coefficient of x is 9.

$\dfrac{1}{9}(9x) = (20)\dfrac{1}{9}$ The reciprocal of 9 is $\dfrac{1}{9}$.

$1 \cdot x = \dfrac{20}{9}$ Multiply both sides of the equation by $\dfrac{1}{9}$.

$x = \dfrac{20}{9}$

Check:
$$\frac{5}{9} \; ? \; \frac{\frac{20}{9}}{4}$$
$$5(4) \; ? \; 9\left(\frac{20}{9}\right)$$
$$20 = 20 \; \checkmark$$

(b) 8 is to y as 10 is to 3.

$$\frac{8}{y} = \frac{10}{3}$$

$$8(3) = 10y \qquad \text{Cross multiply.}$$

$$10y = 24 \qquad \text{The coefficient of } y \text{ is 10.}$$

$$\left(\frac{1}{10}\right)10y = 24\left(\frac{1}{10}\right) \qquad \text{The reciprocal of 10 is } \frac{1}{10}.$$

$$1 \cdot y = \frac{24}{10} \qquad \text{Multiply both sides of the equation by } \frac{1}{10}.$$

$$y = \frac{12}{5} \qquad \text{Reduce to lowest terms.}$$

Check: $\dfrac{8}{\dfrac{12}{5}}$? $\dfrac{10}{3}$

$$8(3) \ ? \ \frac{12}{5}(10)$$

$$24 = 24 \quad \checkmark$$

(c)

$$\frac{\dfrac{z}{3}}{4} = \frac{\dfrac{7}{5}}{6}$$

$$\frac{5}{6}z = \frac{3}{4}(7) \qquad \text{Cross multiply.}$$

$$\frac{5}{6}z = \frac{21}{4} \qquad \text{The coefficient of } z \text{ is } \frac{5}{6}.$$

$$\left(\frac{6}{5}\right)\left(\frac{5}{6}\right)z = \left(\frac{21}{4}\right)\left(\frac{6}{5}\right) \qquad \begin{array}{l}\text{The reciprocal of } \frac{5}{6} \text{ is } \frac{6}{5}.\\ \text{Multiply both sides by } \frac{6}{5}.\end{array}$$

$$1 \cdot z = \left(\frac{21}{\underset{2}{4}}\right)\left(\frac{\overset{3}{6}}{5}\right) \qquad \text{Simplify.}$$

$$z = \frac{63}{10}$$

Check: $\dfrac{\dfrac{63}{10}}{\dfrac{3}{4}}$? $\dfrac{\dfrac{7}{5}}{6}$

$$\left(\frac{\overset{21}{63}}{\underset{2}{10}}\right)\left(\frac{\overset{}{5}}{\underset{2}{6}}\right) \ ? \ \left(\frac{3}{4}\right)(7)$$

$$\frac{21}{4} = \frac{21}{4} \quad \checkmark$$

(d)
$$\frac{2\frac{3}{8}}{x} = \frac{19}{8}$$

$$\left(2\frac{3}{8}\right)(8) = 19x \qquad \text{Cross multiply.}$$

$$\left(\frac{19}{8}\right)(8) = 19x \qquad \text{Change } 2\frac{3}{8} \text{ to an improper fraction.}$$

$$19 = 19x$$
$$19x = 19 \qquad \text{Switch left side and right side. The coefficient of } x \text{ is 19.}$$

$$\left(\frac{1}{19}\right)19x = 19\left(\frac{1}{19}\right) \qquad \text{The reciprocal of 19 is } \frac{1}{19}.$$

$$1 \cdot x = 1$$
$$x = 1 \qquad \text{Multiply both sides by } \frac{1}{19}.$$

Check:
$$\frac{2\frac{3}{8}}{1} \;?\; \frac{19}{8}$$

$$\left(2\frac{3}{8}\right)(8) \;?\; (1)(19)$$

$$\left(\frac{19}{8}\right)(8) \;?\; 19$$

$$19 = 19 \quad \checkmark$$

(e)
$$\frac{1.5}{50} = \frac{0.012}{t}$$

$$1.5t = 50(0.012) \qquad \text{Cross multiply.}$$

$$1.5t = 0.600$$

$$\left(\frac{1}{1.5}\right)(1.5)t = (0.600)\left(\frac{1}{1.5}\right) \qquad \text{Multiply both sides by } \frac{1}{1.5}.$$

$$t = \frac{0.600}{1.5}$$

$$\begin{array}{r} 0.4 \\ 1.5_\wedge \overline{)0.6_\wedge 00} \end{array}$$

$$t = 0.4 \qquad \text{Since the other terms are decimals, express the answer as a decimal.}$$

Check:
$$\frac{1.5}{50} \;?\; \frac{0.012}{0.4}$$

$$(1.5)(0.4) \;?\; (50)(0.012)$$
$$0.60 = 0.60 \quad \checkmark$$

▲

Quick Quiz

Find the value of the unknown term in each
of the following proportions.

Answers

1. $\dfrac{x}{7} = \dfrac{6}{21}$

1. $x = 2$

2. $\dfrac{4\frac{1}{8}}{y} = \dfrac{2\frac{5}{8}}{\frac{6}{11}}$

2. $y = \dfrac{6}{7}$

3. $\dfrac{0.24}{0.014} = \dfrac{3.6}{t}$

3. $t = 0.21$

5.3 Exercises

Indicate which of the following statements are true and which are false. For those that are false, change the italicized expression to make the statement true.

1. A *variable* is a letter that represents an unknown quantity.

2. The coefficient of a variable is the number that is *added* to the variable.

3. The first step in solving a proportion for an unknown is to *reverse multiply*.

4. To solve for the unknown term in a proportion, we *divide* both sides by the reciprocal of the coefficient of the unknown.

Identify the variables and their coefficients in each of the following mathematical expressions.

5. $5x$ 6. $2y$ 7. $17t$ 8. w 9. $\dfrac{5}{8}u$

10. $\dfrac{7}{2}v$ 11. $1.4a$ 12. $2.7b$ 13. $9(2)x$ 14. $4(5)y$

Find the value of the unknown term in each of the following proportions.

15. $\dfrac{3}{5} = \dfrac{9}{x}$ 16. $\dfrac{x}{2} = \dfrac{15}{10}$ 17. $\dfrac{4}{y} = \dfrac{2}{7}$ 18. $\dfrac{5}{8} = \dfrac{w}{9}$

19. $\dfrac{13}{4} = \dfrac{10}{x}$ 20. $\dfrac{6}{x} = \dfrac{15}{8}$ 21. $\dfrac{12}{18} = \dfrac{y}{21}$ 22. $\dfrac{81}{n} = \dfrac{9}{8}$

23. $\dfrac{6}{11} = \dfrac{m}{33}$ 24. $\dfrac{21}{12} = \dfrac{28}{p}$ 25. $\dfrac{54}{q} = \dfrac{9}{13}$ 26. $\dfrac{w}{17} = \dfrac{81}{51}$

27. $\dfrac{18}{23} = \dfrac{3}{y}$ 28. $\dfrac{4}{19} = \dfrac{z}{10}$ 29. $\dfrac{5}{21} = \dfrac{7}{x}$ 30. $\dfrac{x}{69} = \dfrac{4}{3}$

31. $\dfrac{12}{35} = \dfrac{y}{10}$

32. $\dfrac{9}{x} = \dfrac{25}{11}$

33. $\dfrac{18}{54} = \dfrac{81}{z}$

34. $\dfrac{w}{63} = \dfrac{42}{28}$

35. $\dfrac{96}{y} = \dfrac{60}{24}$

36. $\dfrac{36}{21} = \dfrac{t}{39}$

37. $\dfrac{60}{28} = \dfrac{q}{42}$

38. $\dfrac{18}{120} = \dfrac{24}{r}$

39. $\dfrac{\frac{4}{5}}{\frac{3}{8}} = \dfrac{x}{\frac{15}{16}}$

40. $\dfrac{\frac{7}{8}}{x} = \dfrac{\frac{2}{9}}{\frac{4}{21}}$

41. $\dfrac{\frac{5}{11}}{s} = \dfrac{\frac{10}{27}}{\frac{4}{15}}$

42. $\dfrac{u}{\frac{7}{16}} = \dfrac{\frac{12}{25}}{\frac{3}{5}}$

43. $\dfrac{\frac{4}{21}}{x} = \dfrac{2\frac{1}{7}}{4\frac{7}{8}}$

44. $\dfrac{1\frac{5}{6}}{\frac{5}{12}} = \dfrac{y}{3\frac{1}{3}}$

45. $\dfrac{6\frac{2}{3}}{1\frac{1}{5}} = \dfrac{4\frac{1}{6}}{w}$

46. $\dfrac{z}{3\frac{1}{8}} = \dfrac{7\frac{3}{5}}{2\frac{3}{4}}$

47. $\dfrac{9\frac{4}{9}}{7\frac{1}{2}} = \dfrac{v}{2\frac{1}{4}}$

48. $\dfrac{4\frac{4}{5}}{6\frac{1}{4}} = \dfrac{r}{5\frac{5}{8}}$

49. $\dfrac{3.2}{0.08} = \dfrac{x}{0.002}$

50. $\dfrac{x}{0.56} = \dfrac{40}{0.07}$

51. $\dfrac{0.6}{y} = \dfrac{12}{0.05}$

52. $\dfrac{0.81}{0.27} = \dfrac{0.09}{z}$

53. $\dfrac{5.4}{w} = \dfrac{0.21}{0.007}$

54. $\dfrac{0.036}{2.4} = \dfrac{u}{50}$

55. $\dfrac{7.2}{30} = \dfrac{0.006}{t}$

56. $\dfrac{400}{12} = \dfrac{r}{0.09}$

57. $\dfrac{0.056}{4.2} = \dfrac{0.08}{y}$

58. $\dfrac{x}{0.0048} = \dfrac{200}{3.2}$

59. $\dfrac{1.2}{z} = \dfrac{7,200}{0.003}$

60. $\dfrac{s}{4,200} = \dfrac{120}{0.0036}$

61. $\dfrac{0.0008}{0.024} = \dfrac{0.002}{t}$

62. $\dfrac{0.03}{0.00016} = \dfrac{u}{0.64}$

 63. $\dfrac{168}{x} = \dfrac{252}{216}$

 64. $\dfrac{y}{288} = \dfrac{192}{576}$

 65. $\dfrac{1,344}{4,256} = \dfrac{w}{2,128}$

 66. $\dfrac{3,402}{1,989} = \dfrac{4,536}{z}$

 67. $\dfrac{4.71}{x} = \dfrac{0.94}{0.038}$

Round off x to the nearest thousandth.

 68. $\dfrac{0.079}{6.21} = \dfrac{y}{84.5}$

Round off y to the nearest thousandth.

 69. $\dfrac{3.54}{0.091} = \dfrac{862.7}{w}$

Round off w to the nearest whole number.

 70. $\dfrac{z}{0.326} = \dfrac{53.97}{0.00248}$

Round off z to the nearest tenth.

5.4

Applications of Proportions

Proportions provide us with a powerful tool for solving a wide variety of word problems that express a relationship between two quantities. The procedure outlined on the next page can be used to solve many word problems using proportions. Example 1 illustrates how to apply this procedure.

<table>
<tr><td>

**To Solve Word
Problems Using
Proportions**

</td><td>

1. Determine the unknown quantity you are asked to find and assign it a letter, such as *x*. State what quantity *x* represents using the format,

Let *x* = unknown quantity

This step is called *defining the variable*.

2. Set up a ratio that expresses the relationship between two known quantities given in the problem.

3. Set this ratio equal to a corresponding equivalent ratio that contains the unknown.

4. Solve the resulting proportion for the unknown.

5. Check your answer by substituting your answer for the variable in the original equation. Also check that your answer makes sense as an answer to the question asked in the problem, and label it with the appropriate units. ▲

</td></tr>
</table>

EXAMPLE 1   In a chemistry class, the ratio of men to women is 3 to 2. If there are 21 men in the class, how many women are there?

Solution   Let w = the number of women in the class.   Define the variable w.
Ratio of men to women is $3:2$ or $\frac{3}{2}$.   Set up a ratio between men and women.
Ratio of men to women is $21:w$ or $\frac{21}{w}$.

$$\text{men} \rightarrow \quad \frac{3}{2} = \frac{21}{w} \quad \leftarrow \text{men}$$
$$\text{women} \rightarrow \qquad\qquad \leftarrow \text{women}$$

Set this equal to a corresponding ratio containing w.

$$3w = 2(21)$$ Solve the proportion for w.
$$3w = 42$$

$$w = \frac{42}{3}$$

$$w = 14 \text{ women}$$ Label the answer.

Check: $$\frac{3}{2} \; ? \; \frac{21}{14}$$ Check by substituting back into the original equation.

$$3(14) \; ? \; 2(21)$$
$$42 = 42 \quad \checkmark$$

There are 14 women in the class.   ▲

In Example 1, the result of 14 is a reasonable answer, because if the ratio of men to women is 3 to 2, there must be fewer women than men in the class. An answer that is not a whole number would not make sense, since we do not refer to fractions of a person. Therefore an answer such as $14\frac{1}{2}$ would be a clue that there was probably a mistake made in performing the calculation.

EXAMPLE 2 If 8 pencils sell for 60¢, how much would 12 pencils cost?

Let x = cost of 12 pencils.   Define the variable x.

Ratio of pencils to cents is $\frac{8}{60}$.   Set up a ratio between pencils and cents.

Ratio of pencils to cents is $\frac{12}{x}$.

$$\text{pencils} \rightarrow \quad \frac{8}{60} = \frac{12}{x} \quad \leftarrow \text{pencils}$$
$$\text{cents} \rightarrow \qquad\qquad\qquad \leftarrow \text{cents}$$

Set this equal to a corresponding ratio containing x.

$$8x = 60(12)$$ Solve the proportion for x.

$$\left(\frac{1}{8}\right)8x = 720\left(\frac{1}{8}\right)$$

$$x = \frac{720}{8}$$

$$x = 90¢$$ Label the answer.

Check: $\dfrac{8}{60} \; ? \; \dfrac{12}{90}$ Check your answer.

$$8(90) \; ? \; 60(12)$$
$$720 = 720 \; \checkmark$$

The cost of 12 pencils is 90¢.   ▲

Notice that in Example 2 we could have used a different proportion to solve the problem. We can set up a ratio between any two known quantities, as long as we also can set up a corresponding equivalent ratio that contains the unknown. For example,

$$\text{pencils} \rightarrow \quad \frac{8}{12} = \frac{60}{x} \quad \leftarrow \text{cents}$$
$$\text{pencils} \rightarrow \qquad\qquad\qquad \leftarrow \text{cents}$$

$$8x = 12(60)$$

$$\left(\frac{1}{8}\right)8x = 720\left(\frac{1}{8}\right)$$

$$x = 90¢$$

In order to clearly indicate all of the correct proportions that could be used to solve the problem in Example 2, we need to distinguish the first quantities compared from the second quantities compared. In this case, the first quantities compared are 8 pencils and 60¢, and the second quantities compared are 12 pencils and x¢; that is,

$$\begin{array}{cc} \text{first} \\ \text{quantities} \end{array} \quad \frac{8 \text{ pencils}}{60¢} = \frac{12 \text{ pencils}}{x¢} \quad \begin{array}{c} \text{second} \\ \text{quantities} \end{array}$$

which is the proportion we used to solve Example 2. All of the following proportions could be used also to correctly solve the problem. In each case, the first quantities compared appear in color.

1. $\dfrac{\text{pencils}}{\text{cents}} = \dfrac{\text{pencils}}{\text{cents}}$ Pencils are in both numerators.
First quantities appear in the same ratio.

2. $\dfrac{\text{cents}}{\text{pencils}} = \dfrac{\text{cents}}{\text{pencils}}$ Cents are in both numerators.
First quantities appear in the same ratio.

3. $\dfrac{\text{pencils}}{\text{pencils}} = \dfrac{\text{cents}}{\text{cents}}$ Pencils appear in the same ratio.
First quantities are in both numerators.

4. $\dfrac{\text{pencils}}{\text{pencils}} = \dfrac{\text{cents}}{\text{cents}}$ Pencils appear in the same ratio.
Second quantities are in both numerators.

Thus, a proportion is set up correctly to solve a word problem if it is set up in either of the following ways.

The same units appear in both numerators, and both first quantities appear in the same ratio (Cases 1 and 2).

OR

The same units appear in the same ratio, and both first or both second quantities appear in the numerators (Cases 3 and 4).

EXAMPLE 3 Set up four correct proportions that could be used to solve the following word problem. Solve one of these proportions to find the answer to the problem.

In a bag of marbles, the ratio of green marbles to yellow ones is 7 to 4. If the bag contains 12 yellow marbles, how many marbles are green?

Solution Let $x =$ the number of green marbles.

The first quantities compared are 7 green to 4 yellow, and the second quantities compared are 12 yellow to x green. The following four proportions could be used to correctly solve this problem. In each case, the first quantities compared appear in color.

1. $\dfrac{\text{green}}{\text{yellow}} = \dfrac{\text{green}}{\text{yellow}}$ $\dfrac{7}{4} = \dfrac{x}{12}$

2. $\dfrac{\text{yellow}}{\text{green}} = \dfrac{\text{yellow}}{\text{green}}$ $\dfrac{4}{7} = \dfrac{12}{x}$

3. $\dfrac{\text{green}}{\text{green}} = \dfrac{\text{yellow}}{\text{yellow}}$ $\dfrac{7}{x} = \dfrac{4}{12}$

4. $\dfrac{\text{green}}{\text{green}} = \dfrac{\text{yellow}}{\text{yellow}}$ $\dfrac{x}{7} = \dfrac{12}{4}$

Let us solve the fourth proportion for x.

$$\begin{array}{l}\text{green} \rightarrow \\ \text{green} \rightarrow\end{array} \dfrac{x}{7} = \dfrac{12}{4} \begin{array}{l}\leftarrow \text{yellow} \\ \leftarrow \text{yellow}\end{array}$$

$$\dfrac{x}{7} = \dfrac{3}{1} \qquad \text{Simplify terms.}$$

$$x = 7 \cdot 3 \qquad \text{Cross multiply.}$$

$$x = 21 \text{ green marbles} \qquad \text{Label the answer.}$$

Check: $\dfrac{21}{7} \; ? \; \dfrac{12}{4}$ Check the answer.

$(21)4 \; ? \; 7(12)$

$84 = 84 \; \checkmark$ ▲

EXAMPLE 4   In a city in southern Arizona, all of the 6,300 registered voters are either Republicans or Democrats. If 2 out of 7 registered voters are Democrats, what is the number of registered Republican voters?

Solution   Since we know the number of registered voters, and we are asked to find the number of Republicans, we must first determine the ratio of Republicans to all voters. If all the voters are either Republicans or Democrats, and 2 out of 7 voters are Democrats, then

$$7 \text{ voters} - \underbrace{2 \text{ voters}}_{\text{Democrats}} = \underbrace{5 \text{ voters}}_{\text{Republicans}}$$

Thus, for every 7 voters, 2 are Democrats and 5 are Republicans.
The ratio of Republicans to voters is 5 to 7.
Let r = the number of Republicans.   Define the variable.

$\begin{array}{l}\text{Republicans} \to \\ \text{voters} \to \end{array} \dfrac{5}{7} = \dfrac{r}{6{,}300} \begin{array}{l} \leftarrow \text{Republicans} \\ \leftarrow \text{voters}\end{array}$ Set up a proportion with one known ratio and an equivalent ratio that contains the unknown.

$5(6{,}300) = 7r$ Cross multiply.

$\dfrac{1}{7}(5)(6{,}300) = \dfrac{1}{7}(7r)$ Multiply both sides by the reciprocal of 7.

$\dfrac{5(6{,}300)}{7} = r$ Transpose left side and right side.

$r = \dfrac{5(\overset{900}{\cancel{6{,}300}})}{7}$ Simplify.

$r = 4{,}500 \text{ Republicans}$

     Label the answer.

Check: $\dfrac{5}{7} \; ? \; \dfrac{4{,}500}{6{,}300}$ Check the answer.

$5(6{,}300) \; ? \; 7(4{,}500)$

$31{,}500 = 31{,}500 \; \checkmark$

There are 4,500 registered Republicans.   ▲

EXAMPLE 5   A basket of fruit contains only apples, oranges, and bananas. The ratio of oranges to apples is 3 to 2, and the ratio of bananas to apples is 4 to 3. There are 18 oranges in the basket.

(a) How many apples are in the basket?

We know that there are 18 oranges in the basket and that the ratio of oranges to apples is 3 to 2. Let us set up a proportion using oranges and apples.

Let a = the number of apples in the basket.   Define the variable.

$$\text{oranges} \rightarrow \frac{3}{2} = \frac{18}{a} \leftarrow \text{oranges} \atop \leftarrow \text{apples}$$

Set up a proportion with one known ratio and an equivalent ratio that contains the unknown.

$$3a = 2(18)$$ Cross multiply.

$$\left(\frac{1}{3}\right)3a = 36\left(\frac{1}{3}\right)$$ Multiply both sides by the reciprocal of 3.

$$1a = \frac{36}{3}$$ Simplify.

$$a = 12 \text{ apples}$$ Label the answer.

Check: $\dfrac{3}{2} \ ? \ \dfrac{18}{12}$ Check.

$$3(12) \ ? \ 2(18)$$
$$36 = 36 \ \checkmark$$

There are 12 apples in the basket.

(b) How many bananas are in the basket?

We know that there are 12 apples in the basket and that the ratio of bananas to apples is 4 to 3. Let us set up a proportion using bananas and apples.

Let b = the number of bananas in the basket.   Define the variable.

$$\text{bananas} \rightarrow \frac{4}{3} = \frac{b}{12} \leftarrow \text{bananas} \atop \leftarrow \text{apples}$$

Set up the proportion.

$$4(12) = 3b$$ Solve for b.
$$3b = 48$$

$$\left(\frac{1}{3}\right)3b = 48\left(\frac{1}{3}\right)$$

$$1b = \frac{48}{3}$$

$$b = 16 \text{ bananas}$$ Label the answer.

Check: $\dfrac{4}{3} \ ? \ \dfrac{16}{12}$ Check.

$$4(12) \ ? \ 3(16)$$
$$48 = 48 \ \checkmark$$

There are 16 bananas in the basket.

(c) How many pieces of fruit are in the basket?

Since there are only oranges, apples, and bananas in the basket, add the number of pieces of each kind of fruit to find the total amount of fruit in the basket.

$$18 + 12 + 16 = 46 \text{ fruits}$$

There are 18 oranges, 12 apples, and 16 bananas.

There are 46 pieces of fruit in the basket. ▲

The technique we have established for solving word problems using proportions requires that we use only two different units. If three different units are mentioned in a problem, we must convert one of the units to one of the other two units.

EXAMPLE 6 A car traveling at a constant speed takes 25 minutes to travel 20 miles. How far does the car travel in 2 hours?

Solution In this problem, distance is given in miles, and time is given in both minutes and hours. Therefore, let us convert 2 hours to its equivalent in minutes, so that the problem contains only two different units.

$$2 \text{ hr} = 2 \text{ hr} \cdot \frac{60 \text{ min}}{1 \text{ hr}} = 120 \text{ min}$$

Since 1 hr = 60 min.

Let $x =$ distance traveled in 120 min.

Define the variable.

$$\begin{array}{l} \text{minutes} \rightarrow \\ \text{miles} \rightarrow \end{array} \frac{25}{20} = \frac{120}{x} \begin{array}{l} \leftarrow \text{minutes} \\ \leftarrow \text{miles} \end{array}$$

Set up the proportion.

$$25x = 120(20)$$

Solve for x.

$$\frac{1}{25}(25x) = \frac{1}{25}(2{,}400)$$

$$x = \frac{2{,}400}{25}$$

$$x = 96 \text{ mi}$$

Check: $\frac{25}{20} \; ? \; \frac{120}{96}$

$$25(96) \; ? \; 20(120)$$
$$2{,}400 = 2{,}400 \; \checkmark$$

The car can travel 96 miles in 2 hours. ▲

5.4 Exercises

Indicate which of the following statements are true and which are false. For those that are false, change the italicized expression to make the statement true.

1. Proportions can be used to solve problems that express a relationship between *two* quantities.

2. Assigning a letter to an unknown quantity is called *depicting the variance*.

3. There *is* more than one correct way to set up a proportion to solve a given problem.

4. If a proportion is set up correctly and both first quantities appear in the same ratio, *different* units appear in the numerators of the ratios.

Set up four different proportions that can be used to solve each of the following word problems. Solve one of the proportions in each problem to answer the question asked in the problem.

5. The ratio of a quarterback's attempted passes to completed passes is 7 to 2. If he completed 12 passes, how many did he attempt?

6. Two men have the same ratio of shirts to ties. The first man has 4 ties and 18 shirts. If the second man has 27 shirts, how many ties does he have?

7. If it takes a secretary 20 minutes to type four pages, how many pages can he type in 45 minutes?

8. A motorboat travels 8 miles in 15 minutes. How long does it take for the boat to travel 24 miles while maintaining this same speed?

9. If a 10-inch candle burns for 4 hours, how long would a 6-inch candle burn?

10. Facial tissues are on sale for a price of 3 boxes for $2.00. How much would 5 boxes of tissues cost?

Solve each of the following word problems.

11. If a 27-foot tree casts an 18-foot shadow, how long is the shadow cast by a 12-foot tree?

12. Seven bags of sand weigh 18 pounds. How much does 21 bags of sand weigh?

13. A softball player makes 3 hits in every 8 times at bat. If she is up at bat 40 times in one week, how many hits would you expect her to make?

14. If it takes a painter 2 hours to paint 36 square feet of wall space, how long would you expect it to take her to paint 63 square feet of wall space?

15. A survey reveals that 4 out of 5 dentists recommend sugarless gum to their patients who chew gum. If 300 dentists were surveyed, how many recommended sugarless gum?

16. Notebooks are on sale for a price of 3 for $3.57. How much does 5 notebooks cost?

17. For every 35 people who walk into a car dealership, 2 buy a car there. If 245 people visited a car dealership during the month of March, how many car sales were made that month at that dealership?

18. Yarn sells for a price of 6 skeins for $4.98. How much does 20 skeins of yarn cost?

19. If 3 bottles of shampoo sell for $2.55, how much would 2 bottles cost?

20. How many calories are in 12 fluid ounces of beer if 8 fluid ounces contain 116 calories?

21. One out of 3 seniors from a state university plan to go directly to graduate or professional school upon graduation. If 821 students plan to go directly to graduate or professional school, how many seniors are in the class?

22. A car can travel 82 miles on 4 gallons of gas. How far can the car travel on 16.4 gallons of gas?

23. At a tennis ball factory, for every 50 balls produced, 3 are defective. If 7,250 tennis balls are produced, how many would you expect to be defective?

24. A midwestern law school accepted 2 out of every 5 applicants to its entering class. If the school received 435 applications, how many students were accepted?

25. Eleven out of every 20 people living in an apartment building are single. If 140 people live in that building, how many are single?

26. Two out of 5 books on a student's bookshelf are mathematics books. If she has 25 books on her bookshelf, how many are math books?

27. A 6-foot-tall man casts a shadow $3\frac{1}{2}$ feet long. His daughter casts a shadow 14 inches long. How tall is his daughter?

28. A survey conducted at a liberal arts college indicates that 3 out of 8 students smoke cigarettes. If 4,000 students attend the college, how many smoke cigarettes?

29. In a class of 54 students, 2 out of 9 wear glasses. How many students in that class do not wear glasses?

30. Four out of 7 people ice skating at a local rink cannot skate backwards. If 42 skaters are on the ice, how many can skate backwards?

31. A sugar solution is made by dissolving 1.5 grams of sugar in 18 grams of water. How many grams of sugar must be dissolved in 12 grams of water to make a sugar solution of the same concentration?

32. If you paid $12 for 6.8 gallons of gas, how much would 17 gallons of gas cost?

33. Three out of 5 people who participated in a taste test preferred regular cola over diet cola. If 125 people took the taste test, how many preferred diet cola?

34. Five out of 24 calls that a local fire station receives are false alarms. If the station received 120 calls during a given month, how many were real emergencies?

35. On a map of California, $\frac{1}{2}$ inch represents 30 miles. If two cities on this map are $5\frac{1}{4}$ inches apart, what is the actual distance between them?

36. Freshly ground coffee sells for $2.28 per pound. How much does $\frac{1}{4}$ pound cost?

37. If there are 220 calories in a pint of orange juice, how many calories are in a 12-fluid-ounce serving of orange juice?

38. If ribbon costs $1.86 per yard, how much does $2\frac{1}{2}$ feet of ribbon cost?

39. Ground beef sells for $1.68 per pound. How much does 10 ounces of ground beef cost?

40. The scale on a campus map states that 1 inch represents 500 feet. If the distance from your dorm to the location of your first class is $10\frac{1}{2}$ inches on this map, do you have to walk more than a mile or less than a mile to get to your first class? By how many feet?

41. A man wants to bake a cake for his wife's birthday and decides to use a recipe that calls for 5 eggs and $1\frac{1}{4}$ cups of sugar. If he discovers that there are only 4 eggs in the refrigerator, how much sugar should he use so that these ingredients are mixed in the right proportions?

42. A carpenter earns $117 for a 6-hour job. How much would she receive for a job that takes $3\frac{1}{3}$ hours to complete?

43. Three yards of cotton fabric cost $3.88. What is the cost of $6\frac{3}{4}$ yards of the same fabric?

44. A recipe that yields 4 dozen two-inch cookies calls for $\frac{2}{3}$ cup of sugar and $1\frac{1}{2}$ cups of flour. How much sugar and how much flour would you use to make 3 dozen two-inch cookies?

45. On prime-time television, there are $3\frac{1}{2}$ minutes of commercials for every 15 minutes of broadcast time. How many minutes of commercials would one see while watching an hour of television during prime time?

46. If $\frac{3}{4}$ gallon of paint covers 250 square feet of wall space, how much paint would be needed for a room with 510 square feet of wall space?

47. A jar contains only nickels, dimes, and quarters. The ratio of nickels to dimes is 5 to 3, and the ratio of nickels to quarters is 2 to 3. There are 12 dimes in the jar.
    a. How many nickels are in the jar?
    b. How many quarters are in the jar?
    c. What is the total number of coins in the jar?

48. A drawer contains only knives, forks, and spoons. The ratio of knives to forks is 7 to 9, and the ratio of forks to spoons is 6 to 5. There are 15 spoons in the drawer.
    a. How many forks are in the drawer?
    b. How many knives are in the drawer?
    c. How many pieces of silverware are in the drawer?

49. In a class of 84 students composed of only sophomores, juniors, and seniors, 2 out of 7 students are juniors, and the ratio of sophomores to juniors is 3 to 4.
    a. How many students are juniors?
    b. How many students are sophomores?
    c. How many students are seniors?

50. In a bowl that contains only walnuts, brazil nuts, and pecans, the ratio of walnuts to pecans is 2 to 3. There is a total of 27 nuts in the bowl, and 4 out of 9 are pecans.
    a. How many pecans are in the bowl?
    b. How many walnuts are in the bowl?
    c. How many brazil nuts are in the bowl?

51. A chemist made a solution by dissolving 7 mg of sodium chloride in 500 ml of water. How much sodium chloride should be added to 800 ml of water to make a solution of the same strength?

52. A nurse earned $105 for working a 6-hour shift. How much would he earn for working an 8-hour shift at the same rate of pay?

53. Four out of 25 women in the labor force last year were professional or technical workers. If there were 38,414 women in the labor force last year, how many were professional or technical workers? Round off your answer to the nearest whole number.

54. If 13.7 gallons of gas cost $25.89, how much gas can you buy for $5.00? Round off your answer to the nearest tenth of a gallon.

55. The ratio of the area of Japan to that of Peru is 9 to 31. If Peru covers an area of 496,222 square miles, how large is Japan? Round off your answer to the nearest thousand miles.

56. Eight out of 25 babies born in the United States last year had mothers between the ages of 20 and 24. If 1,155,167 mothers in that age group had babies, what was the total number of babies born in the United States last year? Round off your answer to the nearest thousand.

5.5
Summary and Review

Key Terms

(5.1) A **ratio** is a fraction that expresses a relationship between two quantities.

A **unit fraction** is a fraction equivalent to 1.

A **rate** is a ratio that compares two different quantities and has a denominator equal to 1.

(5.2)   A **proportion** is a statement of equality between two ratios. If the proportion $\frac{a}{b} = \frac{c}{d}$, the numbers a, b, c, and d are called the **terms** of the proportion. The terms a and d are called the **extremes**. The terms b and c are called the **means**.

In any proportion, **the product of the extremes equals the product of the means**. If $\frac{a}{b} = \frac{c}{d}$, then $ad = bc$. The products ad and bc are called **cross products**.

(5.3)   A **variable** is a letter that represents an unknown quantity.

The **coefficient** of a variable is the number that is multiplied by the variable.

Calculations

(5.1)   **To compare two quantities using a ratio**, put the first quantity in the numerator, and put the second quantity in the denominator.

(5.1)   **To convert from one unit of measurement to another**, multiply by the unit fraction that has the old unit in the denominator and the new unit in the numerator.

(5.2)   **To show that two ratios are equal**, show that the product of the extremes is equal to the product of the means.

To show that the first ratio is greater than the second ratio, show that the product of the extremes is greater than the product of the means.

To show that the first ratio is less than the second ratio, show that the product of the extremes is less than the product of the means.

(5.3)   **To solve a proportion for an unknown term**, cross multiply to set the product of the extremes equal to the product of the means. Then multiply both sides of the equation by the reciprocal of the coefficient of the unknown.

(5.4)   **To solve a word problem using a proportion**, first define the variable. Then set up a ratio that expresses the relationship between two known quantities given in the problem. Set this ratio equal to a corresponding equivalent ratio that contains the unknown. Solve the resulting proportion for the unknown variable, and answer the question asked in the problem.

To determine if a proportion is set up correctly to solve a word problem, check that either: the same units appear in both numerators, and both first quantities appear in the same ratio,

OR

the same units appear in the same ratio, and both first or both second quantities appear in the numerators.

Chapter **5**
Review Exercises

Indicate which of the following statements are true and which are false. For those that are false, change the italicized expression to make the statement true.

1.   A *rate* is a ratio that compares two different quantities and has a denominator of 1.

2.   When comparing two ratios, if the product of the extremes exceeds the product of the means, the second ratio is *greater* than the first.

3.   The coefficient of a variable is the number that is *added to* the variable.

4.   The price of a single item is called a *unit price*.

5. A proportion is set up *correctly* if the same units appear in both numerators, and both first quantities appear in the same ratio.

6. In a proportion, the product of the means and the product of the extremes are sometimes called *reverse products*.

(5.1) **Express each of the following comparisons as a ratio with the same units in the numerator and denominator. Reduce to lowest terms.**

7. 4 feet to 2 yards

8. 18 inches to 3 feet

9. 6 hours to 45 minutes

10. 30 days to 5 weeks

11. 10 quarts to 3 gallons

12. 2 pints to 24 fluid ounces

13. 12 nickels to 2 quarters

14. 74 dimes to 5 dollars

15. $3\frac{1}{2}$ pounds to 18 ounces

16. 54 items to $2\frac{1}{3}$ dozen

17. $\frac{3}{4}$ mile to 900 feet

18. $5\frac{1}{2}$ years to 30 months

(5.2) **Determine which of the following ratios is greater, and place the appropriate sign (> or <) between them.**

19. $\frac{3}{8}$? $\frac{4}{11}$

20. $\frac{2}{7}$? $\frac{3}{10}$

21. $\frac{4}{5}$? $\frac{5}{6}$

22. $\frac{5}{9}$? $\frac{4}{7}$

23. $\frac{1}{3}$? $\frac{8}{25}$

24. $\frac{4}{21}$? $\frac{1}{5}$

25. $\frac{2\frac{1}{4}}{6}$? $\frac{\frac{2}{3}}{\frac{3}{4}}$

26. $\frac{7}{2\frac{3}{4}}$? $\frac{8}{3\frac{2}{7}}$

27. $\frac{\frac{5}{3}}{\frac{3}{4}}$? $\frac{2\frac{6}{7}}{\frac{3}{8}}$

28. $\frac{\frac{2}{8}}{\frac{3}{4}}$? $\frac{\frac{5}{6}}{2\frac{6}{7}}$

29. $\frac{0.03}{1.1}$? $\frac{0.002}{0.07}$

30. $\frac{0.13}{0.0009}$? $\frac{0.4}{0.003}$

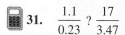 31. $\frac{1.1}{0.23}$? $\frac{17}{3.47}$

32. $\frac{8{,}021}{3.19}$? $\frac{16.3}{0.0065}$

(5.3) **Identify the variables and their coefficients in each of the following mathematical expressions.**

33. $8x$ 34. $20y$ 35. z 36. $4.9w$ 37. $\frac{3}{5}t$ 38. $\frac{2}{9}a$ 39. $(3)(5)b$ 40. $(4)(7)u$

(5.3) **Find the value of the unknown term in each of the following proportions.**

41. $\frac{3}{5} = \frac{x}{40}$

42. $\frac{w}{60} = \frac{4}{3}$

43. $\frac{4}{x} = \frac{2}{7}$

44. $\frac{6}{8} = \frac{9}{y}$

45. $\frac{9}{12} = \frac{t}{6}$

46. $\frac{6}{u} = \frac{9}{3}$

47. $\frac{24}{36} = \frac{16}{n}$

48. $\frac{28}{p} = \frac{21}{33}$

49. $\frac{\frac{2}{3}}{8} = \frac{x}{\frac{6}{7}}$

50. $\frac{y}{\frac{3}{4}} = \frac{\frac{2}{5}}{9}$

51. $\frac{2\frac{5}{8}}{5\frac{5}{6}} = \frac{y}{8\frac{1}{3}}$

52. $\frac{3\frac{3}{4}}{2\frac{1}{7}} = \frac{5\frac{4}{9}}{x}$

53. $\frac{2.7}{t} = \frac{0.09}{0.2}$

54. $\frac{w}{0.004} = \frac{54}{0.8}$

55. $\frac{0.48}{0.064} = \frac{n}{3.2}$

56. $\frac{56}{p} = \frac{4.9}{0.042}$

57. $\dfrac{2.73}{0.512} = \dfrac{x}{0.0081}$

Round off x to the nearest ten-thousandth.

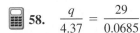 **58.** $\dfrac{q}{4.37} = \dfrac{29}{0.0685}$

Round off q to the nearest whole number.

(5.4) Solve each of the following word problems.

59. A regular deck of 52 cards contains 4 queens. What ratio represents the fraction of the cards that are queens?

60. A calculus class has 225 students, of which 175 are engineering majors. What ratio represents the fraction of the students that are engineering majors?

61. A 5-pound bag of flour sells for $1.29. What is its unit price per pound?

62. A designer bought $3\frac{1}{4}$ yards of plaid fabric and 10 feet of striped fabric. Write a fraction that has the same units in the numerator and denominator that compares these two amounts. If both pieces of material have the same width, did she buy more of the plaid material or striped material?

63. A student spent $\frac{7}{8}$ hour on his math homework and 50 minutes on his chemistry homework. Write a fraction with the same units in the numerator and denominator that compares these two times. On which assignment did he spend more time?

64. A $1\frac{1}{2}$-cup serving of green beans contains 13 calories. A $\frac{2}{3}$-cup serving of broccoli contains 29 calories. For the same size serving, which contains more calories, broccoli or green beans?

65. The pitcher on a baseball team made 27 hits in 62 times at bat. The shortstop made 19 hits in 43 times at bat. Who has the higher batting average?

66. Four out of 7 sophomores at a state university have not yet decided their major. If there are 2,149 sophomores at that school, how many have decided their major?

67. If 4 dining room chairs cost $164, how much would 6 chairs cost?

68. If 3 tires weigh 57 pounds, how much do 4 tires weigh?

69. Three out of 8 working mothers in a major U.S. city have at least one child under the age of 5. If there are 1,816 mothers employed in that city, how many have at least one child under the age of 5?

70. The scale on a map states that $1\frac{3}{4}$ inches represents 280 miles. If two cities are $2\frac{1}{2}$ inches apart on the map, what is the distance between them?

71. To make a glass of chocolate milk, you dissolve 2 teaspoons of chocolate mix in 1 cup of milk. How much chocolate mix would you need to make 1 quart of chocolate milk?

72. Two out of every 15 items sold at a local pharmacy are prescription drugs. If 8,265 items were sold last week, how many were prescription medicines?

73. For every dollar earned, a computer programmer must pay 35¢ in taxes. If her yearly salary is $37,000, how much does she pay in taxes for one year?

74. Two out of 9 employees of a plastics company make over $35,000 a year. If 3,483 people are employed at that company, how many make over $35,000 a year?

75. The state of Montana has a population of 694,409. If Montana covers 145,587 square miles, what is its average population density per square mile? Round your answer to the nearest hundredth.

76. At an international competition, a Belgian swam 500 yards in 6 minutes 35.24 seconds, and a Canadian swam 200 yards in 2 minutes 37.82 seconds. Who swam faster, the Belgian or the Canadian?

77. In the United States, the ratio of dollars spent on transportation to dollars spent on food is 2 to 3. If $184.7 billion is spent on food, how much is spent on transportation? Round off your answer to the nearest tenth of a billion dollars.

 78. The ratio of the number of people who watch prime-time television on Friday night to the number who watch on Sunday night is 17 to 21. If 84,690,000 watch prime-time television on Friday night, how many people watch prime-time television on Sunday night? Round off your answer to the nearest ten thousand.

Chapter **5**
Explain It in Words

79. What is a ratio? Give an example.

80. What is a unit fraction? Give an example.

81. Explain what is meant by "the fundamental property of proportions." Give an example.

82. How can you show that one ratio is greater than another? Give an example.

83. What kinds of word problems can be solved using a proportion? Give an example.

84. Why is it that you can set up a proportion more than one way and still get the correct answer? Give an example.

Chapter **5**
Chapter Test

[5.1] **Express each of the following as a ratio having the same units in the numerator and denominator.**

1. 3 feet to 28 inches

2. 90 minutes to 2 hours

3. 9 quarters to 15 dimes

4. 2 years to 28 months

5. 2 pounds to 20 ounces

[5.2] **Determine which of the following ratios is greater and place the appropriate sign (> or <) between them.**

6. $\dfrac{7}{9} ? \dfrac{4}{5}$

7. $\dfrac{8}{15} ? \dfrac{4}{7}$

8. $\dfrac{\frac{5}{6}}{\frac{1}{4}} ? \dfrac{\frac{7}{8}}{\frac{1}{3}}$

9. $\dfrac{3\frac{5}{8}}{\frac{8}{9}} ? \dfrac{1\frac{5}{6}}{\frac{3}{8}}$

10. $\dfrac{71}{0.03} ? \dfrac{4.9}{0.002}$

[5.3] **Find the value of the unknown term in each of the following proportions.**

11. $\dfrac{x}{12} = \dfrac{4}{3}$

12. $\dfrac{2}{5} = \dfrac{18}{x}$

13. $\dfrac{3\frac{1}{4}}{w} = \dfrac{2\frac{2}{5}}{8}$

14. $\dfrac{1\frac{7}{8}}{3\frac{3}{4}} = \dfrac{x}{3\frac{4}{5}}$

15. $\dfrac{5.4}{z} = \dfrac{0.9}{0.4}$

[5.4]   Solve each of the following word problems.

16.   If 3 pounds of coffee cost $9.87, what is the unit price per pound?

17.   A station wagon can travel 125 miles on 4 gallons of gas and a compact car can travel 95 miles on 3 gallons of gas. Which car gets the better mileage?

18.   Two out of 3 students surveyed said they watch less than 10 hours of television each week. If 78 students were surveyed, how many watch less than 10 hours of television each week?

19.   If 3 cans of tuna sell for $4.17, how much would 5 cans cost?

20.   If a woman who is 5 feet 4 inches tall casts a shadow 6 feet long, how long of a shadow would be cast by a man standing next to her who is 6 feet 8 inches tall?

Chapter **6**

Percent

6.1
Meaning of Percent

The word *percent* is common in our everyday vocabulary. It is used to refer to such things as taxes, discounts, interest, commissions, and tips. In this chapter, we will discuss how the notion of a percent relates to two mathematical quantities we are already familiar with—fractions and decimals. We will also establish techniques for using percents to solve a variety of problems applicable to our daily life.

▶ **DEFINITION** The word *percent* means "per one hundred," or the ratio of a number to 100.

Since a percent refers to the number of parts out of one hundred, any percent can be rewritten as a fraction with a denominator of 100. For example, 25% means 25 parts out of one hundred, or $\frac{25}{100}$, which reduces to $\frac{1}{4}$. We can rewrite any percent in this manner, as outlined below.

To Convert a Percent to a Fraction

1. Remove the percent sign (%).
2. Put the number over a denominator of 100.
3. Reduce to lowest terms. ▲

EXAMPLE 1 Convert each of the following percents to fractions.

(a) 80%

$$80\% = \frac{80}{100} = \frac{4}{5}$$

Since 80% means 80 parts out of 100, remove % sign, put 80 over denominator of 100, and reduce.

(b) 3%

$$3\% = \frac{3}{100}$$

Remove % sign and put 3 over denominator of 100.

236

(c) 58.25%

$$58.25\% = \frac{58.25}{100}$$ 58.25% means 58.25 hundredths.

$$= \frac{58.25 \times 100}{100 \times 100}$$ Multiply numerator and denominator by 100 to remove decimals.

$$= \frac{5,825}{10,000}$$

$$= \frac{233}{400}$$ Divide numerator and denominator by 25 to reduce to lowest terms.

(d) 7.5%

$$7.5\% = \frac{7.5}{100}$$ Remove % sign and put 7.5 over denominator of 100.

$$= \frac{7.5 \times 10}{100 \times 10} = \frac{75}{1,000} = \frac{3}{40}$$ Simplify to remove decimal and reduce.

(e) 100%

$$100\% = \frac{100}{100} = 1$$

Notice that 100% is equal to 1.

(f) 225%

$$225\% = \frac{225}{100} = \frac{9}{4} = 2\frac{1}{4}$$

Notice that any percent greater than 100% is greater than 1.

(g) $\frac{2}{3}\%$

$$\frac{2}{3}\% = \frac{\frac{2}{3}}{100} = \frac{\frac{2}{3} \times 3}{100 \times 3} = \frac{2}{300}$$ Multiply numerator and denominator by 3.

(h) $33\frac{1}{3}\%$

$$33\frac{1}{3}\% = \frac{33\frac{1}{3}}{100}$$ Rewrite as a complex fraction with a denominator of 100.

$$= \frac{\frac{100}{3}}{100}$$ Convert the numerator to an improper fraction.

$$= \frac{\frac{100}{3} \times 3}{100 \times 3}$$ Multiply the numerator and denominator by 3 to simplify the complex fraction.

$$= \frac{100}{300}$$

$$= \frac{1}{3}$$ Reduce to lowest terms. ▲

Any fraction that has a denominator of 100 can be rewritten as a percent by removing the denominator and placing a percent sign after the numerator. For example,

$$\frac{47}{100} = 47\%$$

If a fraction does not have a denominator of 100, we first rewrite it as an equivalent fraction that does have a denominator of 100, and then convert it to a percent. For example, to convert $\frac{3}{20}$ to a percent, first multiply numerator and denominator by 5 to obtain an equivalent fraction with a denominator of 100.

$$\frac{3}{20} = \frac{3 \times 5}{20 \times 5} = \frac{15}{100} = 15\%$$

EXAMPLE 2   Rewrite each of the following fractions as a percent.

(a) $\dfrac{7}{100}$

$$\frac{7}{100} = 7\%$$ The denominator is 100, so remove it and place a % sign after the numerator, 7.

(b) $\dfrac{66\frac{2}{3}}{100}$

$$\frac{66\frac{2}{3}}{100} = 66\frac{2}{3}\%$$ The denominator is 100, so remove it and place a % sign after the numerator, $66\frac{2}{3}$.

(c) $\dfrac{8.1}{10}$

$$\frac{8.1}{10} = \frac{8.1 \times 10}{10 \times 10} = \frac{81}{100} = 81\%$$ Multiply numerator and denominator by 10 to get an equivalent fraction with a denominator of 100.

(d) $\dfrac{7}{25}$

$$\frac{7}{25} = \frac{7 \times 4}{25 \times 4} = \frac{28}{100} = 28\%$$ Multiply numerator and denominator by 4 to get an equivalent fraction with a denominator of 100.

(e) $\dfrac{129}{1,000}$

$$\frac{129}{1,000} = \frac{129 \times \frac{1}{10}}{1,000 \times \frac{1}{10}}$$ Multiply numerator and denominator by $\frac{1}{10}$ to get an equivalent fraction with a denominator of 100.

$$= \frac{\frac{129}{10}}{100} = \frac{12.9}{100} = 12.9\%$$

We can also use a proportion to find an equivalent fraction with a denominator of 100 by writing a variable in place of the unknown numerator.

EXAMPLE 3   Convert $\frac{3}{8}$ to a percent.

Solution $\dfrac{3}{8} = \dfrac{x}{100}$

First set up a proportion to find a fraction equivalent to $\frac{3}{8}$ with a denominator of 100.

$3 \cdot 100 = 8 \cdot x$

Cross multiply and solve for x.

$8x = 300$

$\left(\dfrac{1}{8}\right)8x = 300\left(\dfrac{1}{8}\right)$

$x = \dfrac{300}{8}$

$x = 37.5$

$$\begin{array}{r} 3\,7.5 \\ 8\,\overline{\smash{)}\,3\,0\,0.0} \\ 2\,4 \\ \hline 6\,0 \\ 5\,6 \\ \hline 4\,0 \\ 4\,0 \\ \hline 0 \end{array}$$

$\dfrac{3}{8} = \dfrac{37.5}{100} = 37.5\%$

Substitute 37.5 for x in the proportion and change $\frac{37.5}{100}$ to a percent.   ▲

The procedures that we established for converting fractions to percents are summarized below.

To Convert a Fraction to a Percent

1. Find an equivalent fraction with a denominator of 100, using a proportion if necessary.

2. Remove the denominator of 100 and write a percent sign after the numerator.   ▲

EXAMPLE 4   Convert each of the following fractions to percents.

(a) $\dfrac{5}{16}$

First set up a proportion to find an equivalent fraction with a denominator of 100.

$\dfrac{5}{16} = \dfrac{x}{100}$

$5 \cdot 100 = 16 \cdot x$

$16x = 500$

$\left(\dfrac{1}{16}\right)16x = 500\left(\dfrac{1}{16}\right)$

$x = \dfrac{500}{16}$

$x = 31.25$

$$\begin{array}{r} 3\,1.2\,5 \\ 16\,\overline{\smash{)}\,5\,0\,0.0\,0} \\ 4\,8 \\ \hline 2\,0 \\ 1\,6 \\ \hline 4\,0 \\ 3\,2 \\ \hline 8\,0 \\ 8\,0 \\ \hline 0 \end{array}$$

$\dfrac{5}{16} = \dfrac{31.25}{100} = 31.25\%$

Substitute 31.25 for x in the proportion and change $\frac{31.25}{100}$ to a percent.

(b) $\dfrac{9}{8}$

$$\frac{9}{8} = \frac{x}{100}$$

$$9 \cdot 100 = 8 \cdot x$$

$$8x = 900$$

$$\left(\frac{1}{8}\right)8x = 900\left(\frac{1}{8}\right)$$

$$x = \frac{900}{8}$$

$$x = 112.5$$

$$\frac{9}{8} = \frac{112.5}{100} = 112.5\%$$

First set up a proportion to find an equivalent fraction with a denominator of 100.

```
      1 1 2.5
  8 | 9 0 0.0
      8
      ─────
      1 0
        8
      ─────
        2 0
        1 6
      ─────
          4 0
          4 0
      ─────
            0
```

Now change $\frac{112.5}{100}$ to a percent.

(c) $\dfrac{1}{3}$

$$\frac{1}{3} = \frac{x}{100}$$

$$1 \cdot 100 = 3 \cdot x$$

$$3x = 100$$

$$\left(\frac{1}{3}\right)3x = 100\left(\frac{1}{3}\right)$$

$$x = \frac{100}{3}$$

$$x = 33\frac{1}{3}$$

$$\frac{1}{3} = \frac{33\frac{1}{3}}{100} = 33\frac{1}{3}\%$$

First set up a proportion to find an equivalent fraction with a denominator of 100.

```
      3 3 R1
  3 | 1 0 0      = 33 1/3
      9
      ─────
      1 0
        9
      ─────
        1
```

Notice that when the long division process continues to yield a remainder, we express the answer as a mixed number.

Now change $\dfrac{33\frac{1}{3}}{100}$ to a percent.   ▲

Quick Quiz

Change each of the following percents to fractions reduced to lowest terms.

Answers

1. 60%
2. 5%

Change each of the following fractions to percents.

3. $\dfrac{73}{100}$

4. $\dfrac{9}{25}$

1. $\dfrac{3}{5}$

2. $\dfrac{1}{20}$

3. 73%

4. 36%

6.1 Exercises

Indicate which of the following statements are true and which are false. For those that are false, change the italicized or underlined expression to make the statement true.

1. Percent means *per one hundred*.

2. Any percent can be rewritten as a fraction with a *numerator* of 100.

3. 20% *is greater* than 2.

4. Percents greater than 100% are always equivalent to numbers greater than <u>10</u>.

Convert each of the following percents to a fraction reduced to lowest terms.

5. 75%	6. 30%	7. 2%	8. 6%	9. 57%	10. 93%
11. 64%	12. 88%	13. 120%	14. 250%	15. 500%	16. 1,800%
17. 4.4%	18. 7.6%	19. 0.45%	20. 0.28%	21. 3.07%	22. 60.1%
23. 17.3%	24. 54.9%	25. 18.75%	26. 56.25%	27. 93.75%	28. 62.50%

29. $\frac{1}{2}\%$ 30. $\frac{3}{5}\%$ 31. $31\frac{1}{4}\%$ 32. $87\frac{1}{2}\%$ 33. $70\frac{1}{5}\%$ 34. $2\frac{3}{4}\%$

35. $66\frac{2}{3}\%$ 36. $8\frac{1}{3}\%$

Convert each of the following fractions to a percent.

37. $\frac{13}{100}$ 38. $\frac{88}{100}$ 39. $\frac{1}{100}$ 40. $\frac{49}{100}$ 41. $\frac{3.2}{100}$ 42. $\frac{0.73}{100}$

43. $\frac{5\frac{1}{3}}{100}$ 44. $\frac{72\frac{2}{9}}{100}$ 45. $\frac{3}{4}$ 46. $\frac{8}{5}$ 47. $\frac{2}{5}$ 48. $\frac{3}{2}$

49. $\frac{7}{25}$ 50. $\frac{9}{50}$ 51. $\frac{13}{20}$ 52. $\frac{11}{25}$ 53. $\frac{8}{10}$ 54. $\frac{15}{10}$

55. $\frac{27}{1,000}$ 56. $\frac{318}{1,000}$ 57. $\frac{318}{10,000}$ 58. $\frac{64}{10,000}$ 59. $\frac{7}{40}$ 60. $\frac{1}{16}$

61. $\frac{5}{8}$ 62. $\frac{3}{16}$ 63. $\frac{1}{9}$ 64. $\frac{2}{3}$ 65. $\frac{7}{15}$ 66. $\frac{11}{30}$

Answer each of the following questions.

67. At a state university, $\frac{3}{5}$ of all students work part-time while attending school. What percentage of the student body has part-time jobs?

68. At a large computer rm, $\frac{1}{4}$ of all employees commute more than 10 miles to work. What percent of the company's employees commutes more than 10 miles to work?

69. A 2-ounce serving of spaghetti contains 35% of the U.S. recommended daily allowance of thiamine. Express this percent as a fraction reduced to lowest terms.

70. Of all persons surveyed, 72% said they would prefer to be compensated for overtime hours by receiving extra vacation days instead of being paid for the additional hours worked. What fraction of those surveyed does this percent represent?

6.2
Fractions, Decimals, and Percents

Since any percent can be rewritten as a fraction with a denominator of 100, we can easily convert any percent to a decimal. Consider the following examples.

$$32\% = \frac{32}{100} = 0.32$$

$$6.7\% = \frac{6.7}{100} = \frac{6.7 \times 10}{100 \times 10} = \frac{67}{1,000} = 0.067$$

$$125\% = \frac{125}{100} = 1.25$$

These examples illustrate the procedure stated below.

To convert a percent to a decimal, remove the percent sign and move the decimal point two places to the left.   ▲

EXAMPLE 1   Convert each of the following percents to a decimal.

(a) 19.2%
$$19.2\% = 19.2 = 0.192$$
Remove % sign and move decimal point 2 places to left.

(b) 3%
$$3\% = 03. = 0.03$$
Remove % sign and move decimal point 2 places to left.

(c) 0.54%
$$0.54\% = 00.54 = 0.0054$$
Remove % sign and move decimal point 2 places to left.

(d) $15\frac{3}{4}\%$

$$15\frac{3}{4}\% = 15.75\% = 15.75\% = 0.1575$$
First rewrite mixed number as a decimal.   ▲

We can use a similar technique to rewrite a decimal as a percent. Consider the following examples.

$$0.95 = \frac{95}{100} = 95\%$$

$$6.02 = \frac{602}{100} = 602\%$$

$$0.007 = \frac{7}{1,000} = \frac{0.7}{100} = 0.7\%$$

These examples illustrate the procedure stated below.

To convert a decimal to a percent, move the decimal point two places to the right and attach a percent sign.   ▲

EXAMPLE 2   Convert each of the following decimals to a percent.

(a)   0.88

$$0.88 = 0.88_\frown = 88\%$$

Move decimal point 2 places to right and add % sign.

(b)   0.0503

$$0.0503 = 0.05_\frown 03 = 5.03\%$$

Move decimal point 2 places to right and add % sign.

(c)   2

$$2 = 2.00_\frown = 200\%$$

Move decimal point 2 places to right and add % sign.   ▲

In Section 6.1, we learned a technique for converting a fraction to a percent by using a proportion. Now that we know how to change a decimal to a percent, we can also use the additional technique stated below.

To Convert a Fraction to a Percent

1. Change the fraction to a decimal by using long division.

2. Rewrite the decimal as a percent.   ▲

EXAMPLE 3   Convert each of the following fractions to a percent.

(a) $\dfrac{5}{8}$

$$\dfrac{5}{8} = 5 \div 8 = 0.625$$

First change $\frac{5}{8}$ to a decimal using long division.

$$
\begin{array}{r}
.625 \\
8\overline{\smash{)}5.000} \\
\underline{4\,8} \\
2\,0 \\
\underline{1\,6} \\
4\,0 \\
\underline{4\,0} \\
0
\end{array}
$$

$$0.625 = 0.62_\frown 5 = 62.5\%$$

Now convert 0.625 to a percent.

(b) $3\dfrac{1}{8}$

$$3\dfrac{1}{8} = \dfrac{25}{8} = 25 \div 8 = 3.125$$

Rewrite $3\dfrac{1}{8}$ as an improper fraction, and then convert it to a decimal.

$$
\begin{array}{r}
3.1\,2\,5 \\
8\,\overline{\big)\,2\,5.0\,0\,0} \\
2\,4 \\
\hline
1\,0 \\
8 \\
\hline
2\,0 \\
1\,6 \\
\hline
4\,0 \\
4\,0 \\
\hline
0 \\
\end{array}
$$

$3.125 = 3.12\underset{\curvearrowright}{\,}5 = 312.5\%$

Now convert 3.125 to a percent.

(c) $\dfrac{1}{6}$

$$\dfrac{1}{6} = 1 \div 6 = 0.166\ldots$$

It is apparent that the decimal 0.166 . . . will not terminate. In a case like this, we generally carry out the division to the hundredths place, and note the place value of the remainder. We then express the quotient as the sum of a decimal and a fraction whose numerator is the remainder and whose denominator is the divisor. That is,

$$\dfrac{1}{6} = 0.16\dfrac{4}{6} = 0.16\dfrac{2}{3}$$

$$0.16\dfrac{2}{3} = \dfrac{16\dfrac{2}{3}}{100} = 16\dfrac{2}{3}\%$$

First change $\dfrac{1}{6}$ to a decimal using long division.

$$
\begin{array}{r}
.166\ldots \\
6\,\overline{\big)\,1.000} \\
6 \\
\hline
4\,0 \\
3\,6 \\
\hline
4\,0 \\
3\,6 \\
\hline
4 \\
\end{array}
$$

$$
\begin{array}{r}
.16\tfrac{4}{6} \\
6\,\overline{\big)\,1.00} \\
6 \\
\hline
4\,0 \\
3\,6 \\
\hline
4 \leftarrow \text{hundredths place} \\
\end{array}
$$

Convert $0.16\dfrac{2}{3}$ to a percent. ▲

Notice that $16\dfrac{2}{3}\%$ is the exact answer to the problem in Example 3(c). To express $\dfrac{1}{6}$ as an approximate percent correct to the nearest tenth of a percent, we would carry out the long division to the nearest ten-thousandth and then round off to the nearest thousandth before converting the decimal to a percent. That is,

$$\dfrac{1}{6} = 0.16\overline{66} \cong 0.167 = 0.16\underset{\curvearrowright}{\,}7 = 16.7\%$$

Quick Quiz

Answers

1. Convert 7.8% to a decimal.
2. Convert 0.153 to a percent.
3. Convert $\frac{3}{5}$ to a percent.

1. 0.078
2. 15.3%
3. 60%

6.2 Exercises

Indicate which of the following statements are true and which are false. For those that are false, change the italicized expression to make the statement true.

1. To convert a percent to a decimal, remove the percent sign, and move the decimal point two places to the *right*.

2. To convert a decimal to a percent, move the decimal point two places to the *left*, and attach a percent sign.

3. To express a fraction as a percent correct to the nearest tenth of a percent, we first convert it to a decimal and round off to the nearest *thousandth*.

4. A repeating decimal can be expressed as an *exact* percent that contains a fraction.

Convert each of the following percents to a decimal.

5. 71%	**6.** 12%	**7.** 5%	**8.** 1%	**9.** 24.8%	**10.** 99.9%
11. 4.09%	**12.** 7.63%	**13.** 0.28%	**14.** 0.06%	**15.** 225%	**16.** 170%
17. 3,684%	**18.** 1,000%	**19.** 5.529%	**20.** 3.3%	**21.** 18.03%	**22.** 181.7%
23. $8\frac{1}{2}$%	**24.** $2\frac{3}{4}$%	**25.** $45\frac{3}{5}$%	**26.** $72\frac{1}{10}$%	**27.** $130\frac{1}{8}$%	**28.** $205\frac{5}{16}$%

29. $2\frac{1}{7}$% Round off to the nearest ten-thousandth. **30.** $77\frac{2}{3}$% Round off to the nearest ten-thousandth.

Convert each of the following decimals to a percent.

31. 0.37	**32.** 0.18	**33.** 0.04	**34.** 0.02	**35.** 0.66	**36.** 0.95	**37.** 0.613
38. 0.809	**39.** 0.009	**40.** 0.045	**41.** 0.6	**42.** 0.1	**43.** 0.0675	**44.** 0.0023
45. 4.5	**46.** 3.6	**47.** 3	**48.** 10	**49.** 13.2	**50.** 65	**51.** 0.813
52. 0.040	**53.** 5.07	**54.** 0.301	**55.** 0.0605	**56.** 0.0012		

Convert each of the following fractions to a percent.

57. $\frac{1}{2}$	**58.** $\frac{3}{5}$	**59.** $\frac{7}{8}$	**60.** $\frac{9}{20}$	**61.** $\frac{1}{9}$	**62.** $\frac{2}{3}$	**63.** $\frac{3}{16}$	**64.** $\frac{8}{25}$
65. $\frac{27}{40}$	**66.** $\frac{51}{75}$	**67.** $\frac{5}{12}$	**68.** $\frac{5}{6}$	**69.** $\frac{2}{15}$	**70.** $\frac{11}{20}$	**71.** $\frac{2}{7}$	**72.** $\frac{5}{9}$
73. $\frac{5}{13}$	**74.** $\frac{4}{11}$	**75.** $\frac{18}{5}$	**76.** $\frac{7}{4}$	**77.** $\frac{11}{2}$	**78.** $\frac{9}{8}$	**79.** $2\frac{3}{4}$	**80.** $5\frac{1}{8}$
81. $1\frac{1}{2}$	**82.** $3\frac{1}{3}$						

83. $\dfrac{7}{54}$ Round off to the nearest tenth of a percent.

84. $\dfrac{4}{77}$ Round off to the nearest tenth of a percent.

85. $\dfrac{87}{144}$ Round off to the nearest tenth of a percent.

86. $\dfrac{62}{111}$ Round off to the nearest tenth of a percent.

87. $\dfrac{485}{588}$ Round off to the nearest tenth of a percent.

88. $\dfrac{307}{922}$ Round off to the nearest tenth of a percent.

Complete the following chart so that each row contains an equivalent fraction, decimal, and percent.

	Fraction	Decimal	Percent
89.	$\dfrac{1}{20}$		
90.		0.125	
91.	$\dfrac{3}{16}$		
92.			25%
93.		0.33 . . .	
94.	$\dfrac{2}{5}$		
95.			$66\frac{2}{3}\%$
96.		0.7	
97.	$\dfrac{5}{6}$		
98.			275%

Answer each of the following questions.

99. Three out of 8 students in a general chemistry class are also taking calculus. What percent of the students in this chemistry class is also taking calculus?

100. Of the people living in a midwestern town who voted in the last election, 36% voted Republican. What fraction of the voters in this town voted Republican?

101. One pound of a certain alloy contains 0.48 pound of pure zinc. What percent of the alloy is pure zinc?

102. A runner completed 72% of a 1-mile race 3 minutes after the race began. Express as a decimal the distance he ran during the first 3 minutes of the race.

6.3
Equations Involving Percents

Many applications using percents require us to solve an equation to find an unknown quantity. In this section, we will learn how to translate an expression that contains a percent into a mathematical equation that can be solved for an unknown quantity.

EXAMPLE 1   What number is 20% of 18?

Solution   Let n = unknown number.   Define the variable.

What number is 20% of 18
$$\downarrow \quad \downarrow \downarrow \quad \downarrow \downarrow$$
$$n \; = 20\% \; \cdot \; 18 \qquad \qquad \text{Translate into an equation.}$$
$$n \; = (0.20)(18) \qquad \qquad \text{Convert the percent to a decimal.}$$
$$n \; = 3.6 \qquad \qquad \text{Solve for the unknown.}$$

3.6 is 20% of 18. ▲

When solving this problem, we chose to convert 20% to a decimal instead of a fraction to simplify the arithmetic.

Look again at the answer to Example 1: 3.6 is 20% of 18. The three numbers in this statement are often referred to as the percentage, the rate, and the base.

$$\text{3.6 is 20\% of 18}$$
$$\uparrow \qquad \uparrow \qquad \uparrow$$
Percentage = Rate × Base

Note that the **rate** is a percent and the **percentage** is a number. The **base** is that number following the word *of*. The percentage, rate, and base can be identified in any problem that contains a percent.

For *all* problems involving percents, the following formula holds true.

Percentage = Rate × Base ▲

EXAMPLE 2   12 is 75% of what number?

Solution   Let x = unknown number (base).   Define the variable.

12 is 75% of what number
$$\downarrow \downarrow \downarrow \quad \downarrow \quad \downarrow$$
$$12 = 75\% \; \cdot \quad x \qquad \qquad \text{Translate into an equation.}$$

$$12 \; = \frac{75}{100}x \qquad \qquad \text{Convert the percent to a fraction.}$$

$$12 \; = \frac{3}{4}x$$

$$\left(\frac{4}{3}\right)\frac{3}{4}x \; = \; 12\left(\frac{4}{3}\right) \qquad \qquad \text{Solve for the unknown by multiplying both sides by the reciprocal of the coefficient}$$
$$x \; = \; 16 \qquad \qquad \text{of } x.$$

12 is 75% of 16.
$$\uparrow \qquad \uparrow \qquad \uparrow$$
Percentage   Rate   Base ▲

When solving Example 2, we chose to convert 75% to a fraction instead of a decimal to simplify the arithmetic.

EXAMPLE 3   8 is what percent of 25?

Solution   Let r = unknown rate.   Define the variable.

8 is what percent of 25
↓ ↓   ↓   ↓ ↓
8 = r ·   25   Translate into an equation.

$$8 = 25r$$

$$\left(\frac{1}{25}\right)25r = 8\left(\frac{1}{25}\right)$$ Solve for the unknown by multiplying both sides by the reciprocal of the coefficient of r.

$$r = \frac{8}{25}$$

$$r = 0.32$$ Convert the answer to a percent.

$$r = 32\%$$

8 is 32% of 25.
↑   ↑   ↑
Percentage Rate   Base   ▲

Using Proportions to Solve Percent Problems

Each of the previous examples also could have been solved by setting up a proportion that expresses a relationship involving a percent. The first ratio in the proportion is obtained by rewriting the percent (rate) as a fraction. This ratio is set equal to the ratio of the percentage to the base. The base, which is the number following the word *of*, appears in the denominator of the second ratio. The procedure for using a proportion to solve a percent problem is summarized here.

To Use a Proportion to Solve a Percent Problem

1. Rewrite the percent (rate) as a fraction. This is the first ratio in the proportion.

2. Set this first ratio equal to the ratio of the percentage to the base.

3. Solve for the unknown.   ▲

EXAMPLE 4   What is 85% of 260?

Solution   Let n = unknown number (percentage).   Define the variable.

$$n \text{ is } 85\% \text{ of } 260$$

Identify the percentage, rate, and base.

Percentage Rate Base

$$\text{Rate} \rightarrow \frac{85}{100} = \frac{n}{260} \begin{array}{l} \leftarrow \text{Percentage} \\ \leftarrow \text{Base} \end{array}$$

Express the rate as a fraction. Set this equal to the ratio of the percentage to the base.

$$85(260) = 100n$$

$$\left(\frac{1}{100}\right)100n = 85(260)\left(\frac{1}{100}\right)$$

Solve for the unknown.

$$n = \frac{22{,}100}{100}$$

$$n = 221$$

221 is 85% of 260.

▲

EXAMPLE 5 17 is $33\frac{1}{3}\%$ of what number?

Solution Let y = unknown number (base).

Define the variable.

$$17 \text{ is } 33\tfrac{1}{3}\% \text{ of } y$$

Identify the percentage, rate, and base.

Percentage Rate Base

$$\frac{33\frac{1}{3}}{100} = \frac{17}{y}$$

Express the rate as a fraction. Set this equal to the ratio of the percentage to the base.

$$33\frac{1}{3}y = 100 \cdot 17$$

$$\frac{100}{3}y = 100(17)$$

Solve for the unknown.

$$\left(\frac{3}{100}\right)\frac{100}{3}y = 100(17)\left(\frac{3}{100}\right)$$

$$y = (17)(3)$$

$$y = 51$$

17 is $33\frac{1}{3}\%$ of 51.

▲

EXAMPLE 6 4.5 is what percent of 30?

Solution Let p = unknown percent.

Define the variable.

$$4.5 \text{ is } p\% \text{ of } 30$$

Identify the percentage, rate, and base.

Percentage Rate Base

$$\frac{p}{100} = \frac{4.5}{30}$$

Express the rate as a fraction. Set this equal to the ratio of the percentage to the base.

$$30p = (100)(4.5)$$

$$\left(\frac{1}{30}\right)30p = 450\left(\frac{1}{30}\right)$$

Solve for the unknown.

$$p = \frac{450}{30}$$

$$p = 15$$

4.5 is 15% of 30. ▲

Word Problems Involving Percents

We will now examine some word problems that can be solved by using equations that contain percents.

EXAMPLE 7 In a class of 44 students, 25% received a final grade of A. How many students got A's?

Solution Since 25% of the class of 44 students got A's, the problem asks us to determine the following:

What is 25% of 44?

We will solve this problem using the first technique we discussed, in which we translate the problem directly into a mathematical equation.

Let a = number of students getting A's.   Define the variable.

a is 25% of 44
↓ ↓   ↓   ↓ ↓
$a = 25\% \cdot 44$ Translate into an equation.

$a = \left(\frac{25}{100}\right)(44)$ Change the rate to a fraction.

$a = \left(\frac{1}{4}\right)(44)$ Solve.

$a = 11$ students
Thus, 11 students got A's.

Think about whether this answer seems reasonable. ▲

EXAMPLE 8 At a major university 19.3% of all seniors are undecided as to whether they should look for a job or apply to graduate school. If 386 seniors are undecided, how many students are in the senior class?

Solution Since 386 undecided seniors are 19.3% of all seniors, the problem asks us to determine the following:

386 is 19.3% of what number?

We will solve this problem by setting up a proportion.

Let s = number of seniors.   Define the variable.

$\underset{\text{Percentage}}{386}$ is $\underset{\text{Rate}}{19.3\%}$ of $\underset{\text{Base}}{s}$ Identify percentage, rate, and base.

$$\frac{19.3}{100} = \frac{386}{s}$$ Express the rate as a fraction. Set it equal
to the ratio of the percentage to the base.

$$(19.3)s = (100)(386)$$ Solve for the unknown.

$$\left(\frac{1}{19.3}\right)(19.3)s = (38,600)\left(\frac{1}{19.3}\right)$$

$$s = \frac{38,600}{19.3}$$

$$s = 2,000 \text{ seniors}$$

Thus, there are 2,000 seniors in the class.   Think about whether this answer seems
reasonable.   ▲

EXAMPLE 9   In a recent survey, 62 out of 300 drivers responded that they do not regularly use safety belts. What percent of the drivers surveyed does not regularly use safety belts?

Solution   The problem asks us to determine the following:
62 is what percent of 300?
Let r = unknown rate.

$\underset{62}{62}$ is $\underset{}{\text{what}}$ percent of $\underset{300}{300}$ We will solve this problem by translating it
$62 = \quad r \quad \cdot \quad 300$ directly into an equation.

$$62 = 300r$$

$$\left(\frac{1}{300}\right)300r = 62\left(\frac{1}{300}\right)$$

$$r = \frac{62}{300}$$ $$\begin{array}{r} .20\frac{200}{300} \\ 300\overline{\smash)62.00} \\ \underline{600} \\ 200 \end{array}$$

$$r = 0.20\frac{200}{300}$$ Carry out the division to the hundredths
place. Express the quotient as the sum of a
decimal and a fraction whose numerator is
the remainder and whose denominator is
the divisor.

$$r = 0.20\frac{2}{3}$$

$$r = 20\frac{2}{3}\%$$ Convert the decimal to a percent.

Thus, $20\frac{2}{3}\%$ of the people surveyed do not
use safety belts.   Think about whether this answer seems
reasonable.   ▲

Quick Quiz

Find the unknown in each of the following word problems.

		Answers
1.	What number is 12% of 700?	**1.** 84
2.	0.08 is what percent of 32?	**2.** 0.25%
3.	66 is 1.5% of what number?	**3.** 4,400

6.3 Exercises

Indicate which of the following statements are true and which are false. For those that are false, change the italicized expression to make the statement true.

1. The product of the rate and the base equals the *percentage*.

2. In setting up an equation to solve a percent problem, the word "of" translates to the operation *division*.

3. The number following the word "of" is called the *percentage*.

4. In a percent problem, the *rate* is usually expressed as a percent.

Find the unknown in each of the following problems.

5. What number is 80% of 65?

6. What number is 45% of 30?

7. 18 is 24% of what number?

8. 9 is 72% of what number?

9. 14 is what percent of 56?

10. 98 is what percent of 200?

11. 3.8 is 16% of what number?

12. What number is 3.5% of 78?

13. 0.017 is what percent of 5?

14. 0.008 is 0.025% of what number?

15. 5.6 is what percent of 64?

16. What number is 1.6% of 85?

17. 19 is 19% of what number?

18. 67 is 100% of what number?

19. 31 is 200% of what number?

20. What number is 150% of 21?

21. 85 is what percent of 25?

22. 24 is what percent of 8?

23. 13 is 16% of what number?

24. What number is 84% of 73?

25. 1.8 is 0.06% of what number?

26. What number is 0.053% of 26?

27. 7.5 is what percent of 400?

28. 2.2 is what percent of 50?

29. What number is $33\frac{1}{3}$% of 63?

30. What number is $9\frac{1}{11}$% of 88?

31. 72 is what percent of 28.8?

32. 37 is what percent of 5.92?

33. 45 is 72% of what number?

34. What number is 47% of 29?

35. $5\frac{5}{8}$ is 30% of what number?

36. What number is 60% of $9\frac{1}{4}$?

37. $8\frac{5}{9}$ is what percent of 11?

38. 6 is what percent of $7\frac{1}{5}$?

39. 12 is $66\frac{2}{3}$% of what number?

40. What number is $8\frac{1}{7}$% of 49?

41. What number is 46.92% of 1.837?

42. 6.3841 is what percent of 51.0728?

43. 36.84 is 9.276% of what number?
Round to the nearest hundredth.

44. 2.843 is what percent of 52.89?
Round to the nearest hundredth of a percent.

Solve each of the following word problems.

45. It was found that 12% of all eighth graders who took a standardized reading test scored below a sixth grade reading level. If 150 students took the test, how many scored below a sixth grade reading level?

46. Last summer, a softball team won 5 out of the 25 games that were played. What percent of the games played did they win?

47. Fifteen percent of the teenagers attending a suburban high school smoke cigarettes. If 72 students smoke cigarettes, what is the school's total enrollment?

48. Of all dentists responding to a recent survey, 84% said that they offer nitrous oxide (laughing gas) to patients who need to have cavities filled. If 225 dentists were surveyed, how many use nitrous oxide?

49. Of the people employed at an automobile plant, 14% filed their income tax returns late. If 84 people filed late tax returns, how many are employed at that plant?

50. Twenty-eight of 32 members of a football team weigh over 200 pounds. What percent of the team weighs over 200 pounds?

51. Of the students enrolled in a 9 A.M. calculus class, 72% eat breakfast before coming to class. If there are 200 students in the class, how many eat breakfast before class?

52. Nine families in a suburban neighborhood moved away within the last two years. If 60 families lived in that neighborhood, what percent has moved elsewhere?

53. Twenty-six out of 160 students failed general chemistry. What percent of the class failed?

54. Of the people working for a downtown bank, 88% take public transportation to work. If 550 people are employed at the bank, how many take public transportation?

55. Eighty-two percent of the people attending a banquet prefer coffee to tea. If 287 people prefer coffee, how many people are at the banquet?

56. In a second grade class, 12 out of 32 students come from families in which the parents are divorced or separated. What percent of the class does this represent?

57. On a test worth a total of 60 points, a student obtained a score of 45. What is his grade expressed as a percent?

58. A physician reported that 52% of a company's employees are overweight. If 450 people work at that company, how many are overweight?

59. A 250-milliliter acid solution contains 5 milliliters of pure acid. What percent of the solution is pure acid?

60. Of the students in a university, 14% participate in intercollegiate sports. If 1,250 students are enrolled at the school, how many participate in intercollegiate sports?

61. A well-known business school accepted 185 students out of an applicant pool of 592. What percent of all applicants was accepted?

62. An engineer earning a salary of $38,500 received a 9% raise. What is the dollar value of the raise?

63. In a town of 13,000 eligible workers, the unemployment rate is 8.5%. How many eligible workers are without jobs?

64. A social worker's monthly gross salary is $1,200, of which 6.8% is deducted every month for social security. What is the dollar value of this deduction?

65. One biscuit of shredded wheat cereal combined with $\frac{1}{2}$ cup of whole milk contains 10% of the U.S. recommended daily allowance (RDA) of protein. If each biscuit contains 2 grams of protein, and $\frac{1}{2}$ cup of whole milk contains 4 grams of protein, what is the recommended daily allowance of protein?

66. At a private college, 60 students out of a junior class of 1,250 studied abroad during their junior year. What percent of the junior class does this represent?

67. An airline estimates that 5.6% of the people making flight reservations will not keep them. If 125 people book flights, how many people does the airline expect will cancel or not show up?

68. A four-year study done by the Justice Department shows that 21% of all violent crimes committed in the United States happen between friends or relatives. If 17.9 million cases were studied, how many involved friends or relatives?

69. The Internal Revenue Service examined 1,844,986 individual federal income tax returns. If 87,338,611 returns were filed, what percent was examined by the IRS? Round off your answer to the nearest tenth of a percent.

70. Last fall, 28.8% of all households with televisions in the United States watched the top-rated prime-time program. If there were a total of 76,299,500 households with television in the United States last fall, how many watched the number one prime-time show?

71. The hydroelectric plant at the Grand Coulee Dam currently is producing 6,263 megawatts of power. Its ultimate capacity is 10,080 megawatts of power. What percent of its ultimate capacity is its current level of power production? Round off your answer to the nearest tenth of a percent.

72. Fresh water covers 68,490 square miles of the Canadian province of Ontario. If this represents 16.6% of the total area of Ontario, what is the total area of the province? Round off your answer to the nearest square mile.

6.4
Applications of Percent

We will now examine problems that illustrate some of the most common applications of percent. These include calculating amounts earned in commissions, figuring out taxes, determining interest on money invested or borrowed, and calculating how much to tip. In these problems that involve percent, as in all others, the formula is

$$\text{Percentage} = \text{Rate} \times \text{Base}$$

Commission

A **commission** is a bonus paid to a salesperson as an incentive to maximize individual sales productivity. Commission is expressed as a percent of sales and can be calculated by multiplying the percent by the sales (see below). In commission problems, the percentage is the amount of the commission, the rate is a percent, and the base is the value of sales.

To calculate a commission, multiply the commission rate by the sales.

Commission = Rate × Sales ▲

EXAMPLE 1   A salesman earns a weekly commission of 11% on all sales exceeding $1,500. Last week he sold $3,200 worth of goods. What was his commission for the week?

Solution   First determine the value of the sales that will earn a commission, that is, the amount exceeding $1,500.

$3,200   value of total sales
$\underline{-\ 1,500}$ value of sales not subject to commission
$1,700   value of sales subject to commission

To determine the commission, find 11% of $1,700.
Let c = commission
Commission = % of Sales
$c = 11\%$ of $1,700
$c = 0.11 \times 1,700$
$c = 187$

$$
\begin{array}{r}
1{,}7\,0\,0 \\
\times\ \ \ 0.1\,1 \\
\hline
1\,7\,0\,0 \\
1\,7\,0\,0\ \ \ \\
\hline
1\,8\,7.0\,0
\end{array}
$$

Thus, his commission is $187.   ▲

EXAMPLE 2   A shoe saleswoman is given the option of either working for a straight monthly salary of $1,150 or working for a monthly salary of $850 plus a 6% commission on all shoes she sells. If she chooses to work for a commission, what is the minimum amount of merchandise she must sell per month in order to make at least as much as she would if working for a straight salary?

Solution   First determine the value of her minimum monthly commission in order for both salaries to be equal.

$1,150   straight monthly salary
$\underline{-\ \ \ 850}$ base salary before commission
$\ \ 300   minimum monthly commission

Now determine the sales necessary to generate a $300 commission.
Let s = sales
Commission = % of Sales
$$\$300 = 6\% \text{ of } s$$

$$300 = \frac{6}{100} \cdot s$$

$$\frac{6}{100}s = 300$$

$$\left(\frac{100}{6}\right)\frac{6}{100}s = \overset{50}{\cancel{300}}\left(\frac{100}{\cancel{6}}\right)$$

$$s = 5,000$$

She therefore must sell at least $5,000 worth of shoes every month.   ▲

Taxes

Taxes are one of the most frequently occurring examples of percents in our lives. They appear as a percent of money earned (income tax), a percent of the value of goods purchased (sales tax), a percent of assessed property value (property tax), a percent of the value of imported products (tariff), and a percent of an inheritance (inheritance tax).

Figure 6.1 (opposite) is a copy of the tax rate schedules published by the Internal Revenue Service for determining federal income tax on taxable incomes greater than $50,000. The tax on incomes less than that can be read directly from other tax tables. Schedule X is for single taxpayers, Schedule Y-1 is for married persons filing jointly, Schedule Y-2 is for married persons filing separately, and Schedule Z is for unmarried head of household. Example 3 illustrates how to use these schedules to determine federal income tax.

EXAMPLE 3 Use Figure 6.1 to determine how much income tax each of the following people must pay given that the following amounts appear on line 37 of Form 1040.

(a) $27,000 head of household

We look at Schedule Z and find the line that represents a taxable income of $27,800. The tax is 15% of the amount over $0.

15% of $27,800   Calculate 15% of $27,800.

$= 0.15 \times \$27,800$

$= \$4,170$

The income tax is $4,170.

(b) $46,812 married, filing jointly

We look at Schedule Y-1 and find the line that represents a taxable income of $46,812.

The tax is $5,535.00 + 28% of the amount over $36,900.

$46,810   First determine the amount over $36,900.

$- \;\; 36,900$
_____

$\$\; 9,912$ excess

28% of $9,912   Then calculate 28% of the excess.

$= 0.28 \times \$9,912$

$= \$2,775.36$

$5,535.00   Then add $2775.36 to $5535.00 to find the

$+ \;\; 2,775.36$ total income tax.
_____

$8,310.36

The income tax is $8,310.36

(c) $121,585 single

We look at Schedule X and find the line that represents a taxable income of $121,585. The tax is $31,172.00 + 36% of the amount over $115,000.

$121,585   First determine the amount over $115,000.

$- \;\; 115,000$
_____

$\$\;\; 6,585$ excess

36% of $6,585   Then calculate 36% of the excess.

$= 0.36 \times \$6,585$

$= \$2,370.60$

$$\begin{array}{r} \$31,172.00 \\ +\quad 2,370.60 \\ \hline \$33,542.60 \end{array}$$

Then add $2,370.60 to $31,172.00 to find
the total income tax.

The income tax is $33,542.60.   ▲

Figure 6.1
Tax Rate Schedules

Schedule X—Use if your filing status is **Single**

If the amount on Form 1040, line 37, is: Over—	But not over—	Enter on Form 1040, line 38	of the amount over—
$0	$22,100	. . . . 15%	$0
22,100	53,500	$3,315.00 + 28%	22,100
53,500	115,000	12,107.00 + 31%	53,500
115,000	250,000	31,172.00 + 36%	115,000
250,000	. . . . . .	79,772.00 + 39.6%	250,000

Schedule Y-1—Use if your filing status is **Married filing jointly** or **Qualifying widow(er)**

If the amount on Form 1040, line 37, is: Over—	But not over—	Enter on Form 1040, line 38	of the amount over—
$0	$36,900	. . . . 15%	$0
36,900	89,150	$5,535.00 + 28%	36,900
89,150	140,000	20,165.00 + 31%	89,150
140,000	250,000	35,928.50 + 36%	140,000
250,000	. . . . . .	75,528.50 + 39.6%	250,000

Schedule Y-2—Use if your filing status is **Married filing separately**

If the amount on Form 1040, line 37, is: Over—	But not over—	Enter on Form 1040, line 38	of the amount over—
$0	$18,450	. . . . 15%	$0
18,450	44,575	$2,767.50 + 28%	18,450
44,575	70,000	10,082.50 + 31%	44,575
70,000	125,000	17,964.25 + 36%	70,000
125,000	. . . . . .	37,764.25 + 39.6%	125,000

Schedule Z—Use if your filing status is **Head of household**

If the amount on Form 1040, line 37, is: Over—	But not over—	Enter on Form 1040, line 38	of the amount over—
$0	$29,600	. . . . 15%	$0
29,600	76,400	$4,440.00 + 28%	29,600
76,400	127,500	17,544.00 + 31%	76,400
127,500	250,000	33,385.00 + 36%	127,500
250,000	. . . . . .	77,485.00 + 39.6%	250,000

In sales tax problems, the percentage is the amount of the tax, the rate is a percent, and the base is a price (see below).

To calculate a sales tax, multiply the tax rate by the price.

Sales tax = Rate × Price ▲

EXAMPLE 4 A couple wants to buy a sofa that sells for $399. What is the total amount they must pay if the sales tax is 5%?

Solution First determine the sales tax.

Let t = sales tax.   Define the variable.

Sales tax = % of Price
$$t = 5\% \text{ of } \$399$$
$$t = 0.05 \times 399 \qquad \text{Translate into an equation.}$$
$$t = 19.95 \qquad\qquad \text{Sales tax = Rate × Price}$$
Solve for t.

Now find the total amount they must pay.

$399.00	sofa	Add the sales tax of $19.95 to $399, the
+ 19.95	tax	cost of the sofa.
$418.95	total	

Thus, the total cost is $418.95 including tax. ▲

Interest

Interest is an amount you pay to borrow money, or an amount you earn for allowing someone else to use your money, for example, when you deposit your money in a bank or purchase investments. The amount of money you borrow or deposit on which interest is calculated is called the **principal**. As the amount of time the money is being used increases, so does the interest. To calculate the amount of interest, we multiply the rate by the principal by the time (see below). The percentage is the interest, the rate is a percent, the base is the principal, and there is a new factor—time.

Interest = Rate × Principal × Time

We will use the letter I to represent interest, r to represent rate, p to represent principal, and t to represent the length of time for which the money is borrowed or invested.

To calculate an amount of interest, use the formula
$$I = r \cdot p \cdot t$$
where I = interest, r = rate, p = principal, and t = time in years. ▲

EXAMPLE 5 An insurance agent deposited \$3,500 in a savings account that pays an annual interest rate of $5\frac{1}{2}\%$. How much interest will he earn in 2 years?

Solution $r = 5\frac{1}{2}\%$ Identify the values for the rate, principal, and time, and substitute them into the formula $I = r \cdot p \cdot t$.

$p = \$3,500$

$t = 2 \text{ years}$

$I = r \cdot p \cdot t$

$I = 5\frac{1}{2}\% \cdot 3,500 \cdot 2$ Solve for I.

$I = \dfrac{5\frac{1}{2}}{100} \cdot 3,500 \cdot 2$

$I = \left(5\frac{1}{2}\right)(70)$

$$\begin{array}{r} 35 \\ \times\ 11 \\ \hline 35 \\ 35 \\ \hline 385 \end{array}$$

$I = \left(\dfrac{11}{2}\right)(70)$

$I = 385$

Thus, he would earn \$385 in interest in 2 years. ▲

Example 5 illustrates how much interest would be earned if the interest were calculated, or *compounded*, just once, at the end of 2 years. Banks used to advertise that their interest was compounded semi-annually (twice a year), quarterly (four times a year), monthly, or even daily. You earn more interest the more frequently it is compounded, because the interest earned in the first period (the length of time over which the interest is calculated) becomes part of the principal for the next interest period, and you begin earning interest on interest.

For example, suppose you deposit \$10,000 at an annual interest rate of 5% for 1 year. If interest is compounded only at the end of the year (simple interest), you compute the interest as follows:

$I = r \cdot p \cdot t$

$I = (5\%)(10,000)(1)$

$I = (0.05)(10,000)(1)$

$I = 500$

You earn \$500 interest for the entire year.

If interest is compounded semi-annually, then the interest is computed twice. Let us calculate the interest for the first half of the year ($\frac{1}{2}$ yr).

$I = r \cdot p \cdot t$

$I = (0.05)(10,000)\left(\dfrac{1}{2}\right)$

$I = 250$

$$\begin{array}{r} 5,000 \\ \times\ 0.05 \\ \hline 250.00 \end{array}$$

Now we add the $250 interest to the principal of $10,000 to obtain a new principal of $10,250. The interest for the second half of the year is computed on this new principal.

$$I = r \cdot p \cdot t$$

$$I = (0.05)(10,250)\left(\frac{1}{2}\right)$$

$$I = (0.05)(5,125)$$

$$I = 256.25$$

$$\begin{array}{r} 12 \\ 5,125 \\ \times \quad .05 \\ \hline 256.25 \end{array}$$

Thus, the total interest for the year is calculated as follows:

$$\begin{array}{rl} \$250.00 & \text{first half} \\ + \ 256.25 & \text{second half} \\ \hline \$506.25 & \text{interest for the entire year} \end{array}$$

Notice that you would earn $6.25 more in interest if it were compounded twice a year instead of once a year ($506.25 − $500.00 = $6.25).

Today most banks use high-speed computers to compound interest continuously to give you the highest yield possible. This is why banks now advertise both the annual interest rate and the effective annual yield to tell you the actual amount of money earned for a given time period.

Throughout this book, assume that interest is compounded just once during the time period specified, unless you are told otherwise.

EXAMPLE 6 An electrician wishes to take out a home improvement loan for $5,000 at an annual interest rate of 18%. If he pays back the loan 9 months later, what is the total amount he will have to pay back?

Solution First determine the interest on the loan after 9 months.

$r = 18\% = 0.18$
$p = \$5,000$

Identify the values for rate, principal, and time, and substitute them into the formula $I = r \cdot p \cdot t$.

$$t = 9 \text{ mo} = 9 \text{ mo} \times \frac{1 \text{ yr}}{12 \text{ mo}}$$

$$= \frac{9}{12} \text{ yr} = \frac{3}{4} \text{ yr}$$

Since 18% is an annual interest rate, meaning for 1 year, we must express the time period for the loan in the same units, so convert 9 months to $\frac{3}{4}$ year.

$$I = r \cdot p \cdot t$$

$$I = (0.18)(5,000)\left(\frac{3}{4}\right)$$

$$I = \frac{(0.18)(5,000)(3)}{4}$$

$$I = 675$$

Solve for I.

$$\begin{array}{r} 1,2\,5\,0 \\ \times \qquad 3 \\ \hline 3\,7\,5\,0 \\ \times \quad .1\,8 \\ \hline 3\,0\,0\,0\,0 \\ 3\,7\,5\,0 \\ \hline 6\,7\,5.0\,0 \end{array}$$

The interest is therefore $675.

Now determine the total amount he will have to pay back.

$5,000   Add the interest to the amount borrowed.
+   675
─────
$5,675

The electrician must pay back $5,675.   ▲

Tipping

A *tip* is an extra amount of money paid to someone in return for a service. People we usually tip include waiters, cab drivers, hairdressers, and porters. The amount we tip is often considered to be a standard percent of the cost of the service. For example, in a restaurant, the accepted practice is to leave a tip equivalent to 15% of the bill.

EXAMPLE 7   Suppose your bill in a restaurant amounts to $6.83. How much money should you leave for a tip of 15%?

Solution   Find 15% of $6.83.

Tip = 15% × 6.83

Tip = (0.15)(6.83)

$$\begin{array}{r} 6.83 \\ \times\ \ .15 \\ \hline 3415 \\ 683\ \ \\ \hline 1.0245 \end{array}$$

Tip = 1.0245 ≅ 1.02   Round off to the nearest cent.

You should leave a tip of $1.02.   ▲

It is unnecessary, and often impractical, to figure out an exact tip. To estimate a 15% tip, we can use the following procedure shown below.

1.   Round off the bill to the nearest dollar.
2.   Calculate 10% of the amount in step 1 by moving the decimal point one place to the left.
3.   Calculate 5% of the bill by finding one-half of the amount in step 2.
4.   Add the results of steps 2 and 3 to obtain the amount of a 15% tip.

Let us now use this procedure to estimate the tip for a bill of $6.83.

$6.83 ≅ $7.00   Round off to the nearest dollar.

10% × $7.00 = $0.70   Calculate 10% of $7.00.

$\frac{1}{2}$($0.70) = $0.35   Calculate 5% of $7.00.

$0.70 + $0.35 = $1.05   Add to find the 15% tip.

The tip estimate is $1.05. Notice that this is quite close to the exact answer of $1.02 that we obtained in Example 7.

EXAMPLE 8   Estimate the tip for a bill of $17.59.

Solution $17.59 \cong \$18$ Round off to the nearest dollar.

$10\% \times \$18 = \$1.80$ Calculate 10% of $18.

$\dfrac{1}{2}(\$1.80) = \$0.90$ Calculate 5% of $18.

$\$1.80 + \$0.90 = \$2.70$ Add to find the 15% tip.

The tip is $2.70.   ▲

EXAMPLE 9   At a restaurant, a table of five people receives a bill for $81.07, not including tax or tip. If they agree to split the cost evenly, how much should each person contribute? Assume a 7% tax and a 15% tip.

Solution   First determine the tax and tip.

$7\% + 15\% = 22\%$ Add the rates for tax and tip to determine the total rate that the bill does not include.

$\$81.07 \cong \$81$ Round off bill.
Find 22% of $81.

$\quad 20\% \text{ of } \$81 = 0.20(81) = \$16.20$ Find 20% of $81.
$\underline{+\ 2\% \text{ of } \$81 = 0.02(81) = \$\ \ 1.62}$ Find 2% of $81.

$22\% \text{ of } \$81 \qquad = \qquad \$17.82$ Add to find 22% of $81.

$\$17.82 \cong \$18.00$ Round off answer.

Find the total cost of the meal.

$\$81 + \$18 = \$99$ Add $18 to bill for tax and tip.

Determine how much each person should pay.

$\$99 \cong \$100 = \$100 \div 5 = \$20$ To split the bill 5 ways, round off and divide by 5.

Each person should contribute $20.   ▲

It is worthwhile to practice doing an exercise like this in your head.

6.4 Exercises

Indicate which of the following statements are true and which are false. For those that are false, change the italicized or underlined expression to make the statement true.

1.   Interest is calculated by finding a percentage of the *principal*, which is the amount of money borrowed or invested.

2.   A bonus that is calculated to be a percentage of a salesperson's sales is called a *commissary*.

3.   The standard tip in a restaurant is <u>9</u>%.

4.   To calculate the tax on an item sold in a city where the sales tax is 5%, you multiply the price of the item by <u>0.005</u>.

Use Figure 6.1 to determine how much income tax each of the following people must pay given that the following amounts appear on line 37 of Form 1040.

5. $21,700   single

6. $36,200   married, filing jointly

7. $72,820   married, filing separately

8. $58,360   single

9. $87,485   head of household

10. $152,764   married, filing jointly

11. $283,465   head of household

12. $205,975   married, filing separately

Estimate a 15% tip for each of the following restaurant bills.

13. $15.13

14. $20.78

15. $3.92

16. $8.04

17. $59.62

18. $80.19

19. $174.71

20. $206.21

Solve each of the following word problems.

21. A comedian bought a new car that cost $7,550. If sales tax is 7%, what is the total amount that the comedian must pay?

22. A teenager bought a stereo system on sale for $649.00 and paid a total of $681.45. How much sales tax did the teenager pay? What is the tax rate for that city?

23. A New England state has an automobile excise tax that is 13% of the market value of any car registered in that state. How much excise tax would a person pay on a car with a market value of $2,350?

24. All automobiles imported into the United States from Japan are subject to an 11% tariff. If a car dealer paid a $37,840 tariff on a shipment of new cars, what is the value of the cars that the dealer received?

25. A doctor who earns a salary of $112,060 claims to be in a 33% tax bracket. How much income tax does the doctor pay?

26. An artist bought art supplies totaling $87.25 and had to pay a sales tax of $3.49. What is the tax rate?

27. A man bought a new shirt for $17.99 in a state with a 5% sales tax on clothes. How much change will he get from a $20 bill?

28. An architect owns a house that has a market value of $120,000. The county has set the assessed value of a house to be 25% of the market value. If the property tax is 4% of the assessed value, compute the property tax on this house.

29. A used car salesman makes a 9% commission on all cars he sells. If he sells a car for $4,500, what is his commission?

30. A real estate agent earns a 6% commission. If she sells a house for $125,700, what is her commission?

31. A salesclerk for a men's clothing store earned $437.78 in commissions for $6,254 worth of sales. What is the salesclerk's commission rate?

32. A cosmetics saleswoman earned a commission of $75 for selling $1,250 worth of merchandise. What is her commission rate?

33. A woman earns a $12\frac{1}{2}$% commission on all housewares that she sells by organizing and conducting home sales parties. If one year she earned $150 in commissions, what was the value of the housewares that she sold?

34. A textbook sales representative who earns a 4% commission sold a text to a major university that will order 2,000 copies. If the bookstore pays $29.95 for each book, what is the sales representative's commission?

35. A salesman for a machine tool company has a choice of working for a salary of $11,000 plus a 12% commission or for a salary of $17,000 plus an 8% commission. He anticipates that he can sell $72,000 worth of heavy machinery that year. Which option should he choose to earn the most money?

36. An encyclopedia salesman decides to work solely on a commission basis at a rate of 25% instead of working for a straight salary of $14,000 a year. If encyclopedias sell for $350 a set, how many sets must he sell in order for him to make at least as much money as he would if working for a straight salary?

37. An engineer wishes to take out an $8,900 loan at an annual interest rate of $13\frac{1}{2}$%. How much will she have to repay after 3 years?

38. Suppose you deposit $6,400 in a two-year account that has an annual interest rate of 13.45%. How much money will be in the account at the end of 2 years?

39. A woman deposited $200 in a savings account exactly 1 year ago, and now has $211 in her account. What is the annual interest rate?

40. After 1 year, a pianist had to pay $325 in interest on a loan of $2,500. What is the interest rate for the loan?

41. An electrician wishes to buy a $6,800 car and make a down payment of $1,500. If he borrows the remainder at an annual interest rate of 12%, how much interest will he owe after 1 year?

42. Interest on a $5,000 investment is compounded semi-annually. The annual interest rate is 18%. How much interest is earned in 1 year?

43. After 2 years, a businesswoman earned $6,800 in dividends, which is equivalent to an annual rate of return of 17%. What is the value of her original investment?

44. A couple buys a $135,000 house with a down payment of $18,000. They take out a 14% mortgage for the remainder. After 1 month, in addition to the principal, how much interest will they owe?

45. Cab fare to the airport is $16.50. How much is a 20% tip on the fare?

46. An actress gets her hair done at a beauty salon for $35. If she wants to leave a 20% tip, how much will it cost her?

47. A businessman left an $8 tip for the desk clerk at a hotel where he stayed. The tip amounted to 5% of his bill. What was his hotel bill?

48. A lawyer takes a cab home from work and the fare amounts to $4.50. If she wants to give the driver a 20% tip, how much change should she ask for from a $10 bill?

49. A party of 8 people gets a bill of $119.56 at a restaurant, not including tax and tip. The meal tax is 7%, and they agree to tip 15%. Estimate how much each person should pay if the bill is split evenly.

50. Three people at a restaurant receive a bill of $23.72, not including tax and tip. They agree to tip 20%, and the meal tax is 5%. If they split the bill evenly, estimate how much each person owes.

51. A student went to the bookstore and purchased books and supplies for the following amounts: $1.89, $0.49, $3.57, $35.99, $21.99, and $0.63. If the sales tax is 6%, what was his total bill? Round off your answer to the nearest cent.

52. A stockbroker sold investments totaling $44,178 and received $7,687 in commissions. What was his rate of commission? Round off your answer to the nearest tenth of a percent.

 53. An advertising executive invests $15,000 in a money market fund that accumulates dividends at an annual rate of 16% compounded quarterly. How much will she earn in dividends after 1 year? Round off your answer to the nearest cent.

 54. A couple left a $3.00 tip on a restaurant bill totaling $17.82. What percent did they tip? Round off your answer to the nearest tenth of a percent.

6.5
Percent Increase and Decrease

Some of the most common applications of percent involve expressing an increase or decrease in a particular quantity as a percent. In order to find the amount of increase or decrease, we multiply the rate by the original amount (see below).

To calculate an amount of increase or decrease, multiply the rate by the original amount.

Amount of increase or decrease = Rate × Original amount ▲

EXAMPLE 1 A college admissions officer reported an 8% decrease in freshman applications from the previous year. If 3,850 applications were received last year, how many fewer were received this year?

Solution The rate of decrease is 8%.
3,850 is the original amount.
Amount of decrease = Rate × Original amount
= 8% × 3,850
= (0.08)(3,850) = 308 fewer
The college received 308 fewer applications. ▲

EXAMPLE 2 The U.S. Department of Labor reported that unemployment during the month of May rose one-tenth of 1 percent. If 14,500,000 were unemployed at the end of April, how many were unemployed at the end of May?

Solution One tenth of 1 percent = 0.1%.
This is the rate of increase.
The original amount is 14,500,000.
Amount of increase = Rate × Original amount
= 0.1% × 14,500,000
= (0.001)(14,500,000) = 14,500 more
14,500 more people were unemployed at the end of May.

Find the total number of unemployed at the end of May.

14,500,000	unemployed in April	Add the increase in May to the number of
+ 14,500	increase	unemployed in April.
14,514,500	unemployed in May	

Thus, 14,514,500 persons were unemployed at the end of May. ▲

A common example of a percent decrease is a discount, or markdown. A **discount (markdown)** is a percentage deducted from the original price when an item goes on sale. A discount is calculated by multiplying the discount rate by the original price.

Discount = Rate × Original price

The sale price is calculated by subtracting the discount from the original price.

To determine a sale price, subtract the discount from the original price.

Sale price = Original price − Discount ▲

EXAMPLE 3 Sweaters that normally sell for $35 each are marked down 15%. What is the sale price?

Solution First calculate the discount.

Discount = Rate × Original price
= 15% × 35
= (0.15)(35) = $5.25 discount

Now find the sale price.

Sale price = Original price − Discount

$35.00	original price
− 5.25	discount
$29.75	sale price

The sale price is therefore $29.75. ▲

EXAMPLE 4 A pair of jeans marked down 25% is on sale for $12. What was the original price?

Solution In this problem, we know the discount rate and the sale price.

Sale price = Original price − Discount   Since the discount is 25% of the original
75%   =   100%   −   25%   price, the sale price is 75% of the original
price.
Therefore, 75% of original price = sale price
75% of original price = 12   We know the sale price is $12.
Let x = original price
75% of x = 12

$$\frac{75}{100} \cdot x = 12$$

$$\frac{3}{4}x = 12$$

$$\left(\frac{4}{3}\right)\frac{3}{4}x = \overset{4}{\cancel{12}}\left(\frac{4}{\cancel{3}}\right)$$

$$x = 16$$

Therefore, the original price was $16. ▲

A common example of a percent increase is a profit, or markup. A **profit (markup)** is a percentage added to the cost of a product before it is sold to the consumer. A profit is calculated by multiplying the profit rate by the cost.

Rate $\times$ Cost = Profit

The selling price is determined by adding the profit to the cost (see below).

To determine a selling price, add the profit to the cost.

Cost + Profit = Selling price   ▲

EXAMPLE 5   A camera that sells for \$228 costs the dealer \$120. What percent profit does the dealer make from selling this camera?

Solution   First determine the profit in dollars.

$$\text{Selling price} = \text{Cost} + \text{Profit}$$
$$\text{Selling price} - \text{Cost} = \text{Profit}$$

Rearrange the formula to find the difference between the selling price and the cost.

\$228	selling price
− 120	cost
\$108	profit

Now determine the profit rate.
Let r = profit rate

$$\text{Profit} = \text{Rate} \times \text{Cost}$$
$$108 = r \times 120$$
$$120 \times r = 108$$
$$\left(\frac{1}{120}\right)120r = 108\left(\frac{1}{120}\right)$$
$$r = \frac{108}{120} = 0.90$$
$$r = 90\% \text{ profit}$$

Therefore, the dealer will make a 90% profit.   ▲

6.5 Exercises

Indicate which of the following statements are true and which are false. For those that are false, change the italicized expression to make the statement true.

1. The selling price is determined by finding the *difference between* the cost and the profit.

2. When an item goes on sale, its price is *decreased.*

3. Profit is a percentage of the *cost*.

4. A discount is *subtracted from* the original price to obtain the sale price.

Solve each of the following word problems.

5. If bus fares increase from 50¢ to 65¢, what is the percent increase?

6. At a small college, the number of students majoring in English dropped from 25 to 18. What was the percent decrease in English majors?

7. During the first year, a new car that cost $8,995 depreciates in value 30%. What is the market value of this car when it is 1 year old?

8. The number of people buying new cars at a local dealership dropped 40% from last year. If 325 new cars were sold last year, how many were sold this year?

9. A new car with an original sticker price of $7,599 is marked down 10%. What is its sale price?

10. A department store is advertising a 25% sale on all carpeting. If the carpeting you want originally costs $349, how much would you pay for it during the sale?

11. A bicycle store manager wants to sell a ten-speed bike at a 21% profit. If his cost for the bike is $254, what should its selling price be?

12. The cost of manufacturing an electric typewriter is $116. What should its selling price be to ensure a 55% profit?

13. A stock that sold for $40 a share last month is now selling for $56 a share. What is the percent increase in the price of the stock?

14. A VCR that was selling for $350 is now on sale for $280. What is the discount rate?

15. A bank that employed 7,000 people cut its staff by 15%. How many people are employed by the bank after the staff reduction?

16. A bookstore makes a profit of 30% on all dictionaries it sells. If the store's cost for a dictionary is $27, what is its selling price?

17. The phone company has been granted permission to increase monthly service charges 2.5%. If your monthly service charge is now $18, what will it be after the increase?

18. A mechanic making $16.50 per hour received an 8% raise. What is his new hourly wage?

19. A furniture store is advertising a sale of 15% off on all merchandise. What is the sale price of a sofa that originally sells for $499?

20. A coat originally selling for $250 is on sale for $197.50. What is the discount rate?

21. It costs a donut shop $1.60 to make a dozen donuts that sell for $2.60. What percent profit is made on one dozen donuts?

22. A jeweler is selling for $39 a quartz watch that costs him $24. What is his rate of profit?

23. During one year, doctors at a large city hospital reported a 12% increase in heart attack patients. If 125 heart attack victims were admitted during the previous year, how many were treated at that hospital this year?

24. During the month of December, the price of home heating oil rose 3.5%. If you pay $96 a month for heating oil, what will your monthly bill be after the price increase takes effect?

25. The price of beef rose 2.5% last month. If you paid $4.80 for a piece of beef last month, how much would it cost now?

26. The number of applicants to a well-known law school dropped 4% from last year. If 575 people applied last year, how many applied this year?

27. A car dealer is advertising a sports car for $8,995, which is 20% off the sticker price. What is the sticker price?

28. A new refrigerator originally priced at $600 is on sale for $459. What is the discount rate?

29. An appliance store makes a 38% profit on all washing machines it sells. If its cost for one machine is $175, what is the selling price?

30. A boat dealer makes a 22% profit on all sailboats. If the cost of a sailboat is $3,200, what is its selling price?

31. This past year, the market value of a house that was worth $142,000 increased 17%. What is the present market value of the house?

32. Tuition at a private college is scheduled to increase 14% next year. If tuition is now $6,200, what will it be next year?

33. Last year, the price of gasoline rose from $1.80 per gallon to $1.89 per gallon. What was the percent increase?

34. If the price of a head of lettuce rose from 64¢ to $1.12, what was the percent increase?

35. A dress marked down 20% is on sale for $24. What was the original price?

36. A sports jacket originally selling for $89.95 is marked down 20%. If the sales tax is 5%, how much will it cost to buy this sports jacket?

37. This year, there are 16% more students enrolled in a local public high school than in a private high school. If there are 550 students in the private high school, how many attend the public high school?

38. Next year's defense budget is scheduled to be 18% more than this year's budget. If this year we are spending $220 billion on defense, how much will we be spending next year?

39. Last week the value of a utility stock rose $62\frac{1}{3}$%. If the stock sold for $33 a share at the beginning of the week, what was its price at the end of the week?

40. A real estate agency reported a $6\frac{1}{4}$% decrease in the number of people buying homes this year through its agency. If 112 homes were sold through the agency last year, how many were sold this year?

41. Last year the expenditures of U.S. travelers to foreign countries increased 10% over expenditures the previous years. If U.S. travelers spent $11,930,000 two years ago, how much did they spend last year?

42. This past year, dental fees increased an average of 7.5%. If your dentist charged $20 for a check-up last year, how much would you expect her to charge this year?

43. The median annual income for women is 40.5% less than that for men. If the median salary for men is $15,730, what is it for women?

44. The number of patents issued for inventions increased 22% from 1920 to 1930. If 37,050 patents were issued in 1920, how many were issued in 1930?

45. A boat that retails for $4,299 is marked down 25%. If the sales tax is $6\frac{1}{2}$%, what is the cost of the boat? Round off to the nearest cent.

46. During a 10-year period, the population of the United States increased from 180,671,000 to 204,879,000. By what percent did the population increase? Round off to the nearest tenth of a percent.

47. During a 5-year period, the number of farms in the United States decreased from 2,954,000 to 2,808,000. By what percent did the number of farms in the United States decrease? Round off your answer to the nearest hundredth of a percent.

48. The average number of daily telephone conversations increased 72% from 1940 to 1950. If the average number of daily telephone conversations was 98,775 in 1940, what was it in 1950?

6.6
Reading Graphs

A **graph** is a diagram used to represent numerical data so that it can be easily interpreted. It is easier to compare different quantities if the information is presented in a graph rather than in a table containing long columns of numbers. Usually, the numbers in a graph cannot be represented to a high degree of accuracy, but they are sufficient to give us an overall picture of the relationships between these quantities.

Graphs appear frequently in newspapers and magazines and are widely used by businesses, governmental agencies, and educational institutions to display large amounts of numerical data. Different kinds of graphs are used to emphasize different characteristics of the information being presented. The most common types are bar graphs, pictograms, line graphs, and circle graphs. These are all illustrated in the examples that follow.

A **bar graph** is used to show comparisons between different quantities. In a bar graph, numerical quantities are represented by thick bars, which run either horizontally or vertically. The length (or height) of the bar is proportional to the size of the quantity represented. The information contained in the graph can be determined by reading the title of the graph and the labels for the horizontal and vertical axes, which are the lines that run along the bottom and the left-hand side, respectively, of the graph. Each axis is marked off in segments that represent a range of numerical values. Example 1 illustrates a bar graph (page 271, opposite) representing the rainfall in inches for a 1-year period. The questions that follow refer to that graph.

EXAMPLE 1

(a) During which month did the most amount of rain fall? How much rain was recorded that month?

The most amount of rain fell in March and amounted to 10.5 inches.

(b) During which month did the least amount of rain fall? How much rain was recorded that month?

The least amount of rain fell in August and amounted to 0.5 inch.

(c) How much more rain fell in May than in October?

8.5 inches	May
3.5 inches	October
5.0 inches	Difference

5.0 more inches of rain fell in May than in October.

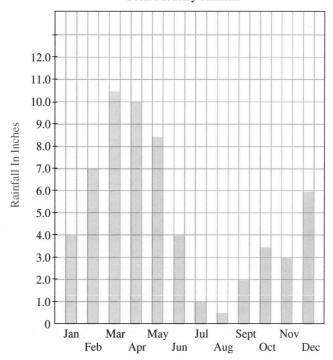

Total Monthly Rainfall

(d) What was the average monthly rainfall for that year? Add the total rainfall for each month and divide the sum by 12:

$4.0 + 7.0 + 10.5 + 10.0 + 8.5 + 4.0 + 1.0 + 0.5$
$+ 2.0 + 3.5 + 3.0 + 6.0 = 60.0$ total inches of rain
$60.0 ÷ 12 = 5.0$ inches per month

The average monthly rainfall was 5.0 inches.   ▲

Another type of graph, which is similar to a bar graph, is a **pictogram**. Instead of using a bar to represent a numerical quantity, pictograms use a series of pictures or symbols, which usually run horizontally across the graph. The number of symbols is proportional to the size of the quantity. A **key** is used to indicate the number represented by a single symbol. A part of a symbol is often used to stand for a fractional part of a given quantity. Even though the numbers portrayed in pictograms are generally not as precise as those in other graphs, pictograms convey a lot of information at a quick glance. Example 2 illustrates a pictogram showing population estimates for a few major cities.

EXAMPLE 2   This example utilizes the graph shown on page 272.

(a)   What is the approximate population of Cairo?

Cairo is represented by five and one-half ⚤. Therefore, its approximate population is
$5.5 × 1,000,000 = 5,500,000$

Estimated Populations of Major World Cities

Source: United Nations Demographic Yearbook

(b) Which city has a population of 2,914,600?

Since each ☂ represents 1,000,000 people, we are looking for a city represented by three ☂. This city is Rome.

(c) Which two cities are closest in population and by approximately how much do they differ in size?

Moscow is represented by eight ☂, indicating a population of 8,000,000. Tokyo is represented by eight and one-half ☂, indicating a population of 8,500,000. Therefore, the cities Moscow and Tokyo are closest in population and differ in size by approximately 500,000 people. ▲

The next type of graph we will consider is a **line graph**. A line graph is used to show how a given quantity changes over time. Each numerical value in a line graph is represented by a dot on the graph, and the dots plotted are connected by straight lines. When information is displayed using a line graph, it is easy to compare rates at which quantities increase or decrease during specified time intervals.

Line graphs often use more than one line to compare trends for different items. The meaning of each line is given in the key. Example 3 illustrates a line graph comparing the quiz grades for two students over a 1-week period. Notice the break in the vertical axis indicating that there are no quiz grades between 0 and 10.

EXAMPLE 3 This example utilizes the graph shown on page 273.

(a) Which student showed the sharpest decline in quiz scores and when did it occur?

Jennifer showed the sharpest decline from Tuesday to Wednesday.

(b) What was the percent increase in Mark's score from Wednesday to Thursday?

To find the percent increase in score, we find the amount of increase from Wednesday

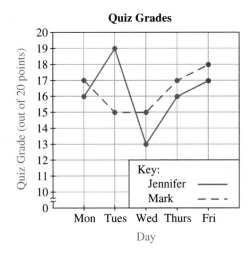

Quiz Grades

to Thursday and then divide by the original score, which is Wednesday's:

$$
\begin{array}{rl}
17 & \text{Thursday} \\
-\,15 & \text{Wednesday} \\
\hline
2 \text{ points} & \text{Amount of increase}
\end{array}
$$

$$\frac{\text{Amount of increase}}{\text{Wednesday's score}} = \frac{2}{15} = 0.13333\ldots = 13\tfrac{1}{3}\%$$

Mark's score increased $13\tfrac{1}{3}\%$ from Wednesday to Thursday.

(c) Which student has the highest quiz average for the week? By how much do their averages differ?

Find the sum of the quiz grades for Mark and Jennifer and divide each sum by 5:

Mark:

$17 + 15 + 15 + 17 + 18 = 82$
$82 \div 5 = 16.4$

Jennifer:

$16 + 19 + 13 + 16 + 17 = 81$
$81 \div 5 = 16.2$

Mark has the highest quiz average. The averages differ by 0.2 point.   ▲

The final type of graph we will consider is a **circle graph**. Circle graphs are used to show the relationship between parts of a quantity and the whole. Circle graphs are also known as **pie charts**. A circle graph is divided into **sectors**, which resemble pieces of pie. The size of each sector is directly proportional to the size of the quantity represented. Each piece of the circle has a percent associated with it, and the sum of all the pieces is 100%. Example 4 illustrates a circle graph that depicts the budget for federal expenditures.

EXAMPLE 4 **Budget for Federal Expenditures**

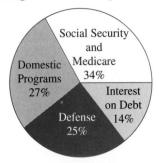

Source: U.S. Office of Management and Budget

(a) If the total budget for federal expenditures was $1,381 billion, what amount was spent in each category?

Social Security and Medicare:

34% of $1,381 billion = (0.34)(1,381) = $469.54 billion

Domestic Programs:

27% of $1,381 billion = (0.27)(1,381) = $372.87 billion

Defense:

25% of $1,381 billion = (0.25)(1,381) = $345.25 billion

Interest on Debt:

14% of $1,381 billion = (0.14)(1,381) = $193.34 billion

(b) How much more money was spent on domestic programs than on defense?

$372.87 billion   Domestic Programs
− 345.25 billion   Defense
$ 27.62 billion

We spent $27.62 billion more on domestic programs than on defense.

(c) What percent of the budget was spent on items other than interest on the debt? How much did this amount to?

Find the difference between 100% and 14% and multiply the result by $1,381 billion:

100% − 14% = 86%

86% of $1,381 billion = (0.86)(1,381) = $1,187.66 billion

Items other than interest on the debt used 86% of the budget. This amounted to $1,187.66 billion. ▲

6.6 Exercises

For Exercises 1–15, answer each of the following questions by reading the corresponding graph for each exercise.

1. a. What grade has the largest number of students? How many students are in that grade?
   b. What is the total number of students represented by this graph?
   c. What percent of the students is enrolled in the fourth grade?

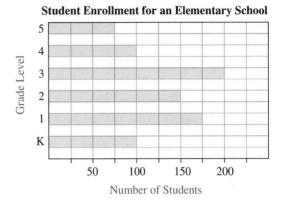

Student Enrollment for an Elementary School

2. a. What school within the university has the smallest number of faculty? How many faculty members are employed by that school?
   b. How many more faculty members are in the school of liberal arts than are in the law school?
   c. What is the total number of faculty in the university?
   d. What percent of the faculty is employed by the school of engineering?

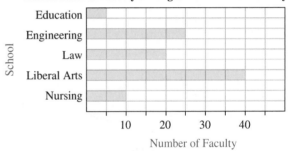

Distribution of Faculty among Schools of a University

3. a. Which city has the highest mean annual snowfall? What does it amount to?
   b. Which city has a mean annual snowfall of 55 inches?
   c. On the average, how much more snow falls per year in Juneau than in Boston?

Mean Annual Snowfall

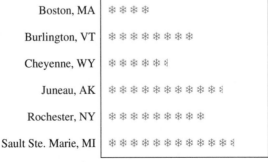

Source: National Oceanic and Atmospheric Administration, U.S. Commerce Department

4. a. From which country did the greatest number of immigrants come? How many persons came from that country?

b. From which country did 150,000 immigrants come?

c. How many more immigrants came from Cuba than from Vietnam?

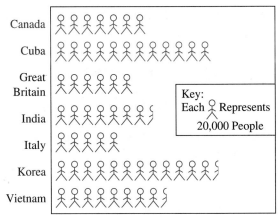

Immigration by Country of Last Residence

Source: U.S. Immigration and Naturalization Service

5. a. During what month were interest rates for automobile loans the highest?

b. By how much did the interest rate increase from January to June?

c. What is the average interest rate for the period from January to June?

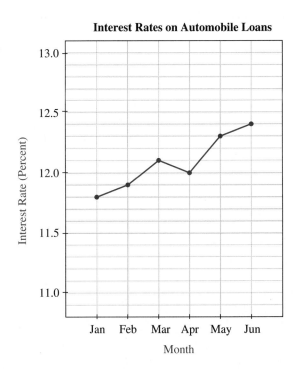

Interest Rates on Automobile Loans

6. **a.** What is the lowest average monthly temperature for New York City? During which month does it occur?
    b. What is the difference between the highest average monthly temperature and the lowest average monthly temperature?
    c. Which two months show the smallest change in temperature?
    d. By how many degrees does the average monthly temperature increase from May to June?

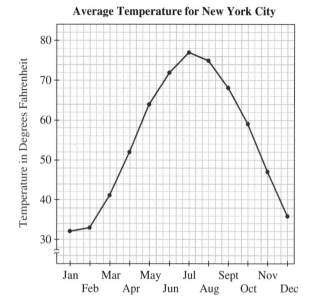

Average Temperature for New York City

Month

Source: National Oceanic and Atmospheric Administration, U.S. Commerce Department.

7. **a.** If a total of 10,851,000 people were living alone, how many in each age group were living alone?
    b. How many more people living alone were 65 and over than were aged 45–64?
    c. What percent of the population living alone was not in the 25–44 age bracket? How many people does this amount to?

Persons Living Alone by Age

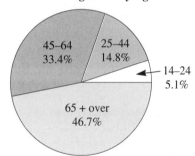

Source: Department of Commerce, Bureau of the Census

8. **a.** If 986,610 people died of heart disease, how many of these deaths were due to heart attacks?
    b. How many more deaths were due to strokes than were due to hypertension?

Deaths Due to Heart Disease

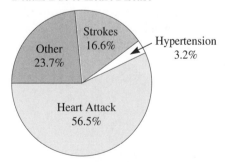

Source: National Center for Health Statistics, U.S. Department of Health and Human Services

9. a. During what 5-year period did the unemployment rate for men increase the most?
   b. What was the average unemployment rate for men during the period 1970–1985?
   c. If the female full-time labor force was 38 million in 1985, how many women were unemployed that year?
   d. How many times greater was the unemployment rate for women in 1975 than that in 1970?

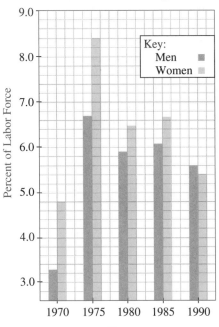

Average Unemployment Rates for Men and Women Working Full-Time

Source: Bureau of Labor Statistics, U.S. Labor Department

10. a. In what year did the smallest number of deaths occur? How many deaths occurred that year?
    b. During what 5-year interval did the number of births decline most severely? Find the percent decrease in births over this 5-year interval. Round off your answer to the nearest percent.
    c. How many more people were born than died in 1965?
    d. What was the average number of births per year for the period 1955–1980? Round off your answer to the nearest tenth of a million.

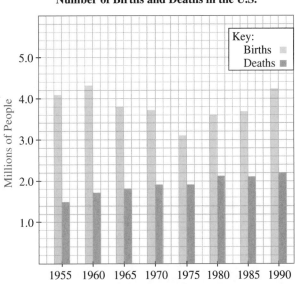

Number of Births and Deaths in the U.S.

Source: National Center for Health Statistics, U.S. Department of Health and Human Services

11.

a. What states produce the same amount of cotton? How much cotton is produced by each of these states?

b. How much more cotton is produced by Mississippi than by Alabama?

c. What state produced approximately 900,000 bales of cotton?

d. What is the average yearly production of cotton for these seven states? Round off your answer to the nearest hundred thousand.

Annual Production of Cotton

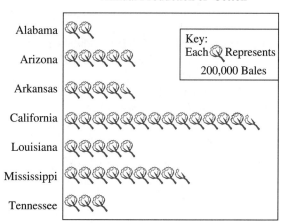

Source: Economic Research Service, U.S. Agriculture Department

12.

a. Approximately how many workers went on strike in 1950?

b. In what year were the greatest number of workers on strike? How many workers were on strike that year?

c. In what year were approximately 900,000 workers on strike?

d. How many more workers were on strike in 1955 than in 1985?

Major Strikes in U.S.

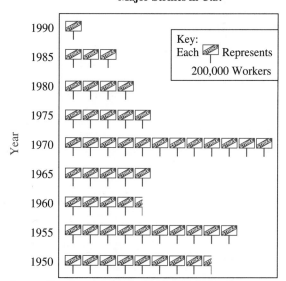

Source: Bureau of Labor Statistics, U.S. Labor Department

13. a. In what month was the price of gas the lowest? What was the price per gallon during this month?
     b. During what 2-month period did the greatest increase in the price of gas occur? How much was this increase?
     c. What was the percent increase in the price of gas from August to December?
     d. What is the average price of gas for the period from February to December?

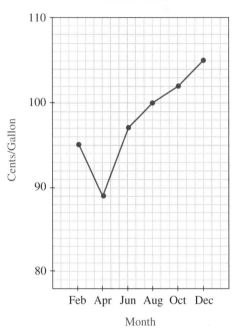

Average Cost of Regular Unleaded Gasoline

14. a. In what year was the marriage rate the highest? What was this rate?
     b. During what 5-year period did the sharpest decrease in the marriage rate occur?
     c. What is the average marriage rate for the years 1950–1975?
     d. What is the percent increase in the marriage rate from 1965 to 1970?

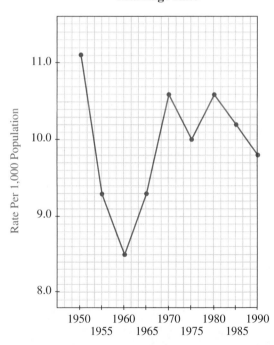

Marriage Rate

Source: National Center for Health Statistics, Public Health Service

15.　a.　If the student's total budget was $12,500, how much was allocated to each item?

　　b.　How much more money was allocated to tuition than to room and board?

　　c.　What percent of the student's budget was not allocated to tuition or room and board? How much does this amount to?

Budget for a College Student

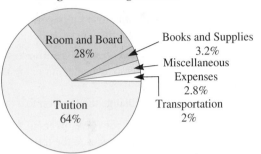

16.　a.　If the total income received by the federal government was $1,091 billion, how much came from individual income taxes? How much came from social insurance taxes?

　　b.　How much more money did the federal government receive from individual income taxes than from corporate income taxes?

　　c.　What percent of the money received by the federal government came from sources other than corporate income taxes? How much does this amount to?

Budget for Federal Receipts

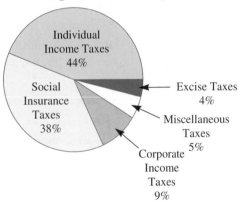

Source: U.S. Office of Management and Budget

6.7
Summary and Review

Key Terms

(6.1)　The word **percent** means per one hundred, or parts of one hundred.

(6.4)　A **commission** is a bonus paid to a salesperson as an incentive to maximize individual sales productivity.

　　　　Interest is an amount you pay to borrow money, or an amount you earn for allowing someone else to use your money when you deposit it in a bank or purchase investments.

　　　　The **principal** is the amount of money you borrow or deposit, upon which interest is calculated.

(6.5)　A **discount**, or **markdown**, is a percentage deducted from the original price when an item goes on sale.

　　　　A **profit**, or **markup**, is a percentage added to the cost of a product before it is sold to the customer.

(6.6)　A **graph** is a diagram used to represent numerical data so that it can be easily interpreted.

Calculation

(6.1) **To convert a percent to a fraction**, remove the percent sign (%), put the number over a denominator of 100, and reduce to lowest terms.

To convert a fraction to a percent, find an equivalent fraction with a denominator of 100, remove the denominator, and write a percent sign after the numerator,

OR

Change the fraction to a decimal and rewrite the decimal as a percent.

(6.2) **To convert a percent to a decimal**, remove the percent sign and move the decimal point two places to the left.

To convert a decimal to a percent, move the decimal point two places to the right and attach a percent sign.

(6.3) **To calculate the percentage or value of a percent**, multiply the rate by the base.

Percentage = Rate × Base

To use a proportion to solve a percent problem, rewrite the percent as a fraction (the first ratio) and set this equal to the ratio of the percentage to the base. Solve for the unknown.

(6.4) **To calculate a sales tax**, multiply the tax rate by the price.

Sales tax = Rate × Price

To calculate a commission, multiply the commission rate by the sales.

Commission = Rate × Sales

To calculate an amount of interest, multiply the rate by the principal by the time.

$I = r \cdot p \cdot t$

(6.5) **To calculate an amount of increase or decrease**, multiply the rate by the original amount.

Amount of increase or decrease = Rate × Original amount

To determine a sale price, subtract the discount from the original price.

Sale price = Original price − Discount

To determine a selling price, add the profit to the cost.

Selling price = Cost + Profit

Chapter 6
Review Exercises

Indicate which of the following statements are true and which are false. For those that are false, change the italicized or underlined expression to make the statement true.

1. One hundred percent is equivalent to the number <u>1</u>.
2. To determine the profit, find the *sum of* the selling price and the cost.
3. A commission is paid to a *consumer*.
4. A discount always *reduces* the original price.
5. When using a proportion to solve a percent problem, the fractional equivalent of the percent is set equal to a ratio with the base in the *numerator*.

6. To convert a fraction to a percent, change the fraction to a decimal, move the decimal point two places to the *left*, and attach a % sign.

(6.1) **Convert each of the following percents to a fraction reduced to lowest terms.**

7. 57%   8. 4%   9. 8.2%   10. 0.9%   11. 180%

12. 72.25%   13. $\frac{3}{4}\%$ 14. $7\frac{1}{2}\%$ 15. $83\frac{1}{3}\%$ 16. $12\frac{1}{7}\%$

(6.1–6.2) **Convert each of the following fractions to a percent.**

17. $\frac{9}{100}$ 18. $\frac{3}{10}$ 19. $\frac{4}{5}$ 20. $\frac{7}{25}$ 21. $2\frac{1}{2}$

22. $1\frac{3}{4}$ 23. $\frac{7}{16}$ 24. $\frac{1}{8}$ 25. $\frac{5}{13}$ 26. $\frac{7}{11}$

(6.1–6.2) **Round off to the nearest tenth of a percent.**

 27. $\frac{107}{576}$ 28. $\frac{348}{795}$

(6.2) **Convert each of the following percents to a decimal.**

29. 18%   30. 80%   31. 2%   32. 4.4%   33. 123%

34. 2,000%   35. 0.09%   36. 6.831%   37. $18\frac{1}{5}\%$ 38. $92\frac{3}{8}\%$

(6.2) **Convert each of the following decimals to a percent.**

39. 0.28   40. 0.07   41. 0.319   42. 0.602   43. 0.8

44. 5.0   45. 0.0043   46. 0.0001   47. 7.2   48. 26.8

(6.3) **Find the unknown in each of the following questions.**

49. What number is 60% of 154?   50. What number is 25% of 76?

51. 9 is 8% of what number?   52. 0.065 is 65% of what number?

53. 7.2 is what percent of 45?   54. 0.054 is what percent of 36?

55. What number is 250% of 6.2?   56. 0.98 is 0.32% of what number?

57. 670 is what percent of $22\frac{1}{3}$?   58. What number is $33\frac{1}{3}\%$ of 267?

59. 3.6289 is what percent of 267.9? Round off to the nearest hundredth of a percent.   60. 0.7634 is 82.719% of what number? Round off to the nearest ten-thousandth.

(6.3–6.6) **Solve each of the following word problems.**

61. Three out of 5 people working for a small computer company chose a 4-day workweek over the traditional 5-day workweek. What percent of the company's employees opted for a 4-day workweek?

62. One out of every 3 students at a local community college is a part-time student. What percent of the student body is part-time?

63. Of the people who flew the 9 A.M. shuttle from Boston to New York, 88% were traveling on business. If 125 people were on the flight, how many were flying for business purposes?

64. Sixteen percent of the people in a part-time M.B.A. program were working on their second masters degree. If 28 people were working on their second degree, how many people were in the program?

65. A photographer bought some darkroom supplies totaling $37.80 before tax. If sales tax is 5%, how much will the supplies cost her?

66. A computer salesman earns a weekly salary of $172 plus an 8% commission on all hardware he sells. If he sold $13,700 worth of computer hardware last week, what was his total income last week?

67. In September the number of students enrolled in a local elementary school was 12% less than last year. If 450 pupils attended the school last year, how many are enrolled this year?

68. Last week, the price of broccoli rose from 75¢ a bunch to 99¢ a bunch. What was the percent increase in price?

69. Sixteen out of 36 people who belong to a photography club said that they became interested in photography as teenagers. What percent of the members does this represent?

70. Out of 432 people staying at a hotel, 72 registered to stay more than two nights. What percent of the hotel's occupants are booked for more than two nights?

71. If you deposit $600 in a bank account at an annual interest rate of $5\frac{3}{4}\%$, how much money will be in your account 4 months later?

72. A party of 6 accumulates a bill of $59.75 at a restaurant, not including tax and tip. Tax is 5%, and they decide to leave a 15% tip. Estimate how much each person should contribute if the bill is split evenly.

73. A pair of shoes originally selling for $39.95 is on sale for 20% off. What is the sale price?

74. A television set costs a wholesale appliance store $80. If the television is sold for $122, what percent profit does the store make?

75. A computer operator deposited $750 in a money market account that pays 5% interest. How much money will be in the account after 2 years?

76. An electrical engineer took out a car loan for $8,000 at an annual interest rate of $7\frac{1}{2}\%$. How much interest will she have paid after 6 months?

77. A money manager invested $12,000 in a bond fund. If after 1 year the investment was worth $13,200, what is the annual rate of interest?

78. After 3 years, an auto mechanic had to pay $4,800 in interest on a $20,000 home improvement loan. What is the annual rate of interest?

79. Of all the people surveyed at a local supermarket, $66\frac{2}{3}\%$ said that they prefer brand-name products instead of generic brands. If 822 people were surveyed, how many prefer brand-name products?

80. Of all the people in a small town in New Mexico, $14\frac{2}{7}\%$ who voted in the last gubernatorial election voted for the Democratic candidate. If the Democratic candidate received 216 votes from that town, how many people from that town voted in the election?

81. During a 10-year period, the number of families in the United States increased from 45,111,000 to 51,586,000. What was the percent increase? Round off your answer to the nearest hundredth of a percent.

82. Last year, the United States spent $286,547,000,000 on social welfare programs. If this figure represented 18.9% of the gross national product (GNP) for that year, what was the GNP? Round off your answer to the nearest billion dollars.

83. **a.** If there were a total of 50,910,000 families, how many had an income of $25,000 or over?

b. How many more families had an income under $5,000 than had an income of $5,000 to $6,999?

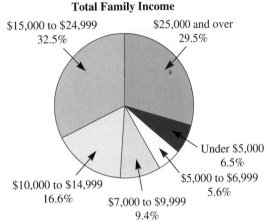

Total Family Income

$15,000 to $24,999
32.5%

$25,000 and over
29.5%

Under $5,000
6.5%

$5,000 to $6,999
5.6%

$7,000 to $9,999
9.4%

$10,000 to $14,999
16.6%

Source: Department of Commerce, Bureau of the Census

84. **a.** During what year was 22% of the female population in the labor force?

b. During what 10-year period did the percentage of women in the labor force increase the most?

c. By how much did the percent of female population in the working force increase from 1960 to 1970?

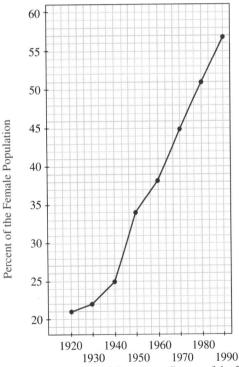

Women in the Labor Force

Percent of the Female Population

Source: Department of Commerce, Bureau of the Census, and Department of Labor, Bureau of Labor Statistics

Chapter **6**
Explain It in Words

85. How do you convert a percent to a decimal? Use an example to explain why this technique works.

86. How do you convert a decimal to a percent? Use an example to explain why this technique works.

87. How do you convert a fraction to a percent? Use an example to explain why this technique works.

88. How do you convert a percent to a fraction? Use an example to explain why this technique works.

89. When would you use a pie chart instead of a bar graph to display data?

90. How do you calculate in your head a 15% tip on a restaurant bill? Give an example.

91. How do you calculate the amount of interest paid on an amount of money that has been borrowed? Give an example.

92. How do you calculate a sale price, given the original price and a discount rate? Give an example.

Chapter **6**
Chapter Test

[6.1] **Convert each of the following percents to a fraction reduced to lowest terms.**

1. 52%   2. 0.4%   3. 8.25%

[6.1, 6.2] **Convert each of the following fractions to a percent.**

4. $\dfrac{19}{50}$ 5. $1\dfrac{3}{8}$ 6. $\dfrac{3}{40}$

[6.2] **Convert each of the following percents to a decimal.**

7. 64%   8. $2\dfrac{3}{4}\%$

[6.2] **Convert each of the following decimals to a percent.**

9. 0.54   10. 0.0028

[6.3] **Find the unknown in each of the following.**

11. What number is 60% of 95?   12. 24 is 4% of what number?

13. 8.2 is what percent of 41?   14. What number is $33\dfrac{1}{3}\%$ of 84?

[6.4, 6.5] **Solve each of the following word problems.**

15. A real estate agent earns a 6% commission. If he sells a house for $135,000, how much does he earn from the sale?

16. If $3,500 is deposited in a bank account that pays an annual interest rate of 7%, how much interest will the account earn in 2 years?

17. A businesswoman gave a cab driver a $2 tip that was 20% of the cab fare. What was the cab fare?

18. A man bought a number of items at the bakery totaling $10.20. If the sales tax on the purchase was $0.51, what is the tax rate?

19. After 1 year, a writer had to pay $589 in interest on a $6,200 loan. What was the interest rate?

20. A sweater that retails for $35 is marked down 20%. What is the sale price?

21. It costs a manufacturer $75 to produce a portable CD player. What should the selling price be to ensure a 40% profit?

22. If the number of daily customers served by a restaurant increased from 160 to 208, what was the percent increase?

[6.6]   **Composition of a Daily Diet**

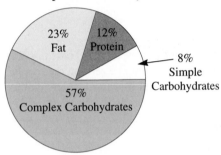

23. If the person whose diet is represented by this graph consumes 1,800 calories a day, how many calories are from protein?

24. How many more calories are from fat than from protein?

25. How many calories are from both simple and complex carbohydrates?

Chapter **7**

Positive and Negative Numbers

7.1
The Number Line

Thus far, our calculations have involved only **positive numbers**, which are numbers greater than zero, located to the right of zero on the number line. In this section, we will also be working with **negative numbers**, which are numbers less than zero, located to the left of zero on the number line. Figure 7.1 shows the relative location of the positive and negative numbers on the number line. The number 0 is considered neither positive nor negative.

Figure 7.1

We use negative numbers in daily life to represent numbers less than zero. For example, if the temperature on a winter day is 6 degrees below zero Fahrenheit, we represent it as $-6°$F. If your checking account is overdrawn by $27.13, your bank statement may show a balance of $-\$27.13$.

Positive numbers are sometimes preceded by a positive sign $(+)$. If no sign precedes a number, it is assumed to be positive. Thus, $+2$ (read as "positive 2") and 2 represent the same number, which is located two units to the *right* of zero on the number line.

Negative numbers are always preceded by a negative sign $(-)$. Thus, -7 (read as "negative 7") represents the number located seven units to the *left* of zero on the number line.

The **integers** consist of $\ldots, -5, -4, -3, -2, -1, 0, 1, 2, 3, 4, 5, \ldots$ The symbol $\ldots$ indicates that there are an infinite number of positive and negative integers.

EXAMPLE 1 Compare each of the following pairs of numbers on this number line. Use the appropriate sign $(=, >, \text{ or } <)$ to indicate the relationship between them.

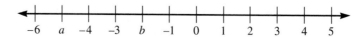

(a) -3 ? 1

$-3 < 1$

Since -3 is to the left of 1 on the number line, -3 is less than 1.

(b) -1 ? -6

$-1 > -6$

Since -1 is to the right of -6 on the number line, -1 is greater than -6.

(c) b ? -2

$b = -2$

Since b is at the same location as -2 on the number line, b is equal to -2.

(d) -4 ? -1

$-4 < -1$

Since -4 is to the left of -1 on the number line, -4 is less than -1.

(e) -3 ? 3

$-3 < 3$

Since -3 is to the left of 3 on the number line, -3 is less than 3.

(f) 0 ? a

$0 > a$

Since 0 is to the right of a on the number line, 0 is greater than a.

(g) b ? a

$b > a$

Since b is to the right of a on the number line, b is greater than a. ▲

Every negative number is the **opposite** of a corresponding positive number that is the same distance away from zero on the number line as is the negative number. For example,

-2 is the opposite of 2.

$-\frac{3}{5}$ is the opposite of $\frac{3}{5}$.

-6.3 is the opposite of 6.3.

Likewise, every positive number is the opposite of a negative number. For example,

8 is the opposite of -8.

$\frac{2}{3}$ is the opposite of $-\frac{2}{3}$.

7.9 is the opposite of -7.9.

The opposite of a number is also called its *additive inverse*. More will be said about additive inverses in Section 7.2.

To indicate the opposite of a number, we write a $-$ sign in front of the number. Thus, the mathematical expression that states that the opposite of negative 5 is positive 5 is written as

$$-(-5) = 5$$

the opposite of   negative

Notice that each $-$ sign in our example has a distinct meaning. You now know three different uses of the $-$ sign. The $-$ sign is used

1. To indicate the operation subtraction.
2. To indicate a negative number.
3. To indicate the opposite of a number.

Whenever a $-$ sign precedes a variable, we interpret it to mean the opposite of that variable. For example, $-a$ is read as "the opposite of a" and *not* as "negative a." The reason for this is that since the value of a is unknown, its opposite could be a negative number or a positive number. For example, if

$$a = -3$$

then

$$-a = -(-3) = 3$$

the opposite of   negative
the opposite of

The statement below summarizes our discussion of opposites. For example, the opposite of 20 is indicated by -20, and the opposite of -8 is indicated by $-(-8) = 8$.

For any number *a*, the *opposite* of *a* is indicated by $-a$.   ▲

EXAMPLE 2 Find the opposite of each of the following numbers and state whether the opposite is a positive or negative number.

(a) 11.2
  The opposite of 11.2 is -11.2.
  -11.2 is a negative number.

(b) $-\dfrac{4}{5}$

  The opposite of $-\dfrac{4}{5}$ is $-\left(-\dfrac{4}{5}\right) = \dfrac{4}{5}$.

  $\dfrac{4}{5}$ is a positive number.

(c) $-(-1)$
  $-(-1) = 1$ The given expression simplifies to 1.
  The opposite of 1 is -1.
  -1 is a negative number.

(d) x if $x > 0$
  The opposite of x is $-x$.
  x is positive.   Since $x > 0$.
  Therefore, $-x$ is negative.

(e) y if $y < 0$
  The opposite of y is $-y$.
  y is negative.   Since $y < 0$.
  Therefore, $-y$ is positive.

(f) $-z$ if $z > 0$

The opposite of $-z$ is $-(-z) = z$.

z is positive. ▲

Since any number and its opposite are both located equal distances away from zero on a number line, they are said to have the same **absolute value**.

▶ **DEFINITION** The *absolute value* of a number is the distance between that number and zero on a number line.

We indicate the absolute value of a number by placing the symbol $|\ \ |$ around the number. For example, the absolute value of 3 is written as $|3|$.

Since we measure distances with positive numbers, the *absolute value* of a number *is never negative*. For example,

$$|-4| = 4 \quad \text{and} \quad |4| = 4$$

The numbers -4 and 4 both have an absolute value of 4 since both numbers are four units away from zero on the number line, as shown in Figure 7.2.

Figure 7.2

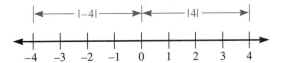

The absolute value of 0 is 0, since 0 is 0 units away from itself.

$$|0| = 0$$

The absolute value of a positive number is equal to the number itself. For example,

$$|9| = 9$$
$$|7.2| = 7.2$$
$$\left|\frac{1}{3}\right| = \frac{1}{3}$$

The absolute value of a negative number is equal to its opposite, which is a positive number. For example,

$$|-12| = 12$$
$$|-0.2| = 0.2$$
$$\left|-\frac{7}{8}\right| = \frac{7}{8}$$

The concept of absolute value is often summarized by the following definition.

▶ **DEFINITION** For any number a, the absolute value of a is defined as

$$|a| = \begin{cases} a & \text{if } a \geq 0 \\ -a & \text{if } a < 0 \end{cases}$$

The first part of the definition,

$$|a| = a \quad \text{if } a \geq 0$$

states that the absolute value of zero or a positive number is equal to itself. Thus,

$$|7| = 7 \quad \text{since } 7 > 0$$

The absolute value of 7 equals 7, since 7 is greater than zero.

The second part of the definition,

$$|a| = -a \quad \text{if } a < 0$$

states that the absolute value of a negative number is equal to its opposite. Thus,

$$|-5| = -(-5) = 5 \quad \text{since } -5 < 0$$

The absolute value of negative 5 equals the opposite of negative 5, which equals 5, since negative 5 is less than zero.

EXAMPLE 3   Simplify each of the following expressions.

(a) $|-6.5|$
       $|-6.5| = -(-6.5) = 6.5$

The absolute value of -6.5 is equal to its opposite, 6.5.

(b) $-|20|$
       $|20| = 20$

First find the absolute value of 20, which is 20.

       $-|20| = -20$

Then find the opposite of 20, which is -20.

Notice that we find the absolute value before we find the opposite. We begin with the innermost expression and work outward.

(c) $-|-73|$
       $|-73| = 73$
       $-|-73| = -73$

The absolute value of -73 is 73.
The opposite of 73 is -73.

(d) $\left| -\left(-\dfrac{1}{2} \right) \right|$

       $-\left(-\dfrac{1}{2} \right) = \dfrac{1}{2}$

The opposite of $-\dfrac{1}{2}$ is $\dfrac{1}{2}$.

       $\left| \dfrac{1}{2} \right| = \dfrac{1}{2}$

The absolute value of $\dfrac{1}{2}$ is $\dfrac{1}{2}$.
Notice that we find the opposite before we find the absolute value. We begin with the innermost expression and work outward. ▲

EXAMPLE 4   Use the $>$, $<$, or $=$ sign to indicate the relationship between each of the following pairs of numbers.

(a) $|-18|$? $|18|$

$|-18| = 18$ Find the absolute value of -18.

$|18| = 18$ Find the absolute value of 18.

$|-18| = |18|$ Since $18 = 18$.

(b) $|-9|$? $|-3|$

$|-9| = 9$ Find the absolute value of -9.

$|-3| = 3$ Find the absolute value of -3.

$|-9| > |-3|$ Since $9 > 3$.

(c) $-|66|$? $|-66|$

$-|66| = -66$ Find the opposite of the absolute value of 66.

$|-66| = 66$ Find the absolute value of -66.

$-|66| < |-66|$ Since $-66 < 66$.   ▲

Quick Quiz

Use the $>$, $<$, or $=$ sign to indicate the relationship between each of the following pairs of numbers.

Answers

1. -5 ? 2 1. $-5 < 2$

2. $|-7|$? $|-3|$ 2. $|-7| > |-3|$

3. $-|-6|$? $|-(-6)|$ 3. $-|-6| < |-(-6)|$

7.1 Exercises

Indicate which of the following statements are true and which are false. For those that are false, change the italicized expression to make the statement true.

1. The opposite of a positive number is *less than* zero.

2. The absolute value of a negative number *is* equal to the opposite of the number.

3. Numbers greater than zero have absolute values *less than* zero.

4. A negative number is located to the *right* of its opposite on a number line.

Compare each of the following pairs of numbers on the following number line. Use the $=$, $>$, or $<$ sign to indicate the relationship between them.

5. -2 ? 2 6. -3 ? -7 7. b ? -4 8. c ? -1

9. a ? -8 10. c ? 0 11. b ? c 12. a ? b

Find the opposite of each of the following numbers and state whether the opposite is a positive or a negative number.

13. -7 14. 3 15. -51 16. -82

17. 5.8 18. 2.1 19. $\dfrac{1}{5}$ 20. $-\dfrac{2}{9}$

21. $-3\dfrac{1}{2}$
22. $7\dfrac{2}{3}$
23. 0.23
24. -0.097

25. $-(-6)$
26. $-(-12)$
27. a if $a < 0$
28. $-b$ if $b < 0$

Simplify each of the following expressions.

29. $|-3|$
30. $|5|$
31. $|51|$
32. $|-27|$
33. $\left|\dfrac{3}{5}\right|$

34. $\left|-\dfrac{7}{8}\right|$
35. $|3.9|$
36. $|-8.2|$
37. $-|62|$
38. $-|-44|$

39. $-\left|-\dfrac{5}{9}\right|$
40. $-\left|\dfrac{2}{3}\right|$
41. $|-82|$
42. $|-(-35)|$
43. $-|52|$

44. $-|-79|$
45. $-\left|8\dfrac{7}{9}\right|$
46. $-\left|-5\dfrac{1}{4}\right|$
47. $-\left|9\dfrac{3}{5}\right|$
48. $\left|-\left(-6\dfrac{1}{2}\right)\right|$

49. $-|5.4|$
50. $-|-6.7|$
51. $-|-1.85|$
52. $-|6.73|$
53. $-|5.48|$

54. $|-(-4.17)|$
55. $-|-(-28)|$
56. $-|-(-61)|$

Use the $=$, $>$, or $<$ sign to indicate the relationship between each of the following pairs of numbers.

57. $|8| \; ? \; |-8|$
58. $-|4| \; ? \; |-4|$
59. $|-9| \; ? \; -|3|$
60. $-|-2| \; ? \; -|5|$

61. $|-8| \; ? \; |2|$
62. $|7| \; ? \; -|5|$
63. $|-23| \; ? \; |23|$
64. $-|41| \; ? \; |-41|$

65. $|-58| \; ? \; -|72|$
66. $|-46| \; ? \; -|27|$
67. $-|3.8| \; ? \; |-5.6|$
68. $-|7.1| \; ? \; -|-9.2|$

69. $-|-0.052| \; ? \; -|0.075|$
70. $-|0.37| \; ? \; -|0.62|$

71. $-\left|-\dfrac{7}{9}\right| \; ? \; -\left|\dfrac{7}{9}\right|$
72. $\left|-\dfrac{3}{5}\right| \; ? \; -\left|\dfrac{3}{5}\right|$
73. $\left|-\left(-\dfrac{5}{8}\right)\right| \; ? \; \left|-\dfrac{1}{2}\right|$
74. $-\left|\dfrac{3}{4}\right| \; ? \; \left|-\dfrac{2}{3}\right|$

7.2
Addition of Signed Numbers

The operation addition can be illustrated by using a number line. Refer to the number line in Figure 7.3 as we work Example 1.

EXAMPLE 1 Add $2 + 5$.

Solution We will use the number line shown in Figure 7.3 to solve this problem. We begin at 0 and move 2 units to the right. This brings us to the first addend, 2. To add 5, we then move 5 additional units to the right. Our final location, 7, is the sum of the two numbers added together.

Figure 7.3

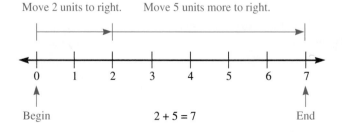

Move 2 units to right.　　Move 5 units more to right.

Begin　　　　$2 + 5 = 7$　　　　End

▲

Thus, to add positive numbers on a number line, begin at zero, and move to the right the distance represented by each addend. In Example 1, the distance we moved along the number line can be represented as the sum of the absolute values of the addends. We can state that

$$2 + 5 = |2| + |5| = 7$$

We therefore have demonstrated the procedure summarized below.

To add two positive numbers, add their absolute values.   ▲

EXAMPLE 2   Add $-3 + (-2)$.

Solution   Notice that whenever a $+$ and $-$ sign appear in succession, parentheses are used to indicate that the sign closest to the number is the sign of the number (-2), and that the other sign is the operation to be performed (addition). Thus, the problem is read as "negative 3 plus negative 2." Using the number line shown in Figure 7.4 to help with our addition, we begin at zero and move 3 units to the left. This brings us to the first addend, -3. To add -2, we then move 2 additional units to the left. Our final location, -5, is the sum of the two numbers.

Figure 7.4

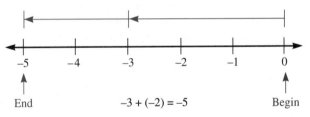

Notice that to solve the problem in Example 2, we moved a total of 5 units, which is the sum of the absolute values of -3 and -2.

$$|-3| + |-2| = 5$$

In Example 2, we found that $-3 + (-2) = -5$. The $-$ sign in front of the 5 indicates we moved a total of 5 units to the left. We therefore have demonstrated the procedure stated below.

To add two negative numbers, add their absolute values and place a negative sign in front of the result.   ▲

EXAMPLE 3   Add each of the following pairs of numbers.

(a) $-4 + (-9)$

$|-4| + |-9| = 4 + 9 = 13$ Find the sum of the absolute values of the addends.

$-4 + (-9) = -13$ Put a $-$ sign in front of the result.

(b) $-8.5 + (-6.2)$

$|-8.5| + |-6.2| = 8.5 + 6.2 = 14.7$ Find the sum of the absolute values of the addends.

$-8.5 + (-6.2) = -14.7$ Put a $-$ sign in front of the result.   ▲

We will now look at some examples of addition in which one addend is negative and the other is positive.

EXAMPLE 4   Add $9 + (-6)$.

Solution   Using the number line shown in Figure 7.5 to help with our addition, we begin at zero and move 9 units to the right. This brings us to the first addend, 9. To add -6, we move 6 units to the left. Our final location, 3, is the sum of the two numbers.

Figure 7.5

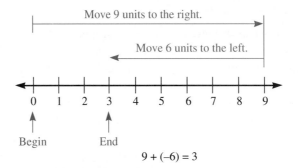

$9 + (-6) = 3$ ▲

Notice that when we added a positive and a negative number together in Example 4, we moved in two different directions, first to the right, and then to the left. To determine the distance between the final location and zero, we can find the difference between the absolute values of the two addends.

$|9| - |-6| = 9 - 6 = 3$

Since we moved farther to the right (9 units) than we moved to the left (6 units), the sum is a positive number, which is located to the right of zero, our starting point. The sign of the sum is therefore the same as the addend with the larger absolute value.

$9 + (-6) = 3$

EXAMPLE 5   Add $-6 + 4$.

Solution   Using the number line shown in Figure 7.6 to help us with our addition, we begin at zero and move 6 units to the left. This brings us to the first addend, -6. To add 4, we move 4 units to the right. Our final location, -2, is the sum of the two numbers.

Figure 7.6

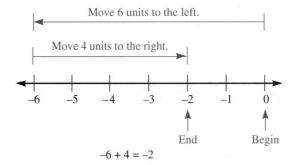

$-6 + 4 = -2$ ▲

Notice in Example 5 that by first moving a distance of 6 units to the left and then 4 units to the right, our final location is 2 units away from zero. Since we moved first to the left, and then to the right, the distance between our final location and zero, our starting point, is the same as the difference between the absolute values of the two addends.

$$|-6| - |4| = 6 - 4 = 2$$

Since we moved farther to the left (6 units) than we did to the right (4 units), the sum is a negative number, which is located to the left of zero. The sign of the sum, therefore, is the same as the addend with the larger absolute value.

$$-6 + 4 = -2$$

We therefore have established the procedure outlined below.

To add a negative number and a positive number, subtract the smaller absolute value from the larger absolute value. The sum will have the same sign as the addend with the larger absolute value. ▲

EXAMPLE 6   Add each of the following pairs of numbers.

(a) $4 + (-17)$

$|-17| - |4| = 17 - 4 = 13$ Subtract the smaller absolute value from the larger.

$|-17| > |4|$ The addend with the larger absolute value is negative.

$4 + (-17) = -13$ The sum is also negative.

(b) $-12 + 36$

$|36| - |-12| = 36 - 12 = 24$ Subtract the smaller absolute value from the larger.

$|36| > |-12|$ The addend with the larger absolute value is positive.

$-12 + 36 = 24$ The sum is also positive.

(c) $-5.7 + 1.3$

$|-5.7| - |1.3| = 5.7 - 1.3 = 4.4$ Subtract the absolute values.

$|-5.7| > |1.3|$ The addend with the larger absolute value is negative.

$-5.7 + 1.3 = -4.4$ The sum is also negative.   ▲

Let us observe what happens when we add a number to its opposite. For example,

$$-8 + 8 = |-8| - |8| = 8 - 8 = 0$$

$$\frac{5}{6} + \left(-\frac{5}{6}\right) = \left|\frac{5}{6}\right| - \left|-\frac{5}{6}\right| = \frac{5}{6} - \frac{5}{6} = 0$$

$$-3.2 + 3.2 = |-3.2| - |3.2| = 3.2 - 3.2 = 0$$

Thus, whenever we add a number to its opposite, we obtain a sum of zero. The opposite of a number is also called its **additive inverse**.

▶ **DEFINITION** The *additive inverse*, or *opposite*, of a number is the number you must add to it to obtain a sum of zero.

EXAMPLE 7 Find the additive inverse of each of the following numbers.

(a) -37

37 is the additive inverse of -37.   Since $-37 + 37 = 0$.

(b) 0.02

-0.02 is the additive inverse of 0.02.   Since $0.02 + (-0.02) = 0$.

(c) $-\frac{1}{8}$

$\frac{1}{8}$ is the additive inverse of $-\frac{1}{8}$.   Since $-\frac{1}{8} + \frac{1}{8} = 0$.   ▲

You now know how to add numbers with the same sign and numbers with different signs. The rules for adding signed numbers are summarized below.

To Add Signed Numbers

1. If both numbers are positive, add their absolute values.

2. If both numbers are negative, add their absolute values and place a negative sign in front of the result.

3. If one number is positive and the other is negative, subtract the smaller absolute value from the larger absolute value. The sum will have the same sign as the addend with the larger absolute value.   ▲

EXAMPLE 8 Perform each of the following additions.

(a) $7 + (-9) + (-8)$

$$7 + (-9) + (-8) = -2 + (-8)$$
$$= -10$$

(b) $-2 + (-4) + 7$

$$-2 + (-4) + 7 = -6 + 7$$
$$= 1$$

(c) $-8 + 4 + (-3) + (-2)$

$$[-8 + 4] + (-3) + (-2) = [-4 + (-3)] + (-2)$$
$$= -7 + (-2)$$
$$= -9 \qquad ▲$$

EXAMPLE 9 A student received a bank statement indicating that he was $29 overdrawn. He then deposited $100 in his account. What was his new balance?

Solution An account that is overdrawn $29 has a balance of $-\$29$. To find the new balance, add $-29 + 100$.

$|100| - |-29| = 100 - 29 = 71$ Find the difference between the absolute values.

$|100| > |-29|$ The number with the larger absolute value is positive.

$-29 + 100 = 71$ The answer is positive.

His new balance is $71.   ▲

Quick Quiz

Perform each of the following additions.   **Answers**

1. $-9 + 3$ **1.** -6

2. $-8 + (-7)$ **2.** -15

3. $3 + (-5) + (-2) + 6$ **3.** 2

4. $-2.3 + 5.7 + (-7.8)$ **4.** -4.4

7.2 Exercises

Indicate which of the following statements are true and which are false. For those that are false, change the italicized expression to make the statement true.

1. The sum of two negative numbers is *positive*.

2. To add a negative and a positive number, find the *difference between* their absolute values.

3. When adding a negative number on a number line, you move to the *left*.

4. The additive inverse of a number is the number you add to it to obtain a sum of *one*.

Find each of the following sums.

5. $7 + (-5)$	**6.** $6 + (-2)$	**7.** $-8 + (-3)$	**8.** $-5 + (-9)$
9. $-3 + 4$	**10.** $-6 + 9$	**11.** $-5 + (-9)$	**12.** $-3 + (-8)$
13. $17 + (-6)$	**14.** $-21 + (-5)$	**15.** $-36 + (-5)$	**16.** $4 + (-23)$
17. $41 + (-20)$	**18.** $-18 + (-55)$	**19.** $-62 + 37$	**20.** $63 + (-25)$
21. $89 + (-45)$	**22.** $-27 + 31$	**23.** $-23 + (-59)$	**24.** $-14 + 85$
25. $-5.3 + 2.8$	**26.** $-2.9 + (-5.7)$	**27.** $7.7 + (-4.8)$	**28.** $-1.9 + 8.5$
29. $7.9 + (-1.3)$	**30.** $-5.8 + 2.2$	**31.** $-0.48 + 0.36$	**32.** $-0.92 + (-0.09)$

33. $367 + (-145)$ **34.** $-185 + 647$ **35.** $\dfrac{5}{9} + \left(-\dfrac{1}{3}\right)$ **36.** $-\dfrac{3}{5} + \left(-\dfrac{1}{2}\right)$

37. $-\dfrac{3}{4} + \dfrac{7}{8}$ **38.** $-\dfrac{1}{6} + \dfrac{2}{3}$ **39.** $-\dfrac{5}{8} + \dfrac{3}{8}$ **40.** $-\dfrac{2}{7} + \left(-\dfrac{5}{7}\right)$

41. $2\dfrac{1}{2} + \left(-1\dfrac{1}{4}\right)$ **42.** $-3\dfrac{5}{6} + \left(-2\dfrac{1}{3}\right)$ **43.** $5 + (-7) + (-3)$ **44.** $-8 + 9 + (-5)$

45. $-4 + 14 + (-9)$ **46.** $11 + (-7) + 18$ **47.** $24 + (-37) + 19$ **48.** $-84 + 69 + 13$

49. $-402 + (-287) + 154$ **50.** $-114 + 567 + (-382)$

51. $\dfrac{1}{3} + \left(-\dfrac{5}{12}\right) + \dfrac{3}{4}$ **52.** $-\dfrac{3}{5} + \left(-\dfrac{7}{15}\right) + \dfrac{3}{4}$

53. $81{,}065{,}050 + (-46{,}729{,}486)$ **54.** $-27{,}364{,}873 + (-48{,}316{,}092)$

55. $-5.00218 + 3.28947$ **56.** $7.21843 + (-8.01241)$

Find the additive inverse of each of the following numbers.

57. 28 **58.** -91 **59.** $-\dfrac{3}{5}$ **60.** $-8\dfrac{3}{4}$

61. 0.675 **62.** 0.003 **63.** $-534{,}987$ **64.** $283{,}005$

Solve each of the following word problems using addition of signed numbers.

65. At 9 A.M. the temperature in Anchorage, Alaska, was $-12°$ F. During the next 3 hours, the temperature rose $5°$. What was the temperature 3 hours later?

66. An elevator in an observation tower at Niagara Falls began at ground level, went down 8 floors to pick up passengers, and then went up 14 floors to an observation deck where they got out. How many floors above ground level is this observation deck?

67. The highest point in California, Mount Whitney, is 14,776 feet higher than the lowest point in California. The lowest point in the state is Death Valley, which is 282 feet below sea level. What is the elevation of Mount Whitney?

68. A tourist on vacation in Las Vegas won $25 at the blackjack table. That same evening, she lost $51 and then won an additional $33. What were her net winnings for that evening?

69. A football team was penalized 15 yards on one play and then gained 8 yards on the next play. If the ball was on the 22-yard line before the penalty, where was the ball after the next play?

70. A stock opened at $30\frac{1}{2}$ points. During that day's trading session, it gained $1\frac{1}{4}$ points, lost $\frac{3}{8}$ point, and then lost an additional $\frac{3}{4}$ point. What was the closing price of the stock?

7.3
Subtraction of Signed Numbers

Since addition and subtraction are inverse operations, any subtraction problem can be expressed as an addition problem. Let us consider the following examples.

$$
\begin{array}{rcll}
7 - 6 &=& 1 & \text{since} & 1 + 6 = 7 \\
3 - 9 &=& -6 & \text{since} & -6 + 9 = 3 \\
-4 - 5 &=& -9 & \text{since} & -9 + 5 = -4 \\
5 - (-2) &=& 7 & \text{since} & 7 + (-2) = 5
\end{array}
$$

Notice that we can obtain the same results by adding the opposite of the quantity being subtracted.

$$
\begin{array}{rcl}
7 - 6 &=& 7 + (-6) = 1 \\
3 - 9 &=& 3 + (-9) = -6 \\
-4 - 5 &=& -4 + (-5) = -9 \\
5 - (-2) &=& 5 + (+2) = 7
\end{array}
$$

Thus, to subtract a number from a given quantity, add its opposite to that quantity.

For any signed numbers a and b,
$$a - b = a + (-b)$$
▲

EXAMPLE 1 Perform each of the following subtractions.

(a) $2 - 8$

$\qquad 2 - 8 = 2 + (-8)$ To subtract 8, add -8.

$\qquad\qquad\quad = -6$

SECTION 7.3
Subtraction of Signed Numbers

(b) $-3 - 9$

$$-3 - 9 = -3 + (-9)$$ To subtract 9, add -9.
$$= -12$$

(c) $5 - (-6)$

$$5 - (-6) = 5 + (+6)$$ To subtract -6, add $+6$.
$$= 11$$

(d) $-78 - (-21)$

$$-78 - (-21) = -78 + (+21)$$ To subtract -21, add $+21$.
$$= -57$$

(e) $-8.7 - 9.5$

$$-8.7 - 9.5 = -8.7 + (-9.5)$$ To subtract 9.5, add -9.5.
$$= -18.2$$

(f) $\dfrac{1}{4} - \left(-\dfrac{7}{8}\right)$

$$\dfrac{1}{4} - \left(-\dfrac{7}{8}\right) = \dfrac{1}{4} + \left(+\dfrac{7}{8}\right)$$ To subtract $-\frac{7}{8}$, add $+\frac{7}{8}$.

$$= \dfrac{2}{8} + \dfrac{7}{8} = \dfrac{9}{8}$$ ▲

To solve problems that involve both addition and subtraction, change each subtraction sign to an addition sign and change the number following that sign to its opposite.

EXAMPLE 2   Perform each of the operations indicated.

(a) $5 - (-7) + (-4)$

$$5 - (-7) + (-4) = 5 + (+7) + (-4)$$ To subtract -7, add $+7$.
$$= 12 + (-4)$$
$$= 8$$

(b) $-6 - 20 - (-47)$

$$-6 - 20 - (-47) = -6 + (-20) - (-47)$$ To subtract 20, add -20.
$$= -6 + (-20) + (+47)$$ To subtract -47, add $+47$.
$$= -26 + 47$$
$$= 21$$

(c) $4.2 + (-9.8) - (-6.5)$

$$4.2 + (-9.8) - (-6.5) = 4.2 + (-9.8) + (+6.5)$$ To subtract -6.5, add $+6.5$.
$$= -5.6 + 6.5$$
$$= 0.9$$ ▲

EXAMPLE 3   Mercury has a melting point of $-38.87°$ C. Helium has a melting point of $-272.20°$ C. What is the difference between the melting point of mercury and the melting point of helium?

Solution Solve the problem using subtraction of signed numbers.

$$-38.87 - (-272.20) = -38.87 + (+272.20) \quad \text{To subtract } -272.20, \text{ add } +272.20.$$
$$|272.20| - |38.87| = 272.20 - 38.87 \quad \text{Subtract the smaller absolute value}$$
$$= 233.33 \quad \text{from the larger.}$$

$$\begin{array}{r} 272.20 \\ -\ 38.87 \\ \hline 233.33 \end{array}$$

$$\underset{\text{mercury}}{-38.87} - \underset{\text{helium}}{(-272.20)} = 233.33$$

Thus, the melting point of helium is 233.33° C less than that of mercury. ▲

Quick Quiz

Perform each of the following subtractions.   **Answers**

1. $-8 - 9$ **1.** -17
2. $4 - (-7)$ **2.** 11

Perform each of the operations indicated.

3. $-2 + (-5) - (-3)$ **3.** -4
4. $3 - 7 - (-9) + (-2)$ **4.** 3

7.3 Exercises

Indicate which of the following statements are true and which are false. For those that are false, change the italicized expression to make the statement true.

1. Subtracting a negative number from a given quantity is the same as *adding* its opposite to that quantity.

2. The operation subtraction is the inverse of *division*.

3. When we subtract a negative number from its opposite, the result is *less than* zero.

4. Adding the opposite of a positive number to a given quantity is the same as subtracting a *negative* number from that quantity.

Perform each of the following subtractions.

5. $5 - 7$ 6. $3 - 8$ 7. $-4 - 2$ 8. $-9 - 1$

9. $0 - 7$ 10. $8 - (-6)$ 11. $5 - (-9)$ 12. $0 - (-4)$

13. $-8 - (-7)$ 14. $-4 - (-4)$ 15. $12 - 35$ 16. $-25 - 41$

17. $-17 - (-53)$ 18. $78 - (-39)$ 19. $3.8 - 5.6$ 20. $-2.9 - 7.1$

21. $5.4 - (-8.7)$ 22. $-6.9 - (-8.3)$ 23. $-453 - 298$ 24. $-307 - (-154)$

25. $-0.036 - 0.298$ 26. $-0.263 - (-0.457)$ 27. $0.059 - 0.584$ 28. $-0.037 - (-0.069)$

29. $1,658 - 3,924$ 30. $-4,781 - 3,742$ 31. $\dfrac{3}{8} - \dfrac{5}{8}$ 32. $-\dfrac{7}{9} - \dfrac{8}{9}$

33. $-\dfrac{4}{5} - \dfrac{7}{10}$ 34. $-\dfrac{5}{12} - \dfrac{3}{4}$ 35. $-3\dfrac{1}{2} - 5\dfrac{1}{8}$ 36. $6\dfrac{2}{3} - 8\dfrac{5}{6}$

37. $7,362,051 - 10,000,000$ **38.** $-29,684,317 - (-28,671,953)$

 39. $-4.80671 - 7.35826$ **40.** $5.09543 - (-6.21985)$

Perform each of the operations indicated.

41. $3 + 5 - 6$ **42.** $2 - 7 + 4$ **43.** $1 + 8 - 9$ **44.** $6 - 9 + 3$

45. $2 - 0 - 7$ **46.** $0 - 5 - 3$ **47.** $3 - (-8) + (-4)$ **48.** $-2 + (-9) - 7$

49. $2.2 - (-1.4) - 7.5$ **50.** $-3.9 - (-5.4) + (-6.1)$

51. $45 - 63 + (-32)$ **52.** $-81 + (-97) + 25$

53. $-\dfrac{1}{2} - \left(-\dfrac{3}{8}\right) + \dfrac{1}{4}$ **54.** $\dfrac{1}{3} - \dfrac{2}{9} + \left(-\dfrac{5}{6}\right)$

55. $-2\dfrac{3}{4} + 1\dfrac{2}{3} - \left(-5\dfrac{7}{12}\right)$ **56.** $-4\dfrac{7}{12} + \left(-3\dfrac{2}{9}\right) - 4\dfrac{5}{6}$

 57. $6,285,736 - 8,943,628 + 4,821,946$ **58.** $7,843,621 + (-5,287,453) - (-2,893,186)$

Solve each of the following word problems using subtraction of signed numbers.

59. On Monday, the average temperature was $-8°\,F$. On Tuesday, the average temperature was $3°\,F$. How much warmer was the average temperature on Tuesday than on Monday?

60. If a bank account is overdrawn $37, how much money must be deposited to bring the balance up to $179?

61. Last year a small business showed a loss of $35,540. This year it showed a profit of $9,380. By how much did the earning power of the business increase from last year to this year?

62. The highest point in the Sahara Desert has an elevation of 11,000 feet. The lowest point is 440 feet below sea level. What is the difference between the highest and lowest points in the Sahara Desert?

7.4
Multiplication of Signed Numbers

In order to determine how to multiply signed numbers, we will use the definition of multiplication as repeated addition. Consider the following examples.

$(3)(4)$ means the sum of three 4's: $4 + 4 + 4 = 12$
$(3)(-4)$ means the sum of three (-4)'s: $(-4) + (-4) + (-4) = -12$
$(-3)(4) = (4)(-3)$ means the sum of four (-3)'s:
$\qquad (-3) + (-3) + (-3) + (-3) = -12$

The procedure we have demonstrated is summarized below.

To multiply a positive number and a negative number, multiply their absolute values and place a negative sign in front of the result.   ▲

EXAMPLE 1   Multiply each of the following numbers.

(a) $(2)(-5)$
$\qquad (2)(-5) = -10$

(b) $\left(-\dfrac{1}{3}\right)\left(\dfrac{5}{8}\right)$

$$\left(-\dfrac{1}{3}\right)\left(\dfrac{5}{8}\right) = -\dfrac{5}{24}$$

▲

In order to determine the product obtained when we multiply two negative numbers together, we will use the following example that illustrates the distributive property.

$$(-7)[3 + (-3)] = (-7)(3) + (-7)(-3)$$
$$(-7)(0) = (-7)(3) + (-7)(-3)$$
$$0 = -21 + (-7)(-3) \qquad \text{The left side simplifies to zero.}$$
$$0 = -21 + ? \qquad \text{The right side must also equal zero.}$$

We can now see that the product $(-7)(-3)$ is equal to some number that when added to -21 produces a sum of zero. Since

$$0 = -21 + 21$$

this number is 21. Therefore,

$$(-7)(-3) = 21$$

We have now shown that the product of two negative numbers is a positive number. When multiplying two negative numbers together, therefore, we can ignore the signs of the factors (see below).

To multiply two negative numbers, find the product of their absolute values. ▲

EXAMPLE 2 Perform each of the following multiplications.

(a) $(-8)(-7)$
$(-8)(-7) = 56$

(b) $(-0.03)(-0.6)$
$(-0.03)(-0.6) = 0.018$

(c) $\left(-\dfrac{3}{5}\right)\left(-\dfrac{2}{7}\right)$

$$\left(-\dfrac{3}{5}\right)\left(-\dfrac{2}{7}\right) = \dfrac{6}{35}$$

▲

Let us now investigate what happens when we multiply more than two signed numbers together.

EXAMPLE 3 Perform each of the following multiplications.

(a) $(-3)(7)(-2)(5)$

$$(-3)(7)(-2)(5) = [(-3)(7)](-2)(5)$$
$$= [(-21)(-2)](5)$$
$$= (42)(5)$$
$$= 210$$

Multiply two numbers at a time.

Notice that there were two negative factors, and the product was positive.

(b) $(2)(-6)(-3)(-4)$
$$(2)(-6)(-3)(-4) = [(2)(-6)](-3)(-4)$$
$$= [(-12)(-3)](-4)$$
$$= (36)(-4)$$
$$= -144$$

Notice that there were three negative factors, and the product was negative.

(c) $(-2)(-4)(-10)(-6)$
$$(-2)(-4)(-10)(-6) = [(-2)(-4)](-10)(-6)$$
$$= [(8)(-10)](-6)$$
$$= (-80)(-6)$$
$$= 480$$

Notice that there were four negative factors, and the product was positive.

(d) $(-1)(-7)(-2)(-3)(-4)$
$$(-1)(-7)(-2)(-3)(-4) = [(-1)(-7)](-2)(-3)(-4)$$
$$= [(7)(-2)](-3)(-4)$$
$$= [(-14)(-3)](-4)$$
$$= (42)(-4)$$
$$= -168$$

Notice that there were five negative factors, and the product was negative. ▲

The problems in Example 3 illustrate the methods outlined below.

To Multiply Signed Numbers

1. Multiply their absolute values.
2. If there is an **even** number of negative factors, the product is **positive**.
3. If there is an **odd** number of negative factors, the product is **negative**. ▲

EXAMPLE 4   Perform each of the following multiplications.

(a) $(-4)(5)(2)(-10)$
$$(-4)(5)(2)(-10) = +[4 \cdot 5 \cdot 2 \cdot 10]$$
$$= 400$$

First determine the sign of the answer. Since there are 2 negative factors, the product is positive. Multiply, ignoring the signs of the factors.

(b) $\left(-\dfrac{3}{5}\right)\left(\dfrac{7}{9}\right)\left(-\dfrac{5}{14}\right)\left(-\dfrac{3}{4}\right)$

$$\left(-\dfrac{3}{5}\right)\left(\dfrac{7}{9}\right)\left(-\dfrac{5}{14}\right)\left(-\dfrac{3}{4}\right) = -\left(\dfrac{3}{5}\cdot\dfrac{7}{9}\cdot\dfrac{5}{14}\cdot\dfrac{3}{4}\right)$$

Since there are 3 negative factors, the product is negative.

$$= -\dfrac{\cancel{3}\cdot 7\cdot\cancel{5}\cdot\cancel{3}}{\cancel{5}\cdot\cancel{9}\cdot\cancel{14}\cdot 4}$$

Divide numerator and denominator by common factors.

$$= -\dfrac{1}{8}$$

Recall that when all the numbers in the numerator are canceled, the result is 1 and not 0.

(c) $(-0.03)(-20)(-0.1)(-7)$

$$(-0.03)(-20)(-0.1)(-7) = +[(0.03)(20)(0.1)(7)]$$
$$= (0.6)(0.7)$$
$$= 0.42$$

Since there are 4 negative factors, the product is positive. ▲

By using the rules for multiplying signed numbers, we can determine what happens when a negative number is raised to a power.

$(-2)^2 = (-2)(-2) = 4$ Two negative factors. Product is positive.

$(-2)^3 = (-2)(-2)(-2) = -8$ Three negative factors. Product is negative.

$(-2)^4 = (-2)(-2)(-2)(-2) = 16$ Four negative factors. Product is positive.

$(-2)^5 = (-2)(-2)(-2)(-2)(-2) = -32$ Five negative factors. Product is negative.

Our observations are stated below.

A negative number raised to an *even* power simplifies to a *positive* number.
A negative number raised to an *odd* power simplifies to a *negative* number.   ▲

EXAMPLE 5   Simplify each of the following expressions.

(a) $(-3)^3$

$$(-3)^3 = -(3\cdot 3\cdot 3)$$
$$= -27$$

Since the power is odd, the result is negative.

(b) $(-10)^6$

$$(-10)^6 = 1{,}000{,}000$$

Since the power is even, the result is positive.

(c) $(-1)^{57}$

$$(-1)^{57} = -1$$

Since the power is odd, the result is negative.   ▲

Be careful not to confuse expressions such as $(-3)^2$ with expressions such as -3^2.

$(-3)^2$ means that negative 3 is used as a factor twice:

$$(-3)^2 = (-3)(-3) = 9$$

-3^2 means take the opposite of 3 squared:

$$-3^2 = -(3 \cdot 3)$$
$$= -9$$

EXAMPLE 6   Simplify each of the following expressions.

(a) $(-7)^2$

      $(-7)^2 = (-7)(-7)$ -7 is used as a factor twice.

           $= 49$

(b) -4^2

      $-4^2 = -(4 \cdot 4)$ Take the opposite of 4 squared.

           $= -16$

(c) -3^4

      $-3^4 = -(3 \cdot 3 \cdot 3 \cdot 3)$ Take the opposite of 3 to the 4th power.

           $= -81$ ▲

Let us now look at some examples that involve simplifying arithmetic expressions using the order of operations rules.

EXAMPLE 7   Simplify each of the following expressions.

(a) $-8(-7 + 3)$

      $(-8)(-7 + 3) = (-8)(-4)$ Simplify what is in ( ) first.

                   $= 32$ Then do multiplication.

(b) $7(-3)^2 - (-2)$

      $7(-3)^2 - (-2) = 7(9) - (-2)$ Raise to powers first.

                   $= 63 - (-2)$ Then do multiplication.

                   $= 63 + (+2)$ Finally, do subtraction.

                   $= 65$

(c) $-2\{[-5 - (-3)]^3 + (-4 - 6)^2\}$

      $-2\{[-5 - (-3)]^3 + (-4 - 6)^2\}$ Rewrite subtraction as

      $= -2\{[-5 + (+3)]^3 + [-4 + (-6)]^2\}$ addition of opposites.

      $= -2[(-2)^3 + (-10)^2]$ Simplify within [ ].

      $= -2(-8 + 100)$ Raise to powers.

      $= -2(92)$ Simplify within ( ).

      $= -184$ Multiply.   ▲

Quick Quiz

Perform each of the following multiplications.

Answers

1. $(-3)(-9)$

   1. 27

2. $(-5)(-2)(-7)$

   2. -70

Perform each of the following operations.

3. $(-3)^2 + (-2)^3$

   3. 1

4. $-5[8 + (-5)]$

   4. -15

7.4 Exercises

Indicate which of the following statements are true and which are false. For those that are false, change the italicized expression to make the statement true.

1. The product of two negative numbers is *always* positive.

2. When multiplying signed numbers, the sign of the product is determined by the number of *positive* factors.

3. A negative number raised to an odd power simplifies to a *negative* number.

4. The product of an even number of negative factors and an odd number of positive factors is *positive*.

Perform each of the following multiplications.

5. $(-3)(7)$

6. $(4)(-9)$

7. $(-6)(-5)$

8. $(-8)(-2)$

9. $(0.7)(-0.3)$

10. $(-0.5)(-0.9)$

11. $(8.2)(-3)$

12. $(-2)(6.1)$

13. $(-4.2)(0.4)$

14. $(-0.3)(-5.1)$

15. $(1.1)(-0.6)$

16. $(0.8)(-0.12)$

17. $\left(-\dfrac{1}{2}\right)\left(-\dfrac{3}{5}\right)$

18. $\left(\dfrac{2}{3}\right)\left(-\dfrac{2}{7}\right)$

19. $\left(-\dfrac{3}{8}\right)\left(-\dfrac{4}{15}\right)$

20. $\left(\dfrac{7}{9}\right)\left(-\dfrac{3}{14}\right)$

21. $\left(-1\dfrac{3}{8}\right)\left(-2\dfrac{2}{11}\right)$

22. $\left(5\dfrac{1}{2}\right)\left(-2\dfrac{2}{3}\right)$

23. $\left(-2\dfrac{2}{5}\right)\left(3\dfrac{3}{4}\right)$

24. $\left(-4\dfrac{4}{7}\right)\left(-2\dfrac{5}{8}\right)$

25. $(-5)(-1)(9)$

26. $(-8)(4)(2)$

27. $(7)(0)(-6)$

28. $(-5)(-4)(-7)$

29. $(-3)(-9)(-2)$

30. $(-1)(8)(6)$

31. $(-3.0)(-0.2)(-40)$

32. $(60)(-0.04)(-0.3)$

33. $(-0.08)(-500)(1.2)$

34. $(0.11)(-0.4)(-20)$

35. $\left(-\dfrac{5}{8}\right)\left(\dfrac{4}{9}\right)\left(-\dfrac{3}{10}\right)$

36. $\left(\dfrac{3}{4}\right)\left(-\dfrac{2}{15}\right)\left(\dfrac{8}{9}\right)$

37. $\left(-\dfrac{7}{8}\right)\left(\dfrac{5}{14}\right)\left(-\dfrac{4}{15}\right)$

38. $\left(-\dfrac{2}{9}\right)\left(-\dfrac{5}{16}\right)\left(-\dfrac{3}{10}\right)$

39. $(-5)(-6)(2)(-7)$

40. $(-7)(8)(0)(-4)$

41. $(-3)(5)(2)(-8)$

42. $(-4)(-6)(-2)(-10)$

43. $(-4)(-6)(-1)(-2)(-3)$

44. $(-5)(4)(-7)(-3)(-2)$

45. $(-35.47)(-2.169)$

46. $(438.7)(-72.45)$

47. $(-4.8)(37.6)(-813)$

48. $(-5.4)(-81.4)(-293)$

Simplify each of the following expressions.

49. $(-3)^4$ 50. $(-6)^2$ 51. $(-1)^{32}$ 52. $(-5)^3$

53. $(-9)^2$ 54. $(-6)^2$ 55. -8^2 56. -5^2

57. $(-10)^4$ 58. -2^4 59. $4[2 + (-7)]$ 60. $6(-3 + 8)$

61. $5(-9 + 2)$ 62. $8[-4 + (-5)]$ 63. $(-3)(-10)^2$ 64. $(-2)^5(6)$

65. $(-4)^2 - (-3)(7)$ 66. $(-8)(-5) + (-6)^2$

67. $(-7 - 3)(-2 + 5)$ 68. $[5 + (-2)][-4 - (-3)]$

69. $4(-3)^2 - (-7)^2(2)$ 70. $6(-10)^3 + (-7)(-10)^2$

71. $[7 - (-2)]^2 - [5 + (-3)]^3$ 72. $(-4 - 6)^3 + [-3 + (-4)]^2$

73. $2\{6 - (-3) + [7 + (-5)]^3\}$ 74. $-5\{(-2 + 5)^3 - [6 - (-1)]\}$

75. $(-15)^3 - (-23)(42)$ 76. $(-51)(39) + (-27)^2$

77. $-843[-341 - (-706)]$ 78. $-675[947 + (-253)]$

7.5

Division of Signed Numbers

Since multiplication and division are inverse operations, we can rewrite any division problem as a multiplication problem. For example,

$$30 \div 6 = (30)\left(\frac{1}{6}\right) = 5$$

$$-72 \div (-9) = (-72)\left(-\frac{1}{9}\right) = 8$$

$$48 \div (-8) = (48)\left(-\frac{1}{8}\right) = -6$$

$$-6 \div \frac{1}{4} = (-6)(4) = -24$$

Therefore, to determine the sign of a quotient, we can use the same rules we established for determining the sign of a product (see below).

To Divide Signed Numbers

1. Find the quotient of their absolute values.

2. If both numbers have the same sign, the quotient is positive.

3. If the numbers have opposite signs, the quotient is negative. ▲

EXAMPLE 1   Find each of the following quotients.

(a) $-144 \div 12$

$-144 \div 12 = -12$ Since one number is negative and the other is positive, the quotient is negative.

(b) $-3.5 \div (-0.7)$

$-3.5 \div (-0.7) = 5$ Since both numbers are negative, the quotient is positive.

(c) $\dfrac{1}{3} \div \left(-\dfrac{5}{9}\right)$

$\dfrac{1}{3} \div \left(-\dfrac{5}{9}\right) = -\left(\dfrac{1}{3} \cdot \dfrac{9}{5}\right)$ Since one number is positive and the other is negative, the quotient is negative.

$= -\left(\dfrac{1}{\cancel{3}} \cdot \dfrac{\overset{3}{\cancel{9}}}{5}\right)$

$= -\dfrac{3}{5}$ ▲

A fraction can also be expressed as a division problem. For example,

$$\frac{-56}{8} = -56 \div 8 = -7$$

$$\frac{56}{-8} = 56 \div (-8) = -7$$

$$-\frac{56}{8} = -(56 \div 8) = -7$$

Notice that regardless of whether the negative sign appears in front of the numerator, the denominator, or the entire fraction, the fraction simplifies to the same negative number. If the numerator and denominator contain the product of signed numbers, we can use the procedure outlined to find the quotient.

| **To Simplify a Fraction That Expresses the Quotient of Products of Signed Numbers** | **1.** First determine the sign of the answer. If there is an even number of negative factors in the numerator and denominator, the result is positive. If there is an odd number of negative factors, the result is negative.

2. Reduce to lowest terms, ignoring the signs of the individual factors.   ▲ |

EXAMPLE 2   Simplify $\dfrac{(-5)(63)}{(7)(-25)}$.

Solution $\dfrac{(-5)(63)}{(7)(-25)} = +\dfrac{5 \cdot 63}{7 \cdot 25}$ First determine the sign of the answer. Since there are 2 negative factors, the answer is positive.

$= \dfrac{\cancel{5} \cdot \overset{9}{\cancel{63}}}{7 \cdot \underset{5}{\cancel{25}}}$

$= \dfrac{9}{5}$ Reduce to lowest terms. Ignore the signs of the individual factors.   ▲

EXAMPLE 3 Simplify $\dfrac{(-2)(8)(27)}{(6)(-9)(-32)}$.

Solution $\dfrac{(-2)(8)(27)}{(6)(-9)(-32)} = -\left(\dfrac{2 \cdot 8 \cdot \overset{3}{\cancel{27}}}{\underset{2}{\cancel{6}} \cdot \cancel{9} \cdot \underset{4}{\cancel{32}}}\right)$ First determine the sign of the answer. Since there are 3 negative factors, the result is negative.
Reduce to lowest terms ignoring the signs of the individual factors.

$\qquad\qquad\qquad = -\dfrac{1}{4}$ ▲

EXAMPLE 4 Simplify $\dfrac{(-64)(-36)(-63)(12)}{(-16)(-21)(48)(-72)}$.

Solution $\dfrac{(-64)(-36)(-63)(12)}{(-16)(-21)(48)(-72)}$

$\qquad = +\dfrac{\overset{4}{\cancel{64}} \cdot 36 \cdot \overset{3}{\cancel{63}} \cdot \cancel{12}}{\cancel{16} \cdot \cancel{21} \cdot \underset{\underset{2}{4}}{48} \cdot \cancel{72}}$ Since there are 6 negative factors, the result is positive.

$\qquad = \dfrac{3}{2}$ ▲

Let us now look at some problems that require us to use the order of operations rules.

EXAMPLE 5 Simplify each of the following expressions.

(a) $12 \div (-4) + (-2)(-7)$

$12 \div (-4) + (-2)(-7) = -3 + 14$ Do multiplication and division first.

$\qquad\qquad\qquad\qquad\qquad = 11$ Then do addition.

(b) $[8 + (-2)]^2 \div (-3)^2 - 5$

$[8 + (-2)]^2 \div (-3)^2 - 5$

$\qquad = 6^2 \div (-3)^2 - 5$ Simplify what is in [ ] first.

$\qquad = 36 \div 9 - 5$ Then raise to powers.

$\qquad = 4 - 5$ Then do division.

$\qquad = 4 + (-5)$ Finally, do subtraction.

$\qquad = -1$

(c) $[5 - (-7)] \div [6(-3) + (-2)(-7)]$

$[5 - (-7)] \div [6(-3) + (-2)(-7)]$ Rewrite subtraction in [ ].

$\qquad = [5 + (+7)] \div [-18 + 14]$ Do multiplication in [ ].

$\qquad = 12 \div (-4)$ Do addition in [ ].

$\qquad = -3$ Finally, do division.   ▲

Problems in which division is the last operation to be performed can also be expressed in fractional form. Thus, Example 5(c) can be written as follows:

$$[5 - (-7)] \div [6(-3) + (-2)(-7)] = \dfrac{5 - (-7)}{6(-3) + (-2)(-7)}$$

To simplify fractions that involve more than one operation, first simplify the numerator and denominator separately using the order of operations rules, and then do the division. ▲

EXAMPLE 6 Simplify $\dfrac{(-6)(7) + (-3)}{-1 - (2)(-5)}$.

Solution

$$\dfrac{(-6)(7) + (-3)}{-1 - (2)(-5)} = \dfrac{-42 + (-3)}{-1 - (-10)} \qquad \text{First do multiplication.}$$

$$= \dfrac{-45}{-1 + (+10)} \qquad \text{Then do addition and subtraction.}$$

$$= \dfrac{-45}{9}$$

$$= -5 \qquad \text{Finally, do division.} \qquad ▲$$

EXAMPLE 7 Simplify $\dfrac{(3 - 8) - (7)(-2)}{9 + (-2) + (-6 + 2)}$.

Solution $\dfrac{(3 - 8) - (7)(-2)}{9 + (-2) + (-6 + 2)}$

$$= \dfrac{-5 - (7)(-2)}{9 + (-2) + (-4)} \qquad \text{Do operations in ( ) first.}$$

$$= \dfrac{-5 - (-14)}{9 + (-2) + (-4)} \qquad \text{Do multiplication.}$$

$$= \dfrac{-5 + (+14)}{7 + (-4)} \qquad \text{Do addition and subtraction.}$$

$$= \dfrac{9}{3}$$

$$= 3 \qquad \text{Finally, do division.} \qquad ▲$$

Quick Quiz

Find each of the following quotients.

		Answers
1.	$-18 \div (-6)$	**1.** $\;3$
2.	$-\dfrac{3}{5} \div \dfrac{3}{4}$	**2.** $\;-\dfrac{4}{5}$
3.	$\dfrac{(-6)(8)}{(12)(-2)}$	**3.** $\;2$
4.	$\dfrac{4(-9 + 2)}{(-3)(-14)}$	**4.** $\;-\dfrac{2}{3}$

7.5 Exercises

Indicate which of the following statements are true and which are false. For those that are false, change the italicized expression to make the statement true.

1. If the divisor and quotient are negative, the dividend is *negative*.

2. When dividing two numbers of opposite signs, the quotient is *positive*.

3. If the numerator and denominator of a fraction contain an *even* number of negative factors, the fraction simplifies to a positive number.

4. To simplify a fraction that involves more than one operation, simplify the numerator and denominator separately *before* doing the division.

Find each of the following quotients.

5. $-32 \div 4$

6. $27 \div (-3)$

7. $-72 \div (-9)$

8. $63 \div (-7)$

9. $400 \div (-80)$

10. $-490 \div 70$

11. $-144 \div (-12)$

12. $121 \div (-11)$

13. $-2.4 \div (-0.8)$

14. $-0.42 \div (-70)$

15. $-8.1 \div 90$

16. $6.3 \div (-90)$

17. $-0.056 \div (0.07)$

18. $-0.048 \div (-0.06)$

19. $-\dfrac{3}{8} \div \dfrac{9}{4}$

20. $\left(-\dfrac{5}{7}\right) \div \left(-\dfrac{10}{21}\right)$

21. $\dfrac{1}{12} \div \left(-\dfrac{5}{6}\right)$

22. $-\dfrac{8}{9} \div \dfrac{2}{3}$

23. $2\dfrac{2}{5} \div \left(-1\dfrac{3}{8}\right)$

24. $\left(-5\dfrac{1}{3}\right) \div \left(-2\dfrac{2}{9}\right)$

25. $\dfrac{(3)(-8)}{(-2)}$

26. $\dfrac{(-7)(-6)}{(-3)}$

27. $\dfrac{(-9)}{(-6)(4)}$

28. $\dfrac{(-2)}{(8)(3)}$

29. $\dfrac{(-4)(9)}{(-6)(-8)}$

30. $\dfrac{(5)(-7)}{(14)(-10)}$

31. $\dfrac{(-21)(-5)}{(45)(-7)}$

32. $\dfrac{(49)(-9)}{(7)(-72)}$

33. $\dfrac{(-25)(-64)}{(-30)(-16)}$

34. $\dfrac{(-42)(24)}{(36)(-56)}$

35. $\dfrac{(-9)(16)(-35)}{(20)(-7)(-36)}$

36. $\dfrac{(54)(-21)(-8)}{(32)(-24)(14)}$

37. $\dfrac{(81)(-45)(8)}{(-18)(24)(-90)}$

38. $\dfrac{(-64)(4)(-27)}{(12)(-72)(-8)}$

39. $\dfrac{(-5)(-3)(-12)(-6)}{(2)(-4)(-15)(-7)}$

40. $\dfrac{(-10)(3)(-21)(9)}{(-18)(5)(-4)(7)}$

41. $\dfrac{(-5)(-18)(3)(-12)}{(36)(-15)(4)(-6)}$

42. $\dfrac{(81)(-7)(5)(-56)}{(-8)(45)(63)(-3)}$

43. $\dfrac{(-145)(31)(102)}{(17)(-29)(-186)}$

44. $\dfrac{(-91)(83)(-228)}{(-332)(-57)(13)}$

Simplify each of the following expressions.

45. $(3 - 7) \div (-4)$

46. $[4 + (-9)] \div 5$

47. $-72 \div (-5 - 3)$

48. $-42 \div [3 - (-4)]$

49. $(-6)^2 \div (-3)^2$

50. $(-4)^3 \div (-2)^4$

51. $-18 \div (-3) + (5)(-7)$

52. $-36 \div (-9) - (4)(-2)$

53. $[(6)(-2) - 3] \div 5$

54. $[(-7)(3) + (-6)] \div (-9)$

55. $[7 + (-2) - (-9)] \div (-2)$

56. $[8 - (-6) - (-2)] \div 4$

57. $\dfrac{(3)(-3) + (-5)}{9 - (-2)(6)}$

58. $\dfrac{(8)(-3) - (-4)}{(-3)(-4) + (-2)}$

59. $\dfrac{(-5 - 3)^2}{-13 - (-5)}$

60. $\dfrac{-9 - (-1)}{(5 - 7)^4}$

61. $\dfrac{[8 + (-2)]^2}{(-7)(-6)}$

62. $\dfrac{(-2)(9)}{(-9 + 6)^3}$

63. $\dfrac{(3 - 7) + (4)(-2)}{(-7)(2) + (-9 + 5)}$

64. $\dfrac{(-5)(-3) + (-5 - 1)}{(-2 + 6) - (-8)(3)}$

7.6
Summary and Review

Key Terms

(7.1) **Positive numbers** are numbers greater than zero, located to the right of zero on a number line.

Negative numbers are numbers less than zero, located to the left of zero on a number line.

The set of **integers** includes $\{\ldots, -3, -2, -1, 0, 1, 2, 3, \ldots\}$.

The **opposite** of a number is a number located on the other side of zero on a number line, the same distance away from zero as the original number. For any number a, the opposite of a is indicated by $-a$.

The **absolute value** of a number is the distance between that number and zero on a number line. For any number a, the absolute value of a is defined as follows:

$$|a| = \begin{cases} a & \text{if } a \geq 0 \\ -a & \text{if } a < 0 \end{cases}$$

The **additive inverse**, or **opposite**, of a number is the number you must add to it to obtain a sum of zero.

Calculations

(7.2) **To add two positive numbers**, add their absolute values.

To add two negative numbers, add their absolute values and place a negative sign in front of the result.

To add a positive number and a negative number, subtract the smaller absolute value from the larger absolute value. The sum will have the same sign as the addend with the larger absolute value.

(7.3) **To subtract a number from a given quantity**, add its opposite to the quantity.

(7.4) **To multiply signed numbers**, multiply their absolute values. If there is an even number of negative factors, the product is positive. If there is an odd number of negative factors, the product is negative.

A negative number raised to an even power simplifies to a positive result.

A negative number raised to an odd power simplifies to a negative result.

(7.5) **To divide signed numbers**, find the quotient of their absolute values. If both numbers have the same sign, the quotient is positive. If the numbers have opposite signs, the quotient is negative.

To simplify a fraction that expresses the quotient of products of signed numbers, first determine the sign of the answer. If there is an even number of negative factors in the numerator and denominator, the result is positive. If there is an odd number of negative factors, the result is negative. Then reduce to lowest terms, ignoring the signs of the factors.

To simplify fractions that involve more than one operation, first simplify the numerator and denominator separately using the order of operations rules. Then do the division.

Chapter **7**
Review Exercises

(7.2–7.5) State which of the following problems are correct and which are incorrect. For those that are incorrect, replace the underlined answer with the correct one.

1. $9 + (-7) = \underline{-2}$
2. $-4 - (-5) = \underline{-1}$
3. $(-7)(-9) = \underline{-63}$
4. $54 \div (-6) = \underline{-9}$
5. $9(-2) - (-5) = \underline{-13}$
6. $-25 - (3)(-4) = \underline{-37}$
7. $(-5)^2 + (-3)^2 = \underline{34}$
8. $(-9)^2 - (-2)^4 = \underline{65}$
9. $-48 \div (-4)(-3) = \underline{4}$
10. $(-6)(4) \div (-8) = \underline{-3}$

(7.1) Use the $=$, $>$, or $<$ sign to indicate the relationship between each of the following pairs of numbers.

11. $-7 \: ? \: 1$
12. $3 \: ? \: -5$
13. $-3 \: ? \: -|3|$
14. $-7 \: ? \: |-7|$
15. $|4| \: ? \: |-9|$
16. $|-2| \: ? \: |-8|$
17. $4 \: ? \: -8$
18. $0 \: ? \: -2$
19. $|-2| \: ? \: 2$
20. $|-23| \: ? \: -|23|$
21. $-|15| \: ? \: |-5|$
22. $-|-6| \: ? \: |-(-6)|$
23. $-|9| \: ? \: -(-9)$
24. $-(-6) \: ? \: -6$

(7.2–7.5) Perform each of the following operations.

25. $8 + (-6)$
26. $-9 + (-8)$
27. $2 - 7$
28. $4 - (-5)$
29. $(-6)(-7)$
30. $(8)(-5)$
31. $24 \div (-8)$
32. $-63 \div (-7)$
33. $48 + (-53)$
34. $-61 + (-25)$
35. $17 - 86$
36. $32 - (-74)$
37. $(0.8)(-0.7)$
38. $(-0.05)(-0.6)$
39. $(-2.8) \div (0.07)$
40. $(-0.72) \div (-0.008)$
41. $\dfrac{1}{2} + \left(-\dfrac{1}{3}\right)$
42. $-\dfrac{5}{8} + \left(-\dfrac{1}{4}\right)$
43. $\dfrac{3}{5} - \dfrac{1}{6}$
44. $\dfrac{2}{9} - \left(-\dfrac{5}{12}\right)$
45. $\left(\dfrac{5}{8}\right)\left(-\dfrac{4}{15}\right)$
46. $\left(-\dfrac{7}{9}\right)\left(-\dfrac{6}{7}\right)$
47. $\left(-\dfrac{7}{8}\right) \div \left(-\dfrac{21}{4}\right)$
48. $\left(-\dfrac{5}{9}\right) \div \left(\dfrac{20}{27}\right)$
49. $-7 + (-3) - (-9)$
50. $-11 + 8 - (-5)$
51. $4 - 9 + (-5)$
52. $9 - (-4) + (-3)$
53. $(-3)(7)(-2)$
54. $(-8)(-3)(-4)$
55. $(-4)(-5)(-9)$
56. $(-2)(-6)(7)$
57. $\dfrac{(-8)(14)}{(-6)(7)}$
58. $\dfrac{(-9)(2)}{(3)(-4)}$
59. $\dfrac{(15)(-21)(6)}{(7)(-5)(-42)}$
60. $\dfrac{(-15)(49)(-6)}{(-56)(-5)(36)}$
61. $-3(-2 + 9)$
62. $5[3 + (-8)]$
63. $(-4)^2 - (2)(-5)$
64. $(7)(-6) + (-3)^3$
65. $-5(-3 - 1)^2$
66. $4[7 + (-2)]^2$
67. $(5)(-10)^3 + (3)(-10)^2$
68. $(-2)(-3)^3 - (8)(-1)^9$
69. $[5 - (-2)](3 - 7)$
70. $[8 + (-2)](-4 - 5)$
71. $63 \div (-5 - 2)$
72. $-48 \div [8 + (-2)]$

73. $\dfrac{(-4) + (6)(-2)}{(5)(-3) - (-3)}$

74. $\dfrac{(4)(-6) - (-3)}{7 - (-2 - 5)}$

75. $-356 + (437)(-192)$

76. $-853 - (-781) + (-349)$

77. $(-38)^3 - (-57)^4$

78. $[-493 + (-918)] \div (-17)$

Chapter **7**
Explain It in Words

79. Explain two different uses of the "$-$" sign and give an example of each.

80. Explain how to determine the absolute value of a number. Give an example.

81. Explain why you have to be more careful when adding positive and negative numbers than when you are just adding positive numbers. Give an example.

82. Explain how to add two negative numbers. Give an example.

83. Explain how to subtract a negative number from a positive number. Give an example.

84. Explain how you determine the sign of the answer when multiplying three signed numbers together. Give an example.

Chapter **7**
Chapter Test

[7.1] **Use the =, >, or < sign to indicate the relationship between each of the following pairs of numbers.**

1. $-7 \, ? \, -1$

2. $|-4| \, ? \, -(-4)$

3. $-9 \, ? \, -|-2|$

[7.2] **Add each of the following.**

4. $-3 + 8$

5. $5.6 + (-7.9)$

6. $-\dfrac{3}{4} + \left(-\dfrac{1}{8}\right)$

[7.3] **Subtract each of the following.**

7. $-5 - (-7)$

8. $-8.1 - 3.5$

9. $\dfrac{2}{3} - \left(-\dfrac{3}{5}\right)$

[7.4] **Multiply each of the following.**

10. $(-8)(-6)$

11. $(-0.3)(4.2)$

12. $\left(-\dfrac{2}{9}\right)\left(-\dfrac{6}{7}\right)$

[7.5] **Divide each of the following.**

13. $-56 \div 8$

14. $-4.8 \div (-0.6)$

15. $\dfrac{2}{3} \div \left(-\dfrac{4}{9}\right)$

[7.2–7.5] **Perform each of the following operations.**

16. $8 + (-5) - (-3)$

17. $(2)(-7)(-5)$

18. $\dfrac{(-9)(2)(5)}{(10)(-18)(3)}$

19. $-6[5 + (-7)]^2$

20. $[9 + (-6)](-4 - 2)$

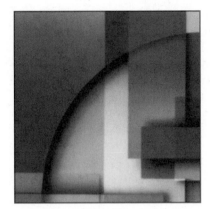

Chapter **8**

Roots of Numbers

8.1
Square Roots

In Chapter 2, we defined the square of a number as the product of that number multiplied by itself. When we square a number, we raise it to the second power. The exponent 2 indicates that the number being squared, or base, appears as a factor twice. For example,

$$7^2 = 7 \cdot 7 = 49$$
$$(-5)^2 = (-5)(-5) = 25$$
$$\left(\frac{1}{3}\right)^2 = \frac{1}{3} \cdot \frac{1}{3} = \frac{1}{9}$$

In this section, we will discuss the inverse of squaring a number, which is taking the **square root** of a number. We indicate the square root of a number by a radical sign, $\sqrt{}$.

▶ **DEFINITION** The *square root* of a given number is a number whose square is equal to the given number. For any number a, the square root of a is indicated by $\sqrt{a}$.

For example,

$\sqrt{4}$ indicates the square root of 4.
$\sqrt{65}$ indicates the square root of 65.

The easiest square roots to determine are square roots of **perfect squares**.

▶ **DEFINITION** A *perfect square* is a number whose square root is a whole number.

Here are some examples of square roots of perfect squares.

$\sqrt{25} = 5$ since $5^2 = 5 \cdot 5 = 25$
$\sqrt{64} = 8$ since $8^2 = 8 \cdot 8 = 64$
$\sqrt{121} = 11$ since $11^2 = 11 \cdot 11 = 121$

If we reexamine closely our definition of square root, we realize that finding the square root of a positive number produces two results—a positive number and its opposite. For example, the square root of 16 is 4 since $4^2 = 16$. The square root of 16 is also -4 since $(-4)^2 = 16$. To indicate the positive square root, or **principal square root**, we use only the radical sign. To indicate the negative square root, we place a negative sign in front of the radical. To indicate both roots, we place a $\pm$ sign in front of the radical. Generally, we are only concerned with the principal square root of a number. If no negative sign appears in front of the radical, find only the positive square root. For example,

$$\sqrt{49} = 7$$
$$-\sqrt{49} = -7$$
$$\pm\sqrt{49} = \pm 7$$

EXAMPLE 1 Simplify each of the following square roots.

(a) $\sqrt{81}$

Find the positive square root of 81. The radical is unsigned.
$\sqrt{81} = 9$ Since $9^2 = 9 \cdot 9 = 81$.

(b) $-\sqrt{144}$

Find the negative square root of 144. There is a $-$ sign in front of the radical.
$-\sqrt{144} = -12$ Since $(-12)^2 = (-12)(-12) = 144$.

(c) $\pm\sqrt{36}$

Find both square roots of 36. There is a $\pm$ sign in front of the radical.
$\pm\sqrt{36} = \pm 6$ Since $6^2 = 36$ and $(-6)^2 = 36$. ▲

The squares of all numbers we have worked with thus far are always positive, so we can find square roots of positive numbers. We cannot take square roots of negative numbers unless we introduce a new system of numbers called imaginary numbers. An **imaginary number** is the square root of a negative number. A detailed discussion of imaginary numbers is beyond the scope of this book, but you will learn more about them in more advanced mathematics courses.

To determine square roots of numbers that lie between two perfect squares, we can use a table that lists square roots accurate to a certain number of decimal places. Table 8.1 lists squares and square roots of the numbers 0 through 99. You should be able to recognize the squares of all numbers between 1 and 20. (In Section 8.3 we will learn the procedure for determining square roots of numbers.)

EXAMPLE 2 Use Table 8.1 to find each of the following square roots.

(a) $\sqrt{68}$

Find 68 in the column labeled "number." Look to the right to the column labeled "square root."
$\sqrt{68} = 8.246$ correct to the nearest thousandth.

TABLE 8.1
Square Root Table

Number n	Square n²	Square Root √n	Number n	Square n²	Square Root √n
0	0	0.000	50	2,500	7.071
1	1	1.000	51	2,601	7.141
2	4	1.414	52	2,704	7.211
3	9	1.732	53	2,809	7.280
4	16	2.000	54	2,916	7.348
5	25	2.236	55	3,025	7.416
6	36	2.449	56	3,136	7.483
7	49	2.646	57	3,249	7.550
8	64	2.828	58	3,364	7.616
9	81	3.000	59	3,481	7.681
10	100	3.162	60	3,600	7.746
11	121	3.317	61	3,721	7.810
12	144	3.464	62	3,844	7.874
13	169	3.606	63	3,969	7.937
14	196	3.742	64	4,096	8.000
15	225	3.873	65	4,225	8.062
16	256	4.000	66	4,356	8.124
17	289	4.123	67	4,489	8.185
18	324	4.243	68	4,624	8.246
19	361	4.359	69	4,761	8.307
20	400	4.472	70	4,900	8.367
21	441	4.583	71	5,041	8.426
22	484	4.690	72	5,184	8.485
23	529	4.796	73	5,329	8.544
24	576	4.899	74	5,476	8.602
25	625	5.000	75	5,625	8.660
26	676	5.099	76	5,776	8.718
27	729	5.196	77	5,929	8.775
28	784	5.292	78	6,084	8.832
29	841	5.385	79	6,241	8.888
30	900	5.477	80	6,400	8.944
31	961	5.568	81	6,561	9.000
32	1,024	5.657	82	6,724	9.055
33	1,089	5.745	83	6,889	9.110
34	1,156	5.831	84	7,056	9.165
35	1,225	5.916	85	7,225	9.220
36	1,296	6.000	86	7,396	9.274
37	1,369	6.083	87	7,569	9.327
38	1,444	6.164	88	7,744	9.381
39	1,521	6.245	89	7,921	9.434
40	1,600	6.325	90	8,100	9.487
41	1,681	6.403	91	8,281	9.539
42	1,764	6.481	92	8,464	9.592
43	1,849	6.557	93	8,649	9.644
44	1,936	6.633	94	8,836	9.695
45	2,025	6.708	95	9,025	9.747
46	2,116	6.782	96	9,216	9.798
47	2,209	6.856	97	9,409	9.849
48	2,304	6.928	98	9,604	9.899
49	2,401	7.000	99	9,801	9.950

(b) $\sqrt{1,849}$

Since the largest number in the column labeled "number" is 99, find 1,849 in the column labeled "square." Since the square root of 1,849 is the number whose square is 1,849, look to the left to the column labeled "number."

$\sqrt{1,849} = 43$ Since $(43)^2 = 43 \cdot 43 = 1,849$.

(c) $\sqrt{5,041}$

Find 5,041 in the column labeled "square." Look to the left to the column labeled "number."

$\sqrt{5,041} = 71$ Since $(71)^2 = 71 \cdot 71 = 5,041$.   ▲

Today, square roots are frequently computed with a pocket calculator. If your calculator has a $\sqrt{x}$ button, simply enter the number whose square root you wish to determine, and then press the $\sqrt{x}$ button.

Quick Quiz

Use Table 8.1 to find each of the following square roots.

                                                Answers

1. $\sqrt{9}$ 1.   3

2. $\pm\sqrt{100}$ 2. ± 10

3. $-\sqrt{169}$ 3. -13

4. $\sqrt{7,921}$ 4.   89

8.1 Exercises

Indicate which of the following statements are true and which are false. For those that are false, change the italicized expression to make the statement true.

1. A number whose square root is a whole number is called a *complete square*.

2. If no sign appears in front of the radical, find only the *positive* square root of the number under the radical.

3. The principal square root of a number is its *negative* square root.

4. The square root of a negative number is *a negative* number.

Find each of the following square roots.

  5. $\sqrt{1}$ 6. $\sqrt{0}$ 7. $\sqrt{49}$ 8. $\sqrt{16}$ 9. $-\sqrt{81}$ 10. $-\sqrt{36}$

11. $\pm\sqrt{64}$ 12. $\pm\sqrt{25}$ 13. $\sqrt{144}$ 14. $\sqrt{100}$ 15. $-\sqrt{9}$ 16. $-\sqrt{121}$

17. $\sqrt{324}$ 18. $\sqrt{256}$ 19. $\sqrt{225}$ 20. $\sqrt{196}$ 21. $\sqrt{289}$ 22. $-\sqrt{289}$

23. $\sqrt{169}$ 24. $\sqrt{400}$ 25. $\sqrt{900}$ 26. $\sqrt{625}$

Use Table 8.1 to find each of the following square roots. Round off square roots of numbers that are not perfect squares to the nearest hundredth.

27. $\sqrt{2}$ 28. $\sqrt{3}$ 29. $\sqrt{5}$ 30. $\sqrt{7}$ 31. $\sqrt{15}$ 32. $\sqrt{37}$

33. $\sqrt{85}$ 34. $\sqrt{28}$ 35. $\sqrt{62}$ 36. $\sqrt{77}$ 37. $\sqrt{63}$ 38. $\sqrt{88}$
39. $\sqrt{576}$ 40. $\sqrt{961}$ 41. $\sqrt{729}$ 42. $\sqrt{676}$ 43. $\sqrt{1,024}$ 44. $\sqrt{3,025}$
45. $\sqrt{4,096}$ 46. $\sqrt{6,084}$ 47. $\sqrt{5,041}$ 48. $\sqrt{1,936}$ 49. $\sqrt{3,844}$ 50. $\sqrt{1,296}$
51. $\sqrt{9,025}$ 52. $\sqrt{6,889}$

Use a pocket calculator to determine each of the following square roots. Round off each answer to the nearest hundredth.

53. $\sqrt{5.4}$ 54. $\sqrt{7.7}$ 55. $\sqrt{61.5}$ 56. $\sqrt{86.3}$ 57. $\sqrt{188}$
58. $\sqrt{524}$ 59. $\sqrt{762}$ 60. $\sqrt{393}$ 61. $\sqrt{554}$ 62. $\sqrt{821}$
63. $\sqrt{638.4}$ 64. $\sqrt{997.2}$ 65. $\sqrt{5,348}$ 66. $\sqrt{6,792}$ 67. $\sqrt{28,427}$
68. $\sqrt{39,615}$ 69. $\sqrt{0.43}$ 70. $\sqrt{0.97}$ 71. $\sqrt{0.0318}$ 72. $\sqrt{0.0689}$
73. $\sqrt{0.00272}$ 74. $\sqrt{0.00546}$ 75. $\sqrt{7,365,217}$ 76. $\sqrt{4,028,641}$

8.2
Properties of Square Roots

Square roots possess two important properties that will enable us to determine the square roots of a wide variety of numbers that are multiples of perfect squares. Consider the following example.

$$\sqrt{4,900} = 70 \text{ since } 70 \cdot 70 = 4,900$$

Notice that 4,900 can be expressed as the product of two perfect squares.

$$4,900 = 49 \cdot 100$$

The square root of 4,900, 70, is the product of the square roots of each of the perfect square factors.

$$\sqrt{4,900} = \sqrt{49 \cdot 100}$$
$$= \sqrt{49} \cdot \sqrt{100}$$
$$= 7 \cdot 10$$
$$= 70$$

We have thus illustrated the property stated below.

For any positive numbers a and b,
$$\sqrt{a \cdot b} = \sqrt{a} \cdot \sqrt{b}$$

In other words, the square root of a product is the product of the square roots of the factors. ▲

For example, $\sqrt{4 \cdot 9} = \sqrt{4} \cdot \sqrt{9}$

Let us now look at some examples of how this property can be used to calculate square roots of numbers.

EXAMPLE 1   Find each of the following square roots.

(a) $\sqrt{900}$

$$\begin{aligned}
\sqrt{900} &= \sqrt{9 \cdot 100} \\
&= \sqrt{9} \cdot \sqrt{100} \\
&= 3 \cdot 10 \\
&= 30
\end{aligned}$$

Express 900 as the product of two perfect squares.

Find the square root of each factor.

Find the product of the square roots.

(b) $\sqrt{14,400}$

$$\begin{aligned}
\sqrt{14,400} &= \sqrt{144 \cdot 100} \\
&= \sqrt{144} \cdot \sqrt{100} \\
&= 12 \cdot 10 \\
&= 120
\end{aligned}$$

Express 14,400 as the product of two perfect squares.

Find the square root of each factor.

Find the product of the square roots.

(c) $\sqrt{360,000}$

$$\begin{aligned}
\sqrt{360,000} &= \sqrt{36 \cdot 10,000} \\
&= \sqrt{36} \sqrt{10,000} \\
&= 6 \cdot 100 \\
&= 600
\end{aligned}$$

If no operation sign appears between two radicals, multiplication is implied.

▲

We can apply this same property to express the square root of a number in simplest form.

To express a square root in simplest form, we determine the square root of the largest perfect square that is a factor of the number under the radical sign. The other factor, which is not a perfect square, remains under the radical sign.   ▲

For example,

$$\begin{aligned}
\sqrt{12} &= \sqrt{4 \cdot 3} \\
&= \sqrt{4} \sqrt{3} \\
&= 2\sqrt{3}
\end{aligned}$$

Notice that even though the multiplication sign does not appear, it is understood to be there.

Thus, $2\sqrt{3}$ is $\sqrt{12}$ expressed in simplest form. We read $2\sqrt{3}$ as "2 times the square root of 3."

EXAMPLE 2   Write each of the following square roots in simplest form.

(a) $\sqrt{18}$

$$\begin{aligned}
\sqrt{18} &= \sqrt{9 \cdot 2} \\
&= \sqrt{9} \sqrt{2} \\
&= 3\sqrt{2}
\end{aligned}$$

9 is the largest perfect square that is a factor of 18.

(b) $\sqrt{32}$

$$\begin{aligned}
\sqrt{32} &= \sqrt{16 \cdot 2} \\
&= \sqrt{16} \sqrt{2} \\
&= 4\sqrt{2}
\end{aligned}$$

16 is the largest perfect square that is a factor of 32.

(c) $\sqrt{250}$

$$\begin{aligned}
\sqrt{250} &= \sqrt{25 \cdot 10} \\
&= \sqrt{25}\,\sqrt{10} \\
&= 5\sqrt{10}
\end{aligned}$$

25 is the largest perfect square that is a factor of 250.

▲

We will now examine a second property of square roots that will enable us to determine square roots of fractions. Consider the following example.

$$\sqrt{\frac{4}{9}} = \frac{2}{3} \quad \text{since} \quad \frac{2}{3} \cdot \frac{2}{3} = \frac{4}{9}$$

Notice that $\frac{4}{9}$ is the quotient of two perfect squares, 4 and 9. The square root of $\frac{4}{9}$, $\frac{2}{3}$, is the quotient of the square root of the numerator 4, and the square root of the denominator 9.

$$\sqrt{\frac{4}{9}} = \frac{\sqrt{4}}{\sqrt{9}} = \frac{2}{3}$$

We have thus illustrated the following property.

For any positive number a and b,

$$\sqrt{\frac{a}{b}} = \frac{\sqrt{a}}{\sqrt{b}}$$

In other words, the square root of a quotient is the quotient of the square roots of the numerator and the denominator. ▲

For example,

$$\sqrt{\frac{9}{25}} = \frac{\sqrt{9}}{\sqrt{25}} = \frac{3}{5}$$

To determine the square roots of certain decimals, we can rewrite the decimal as a fraction, and then use this property to simplify the fraction. For example,

$$\sqrt{0.64} = \sqrt{\frac{64}{100}} = \frac{\sqrt{64}}{\sqrt{100}} = \frac{8}{10} = 0.8$$

EXAMPLE 3 Find each of the following square roots.

(a) $\sqrt{\dfrac{1}{16}}$

$$\sqrt{\frac{1}{16}} = \frac{\sqrt{1}}{\sqrt{16}} = \frac{1}{4}$$

Find the square root of the numerator and denominator and then find the quotient.

(b) $\sqrt{\dfrac{4}{49}}$

$$\sqrt{\frac{4}{49}} = \frac{\sqrt{4}}{\sqrt{49}} = \frac{2}{7}$$

(c) $\sqrt{\dfrac{36}{169}}$

$$\sqrt{\dfrac{36}{169}} = \dfrac{\sqrt{36}}{\sqrt{169}} = \dfrac{6}{13}$$

(d) $\sqrt{0.04}$

$$\sqrt{0.04} = \sqrt{\dfrac{4}{100}} = \dfrac{\sqrt{4}}{\sqrt{100}} = \dfrac{2}{10} = 0.2$$ First, express the decimal as a fraction. Express the answer as a decimal.

(e) $\sqrt{0.0081}$

$$\sqrt{0.0081} = \sqrt{\dfrac{81}{10,000}} = \dfrac{\sqrt{81}}{\sqrt{10,000}} = \dfrac{9}{100} = 0.09$$

(f) $\sqrt{1.44}$

$$\sqrt{1.44} = \sqrt{\dfrac{144}{100}} = \dfrac{\sqrt{144}}{\sqrt{100}} = \dfrac{12}{10} = 1.2$$ ▲

Quick Quiz

Simplify each of the following square roots.

1. $\sqrt{6,400}$
2. $\sqrt{75}$
3. $\sqrt{\dfrac{25}{81}}$
4. $\sqrt{0.36}$

Answers

1. 80
2. $5\sqrt{3}$
3. $\dfrac{5}{9}$
4. 0.6

8.2 Exercises

Indicate which of the following statements are true and which are false. For those that are false, change the italicized word to make the statement true.

1. The square root of a product is the *sum* of the square roots of each factor.

2. The square root of a fraction is equivalent to the square root of the numerator *multiplied* by the square root of the denominator.

3. When a square root is expressed in *simplest* form, the number under the radical does not have any factors that are perfect squares.

4. The number $5\sqrt{3}$ is read as "5 *plus* the square root of 3."

Use the properties of square roots to find each of the following square roots.

5. $\sqrt{400}$ 6. $\sqrt{1,600}$ 7. $\sqrt{2,500}$ 8. $\sqrt{3,600}$ 9. $\sqrt{8,100}$

10. $\sqrt{10,000}$ 11. $\sqrt{12,100}$ 12. $\sqrt{16,900}$ 13. $\sqrt{640,000}$ 14. $\sqrt{490,000}$

15. $\sqrt{\dfrac{1}{25}}$ 16. $\sqrt{\dfrac{1}{49}}$ 17. $\sqrt{\dfrac{4}{81}}$ 18. $\sqrt{\dfrac{9}{64}}$ 19. $\sqrt{\dfrac{36}{121}}$

20. $\sqrt{\dfrac{49}{144}}$ 21. $\sqrt{\dfrac{64}{169}}$ 22. $\sqrt{\dfrac{16}{225}}$ 23. $\sqrt{0.25}$ 24. $\sqrt{0.81}$

25. $\sqrt{0.49}$ 26. $\sqrt{0.16}$ 27. $\sqrt{1.21}$ 28. $\sqrt{2.25}$ 29. $\sqrt{0.09}$

30. $\sqrt{0.01}$ 31. $\sqrt{0.0064}$ 32. $\sqrt{0.0036}$ 33. $\sqrt{0.0009}$ 34. $\sqrt{0.0001}$

Write each of the following square roots in simplest form.

35. $\sqrt{48}$ 36. $\sqrt{27}$ 37. $\sqrt{20}$ 38. $\sqrt{45}$ 39. $\sqrt{50}$ 40. $\sqrt{72}$

41. $\sqrt{80}$ 42. $\sqrt{63}$ 43. $\sqrt{98}$ 44. $\sqrt{28}$ 45. $\sqrt{128}$ 46. $\sqrt{108}$

47. $\sqrt{360}$ 48. $\sqrt{640}$ 49. $\sqrt{288}$ 50. $\sqrt{242}$

8.3
Procedure for Calculating Square Roots (Optional)

Thus far, we have learned how to calculate square roots of numbers that express a product or quotient of perfect squares. In this section, we will learn an arithmetic procedure that can be used to calculate the square root of a number that is not a perfect square to a desired degree of accuracy. This procedure is outlined in the box. The numbered steps below correspond to the numbered steps in Examples 1 and 2.

To Calculate a Square Root

1. Block off every group of two digits to the left and to the right of the decimal point. If there is an odd number of digits to the left of the decimal point, the leftmost block will contain only one digit.

2. Since each block will correspond to one digit in the calculated square root, add an appropriate number of digits to the right of the decimal point so that the number of blocks to the right of the decimal point is one more than the number of decimal places needed for the approximation for the square root. Put a decimal point directly above the decimal point under the radical sign.

3. Find the largest number whose square is less than or equal to the number in the leftmost block. Write it above the block and subtract it from the number in the block.

4. Bring down the next block of two digits.

5. Double the number above the radical, write it to the left of the difference, and attach a blank box to the right of it.

6. Find the largest number to put in the box so that the product of the number with the attached box and the number in the box is less than or equal to the difference on the right.

7. Subtract that product from the difference on the right.

8. Repeat steps 4–7 until a number appears above every block of two digits.

9. Round off the result to obtain the desired square root. ▲

EXAMPLE 1   Calculate $\sqrt{51.7}$ correct to the nearest tenth.

Solution $\sqrt{51.7000}$

$$
\begin{array}{r}
7. \\
\sqrt{51.7000} \\
\underline{49} \\
2
\end{array}
$$

1. Block off groups of two digits.

2. Attach 3 zeros to the right of the decimal point. Put a decimal point above the decimal point under the radical sign.

$$
\begin{array}{r}
7. \\
\sqrt{51.7000} \\
\underline{49} \\
14\boxed{}\ \ 270
\end{array}
$$

3. 49 is the largest perfect square ≤ 51. Put 7 above the first block. Subtract $51 - 49 = 2$.

4. Bring down the next block: 70.

$$
\begin{array}{r}
7. 1 \\
\sqrt{51.7000} \\
\underline{49} \\
14\boxed{1}\ \ 270 \\
\underline{141} \\
129
\end{array}
$$

5. Double the number above the radical: $7 \times 2 = 14$. Write 14 to the left of 270 and attach a box to its right.

6. $\begin{array}{r} 14\,\boxed{2} \\ \times \boxed{2} \\ \hline 28\ \ 4 \end{array} > 270 \qquad \begin{array}{r} 14\,\boxed{1} \\ \times \boxed{1} \\ \hline 14\ \ 1 \end{array} < 270$

Write 1 in the box and above the radical.

7. Subtract $270 - 141 = 129$.

$$
\begin{array}{r}
7. 1\ \ 9 \\
\sqrt{51.7000} \\
\underline{49} \\
14\boxed{1}\ \ 270 \\
\underline{141} \\
142\boxed{9}\ \ 12900 \\
\underline{12861} \\
39
\end{array}
$$

8. Bring down the next block: 00. Double the number above the radical: $71 \times 2 = 142$. Write 142 to the left of 12900 and attach a box to its right.

$\begin{array}{r} 142\,\boxed{9} \\ \times \boxed{9} \\ \hline 12{,}86\ \ 1 \end{array} < 12{,}900$

Write 9 in the box and above the radical.

$7.19 \cong 7.2$

Thus $\sqrt{51.7} \cong 7.2$.

9. Round off the result.

Check:
$$
\begin{array}{r}
7.2 \\
\times \quad 7.2 \\
\hline
144 \\
504 \\
\hline
51.84 \quad \checkmark
\end{array}
$$

Since $51.84 \cong 51.7$, then 7.2 is a reasonable approximation for $\sqrt{51.7}$.   ▲

EXAMPLE 2 Calculate $\sqrt{250}$ correct to the nearest tenth.

Solution $\sqrt{2\,5\,0.0\,0\,0\,0}$

$$\phantom{\sqrt{}}1.$$

$\sqrt{2\,5\,0.0\,0\,0\,0}$
$$\underline{1}$$
$$1$$

1. Block off groups of two digits.

2. Attach 2 blocks of zeros to the right of the decimal point. Put a decimal point above the decimal point under the radical sign.

$$1.$$
$\sqrt{2\,5\,0.0\,0\,0\,0}$
$$1$$
$$\begin{array}{r|l}2\;\square & 1\,5\,0\end{array}$$

3. 1 is the largest perfect square ≤ 2. Write 1 above the number 2 under the radical. Subtract $2 - 1 = 1$.

4. Bring down the next block: 50.

$$15.$$
$\sqrt{2\,5\,0.0\,0\,0\,0}$
$$1$$
$$\begin{array}{r|l}2\;\boxed{5} & 1\,5\,0 \\ & 1\,2\,5 \\ \hline & 2\,5\end{array}$$

5. Double the number above the radical: $1 \times 2 = 2$. Write 2 to the left of 1 and attach a box to its right.

6. $\begin{array}{r}2\,\boxed{5}\\ \times\quad\boxed{5}\\ \hline 12\,5\end{array}$ < 150 $\begin{array}{r}2\,\boxed{6}\\ \times\quad\boxed{6}\\ \hline 15\,6\end{array}$ > 150

Write 5 in the box and above the radical.

$$15.8$$
$\sqrt{2\,5\,0.0\,0\,0\,0}$
$$1$$
$$\begin{array}{r|l}2\;\boxed{5} & 1\,5\,0 \\ & 1\,2\,5 \\ \hline 30\;\boxed{8} & 2\,5\,0\,0 \\ & 2\,4\,6\,4 \\ \hline & 3\,6\end{array}$$

7. Subtract $150 - 125 = 25$.

8. Bring down the next block: 00. Double the number above the radical: $15 \times 2 = 30$. Write 30 to the left of 2500 and attach a box to its right.

$$\begin{array}{r}30\,\boxed{8}\\ \times\quad\boxed{8}\\ \hline 2{,}46\,4\end{array} < 2{,}500$$

Write 8 in the box and above the radical. Subtract $2{,}500 - 2{,}464 = 36$.

$$15.81$$
$\sqrt{2\,5\,0.0\,0\,0\,0}$
$$1$$
$$\begin{array}{r|l}2\;\boxed{5} & 1\,5\,0 \\ & 1\,2\,5 \\ \hline 30\;\boxed{8} & 2\,5\,0\,0 \\ & 2\,4\,6\,4 \\ \hline 316\;\boxed{1} & 3\,6\,0\,0 \\ & 3\,1\,6\,1 \\ \hline & 4\,3\,9\end{array}$$

Bring down next block: 00. Double the number above the radical: $158 \times 2 = 316$. Write 316 to the left of 3600 and attach a box to its right.

$$\begin{array}{r}316\,\boxed{1}\\ \times\quad\boxed{1}\\ \hline 3{,}16\,1\end{array} < 3{,}600$$

Write 1 in the box and above the radical.

9. Round off the result.

$15.81 \cong 15.8$
Thus, $\sqrt{250} \cong 15.8$.

Check:

$$
\begin{array}{r}
1\,5.8 \\
\times \quad 1\,5.8 \\
\hline
1264 \\
790 \\
\hline
158 \\
\hline
249.64 \quad \surd
\end{array}
$$

Square 15.8.

Since $249.64 \cong 250$, then 15.8 is a reasonable approximation for $\sqrt{250}$.   ▲

This procedure is very easy to use for determining the square root of perfect squares, as Example 3 illustrates.

EXAMPLE 3   Calculate $\sqrt{3,240,000}$.

Solution

$$
\begin{array}{r}
1\ \ 8\ \ \ 0\ \ 0\,. \\
\sqrt{3\,2\,4\,0\,0\,0\,0\,.}
\end{array}
$$

$$
\begin{array}{r}
1 \\
\hline
2\,\boxed{8}\ \big|\ 2\,2\,4 \\
2\,2\,4 \\
\hline
0
\end{array}
$$

$$
\begin{array}{r}
2\,\boxed{8} \\
\times \quad \boxed{8} \\
\hline
22\ \ 4\ = 224
\end{array}
$$

Thus, $\sqrt{3,240,000} = 1,800$.

Notice that the 2 zeros are necessary placeholders in the answer.

Check:

$$
\begin{array}{r}
1,8\,0\,0 \\
\times \quad 1,8\,0\,0 \\
\hline
1\,4\,4\,0\,0\,0\,0 \\
1\,8\,0\,0 \\
\hline
3,2\,4\,0,0\,0\,0 \quad \surd
\end{array}
$$

▲

EXAMPLE 4   Calculate $\sqrt{116.2084}$.

Solution

$$
\begin{array}{r}
1\ \ 0\,.\,7\ \ 8 \\
\sqrt{1\,1\,6.2\,0\,8\,4}
\end{array}
$$

$$
\begin{array}{r}
1 \\
\hline
2\,\boxed{0}\ \big|\ 1\,6 \\
0 \\
\hline
2\,0\,\boxed{7}\ \big|\ 1\,6\,2\,0 \\
1\,4\,4\,9 \\
\hline
2\,1\,4\,\boxed{8}\ \big|\ 1\,7\,1\,8\,4 \\
1\,7\,1\,8\,4 \\
\hline
0
\end{array}
$$

$$
\begin{array}{r}
2\,0\,\boxed{8} \\
\times \quad \boxed{8} \\
\hline
1,66\ \ 4\ > 1,620
\end{array}
\qquad
\begin{array}{r}
2\,0\,\boxed{7} \\
\times \quad \boxed{7} \\
\hline
1,44\ \ 9\ < 1,620
\end{array}
$$

$$
\begin{array}{r}
2\,1\,4\,\boxed{8} \\
\times \quad \boxed{8} \\
\hline
17,18\ \ 4\ = 17,184
\end{array}
$$

Thus, $\sqrt{116.2084} = 10.78$.   ▲

EXAMPLE 5 Calculate $\sqrt{0.734}$ correct to the nearest thousandth.

Solution

```
            0. 8  5  6  7
         _____
       √ 0.7 3 4 0 0 0 0 0
           6 4
        _____
  16 5      9 4 0
             8 2 5
       _____
 170 6      1 1 5 0 0
            1 0 2 3 6
       _____
1712 7        1 2 6 4 0 0
              1 1 9 8 8 9
          _____
                6 5 1 1
```

16 6		16 5	
× 6		× 5	
99 6 > 940		82 5 < 940	

170 7		170 6	
× 7		× 6	
11,94 9 > 11,500		10,23 6 < 11,500	

1712 7	
× 7	
119,88 9 < 126,400	

$0.8567 \cong 0.857$

Thus, $\sqrt{0.734} \cong 0.857$.

▲

Quick Quiz

		Answers	
1.	Calculate $\sqrt{34.81}$.	**1.**	5.9
2.	Calculate $\sqrt{752}$ correct to the nearest tenth.	**2.**	27.4

8.3 Exercises

Calculate each of the following square roots. Check your answers by squaring.

1. $\sqrt{27.04}$	2. $\sqrt{90.25}$	3. $\sqrt{44.89}$	4. $\sqrt{75.69}$	5. $\sqrt{985.96}$
6. $\sqrt{561.69}$	7. $\sqrt{1,303.21}$	8. $\sqrt{4,678.56}$	9. $\sqrt{127,449}$	10. $\sqrt{170,569}$
11. $\sqrt{630,436}$	12. $\sqrt{833,569}$	13. $\sqrt{270,400}$	14. $\sqrt{136,900}$	15. $\sqrt{0.5184}$
16. $\sqrt{0.3844}$	17. $\sqrt{0.0841}$	18. $\sqrt{0.9216}$	19. $\sqrt{0.481636}$	20. $\sqrt{0.714025}$
21. $\sqrt{0.004624}$	22. $\sqrt{0.008281}$	23. $\sqrt{1,440,000}$	24. $\sqrt{7,290,000}$	

Calculate each of the following square roots. Round off as indicated.

25. $\sqrt{5.7}$ nearest tenth	26. $\sqrt{3.2}$ nearest hundredth
27. $\sqrt{17.4}$ nearest tenth	28. $\sqrt{82.5}$ nearest tenth
29. $\sqrt{67.9}$ nearest hundredth	30. $\sqrt{43.6}$ nearest hundredth
31. $\sqrt{923}$ nearest tenth	32. $\sqrt{546}$ nearest tenth
33. $\sqrt{372.4}$ nearest hundredth	34. $\sqrt{789.1}$ nearest hundredth
35. $\sqrt{4,317}$ nearest tenth	36. $\sqrt{6,749}$ nearest tenth
37. $\sqrt{58,402}$ nearest whole number	38. $\sqrt{90,357}$ nearest whole number

39. $\sqrt{0.73}$ nearest hundredth

40. $\sqrt{0.45}$ nearest thousandth

41. $\sqrt{0.273}$ nearest hundredth

42. $\sqrt{0.458}$ nearest thousandth

43. $\sqrt{0.013}$ nearest thousandth

44. $\sqrt{0.059}$ nearest thousandth

45. $\sqrt{0.0462}$ nearest hundredth

46. $\sqrt{0.0898}$ nearest hundredth

47. $\sqrt{0.005274}$ nearest hundredth

48. $\sqrt{0.007329}$ nearest thousandth

49. $\sqrt{6,741,308}$ nearest whole number

50. $\sqrt{2,316,452}$ nearest whole number

8.4
Higher Roots of Numbers

Just as we can raise any whole number to an exponent that is a positive integer, we can also take the nth root of a number, where n is a positive integer.

▶ **DEFINITION** The nth root of a number is some number whose nth power is equal to the given number. For any number a, the nth root of a is indicated by $\sqrt[n]{a}$.

For example, $\sqrt[3]{8}$ indicates the cube root of 8. It is equivalent to some number that yields a product of 8 when cubed. Thus,

$$\sqrt[3]{8} = 2 \qquad \text{since } 2^3 = 2 \cdot 2 \cdot 2 = 8$$

In this section, we will deal only with higher roots of numbers that can be calculated on sight or by using a table. Table 8.2, which lists the values obtained when the numbers 1 through 10 are raised to the 2nd, 3rd, 4th, and 5th powers, can be used to calculate the square, cube, 4th, and 5th roots of numbers in the corresponding columns.

TABLE 8.2
Powers of Numbers

n	n^2	n^3	n^4	n^5
1	1	1	1	1
2	4	8	16	32
3	9	27	81	243
4	16	64	256	1,024
5	25	125	625	3,125
6	36	216	1,296	7,776
7	49	343	2,401	16,807
8	64	512	4,096	32,768
9	81	729	6,561	59,049
10	100	1,000	10,000	100,000

EXAMPLE 1 Use Table 8.2 to find each of the following roots.

(a) $\sqrt[4]{81}$

Look down the column labeled n^4 until you find 81.

$\sqrt[4]{81} = 3$ Since $3^4 = 81$.

(b) $\sqrt[3]{216}$

Look down the column labeled n^3 until you find 216.

$\sqrt[3]{216} = 6$ Since $6^3 = 216$.

(c) $\sqrt[5]{100,000}$

$\sqrt[5]{100,000} = 10$ Since $10^5 = 100,000$.   ▲

The same properties that apply to square roots of numbers also apply to higher roots of numbers. We restate these basic properties below.

For any positive numbers a and b and positive integer n,
$$\sqrt[n]{a \cdot b} = \sqrt[n]{a}\,\sqrt[n]{b}$$ ▲

For example,

$$\sqrt[3]{8 \cdot 5} = \sqrt[3]{8}\,\sqrt[3]{5} = 2\sqrt[3]{5}$$

For any positive numbers a and b and positive integer n,
$$\sqrt[n]{\frac{a}{b}} = \frac{\sqrt[n]{a}}{\sqrt[n]{b}}$$ ▲

For example,

$$\sqrt[4]{\frac{16}{81}} \cdot \frac{\sqrt[4]{16}}{\sqrt[4]{81}} = \frac{2}{3}$$

EXAMPLE 2   Simplify each of the following radical expressions. (Use Table 8.2.)

(a) $\sqrt[3]{\dfrac{1}{27}}$

$$\sqrt[3]{\frac{1}{27}} = \frac{\sqrt[3]{1}}{\sqrt[3]{27}} = \frac{1}{3}$$ Find the cube roots of the numerator and denominator.

(b) $\sqrt[5]{96}$

$$\sqrt[5]{96} = \sqrt[5]{32 \cdot 3}$$
$$= \sqrt[5]{32}\,\sqrt[5]{3} = 2\sqrt[5]{3}$$ Find the 5th root of each factor. Simplify.

(c) $\sqrt[4]{0.0016}$

$$\sqrt[4]{0.0016} = \sqrt[4]{\frac{16}{10,000}}$$ Rewrite decimal as a fraction. Find 4th roots of numerator and denominator. Express answer as decimal.

$$= \frac{\sqrt[4]{16}}{\sqrt[4]{10,000}} = \frac{2}{10} = 0.2$$

(d) $\sqrt[3]{54}$

$$\sqrt[3]{54} = \sqrt[3]{27 \cdot 2}$$

Find the cube root of each factor.

$$= \sqrt[3]{27}\ \sqrt[3]{2} = 3\sqrt[3]{2}$$

Simplify.

▲

Recall that a negative number raised to an even power produces a positive result. In this book, we have calculated even roots of only positive numbers, because the calculation of even roots of negative numbers involves working with imaginary numbers. However, a negative number raised to an odd power produces a negative result. Therefore, a negative number has an odd root that is also a negative number. For example,

$$\sqrt[3]{-8} = -2 \quad \text{since} \quad (-2)^3 = -8$$

EXAMPLE 3   Find each of the following roots.

(a) $\sqrt[5]{-1}$

$$\sqrt[5]{-1} = -1$$

Since $(-1)^5 = -1$.

(b) $\sqrt[3]{-125}$

$$\sqrt[3]{-125} = -5$$

Since $(-5)^3 = -125$.

▲

Quick Quiz

Simplify each of the following radical expressions.   **Answers**

1. $\sqrt[4]{16}$ 1.   2

2. $\sqrt[3]{\dfrac{1}{64}}$ 2. $\dfrac{1}{4}$

3. $\sqrt[5]{-10,000}$ 3. -10

8.4 Exercises

Use Table 8.2 to find each of the following roots.

1. $\sqrt[3]{27}$ 2. $\sqrt[5]{32}$ 3. $\sqrt[4]{16}$ 4. $\sqrt[3]{125}$ 5. $\sqrt[4]{256}$

6. $\sqrt[3]{343}$ 7. $\sqrt[4]{625}$ 8. $\sqrt[5]{243}$ 9. $\sqrt[4]{10,000}$ 10. $\sqrt[5]{3,125}$

11. $\sqrt[4]{\dfrac{1}{16}}$ 12. $\sqrt[5]{\dfrac{1}{32}}$ 13. $\sqrt[3]{\dfrac{1}{1,000}}$ 14. $\sqrt[3]{\dfrac{27}{125}}$ 15. $\sqrt[3]{0.008}$

16. $\sqrt[4]{0.0081}$ 17. $\sqrt[5]{0.01024}$ 18. $\sqrt[3]{0.125}$ 19. $\sqrt[5]{-32}$ 20. $\sqrt[3]{-8}$

21. $\sqrt[3]{-512}$ 22. $\sqrt[5]{-1,024}$ 23. $\sqrt[3]{-\dfrac{1}{64}}$ 24. $\sqrt[3]{-\dfrac{8}{27}}$

Write each of the following in simplest form.

25. $\sqrt[3]{16}$ 26. $\sqrt[4]{80}$ 27. $\sqrt[3]{24}$ 28. $\sqrt[5]{64}$

29. $\sqrt[3]{250}$ 30. $\sqrt[4]{162}$ 31. $\sqrt[3]{270}$ 32. $\sqrt[3]{128}$

33. $\sqrt[4]{20,000}$ 34. $\sqrt[3]{3,000}$ 35. $\sqrt[5]{500,000}$ 36. $\sqrt[4]{486}$

8.5
Number Systems Used in Mathematics

Throughout our discussions of various arithmetic operations, we have encountered many different kinds of numbers that are all included in the set of **real numbers**. A **set** is simply a collection or group of objects. Braces { } are used to enclose the members of a set. Each of the number systems we have used thus far are identified below.

Number Systems

1. The set of **natural numbers**, or **counting numbers**, consists of $\{1, 2, 3, 4, 5, 6, \ldots\}$. (Remember that the symbol . . . indicates that there is an infinite number of natural numbers.)

2. The set of **whole numbers** consists of $\{0, 1, 2, 3, 4, \ldots\}$. The whole numbers include all the counting numbers in addition to zero.

3. The set of **integers** consists of $\{\ldots, -4, -3, -2, -1, 0, 1, 2, 3, 4, \ldots\}$. The integers include all the whole numbers and their opposites. Note that there is an infinite number of negative and positive integers.

4. The set of **rational numbers** consists of all numbers that can be written in the form $\frac{a}{b}$, where a and b are integers and $b \neq 0$. For example, $\frac{3}{5}$, $-\frac{9}{8}$, 7, and $-4\frac{1}{2}$ are all rational numbers. All terminating and repeating decimals are included in the set of rational numbers. For example, 0.06, -3.72, and $0.3\overline{3}$ are rational numbers since they can be rewritten as

$$\frac{6}{100}, \quad -\frac{372}{100}, \quad \text{and} \quad \frac{1}{3}, \text{ respectively.}$$

5. The set of **irrational numbers** consists of all numbers that can be expressed as a nonterminating, nonrepeating decimal, but cannot be expressed in the form of a rational number. For example, $\sqrt{2} = 1.414213562 \ldots$ is an irrational number since it cannot be expressed in the form $\frac{a}{b}$, where a and b are integers and $b \neq 0$.

6. The set of **real numbers** consists of all rational and irrational numbers. ▲

Since all of the number systems we described are part of the real number system, they are sometimes called subsets of the real numbers. Set A is a *subset* of set B if every element of set A is also an element of set B. For example, since every whole number is included in the set of rational numbers, the whole numbers are a subset of the rational numbers. However, the rational numbers are not a subset of the whole numbers, because there are many rational numbers that are not whole numbers. Figure 8.1 illustrates which numbers are included in each of the subsets of the real number system.

Figure 8.1

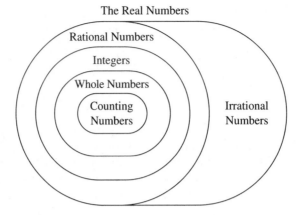

The Real Numbers

Rational Numbers

Integers

Whole Numbers

Counting Numbers

Irrational Numbers

EXAMPLE 1   List all of the number systems in which each of the following numbers is included.

(a) -7

$$-7 = -\frac{7}{1}$$

-7 is an integer, a rational number, and a real number.

(b)   0.28

$$0.28 = \frac{28}{100}$$

0.28 is a rational number and a real number.

(c) $\sqrt{36}$

$$\sqrt{36} = 6 = \frac{6}{1}$$

$\sqrt{36}$ is a counting number, a whole number, an integer, a rational number, and a real number.

(d) $0.454545 \cdots$

$0.454545 \cdots$ is a repeating decimal.

$0.45\overline{45}$ is a rational number and a real number.

(e) $\sqrt{3}$

$\sqrt{3} = 1.73205 \cdots$ is a nonrepeating, nonterminating decimal.

$\sqrt{3}$ is an irrational number and a real number.   ▲

All real numbers can be represented on a number line. Since there is an infinite number of positive and negative real numbers, the number line extends infinitely to the right and to the left. There is also an infinite number of real numbers located between any two real numbers. Figure 8.2 shows the location of a few real numbers on a number line.

Figure 8.2

EXAMPLE 2   Indicate the relationship between each of the following pairs of numbers, using the $>$, $<$, or $=$ sign.

(a) $2 \; ? \; \sqrt{3}$
      $2 > \sqrt{3}$ Since $\sqrt{3} \cong 1.732$.

(b) $-2.4 \; ? \; -\sqrt{9}$
      $-2.4 > -\sqrt{9}$ Since $-\sqrt{9} = -3$.

(c) $\sqrt{\dfrac{1}{4}} \; ? \; \dfrac{1}{2}$

      $\sqrt{\dfrac{1}{4}} = \dfrac{1}{2}$ Since $\dfrac{1}{2} \cdot \dfrac{1}{2} = \dfrac{1}{4}$.

(d) $\sqrt{5} \; ? \; 3$
      $\sqrt{5} < 3$ Since $\sqrt{5} \cong 2.236$.

(e) $\sqrt{0.26} \; ? \; 0.5$
      $\sqrt{0.26} > 0.5$ Since $0.5 = \sqrt{0.25}$.   ▲

Quick Quiz

List all the number systems in which each of the following is included.

Answers

1. $\sqrt{7}$

1. irrational numbers, real numbers

2. $-\sqrt{25}$

2. integers, rational numbers, real numbers

Use the $>$, $<$, or $=$ sign to indicate the relationship between each of the following pairs of numbers.

3. $\sqrt{0.16} \; ? \; 0.4$

3. $\sqrt{0.16} = 0.4$

4. $-\sqrt{49} \; ? \; -7.1$

4. $-\sqrt{49} > -7.1$

8.5 Exercises

Indicate which of the following statements are true and which are false. For those that are false, change the italicized expression to make the statement true.

1.   The set of counting numbers *does not* include zero.

2.   Every integer is also a *whole* number.

3. Every whole number is also a *rational* number.

4. A number cannot belong to both the rational numbers and the *real* numbers.

5. The quotient of any two nonzero integers is always *an integer.*

6. The quotient of any two nonzero rational numbers is always a *rational* number.

7. The irrational numbers are a subset of the *rational* numbers.

8. The counting numbers are a subset of the *integers.*

List all of the number systems in which each of the following numbers is included.

9. -9

10. 4

11. $\sqrt{5}$

12. $-\dfrac{1}{2}$

13. 3.7

14. -6.0

15. $-\sqrt{23}$

16. $11\dfrac{3}{5}$

17. $0.\overline{66}$

18. $\dfrac{7}{8}$

19. -3.15

20. $5\sqrt{2}$

21. $\dfrac{32}{4}$

22. $-\sqrt{49}$

23. -0.4397

24. $-\dfrac{72}{8}$

25. $\sqrt{50}$

26. $-9.484\overline{8}$

27. 0

28. $-\sqrt{1.44}$

Use the $>$, $<$, or $=$ sign to indicate the relationship between each of the following pairs of numbers.

29. $\sqrt{3}\,?\,2$

30. $-\sqrt{2}\,?\,-1$

31. $-3\,?\,-\sqrt{9}$

32. $\sqrt{4}\,?\,2$

33. $\sqrt{0.09}\,?\,0.03$

34. $0.7\,?\,\sqrt{0.49}$

35. $\sqrt{\dfrac{1}{25}}\,?\,\dfrac{1}{5}$

36. $\dfrac{1}{7}\,?\,\sqrt{\dfrac{1}{36}}$

37. $\sqrt{8}\,?\,3$

38. $-4\,?\,-\sqrt{16}$

39. $\sqrt{0.48}\,?\,0.7$

40. $-0.9\,?\,-\sqrt{0.82}$

41. $\sqrt{\dfrac{9}{64}}\,?\,\dfrac{3}{7}$

42. $\dfrac{1}{4}\,?\,\sqrt{\dfrac{3}{16}}$

43. $\sqrt{0.01}\,?\,0.01$

44. $\sqrt{\dfrac{1}{9}}\,?\,\dfrac{1}{9}$

45. $-\sqrt{4}\,?\,-2.1$

46. $-5.9\,?\,-\sqrt{36}$

47. $\sqrt{1.44}\,?\,1.2$

48. $0.08\,?\,\sqrt{0.064}$

8.6
Summary and Review

Key Terms

(8.1) The **square root** of a given number is a number whose square is equal to the given number. For any number a, the square root of a is indicated by $\sqrt{a}$.

A **perfect square** is a number whose square root is a whole number.

The **principal square root** of a positive number is its positive square root.

An **imaginary number** is the square root of a negative number.

(8.4) The **nth root** of a given number is some number whose nth power is equal to the given number. For any number a, the nth root of a is indicated by $\sqrt[n]{a}$.

(8.5) A **set** is a collection or group of objects.

Calculations

(8.2)   **To find the square root of a product**, find the product of the square roots of the factors.

To express a square root in simplest form, determine the square root of the largest perfect square that is a factor of the number under the radical sign. The other factor, which is not a perfect square, remains under the radical sign.

To find the square root of a quotient, find the quotient of the square roots of the numerator and the denominator.

Properties of Roots

(8.4)   For any positive numbers a and b and positive integer n,

$$\sqrt[n]{a \cdot b} = \sqrt[n]{a}\, \sqrt[n]{b}$$

In other words, the nth root of a product is the product of the nth roots of the factors. For any positive numbers a and b and positive integer n,

$$\sqrt[n]{\frac{a}{b}} = \frac{\sqrt[n]{a}}{\sqrt[n]{b}}$$

In other words, the nth root of a quotient is the quotient of the nth roots of the numerator and the denominator.

Number Systems

(8.5)   **Natural Numbers**, or **Counting Numbers**: $\{1, 2, 3, 4, 5, \ldots\}$

Whole Numbers: $\{0, 1, 2, 3, 4, 5, \ldots\}$

Integers: $\{\ldots, -4, -3, -2, -1, 0, 1, 2, 3, 4, \ldots\}$

Rational Numbers: Numbers that can be written in the form $\frac{a}{b}$, where a and b are integers and $b \neq 0$.

Irrational Numbers: Nonterminating, nonrepeating decimals that cannot be expressed in the form of a rational number.

Real Numbers: All rational and irrational numbers.

Chapter **8**
Review Exercises

(8.1–8.3)   Determine each of the following square roots. Round off where indicated.

1. $\sqrt{49}$

2. $-\sqrt{81}$

3. $\pm\sqrt{\dfrac{9}{25}}$

4. $\sqrt{\dfrac{1}{16}}$

5. $-\sqrt{0.36}$

6. $\sqrt{0.64}$

7. $-\sqrt{4,900}$

8. $\pm\sqrt{6,400}$

9. $-\sqrt{0.0009}$

10. $\sqrt{0.0004}$

11. $\sqrt{10,000}$

12. $\sqrt{810,000}$

13. $\sqrt{13.69}$

14. $\sqrt{51.84}$

15. $\sqrt{324}$

16. $\sqrt{4,489}$

17. $\sqrt{16,129}$

18. $\sqrt{383,161}$

19. $\sqrt{0.0784}$

20. $\sqrt{0.5776}$

21. $\sqrt{43,296,400}$

22. $\sqrt{6,604,900}$

23. $\sqrt{0.133225}$

24. $\sqrt{0.471969}$

25. $\sqrt{5.9}$ nearest hundredth
26. $\sqrt{37.1}$ nearest hundredth
27. $\sqrt{637}$ nearest tenth
28. $\sqrt{4,712}$ nearest tenth
29. $\sqrt{38,895}$ nearest whole number
30. $\sqrt{617,204}$ nearest whole number
31. $\sqrt{0.281}$ nearest thousandth
32. $\sqrt{0.0174}$ nearest thousandth
33. $\sqrt{53,742,109}$ nearest whole number
34. $\sqrt{1,347,269}$ nearest whole number
35. $\sqrt{0.634812}$ nearest thousandth
36. $\sqrt{0.004517}$ nearest ten-thousandth

(8.2, 8.4)   Write each of the following radical expressions in simplest form.

37. $\sqrt{200}$
38. $\sqrt{450}$
39. $\sqrt{8}$
40. $\sqrt{54}$
41. $\sqrt{24}$
42. $\sqrt{125}$
43. $\sqrt[4]{16}$
44. $\sqrt[5]{32}$
45. $\sqrt[4]{81}$
46. $\sqrt[3]{125}$
47. $\sqrt[5]{-1}$
48. $\sqrt[3]{-27}$

(8.5)   List all of the number systems in which each of the following numbers is included.

49. $-\sqrt{2}$
50. $\sqrt{25}$
51. 3.78
52. -917
53. $5\frac{7}{8}$
54. $\frac{21}{3}$
55. $\sqrt{0.09}$
56. $0.87\overline{87}$
57. $\sqrt{100}$
58. -3.0
59. $-5\sqrt{15}$
60. $\sqrt{3,600}$

(8.5)   Use the $>$, $<$, or $=$ sign to indicate the relationship between each of the following pairs of numbers.

61. $-\sqrt{3}\ ?\ -2$
62. $2\ ?\ \sqrt{5}$
63. $\sqrt{49}\ ?\ 7$
64. $-\sqrt{37}\ ?\ -6$
65. $\sqrt{0.16}\ ?\ 0.4$
66. $\sqrt{0.9}\ ?\ 0.3$
67. $-\sqrt{\frac{4}{49}}\ ?\ -\frac{3}{7}$
68. $\frac{5}{8}\ ?\ \sqrt{\frac{25}{81}}$
69. $\sqrt{0.049}\ ?\ 0.07$
70. $\sqrt{0.121}\ ?\ 0.11$
71. $\sqrt{65}\ ?\ 8$
72. $-4\ ?\ -\sqrt{17}$

Chapter **8**
Explain It in Words

73. Explain what is meant by a "perfect square" of a number. Give an example.
74. Explain what is meant by the "cube root" of a number. Give an example.
75. Are all integers also rational numbers? Use an example to explain your answer.
76. Are all integers also whole numbers? Use an example to explain your answer.
77. Explain the difference between rational and irrational numbers. Give an example of each.
78. Are all natural numbers also integers? Use an example to explain your answer.

Chapter **8**
Chapter Test

[8.1, 8.2]   Determine each of the following square roots.

1. $\sqrt{25}$
2. $-\sqrt{0.81}$
3. $\sqrt{\frac{9}{64}}$

4. $\pm\sqrt{4,900}$ 5. $\sqrt{0.0036}$

[8.2, 8.4]   **Write each of the following in simplest form.**

6. $\sqrt{72}$ 7. $\sqrt{48}$ 8. $\sqrt[3]{8}$

9. $\sqrt[3]{-27}$ 10. $\sqrt[5]{\dfrac{1}{100,000}}$

[8.5]   **List all of the number systems in which each of the following is included.**

11. $\sqrt{36}$ 12. -8 13. 9.3

14. $\sqrt{5}$ 15. $\dfrac{3}{5}$

[8.5]   **Use the $=$, $>$, or $<$ sign to indicate the relationship between each of the following pairs of numbers.**

16. 0.8 ? $\sqrt{0.64}$ 17. $\sqrt{0.16}$? 0.04 18. $-\sqrt{50}$? -7

19. $\sqrt{7}$? 7 20. $-\sqrt{25}$? -5

Chapter *9*

Measurement in the English and Metric Systems

To **measure** something is to find a number that characterizes one of its properties. To establish a uniform convention for assigning numbers to something we measure, it is necessary to define a **unit of measurement** that indicates the size of a known quantity. Units of measurement are used to indicate the magnitude of various properties such as length, area, weight, volume, and temperature. A *system of measurement* defines the relationships within a set of standard units that are used to describe the various properties and all other units that are derived from them. In this chapter, we will discuss two systems of measurement—the **English system**, which is used primarily in the United States, and the **metric system**, which is used throughout the rest of the world.

9.1
Length

The standard units of length in the English system are the inch, foot, yard, and mile. Table 9.1 summarizes equivalent measures of length in the English system.

TABLE 9.1
English Units of Length

1 foot (ft)	=	12 inches (in.)
1 yard (yd)	=	3 feet (ft)
1 mile (mi)	=	5,280 feet (ft)

In Section 5.1, we showed how to convert from one unit to another by multiplying the quantity to be converted by a unit fraction, that is, a fraction equivalent to 1. Whenever we multiply a unit of measurement by a series of unit fractions to convert it to another unit of measurement, we are using a technique called *dimensional analysis*. This technique is illustrated in Example 1.

EXAMPLE 1 Convert each of the following English units as indicated.

(a) Convert 13,200 feet to miles.

$$13,200 \text{ ft} = 13,200 \cancel{\text{ft}} \times \frac{1 \text{ mi}}{5,280 \cancel{\text{ft}}}$$

$$= 2.5 \text{ mi}$$

Since 1 mi = 5,280 ft, multiply 13,200 ft by the unit fraction $\frac{1 \text{ mi}}{5,280 \text{ ft}}$ so that ft will cancel.

(b) Convert 7 yd to inches.

$$7 \text{ yd} = 7 \cancel{\text{yd}} \times \frac{3 \cancel{\text{ft}}}{1 \cancel{\text{yd}}} \times \frac{12 \text{ in.}}{1 \cancel{\text{ft}}}$$

$$= 252 \text{ in.}$$

Since 1 yd = 3 ft and 1 ft = 12 in., multiply 7 yd by the unit fractions $\frac{3 \text{ ft}}{1 \text{ yd}}$ and $\frac{12 \text{ in.}}{1 \text{ ft}}$ so that yd and ft will cancel. ▲

The standard unit of length in the metric system is the **meter** (m), which is a little longer than a yard. Other units of length in the metric system are derived by multiplying or dividing a meter by powers of 10. These other units are named by attaching the appropriate prefix to the word *meter*. The metric prefixes and their meanings are given in Table 9.2. The equivalent units of length in the metric system are given in Table 9.3.

TABLE 9.2
Metric Prefixes

Prefix	Meaning
milli	thousandth
centi	hundredth
deci	tenth
deka	ten
hecto	hundred
kilo	thousand

TABLE 9.3
Metric Units of Length

1 millimeter (mm) $= \dfrac{1}{1,000}$ meter (m)	or	1,000 mm = 1 m
1 centimeter (cm) $= \dfrac{1}{100}$ meter (m)	or	100 cm = 1 m
1 decimeter (dm) $= \dfrac{1}{10}$ meter (m)	or	10 dm = 1 m
1 dekameter (dam) = 10 meters (m)		
1 hectometer (hm) = 100 meters (m)		
1 kilometer (km) = 1,000 meters (m)		

To facilitate our work with the metric system, it is helpful to be able to visualize the lengths of everyday objects in terms of metric units.

For example,

The length of this book is about 24 cm.
The height of the Empire State Building is 381 m.
The diameter of a penny is about 19 mm.

Figure 9.1 shows how a ruler that is 1 decimeter, or 10 centimeters, in length compares to another ruler that is 4 inches long.

Figure 9.1

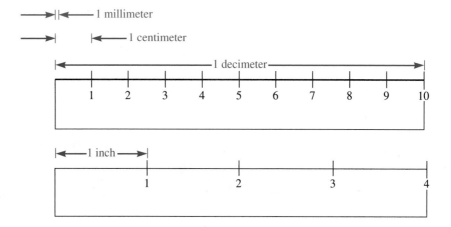

To convert from one unit of length to another in the metric system, we simply multiply or divide the unit to be converted by powers of 10. Therefore, conversion from one unit to another is much simpler in the metric system than in the English system, which is based on fractions.

EXAMPLE 2 Convert each of the following metric units as instructed.

(a) Convert 8 m to centimeters.

$$8 \text{ m} = 8 \text{ m} \times \frac{100 \text{ cm}}{1 \text{ m}} = 800 \text{ cm}$$

Since 1 m = 100 cm, multiply 8 m by the unit fraction $\frac{100 \text{ cm}}{1 \text{ m}}$ so that m cancels.

(b) Convert 1.5 hm to millimeters.

$$1.5 \text{ hm} = 1.5 \text{ hm} \times \frac{100 \text{ m}}{1 \text{ hm}} \times \frac{1{,}000 \text{ mm}}{1 \text{ m}}$$

$$= 150{,}000 \text{ mm}$$

Since 1 hm = 100 m and 1 m = 1,000 mm, multiply 1.5 hm by the unit fractions $\frac{100 \text{ m}}{1 \text{ hm}}$ and $\frac{1{,}000 \text{ mm}}{1 \text{ m}}$ so that hm and m cancel.

To multiply by 100,000, move the decimal point five places to the right.

(c) Convert 82 dm to kilometers.

$$82 \text{ dm} = 82 \text{ dm} \times \frac{1 \text{ m}}{10 \text{ dm}} \times \frac{1 \text{ km}}{1{,}000 \text{ m}}$$

$$= 0.0082 \text{ km}$$

Since 1 m = 10 dm and 1 km = 1,000 m, multiply 82 dm by the unit fractions $\frac{1 \text{ m}}{10 \text{ dm}}$ and $\frac{1 \text{ km}}{1{,}000 \text{ m}}$ so that dm and m cancel.

To divide by 10,000, move the decimal point four places to the left. ▲

The relationship between the units of length in the English and metric systems is given in Table 9.4. Each of the values in Table 9.4 is correct to four places of accuracy, or four significant digits. Therefore, when using Table 9.4 for a calculation, round off your answer so that it contains no more than four significant digits.

TABLE 9.4
Equivalent Units of Length

English to Metric	Metric to English
1 in. = 2.540 cm	1 cm = 0.3937 in.
1 ft = 0.3048 m	1 m = 39.37 in.
1 yd = 0.9144 m	1 m = 1.094 yd
1 mi = 1.609 km	1 km = 0.6214 mi

To convert from one system to another, we can also multiply the unit to be converted by an appropriate unit fraction, as shown in Example 3.

EXAMPLE 3   Convert each of the following as indicated.

(a)   Convert 357 mi to kilometers.

$$357 \text{ mi} = 357 \text{ mi} \times \frac{1.609 \text{ km}}{1 \text{ mi}}$$

Since 1 mi = 1.609 km, multiply 357 mi by the unit fraction $\frac{1.609 \text{ km}}{1 \text{ mi}}$ so that mi cancels.

$$= 574.413 \text{ km} \cong 574.4 \text{ km}$$

Since the conversion factor has only four significant digits, our final answer can have no more than four significant digits. Therefore, we rounded off to the nearest tenth.

(b)   Convert 84 cm to inches.

$$84 \text{ cm} = 84 \text{ cm} \times \frac{0.3937 \text{ in.}}{1 \text{ cm}}$$

Since 1 cm = 0.3937 in., multiply 84 cm by the unit fraction $\frac{0.3937 \text{ in.}}{1 \text{ cm}}$ so that cm cancels.

$$= 33.0708 \text{ in.}$$
$$\cong 33.07 \text{ in.}$$

Since the conversion factor has only four significant digits, our final answer can have no more than four digits. Therefore, we rounded off to the nearest hundredth.   ▲

It is very important to be able to mentally estimate conversions between the English and metric systems based on knowledge of some approximate conversion factors. Since 1 m = 39.37 in. and 1 yd = 36 in., a meter is a little longer than a yard.

$$1 \text{ m} \cong 1 \text{ yd}$$

Since 1 in. = 2.54 cm, we can think of 1 in. as approximately equal to $2\frac{1}{2}$ cm. Therefore,

$$2 \text{ in.} \cong 5 \text{ cm} \quad \text{and} \quad 1 \text{ ft} \cong 30 \text{ cm}$$

Since 1 km = 0.6214 mi, we can think of 1 km as approximately equal to $\frac{5}{8}$ mi ($\frac{5}{8} = 0.625$) or

$$5 \text{ mi} \cong 8 \text{ km}$$

Some approximations that are less exact but easier to remember are

$$1 \text{ mi} \cong 2 \text{ km}$$

$$1 \text{ km} \cong \frac{1}{2} \text{ mi}$$

EXAMPLE 4 Approximate each of the following conversions between English and metric units. Then compare your answers to the values obtained by using Table 9.4.

(a) Convert 15 cm to inches.
Since 5 cm $\cong$ 2 in.,
then 15 cm $\cong$ 6 in.
Use Table 9.4 to find the exact answer.

$\left(\text{Think, } \frac{5 \text{ cm}}{2 \text{ in.}} = \frac{15 \text{ cm}}{?} \right)$

$$15 \text{ cm} = 15 \text{ cm} \times \frac{0.3937 \text{ in.}}{1 \text{ cm}}$$

$$= 5.9055 \text{ in.}$$

Multiply 15 cm by the unit fraction $\frac{0.3937 \text{ in.}}{1 \text{ cm}}$ so that cm cancels.

15 cm = 5.9055 in. $\cong$ 5.906 in. correct to four significant digits. Our approximation is off by about 0.1 in.

6 in. $-$ 5.906 in. $\cong$ 0.1 in.

(b) Convert 5 ft to centimeters.
Since 1 ft $\cong$ 30 cm,
then 5 ft $\cong$ 150 cm
Use Table 9.4 to find the exact answer.

$\left(\text{Think, } \frac{1 \text{ ft}}{30 \text{ cm}} = \frac{5 \text{ ft}}{?} \right)$

$$5 \text{ ft} = 5 \text{ ft} \times \frac{0.3048 \text{ m}}{1 \text{ ft}} \times \frac{100 \text{ cm}}{1 \text{ m}}$$

$$= 152.4 \text{ cm}$$

Multiply 5 ft by the unit fractions $\frac{0.3048 \text{ m}}{1 \text{ ft}}$ and $\frac{100 \text{ cm}}{1 \text{ m}}$ so that ft and m cancel.

5 ft = 152.4 cm correct to four significant digits. Our approximation is off by 2.4 cm.

152.4 $-$ 150 cm = 2.4 cm

(c) Convert 2,500 mi to kilometers.
Since 5 mi $\cong$ 8 km,
2,500 mi $\cong$ 4,000 km
Use Table 9.4.

$\left(\text{Think, } \frac{5 \text{ mi}}{8 \text{ km}} = \frac{2,500 \text{ mi}}{?} \right)$

$$2,500 \text{ mi} = 2,500 \text{ mi} \times \frac{1.609 \text{ km}}{1 \text{ mi}}$$

$$= 4,022.5 \text{ km}$$

Multiply 2,500 mi by the unit fraction $\frac{1.609 \text{ km}}{1 \text{ mi}}$ so that mi cancels.

2,500 mi = 4,022.5 km $\cong$ 4,023 km correct to four significant digits. Our approximation is off by about 23 km.

4,023 km $-$ 4,000 km = 23 km ▲

EXAMPLE 5 The captain of a swim team is 6 ft 2 in. tall, and his coach is 171 cm tall. Who is taller?

Solution Approximate the height of the swim team captain in centimeters.
Since 1 ft $\cong$ 30 cm,
6 ft $\cong$ 180 cm
Since 2 in. $\cong$ 5 cm,
6 ft 32 in. $\cong$ 180 cm + 5 cm = 185 cm
The swim team captain is taller than his coach. Since 185 cm > 171 cm. ▲

Quick Quiz

Perform each of the following conversions.   **Answers**

1. 7 ft = _____ yd   1. $2\frac{1}{3}$

2. 6,000 mm = _____ m   2. 6

3. 5.2 km = _____ cm   3. 520,000

4. 8 in. $\cong$ _____ cm   4. 20

9.1 Exercises

Indicate which of the following statements are true and which are false. For those that are false, change the italicized word to make the statement true.

1. The standard unit of length in the metric system is the *meter*.

2. One yard is about 3 inches *longer* than a meter.

3. One meter is equivalent to 100 *millimeters*.

4. Thirty centimeters is roughly equivalent to 1 *inch*.

Convert each of the following English units as indicated.

5. 8 ft = _____ in.
6. 60 in. = _____ ft
7. 7 ft = _____ yd
8. 528 ft = _____ mi

9. $5\frac{2}{3}$ yd = _____ ft
10. $1\frac{3}{4}$ mi = _____ ft
11. 7 ft 2 in. = _____ in.
12. 3 yd 2 ft = _____ ft

13. 7 yd 1 ft = _____ yd
14. 9 ft 9 in. = _____ ft
15. $\frac{3}{5}$ mi = _____ ft
16. 1,584 ft = _____ mi

17. 23.5 ft = _____ in.
18. 8.5 mi = _____ ft
19. $\frac{3}{4}$ ft = _____ in.
20. $\frac{7}{9}$ yd = _____ in.

21. $\frac{1}{2}$ mi = _____ yd
22. $5\frac{2}{3}$ yd = _____ in.
23. 114 in. = _____ ft
24. 2,256 in. = _____ yd

25. 540 in. = _____ yd
26. 0.05 yd = _____ in.

Convert each of the following metric units as indicated.

27. 5 m = _____ cm
28. 870 cm = _____ m
29. 19 m = _____ mm

30. 6,400 mm = _____ m
31. $3\frac{1}{2}$ km = _____ m
32. $7\frac{3}{4}$ cm = _____ mm

33. 3.8 km = _____ hm
34. 9.5 cm = _____ dm
35. 53 mm = _____ cm

36. 98 dam = _____ hm
37. $\frac{7}{10}$ km = _____ m
38. $\frac{3}{5}$ m = _____ cm

39. 590 cm = _____ m
40. 4.5 hm = _____ m
41. 0.2 km = _____ cm

42. 8,800 mm = _____ m
43. 857 dm = _____ m
44. 631 dam = _____ m

45. 720 cm = _____ dm
46. 3,500 mm = _____ dm
47. 8.1 hm = _____ dam

48. 2.7 cm = _____ hm
49. 0.7 km = _____ mm
50. 0.3 dm = _____ dam

Indicate which measurement most closely approximates each of the following lengths.

51. The length of a playing card: 20 cm   9 cm   9 mm

52. The diameter of a quarter: 3.8 mm   2.5 cm   1.4 cm

53. The length of a football field: 91 m   80 m   107 m

54. The average height of an adult woman: 300 cm   163 cm   250 cm

55. The length of a tennis racquet: 68.5 cm   6.85 m   68.5 mm

56. The width of this book: 19.5 cm   19.5 mm   1.95 m

57. The height of a classroom: 320 mm   3.2 m   0.32 km

58. The height of a 100-story building: 343 km   3.43 km   343 m

59. The length of a pen: 15 cm   150 cm   15 mm

60. The distance between Boston and Philadelphia: 43.6 km   80.5 km   436 km

61. The diameter of the earth: 12,755 m   127,550 m   12,755 km

62. The height of a desk: 77 cm   7 m   77 dm

Approximate each of the following conversions between English and metric units. Then compare your results to the values obtained by using Table 9.4.

63. 35 cm = _____ in.

64. 8 in. = _____ cm

65. 5 ft = _____ cm

66. 210 cm = _____ ft

67. 2 yd = _____ m

68. 8 m = _____ yd

69. 450 mi = _____ km

70. 64 km = _____ mi

71. 5 yd = _____ cm

72. 20 mi = _____ m

Solve each of the following word problems.

73. The distance from San Francisco to Minneapolis is 1,997 mi, and the distance from San Francisco to Chicago is 2,989 km. Which city is farther from San Francisco, Minneapolis or Chicago?

74. A Canadian athlete ran a 100-m race in 10.83 sec. An American athlete ran a 100-yd race in the same time. Who ran faster?

75. A room is 5.98 m long. How many tiles must be laid end to end to reach from one end of the room to the other if each tile is 23 cm in length?

76. A stack of 17 identical books is 76.5 cm high. What is the thickness of each book?

77. A roll of transparent tape is 20.3 m long. How many feet of tape are in this roll? Round off your answer to the nearest tenth of a foot.

78. A speed limit is posted as 55 mi/hr. What is the speed limit in km/hr? Round off your answer to the nearest kilometer.

79. If fabric sells for $4.98 a meter, what is the cost of a piece that measures $3\frac{1}{2}$ yd in length? Round off your answer to the nearest cent.

80. Fencing material sells for $12 a foot. How much would it cost to construct a fence that is 4 m long?

9.2
Weight

The **weight** of an object is a measure of the force of the earth's gravity pulling upon it. As an object moves away from the center of the earth, its weight decreases.

The standard units of weight in the English system are the ounce, pound, and ton. Table 9.5 summarizes equivalent units of weight in the English system.

TABLE 9.5
English Units of Weight

1 pound (lb)	=	16 ounces (oz)
1 ton (T)	=	2,000 pounds (lb)

EXAMPLE 1 Convert each of the following English units of measurement.

(a) Convert $7\frac{5}{8}$ lb to ounces.

$$7\frac{5}{8}\ \text{lb} = 7\frac{5}{8}\ \text{lb} \times \frac{16\ \text{oz}}{1\ \text{lb}}$$

Multiply $7\frac{5}{8}$ lb by the unit fraction $\frac{16\ \text{oz}}{1\ \text{lb}}$ so that lb cancels.

$$= \frac{61}{8} \times 16\ \text{oz} = 122\ \text{oz}$$

(b) Convert 3.2 T to pounds.

$$3.2\ \text{T} = 3.2\ \text{T} \times \frac{2{,}000\ \text{lb}}{1\ \text{T}}$$

Since 1 T = 2,000 lb, multiply 3.2 T by the unit fraction $\frac{2{,}000\ \text{lb}}{1\ \text{T}}$ so that T cancels.

$$= 6{,}400\ \text{lb}$$ ▲

The standard unit of weight in the metric system is the **gram** (g). Other metric units of weight are named by attaching the appropriate prefix to the word *gram*. The equivalent units of weight in the metric system are listed in Table 9.6.

TABLE 9.6
Metric Units of Weight

1 milligram (mg) $= \dfrac{1}{1{,}000}$ gram (g)	or	1,000 mg = 1 g
1 centigram (cg) $= \dfrac{1}{100}$ gram (g)	or	100 cg = 1 g
1 kilogram (kg) = 1,000 grams (g)		
1 metric ton (t) = 1,000,000 grams (g)	or	1 t = 1,000 kg

The units decigram ($\frac{1}{10}$ g), dekagram (10 g), and hectogram (100 g) have been omitted from Table 9.6 since they are rarely used.

The weights of a few well-known objects, in metric units of measurement, are as follows:

An aspirin weighs about 500 mg.
An egg weighs about 50 g.
A football player weighs about 130 kg.

EXAMPLE 2 Convert each of the following metric units as indicated.

(a) Convert 17 mg to grams.

$$17\ \text{mg} = 17\ \text{mg} \times \frac{1\ \text{g}}{1{,}000\ \text{mg}}$$

Since 1 g = 1,000 mg, multiply 17 mg by the unit fraction $\frac{1\ \text{g}}{1{,}000\ \text{mg}}$ so that mg cancels.

$$= 0.017\ \text{g}$$

To divide by 1,000, move the decimal point three places to the left.

(b) Convert 3.5 t to kilograms.

$$3.5\ \text{t} = 3.5\ \text{t} \times \frac{1{,}000\ \text{kg}}{1\ \text{t}}$$

Since 1 t = 1,000 kg, multiply 3.5 t by the unit fraction $\frac{1{,}000\ \text{kg}}{1\ \text{t}}$ so that t cancels.

To multiply by 1,000, move the decimal point 3 places to the right.

$$= 3{,}500\ \text{kg}$$

(c) Convert 4.78 kg to milligrams.

$$4.78 \text{ kg} = 4.78 \text{ k\!\!\!/g} \times \frac{1{,}000 \text{ \!\!\!/g}}{1 \text{ k\!\!\!/g}} \times \frac{1{,}000 \text{ mg}}{1 \text{ \!\!\!/g}}$$

$$= 4{,}780{,}000 \text{ mg}$$

Since 1 kg = 1,000 g and 1 g = 1,000 mg, multiply 4.78 kg by the unit fractions $\frac{1{,}000 \text{ g}}{1 \text{ kg}}$ and $\frac{1{,}000 \text{ mg}}{1 \text{ g}}$ so that kg and g cancel. To multiply by 1,000,000, move the decimal point 6 places to the right. ▲

The relationship between the units of weight in the English and metric systems is given in Table 9.7. Each value is accurate to four significant digits.

TABLE 9.7
Equivalent Units of Weight

English to Metric	Metric to English
1 oz = 28.35 g	1 g = 0.03527 oz
1 lb = 0.4536 kg	1 kg = 2.205 lb
1 T = 0.9072 t	1 t = 1.102 T

From this table, we can derive the following approximations that will enable us to mentally estimate conversions from one system to the other.

$$1 \text{ oz} \cong 30 \text{ g}$$
$$1 \text{ lb} \cong 0.45 \text{ kg}$$
$$1 \text{ kg} \cong 2.2 \text{ lb}$$
$$1 \text{ T} \cong 1 \text{ t}$$

Some approximations that are less exact but easier to remember are

$$1 \text{ lb} \cong \tfrac{1}{2} \text{ kg}$$

$$1 \text{ kg} \cong 2 \text{ lb}$$

EXAMPLE 3 Approximate the following conversions. Then compare your results to the values obtained by using Table 9.7.

(a) Convert 10 lb to kilograms.
Since 1 lb $\cong$ 0.45 kg,
then 10 lb $\cong$ 4.5 kg
Use Table 9.7 to find the exact answer.

$\left(\text{Think, } \frac{1 \text{ lb}}{0.45 \text{ kg}} = \frac{10 \text{ lb}}{?} \right)$

$$10 \text{ lb} = 10 \text{ l\!\!\!/b} \times \frac{0.4536 \text{ kg}}{1 \text{ l\!\!\!/b}}$$

$$= 4.536 \text{ kg}$$

Multiply 10 lb by the unit fraction $\frac{0.4536 \text{ kg}}{1 \text{ lb}}$ so that lb cancels.

10 lb = 4.536 kg correct to four significant digits. Our approximation is off by 0.036 kg.

4.536 kg − 4.5 kg = 0.036 kg

(b) Convert 8 oz to grams.

Since 1 oz ≅ 30 g,

8 oz ≅ 240 g

$$\left(\text{Think, } \frac{1 \text{ oz}}{30 \text{ g}} = \frac{8 \text{ oz}}{?}\right)$$

Use Table 9.7 to find the exact answer.

$$8 \text{ oz} = 8 \text{ oz} \times \frac{28.35 \text{ g}}{1 \text{ oz}} = 226.8 \text{ g}$$

Multiply 8 oz by the unit fraction $\frac{28.35 \text{ g}}{1 \text{ oz}}$ so that oz cancels.

8 oz = 226.8 g, correct to four significant digits. Our approximation is off by 13.2 g.

240 g − 226.8 g = 13.2 g   ▲

EXAMPLE 4   According to an international agreement, an airline is required to compensate a passenger for lost baggage at a rate not exceeding $20 per kilogram. If a 22-lb bag was lost by an airline, what is the maximum amount its owner can expect to receive in compensation?

Solution $22 \text{ lb} = 22 \text{ lb} \times \frac{1 \text{ kg}}{2.2 \text{ lb}} = 10 \text{ kg}$

Convert 22 lb to kg using the conversion factor 1 kg ≅ 2.2 lb.

$$10 \text{ kg} \times \frac{\$20}{\text{kg}} = \$200$$

Then multiply the result by 20.

The maximum amount the owner can expect is $200.   ▲

Quick Quiz

Perform each of the following conversions.   **Answers**

1.   8 oz   = ___ lb   1. $\frac{1}{2}$

2.   20 mg   = ___ g   2.   0.02

3.   9,000 g = ___ kg   3.   9

4.   2 kg   ≅ ___ lb   4.   4.4

9.2 Exercises

Indicate which of the following statements are true and which are false. For those that are false, change the italicized or underlined expression to make the statement true.

1.   One pound is equal to <u>8</u> ounces.

2.   The standard unit of weight in the *English* system is the gram.

3.   One kilogram is roughly equivalent to 2.2 *ounces*.

4.   One kilogram is equivalent to <u>1,000,000</u> milligrams.

Convert each of the following English units as indicated.

5.   32 oz = ___ lb

6.   5 lb = ___ oz

7. $\frac{1}{2}$ T = ___ lb

8.   3,000 lb = ___ T

9. $7\frac{3}{4}$ lb = ___ oz

10.   4.5 T = ___ lb

11.   8 lb 9 oz = ___ oz   12.   20 lb 2 oz = ___ lb   13.   0.053 T = ___ lb
14.   0.0069 T ≅ ___ lb   15. $5\frac{3}{8}$ lb = ___ oz   16.   108 oz = ___ lb
17.   250 lb = ___ T   18.   50,000 lb = ___ T   19.   80 oz = ___ lb
20.   0.72 lb = ___ oz   21. $3\frac{2}{5}$ T = ___ lb   22. $2\frac{7}{8}$ lb = ___ oz
23.   4,800 lb = ___ T   24.   7,200 oz = ___ lb

Convert each of the following metric units as indicated.

25.   5 kg = ___ g   26.   28 g = ___ mg   27.   350 cg = ___ g
28.   9.3 g = ___ cg   29.   5 t = ___ kg   30.   3,800 kg = ___ t
31.   6.7 mg = ___ g   32.   7.2 cg = ___ g   33.   750 mg = ___ g
34.   0.4 g = ___ cg   35.   347 g = ___ mg   36.   218 cg = ___ mg
37.   4.18 kg = ___ g   38.   0.257 g = ___ mg   39.   3 kg = ___ mg
40.   5,000 mg = ___ kg   41.   88,000 g = ___ t   42.   7,000 kg = ___ t
43. $9\frac{1}{2}$ mg = ___ g   44. $8\frac{2}{5}$ g = ___ cg   45. $\frac{5}{8}$ kg = ___ g
46. $3\frac{3}{4}$ g = ___ mg   47.   0.07 t = ___ g   48.   0.02 kg = ___ mg

Indicate which measurement most closely approximates each of the following weights.

49.   The weight of a box of crackers: 454 g   4.54 kg   4,540 mg
50.   The weight of a cold capsule: 650 mg   6.5 g   0.065 kg
51.   The weight of one-half dozen eggs: 1 kg   100 g   300 g
52.   The weight of a newborn baby: 4 kg   40 kg   40 g
53.   The weight of a nickel: 50 mg   5 g   0.5 kg
54.   The weight of an average adult man: 860 g   86 kg   8.6 kg
55.   The weight of an automobile: 500 kg   1 t   50,000 g
56.   The weight of a paper clip: 5 g   50 mg   0.5 g

Approximate each of the following conversions. Then compare your results to the values obtained by using Table 9.7.

57.   30 lb = ___ kg   58.   90 g = ___ oz   59.   4 kg = ___ lb   60.   66 lb = ___ kg
61.   7.2 T = ___ t   62.   23 t = ___ T   63.   700 oz = ___ g   64.   200 kg = ___ lb
65.   3 kg = ___ oz   66.   240 g = ___ lb

Solve each of the following word problems.

67.   If one aspirin weighs 500 mg, how many grams would 100 aspirins weigh?
68.   Which weighs more, 10 lb of potatoes or 5 kg of potatoes?
69.   A chocolate bar that contains nuts weighs 45 g. Another chocolate bar that does not contain nuts weighs 2 oz. Which chocolate bar weighs more?
70.   A 10-year-old boy weighs 60 lb, and a 10-year-old girl weighs 30 kg. Who weighs more?
71.   If a dozen bricks weighs 54 kg, how much does a single brick weigh?
72.   A package containing 16 slices of cheese weighs 340 g. What is the weight of a single slice?

73. A 12-oz package of chocolate chip cookies costs $1.69. A 420-g package of chocolate chip cookies sells for the same price. Which is the better buy?

74. A 2-lb box of sugar sells for $1.29. How much would a box that weighs 2 kg cost? Round off your answer to the nearest cent.

9.3
Capacity

The standard units of capacity (volume) in the English system are the cup, pint, quart, gallon, fluid ounce, tablespoon, and teaspoon. Table 9.8 summarizes equivalent units of capacity in the English system.

TABLE 9.8
English Units of Capacity

1 pint (pt)	=	2 cups (c)
1 quart (qt)	=	2 pints (pt)
1 gallon (gal)	=	4 quarts (qt)
1 pint (pt)	=	16 fluid ounces (fl oz)
1 tablespoon (tbsp)	=	$\frac{1}{2}$ fluid ounce (fl oz)
1 tablespoon (tbsp)	=	3 teaspoons (tsp)

EXAMPLE 1 Convert each of the following English units of measurement.

(a) Convert 22 fl oz to pints.

$$22 \text{ fl oz} = 22 \text{ fl oz} \times \frac{1 \text{ pt}}{16 \text{ fl oz}} = \frac{22}{16} \text{ pt}$$

Multiply 22 fl oz by the unit fraction $\frac{1 \text{ pt}}{16 \text{ fl oz}}$ so that fl oz cancels.

$$= \frac{11}{8} \text{ pt} = 1\frac{3}{8} \text{ pt}$$

(b) Convert 9.5 gal to pints.

$$9.5 \text{ gal} = 9.5 \text{ gal} \times \frac{4 \text{ qt}}{1 \text{ gal}} \times \frac{2 \text{ pt}}{1 \text{ qt}}$$

Multiply 9.5 gal by the unit fractions $\frac{4 \text{ qt}}{1 \text{ gal}}$ and $\frac{2 \text{ pt}}{1 \text{ qt}}$ so that gal and qt cancel.

$$= 9.5 \times 4 \times 2 \text{ pt}$$
$$= 76 \text{ pt}$$

▲

The standard unit of capacity in the metric system is the *liter* (L), which is a little more than a quart. Other metric units of capacity are named by attaching the appropriate prefix to the word *liter*. The equivalent units of capacity in the metric system are listed in Table 9.9.

TABLE 9.9
Metric Units of Capacity

1 milliliter (mL) = $\frac{1}{1,000}$ liter (L)	or	1,000 mL = 1 L
1 hectoliter (hL) = 100 liters (L)		
1 kiloliter (kL) = 1,000 liters (L)		

The units centiliter ($\frac{1}{100}$ L), deciliter ($\frac{1}{10}$ L), and dekaliter (10 L) have been omitted from Table 9.9 since they are rarely used.

The volumes of a few well-known items, in metric units of measurement, are as follows:

A bottle of eyedrops contains 15 mL.
A can of soda contains 355 mL.
A bathtub can hold 150 L.

EXAMPLE 2   Convert each of the following metric units as instructed.

(a)   Convert 3.5 L to milliliters.

$$3.5 \text{ L} = 3.5 \text{ L} \times \frac{1,000 \text{ mL}}{1 \text{ L}}$$

Multiply 3.5 L by the unit fraction $\frac{1,000 \text{ mL}}{1 \text{ L}}$ so that L cancels.

$$= 3,500 \text{ mL}$$

(b)   Convert 80,000 mL to hectoliters.

80,000 mL

$$= 80,000 \text{ mL} \times \frac{1 \text{ L}}{1,000 \text{ mL}} \times \frac{1 \text{ hL}}{100 \text{ L}}$$

Multiply 80,000 mL by the unit fractions $\frac{1 \text{ L}}{1,000 \text{ mL}}$ and $\frac{1 \text{ hL}}{100 \text{ L}}$ so that mL and L cancel.

$$= 0.8 \text{ hL}$$

To divide by 100,000 move the decimal point 5 places to the left. ▲

The relationship between units of capacity in the English and metric systems is given in Table 9.10. Each value is accurate to four significant digits.

TABLE 9.10
Equivalent Units of
Capacity

English to Metric	Metric to English
1 qt = 0.9463 L	1 L = 1.057 qt
1 gal = 3.785 L	1 L = 0.2642 gal
1 fl oz = 29.57 mL	1 mL = 0.03381 fl oz
1 tsp = 4.929 mL	1 mL = 0.2029 tsp
1 tbsp = 14.79 mL	1 mL = 0.06763 tbsp

From this table, we can derive the following approximations that will enable us to mentally estimate conversions from one system to the other.

$$1 \text{ qt} \cong 1 \text{ L}$$
$$1 \text{ fl oz} \cong 30 \text{ mL}$$
$$1 \text{ tsp} \cong 5 \text{ mL}$$
$$1 \text{ tbsp} \cong 15 \text{ mL}$$

Since teaspoons and tablespoons are not measured exactly, the values in Table 9.10 for converting teaspoon and tablespoon to milliliter are given only for the purpose of deriving these approximations.

EXAMPLE 3   Approximate each of the following conversions. Then compare your results to the values obtained by using Table 9.10.

(a) Convert 60 mL to fluid ounces.
Since 30 mL $\cong$ 1 fl oz,
then 60 mL $\cong$ 2 fl oz
Use Table 9.10 to find the exact
answer.

$$60 \text{ mL} = 60 \text{ mL} \times \frac{0.03381 \text{ fl oz}}{1 \text{ mL}}$$

$$= 2.0286 \text{ fl oz}$$

2.0286 fl oz $\cong$ 2.029 fl oz correct to four
significant digits. Our approximation
is off by about 0.03 fl oz.

$\left(\text{Think, } \frac{30 \text{ mL}}{1 \text{ fl oz}} = \frac{60 \text{ mL}}{?}\right)$

Multiply 60 mL by the unit fraction
$\frac{0.03381 \text{ fl oz}}{1 \text{ mL}}$ so that mL cancels.

2.0286 fl oz $-$ 2 fl oz $\cong$ 0.03 fl oz

(b) Convert $2\frac{1}{2}$ gal to liters.

$$2\frac{1}{2} \text{ gal} = \frac{5}{2} \text{ gal} \times \frac{4 \text{ qt}}{1 \text{ gal}}$$

$$= \frac{5}{2} \times 4 \text{ qt} = 10 \text{ qt}$$

Thus, $2\frac{1}{2}$ gal $= 10$ qt $\cong 10$ L.
Use Table 9.10 to find the exact
answer.

$$2\frac{1}{2} \text{ gal} = \frac{5}{2} \text{ gal} \times \frac{3.785 \text{ L}}{1 \text{ gal}}$$

$$= 9.4625 \text{ L}$$

9.4625 L $\cong$ 9.463 L correct to four
significant digits. Our approximation
was off by about 0.5 L.

Since 1 gal $= 4$ qt and 1 qt $\cong 1$ L, we will
first convert $2\frac{1}{2}$ gal to qt.

Multiply $2\frac{1}{2}$ gal by the unit fraction $\frac{3.785 \text{ L}}{1 \text{ gal}}$
so that gal cancels.

10 L $-$ 9.463 L $\cong$ 0.5 L   ▲

EXAMPLE 4   A pitcher of iced tea that holds $3\frac{1}{2}$ L is divided evenly among 14 people. How much iced
tea does each person receive?

Solution $3\frac{1}{2}$ L $= 3.5$ L $\times \dfrac{1,000 \text{ mL}}{1 \text{ L}} = 3,500$ mL

3,500 mL $\div$ 14 people $= 250$ mL/person.
Each person receives 250 mL of iced tea.

Since a serving of iced tea is measured in
mL, first convert $3\frac{1}{2}$ L to mL and then
divide by 14.

▲

Quick Quiz

Perform each of the following conversions.

1. 14 pt $= \underline{\hphantom{XX}}$ gal

2. 50 mL $= \underline{\hphantom{XX}}$ L

3. 8,536 L $= \underline{\hphantom{XX}}$ kL

4. 2 tsp $\cong \underline{\hphantom{XX}}$ mL

Answers

1. $1\dfrac{3}{4}$

2. 0.05

3. 8.536

4. 10

9.3 Exercises

Indicate which of the following statements are true and which are false. For those that are false, change the italicized or underlined expression to make the statement true.

1. A liter is a little *less than* a quart.

2. A teaspoon is a unit of capacity in the *metric* system.

3. A hectoliter is equivalent to $\frac{1}{100}$ liter.

4. One hundred milliliters is *equivalent to* $\frac{1}{10}$ liter.

Convert each of the following English units as instructed.

5. 36 gal = ____ qt

6. 20 pt = ____ c

7. 5 pt = ____ qt

8. 18 qt = ____ gal

9. 28 fl oz = ____ pt

10. $3\frac{1}{2}$ pt = ____ fl oz

11. 5.25 gal = ____ qt

12. 9.5 qt = ____ pt

13. 4 tbsp = ____ tsp

14. 12 tsp = ____ tbsp

15. $5\frac{1}{2}$ qt = ____ pt

16. $8\frac{3}{4}$ gal = ____ qt

17. $4\frac{5}{8}$ pt = ____ fl oz

18. $\frac{3}{4}$ qt = ____ fl oz

19. 1 pt 10 fl oz = ____ pt

20. 9 gal 2 qt = ____ qt

21. 9 pt = ____ gal

22. 20 c = ____ qt

23. 48 c = ____ gal

24. 80 fl oz = ____ gal

Convert each of the following metric units as instructed.

25. 85 L = ____ hL

26. 250 L = ____ kL

27. 62 mL = ____ L

28. 0.9 L = ____ mL

29. 57.6 kL = ____ L

30. 325.8 mL = ____ L

31. 8,800 mL = ____ L

32. 4.37 hL = ____ L

33. 0.897 L = ____ mL

34. 74.8 kL = ____ L

35. $7\frac{1}{2}$ kL = ____ L

36. $\frac{3}{8}$ L = ____ mL

37. $5\frac{3}{4}$ L = ____ mL

38. $\frac{3}{5}$ kL = ____ L

39. 362 L = ____ hL

40. 2.81 L = ____ kL

41. 0.069 L = ____ mL

42. 0.052 hL = ____ mL

43. 92,000 mL = ____ hL

44. 0.008 kL = ____ mL

Indicate which measurement most closely approximates each of the following capacities.

45. A six-pack of beer: 2,100 mL   2,100 L   21 L

46. Capacity of a swimming pool: 75 kL   750 L   7,500 L

47. One gallon of paint: 3.8 L   38 L   380 mL

48. A dosage of cough syrup: 50 mL   5 mL   500 mL

49. Capacity of a gas tank: 20 L   70 L   15 hL

50. A cup of coffee: 2 L   600 mL   250 mL

51. Refrigerator capacity: 40 L   0.4 kL   400 mL

52. A bottle of wine: 1.5 L   15 mL   15 L

Approximate each of the following conversions. Then compare your results to the values obtained by using Table 9.10.

53. 9 qt = ____ L

54. 12 fl oz = ____ mL

55. 25 mL = ____ tsp

56. 2 tbsp = ____ mL

57. 900 mL = ____ fl oz

58. 8 tsp = ____ mL

59. 300 mL = ____ tbsp

60. 7 gal = ____ L

61. 3 L = ____ fl oz

62. 150 fl oz = ____ L

Solve each of the following word problems.

63. A carton of cream contains $\frac{1}{2}$ pt. A pitcher of milk contains 250 mL. Is there more milk or cream?

64. A 32-fl oz bottle of dishwashing detergent sells for $0.69. Would a 1-L bottle cost more or less?

65. A bottle of cough syrup contains 118 mL. If the recommended dosage is 1 tsp, approximately how many doses are contained in this bottle?

66. A carafe containing 1 L of wine was shared equally among five people. How many milliliters of wine did each person get?

67. A half gallon of ice cream is equivalent to how many liters?

68. A cookie recipe calls for $\frac{3}{4}$ tsp of ginger. This amount is equivalent to how many milliliters?

69. If gas sells for $1.89 per gallon, what is the cost of 1 L of gas? Round off your answer to the nearest cent.

70. A can of spray paint that contains 335 mL costs $5.95. Another can that contains 250 mL costs $4.35. Which size is the better buy?

9.4

Temperature

Temperature indicates the degree to which something is either hot or cold. Scales that use the freezing and boiling points of water as reference points have been devised to measure differences in temperature.

The temperature scale used in the English system is the **Fahrenheit scale**, named after Gabriel Fahrenheit, the inventor of the mercury thermometer. On this scale, the freezing point of water is 32° F (read as "32 degrees Fahrenheit"), and the boiling point of water is 212° F.

The temperature scale used in the metric system is the **Celsius** (or **centigrade**) **scale**, which was devised by Anders Celsius. On this scale, the freezing point of water is 0° C (zero degrees Celsius), and the boiling point of water is 100° C.

A comparison of the Celsius and Fahrenheit temperature scales is shown in Figure 9.2 on page 356. Other useful temperatures to know in both Fahrenheit and Celsius are:

Room temperature:	68° F = 20° C
Normal body temperature:	98.6° F = 37° C
Oven temperature for roasting:	350° F = 175° C

By looking at Figure 9.2, we can see that a change in temperature of 1° C is roughly equivalent to a change of 2° F. For example, 10° C (which is 10° C above the freezing point of water) is roughly equivalent to 32° F (the freezing point of water) plus 20° F (a change of 10° C is approximately equal to 20° F since 1° C $\cong$ 2° F), which equals 52° F.

$$10° C \cong 32° F + 20° F$$
$$10° C \cong 52° F$$

Figure 9.2

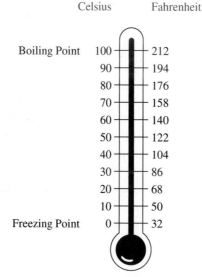

Celsius   Fahrenheit

Boiling Point   100 ─┼─ 212
                  90 ─┼─ 194
                  80 ─┼─ 176
                  70 ─┼─ 158
                  60 ─┼─ 140
                  50 ─┼─ 122
                  40 ─┼─ 104
                  30 ─┼─ 86
                  20 ─┼─ 68
                  10 ─┼─ 50
Freezing Point   0 ─┼─ 32

We also can see that $200°$ F (which is $12°$ F less than the boiling point of water) is roughly equivalent to $100°$ C (the boiling point of water) minus $6°$ C ($12°$ F $\cong 6°$ C since $2°$ F $\cong 1°$ C), which equals $94°$ C.

$$200°\,F \cong 100°\,C - 6°\,C$$
$$200°\,F \cong 94°\,C$$

To determine more exact values when converting from one temperature scale to another, we can use the procedures outlined here and on the next page.

**To Convert from
Celsius to Fahrenheit**

1. Multiply degrees Celsius by $\frac{9}{5}$.
2. Add 32.
   This can be summarized by the formula

   $$F = \frac{9}{5}C + 32$$

   where $F = °F$ and $C = °C$.   ▲

EXAMPLE 1   Convert $25°$ C to $°$ F.

Solution $F = \dfrac{9}{5}C + 32$ Use the formula for converting from Celsius to Fahrenheit.

$F = \dfrac{9}{5}(25) + 32$ Substitute 25 for C.

$F = \dfrac{9}{\overset{}{\underset{1}{5}}}(\overset{5}{25}) + 32$ Perform multiplication before addition. Multiply 25 by $\frac{9}{5}$.

$F = 45 + 32$ Add 32 and 45.

$F = 77$

Thus, $25°$ C $= 77°$ F.   ▲

To Convert from Fahrenheit to Celsius	1. Subtract 32 from degrees Fahrenheit.
	2. Multiply the result by $\frac{5}{9}$. This can be summarized by the formula

$$C = \frac{5}{9}(F - 32)$$

where $C = °C$ and $F = °F$. ▲

EXAMPLE 2 Convert 59° F to ° C.

Solution

$C = \dfrac{5}{9}(F - 32)$ Use the formula for converting from Fahrenheit to Celsius.

$C = \dfrac{5}{9}(59 - 32)$ Substitute 59 for F.

$C = \dfrac{5}{9}(27)$ Do subtraction in ( ) first.

$C = \dfrac{5}{9}(\overset{3}{\cancel{27}})$ Multiply 27 by $\frac{5}{9}$.

$C = 15$
Thus, $59° F = 15° C$.   ▲

EXAMPLE 3 Approximate each of the following conversions. Then determine the exact answers.

(a) Convert 0° F to ° C.

$0° F \cong 0° C - 16° C$ 0° F is 32° F below the freezing point of water. Since $1° C \cong 2° F$, subtract 16° C from 0° C (the freezing point of water).
$0° F \cong -16° C$

Determine the exact answer.

$C = \dfrac{5}{9}(F - 32)$ Use the formula.

$C = \dfrac{5}{9}(0 - 32)$

$C = \dfrac{5}{9}(-32)$

$C = -\dfrac{160}{9} = -17\dfrac{7}{9}$

Thus, $0° F = -17\dfrac{7}{9}° C$

Our approximation is off by $1\dfrac{7}{9}° C$. $-17\dfrac{7}{9}° C - (-16° C) = -1\dfrac{7}{9}° C$

(b) Convert $150°$ C to $°$ F.

$150°$ C $\cong 212°$ F $+ 100°$ F

$150°$ C $\cong 312°$ F

150° C is 50° C above the boiling point of water. Since 1° C $\cong$ 2° F, add 100° F to 212° F (the boiling point of water).

Determine the exact answer.

$$F = \frac{9}{5}C + 32$$

Use the formula.

$$F = \frac{9}{\overset{}{5}}(\overset{30}{\cancel{150}}) + 32$$

$F = 270 + 32$

$F = 302$

Thus, $150°$ C $= 302°$ F.

Our approximation is off by $10°$ F.

$312°$ F $- 302°$ F $= 10°$ F

Quick Quiz

Perform each of the following conversions.

1. $85°$ C $= \underline{\quad}°$ F
2. $5°$ F $= \underline{\quad}°$ C
3. $-10°$ C $= \underline{\quad}°$ F

Answers

1. 185
2. -15
3. 14

9.4 Exercises

Indicate which of the following statements are true and which are false. For those that are false, change the italicized or underlined expression to make the statement true.

1. The freezing point of water is 0° F.

2. The *metric* system uses the Fahrenheit scale.

3. A change of temperature of 2° F is roughly equivalent to a change of 1° C.

4. 120° F is *warmer than* 120° C.

Indicate which measurement most closely approximates each of the following temperatures.

5. Low fever: 99° C   38° C   41° C

6. Ice skating rink: $-1°$ C   32° C   20° C

7. Baking temperature for cookies: 120° C   190° C   375° C

8. Broiling temperature: 190° C   550° C   290° C

9. Hot summer day: 30° C   85° C   20° C

10. Hot coffee: 200° C   40° C   90° C

Convert each of the following temperatures as instructed. Round off to the nearest degree.

11. 68° F $= \underline{\quad}°$ C   12. 95° F $= \underline{\quad}°$ C   13. 23° F $= \underline{\quad}°$ C   14. 41° F $= \underline{\quad}°$ C

15. $35°C =$ ___°F   16. $45°C =$ ___°F   17. $115°C =$ ___°F   18. $95°C =$ ___°F
19. $-4°F =$ ___°C   20. $-22°F =$ ___°C   21. $-5°C =$ ___°F   22. $-20°C =$ ___°F
23. $67°C =$ ___°F   24. $82°C =$ ___°F   25. $24°C =$ ___°F   26. $13°C =$ ___°F
27. $52°F =$ ___°C   28. $89°F =$ ___°C   29. $130°F =$ ___°C   30. $106°C =$ ___°F
31. $-2°C =$ ___°F   32. $-15°F =$ ___°C

Approximate each of the following conversions. Then determine the exact answers.

33. $77°F =$ ___°C   34. $113°F =$ ___°C   35. $65°C =$ ___°F   36. $30°C =$ ___°F
37. $59°F =$ ___°C   38. $239°F =$ ___°C   39. $90°C =$ ___°F   40. $75°C =$ ___°F
41. $-40°F =$ ___°C   42. $-15°C =$ ___°F   43. $28°F =$ ___°C   44. $91°F =$ ___°C
45. $141°C =$ ___°F   46. $-8°C =$ ___°F

Solve each of the following word problems.

47. Following an African safari, an engineer came down with malaria and had a fever of $104°F$. What was her temperature in degrees Celsius?

48. On a cold winter day, the temperature was $-5°C$. What was the temperature in degrees Fahrenheit?

49. Which oven temperature is hotter, $200°C$ or $350°F$?

50. Which freezer temperature is colder, $-10°C$ or $10°F$?

9.5
Summary and Review

Key Terms

(9.1) To **measure** something is to find a number that characterizes one of its properties.

A **unit of measurement** indicates the size of a known quantity.

The **English system** is the system of measurement used in the United States.

The **metric system** is the system of measurement used throughout most of the world.

The **meter** is the standard unit of length in the metric system.

(9.2) The **weight** of an object is a measure of the force of the earth's gravitational pull acting upon it.

The **gram** is the standard unit of weight in the metric system.

(9.3) The **liter** is the standard unit of capacity (volume) in the metric system.

(9.4) **Temperature** indicates the degree to which something is either hot or cold.

The **Fahrenheit** scale is used to measure temperature in the English system.

The **Celsius** (or **centigrade**) scale is used to measure temperature in the metric system.

milli—thousandth	deka—ten
centi—hundredth	hecto—hundred
deci—tenth	kilo—thousand

Calculations

(9.1) **To convert from one unit of measurement to another**, multiply by the appropriate unit fraction(s).

(9.4) **To convert from Celsius to Fahrenheit**, multiply degrees Celsius by $\frac{9}{5}$ and then add 32.

$$F = \frac{9}{5}C + 32$$

To convert from Fahrenheit to Celsius, subtract 32 from degrees Fahrenheit and then multiply the result by $\frac{5}{9}$.

$$C = \frac{5}{9}(F - 32)$$

Chapter 9
Review Exercises

Indicate which of the following statements are true and which are false. For those that are false, change the italicized or underlined expression to make the statement true.

1. The prefix centi means *hundred*.

2. One kilogram is a little *less than* 2 pounds.

3. Four liters is a little *more than* 1 gallon.

4. The boiling point of water is <u>100° F</u>.

(9.1–9.4) **Indicate which measurement most closely approximates each of the following lengths, weights, capacities, or temperatures.**

5. Length of a tennis court: 2.38 m   23.8 m   238 m

6. Length of skis: 175 cm   17.5 m   175 mm

7. Weight of a tube of toothpaste: 2.32 kg   232 g   2.32 g

8. Weight of a vitamin tablet: 250 mg   25 g   2.5 mg

9. Capacity of a tea kettle: 5 L   2 L   200 mL

10. Capacity of a trash can: 14 L   40 L   114 L

11. Length of a birthday candle: 0.6 m   600 mm   6 cm

12. Volume of a small can of frozen orange juice: 177 mL   17 mL   1.7 L

13. Normal body temperature: 37° C   40° C   98.6° C

14. Temperature on a cold winter day: −2° C   20° C   32° C

(9.1–9.4) **Convert each of the following units of measurement as instructed.**

15. 8 in. = ____ ft

16. $2\frac{1}{2}$ mi = ____ ft

17. 36 oz = ____ lb

18. 0.4 lb = ____ oz

19. 6 pt = ____ qt

20. 2 gal = ____ pt

21. 300 m = ____ km

22. 72 mm = ____ cm

23. 8,200 g = ____ kg

24. 6,400 cg = ____ g

25. 53 L = ____ mL

26. 2.7 hL = ____ L

27. 22 yd = ____ ft
28. 40 in. = ____ yd
29. $5\frac{1}{4}$ T = ____ lb
30. 7,500 lb = ____ T
31. 5 tsp = ____ tbsp
32. 60 fl oz = ____ pt
33. 9.2 dam = ____ dm
34. 0.59 hm = ____ mm
35. 0.23 g = ____ mg
36. 75 mg = ____ cg
37. 0.86 kL = ____ L
38. 450 L = ____ kL
39. 50° F = ____ °C
40. 176° F = ____ °C
41. 85° C = ____ °F
42. − 10° C = ____ °F
43. $7\frac{1}{2}$ m = ____ dm
44. $\frac{3}{4}$ km = ____ hm
45. 0.018 kg = ____ mg
46. 5,400 cg = ____ kg
47. 63,000 mL = ____ kL
48. 0.08 hL = ____ mL
49. 21° F = ____ °C
50. 102° C = ____ °F

(9.1–9.4)   Approximate each of the following conversions. Then use tables to determine more precise answers.

51. 150 cm = ____ in.
52. 500 mi = ____ km
53. 44 lb = ____ kg
54. 7 oz = ____ g
55. 3 gal = ____ L
56. 75 mL = ____ tsp
57. 72° F = ____ °C
58. 5° C = ____ °F

(9.1–9.4)   Solve each of the following word problems.

59. A delicatessen uses 250 g of roast beef for every roast beef sandwich it makes. About how many sandwiches can be made from an 8.8-lb piece of roast beef?

60. A box that contains 24 tea bags weighs 37 g. If the box weighs 7 g, what is the weight of a single tea bag?

61. The distance from Tokyo to Berlin is 5,540 mi. What is this distance in kilometers?

62. If an extension cord is 2.4 m long, will it reach from an outlet to a lamp cord that is $8\frac{1}{2}$ ft away?

63. A 237-mL bottle of suntan lotion costs $2.28. A 120-mL bottle costs $1.19. Which size bottle is the better buy?

64. A 2-lb bag of sugar costs $1.89. What is the cost of 1 kg of sugar? Round off your answer to the nearest cent.

65. A nurse found a patient's temperature to be 39° C. What is the equivalent temperature in °F?

66. How much will it cost to fill a 70-L gas tank if gas costs $1.79 per gallon? Round off your answer to the nearest cent.

Chapter 9
Explain It in Words

67. Explain how to convert a measurement from inches to feet. Use an example.

68. Explain how to convert a measurement from grams to kilograms. Use an example.

69. Explain how to convert a measurement from gallons to quarts. Use an example.

70. Explain how to convert a temperature from Fahrenheit to Celsius. Use an example.

71. Would you measure the capacity of a can of soda in milliliters or in hectoliters? Explain why.

72. Would you measure the length of a basketball court in centimeters or in meters? Explain why.

Chapter 9
Chapter Test

[9.1–9.4] **Convert each of the following English units of measurement.**

1. 80 in. = ____ ft
2. 6 lb = ____ oz
3. $2\frac{1}{2}$ gal = ____ qt
4. 0.5 mi = ____ ft
5. 3 pt = ____ fl oz

[9.1–9.3] **Convert each of the following metric units of measurement.**

6. 600 dm = ____ m
7. 70 L = ____ ml
8. 400 g = ____ kg
9. 320 cm = ____ m
10. 5.2 kg = ____ cg

[9.1–9.3] **Approximate each of the following conversions between metric and English units.**

11. 4 kg ≅ ____ lb
12. 2 tbsp ≅ ____ ml
13. 45 cm ≅ ____ in.
14. 3 oz ≅ ____ g
15. 240 km ≅ ____ mi

[9.4] **Convert each of the following temperatures as indicated.**

16. 41°F = ____°C
17. 30°C = ____°F
18. −5°C = ____°F

[9.1–9.2] **Solve each of the following word problems.**

19. The distance between Buffalo and Memphis is 800 mi and the distance between Buffalo and Omaha is 1,420 km. Which city is farther from Buffalo, Memphis or Omaha?

20. If one vitamin weighs 750 mg, how many grams would 100 vitamins weigh?

Chapter *10*

Introduction to Geometry

Geometric Figures

Geometry is the branch of mathematics that deals with measurement. Geometry describes the properties of objects, such as **points**, **lines**, and **planes**. A **point** indicates a location or position and is represented by a dot. A **line** is a set of points extending indefinitely in opposite directions. A **plane** is a set of points that form a flat surface extending indefinitely in all directions. Figure 10.1 illustrates examples of a point, a line, and a plane.

Figure 10.1

We will now learn how to identify geometric figures known as **polygons**.

▶ **DEFINITION** A *polygon* is a figure that lies in a plane and has many sides. A *regular polygon* is a polygon in which all sides are equal.

The names of certain polygons indicate the number of sides that form the figure. For example,

*Tri*angle:	*three*-sided figure	*Hexa*gon:	*six*-sided figure
*Quadri*lateral:	*four*-sided figure	*Octa*gon:	*eight*-sided figure
*Penta*gon:	*five*-sided figure		

Figure 10.2 shows an example of each of these polygons.

Figure 10.2

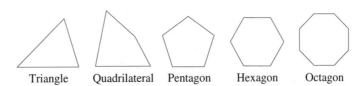

The shape of any polygon is determined by the length of each side and the size of each **angle**.

> ▶ **DEFINITION** An *angle* is formed whenever two lines meet. The two lines are called the *sides* of the angle, and the point at which they meet is called the *vertex* of the angle.

A typical angle is illustrated in Figure 10.3. Notice that each corner of a polygon is the vertex of an angle.

Figure 10.3

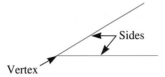

The size of an angle is measured in **degrees** (°), or **degrees of arc**. If you were to stand in one place, but pivot on your feet to turn one complete revolution, you would have turned 360°. One-half of a revolution is equivalent to 180°, and one-quarter of a revolution is equivalent to 90°. Notice that a straight line represents a 180° angle, and the intersection of a horizontal and a vertical line represents a 90° angle, as shown in Figure 10.4.

Figure 10.4

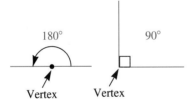

A 90° angle is commonly called a **right angle**. Angles that are greater than 0° and less than 90° are called **acute**, and angles that are greater than 90° and less than 180° are called **obtuse**. These are illustrated in Figure 10.5. Notice that a right angle is indicated by forming a small box at the vertex. Lines that meet in a right angle are said to be **perpendicular**.

Figure 10.5

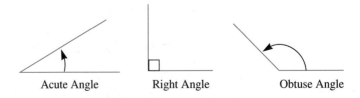

Now that we know the different types of angles, we can classify the different kinds of triangles as follows:

An **equilateral** triangle has three equal sides and three equal angles.
An **isosceles** triangle has two equal sides and two equal angles.
A **right** triangle has a right angle.

These triangles are illustrated in Figure 10.6.

Figure 10.6

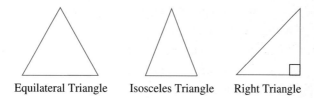

Equilateral Triangle   Isosceles Triangle   Right Triangle

Many quadrilaterals contain lines that are **parallel** and **perpendicular**.

▶ **DEFINITION** *Parallel lines* **are lines in the same plane that never meet.**

▶ **DEFINITION** *Perpendicular lines* **are lines that meet in a right angle.**

Figure 10.7 shows a pair of parallel lines and a pair of perpendicular lines. Notice that parallel lines are always the same distance apart, regardless of how far they are extended.

Figure 10.7

Parallel Lines   Perpendicular Lines

Now that we understand parallel lines, we can classify the different kinds of quadrilaterals as follows:

A **trapezoid** is a quadrilateral that has two parallel sides.
A **parallelogram** is a quadrilateral whose opposite sides are equal and parallel.
A **rectangle** is a parallelogram that has four right angles.
A **square** is a rectangle that has four equal sides.

These quadrilaterals are illustrated in Figure 10.8.

Figure 10.8

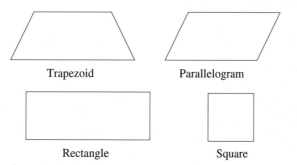

Trapezoid   Parallelogram

Rectangle   Square

10.1 Exercises

Indicate whether each of the following statements are true or false. For those that are false, change the italicized or underlined expression to make the statement true.

1. Lines that meet in a right angle are said to be *parallel*.
2. The sides of a *regular* polygon are all equal.
3. An isosceles triangle has *three* equal sides.
4. A rectangle has four equal *sides*.
5. Perpendicular lines meet in a <u>45°</u> angle.
6. A straight line represents a <u>180°</u> angle.
7. A pentagon has *six* sides.
8. A *trapezoid* is a quadrilateral whose opposite sides are parallel.
9. A right triangle has a <u>45°</u> angle.
10. The point where the two sides of an angle meet is called the *apex*.
11. An obtuse angle is greater than <u>90°</u>.
12. One complete revolution is a <u>180°</u> turn.

Choose the measurement that correctly indicates the size of each of the following angles.

13.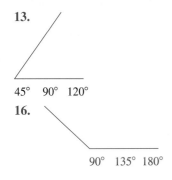
45°   90°   120°

14.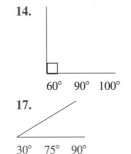
60°   90°   100°

15.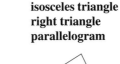
100°   120°   180°

16.
90°   135°   180°

17.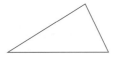
30°   75°   90°

18.
15°   60°   90°

Choose the word from the following list that most completely describes each of the figures in Exercises 19–26.

triangle	hexagon	isosceles triangle	trapezoid
quadrilateral	octagon	right triangle	rectangle
pentagon	equilateral triangle	parallelogram	square

19.

20.

21.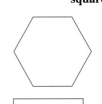

22.

23.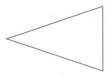

24.

25.

26.

10.2

Perimeter   We will now discuss some characteristics of geometric figures that have widespread applications. The first of these is **perimeter**.

▶ **DEFINITION** *Perimeter* is the distance around a geometric figure.

To find the perimeter of any figure, we find the sum of the lengths of the sides. Since length is measured in linear, or one-dimensional, units, perimeter is also a one-dimensional quantity.

EXAMPLE 1   Find the perimeter of the accompanying figure.

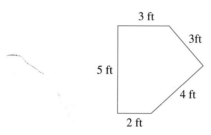

Solution $P = 3 + 3 + 4 + 2 + 5$ Find the sum of the sides.
$P = 17$ ft   Since all dimensions are given in feet, the perimeter is also in feet.   ▲

We will now establish some formulas for finding the perimeters of certain geometric figures. To find the perimeter of a rectangle whose length is indicated by the letter *l*, and whose width is indicated by the letter *w*, as shown in Figure 10.9, we can find the sum of the sides.

Figure 10.9

$$P = l + w + l + w$$
$$P = l + l + w + w$$
$$P = 2l + 2w$$

We have thus established the formula below.

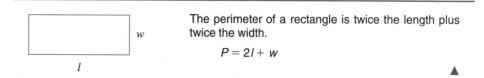

The perimeter of a rectangle is twice the length plus twice the width.

$$P = 2l + w$$

▲

EXAMPLE **2** Find the perimeter of a rectangle whose length is 7 m and whose width is 3 m.

3 m

7 m

Solution $P = 2l + 2w$ Use the formula for finding the perimeter
                                              of a rectangle.

$P = 2(7) + 2(3)$ Substitute 7 for l and 3 for w.

$P = 14 + 6$ Do multiplication before addition.

$P = 20$ m   Since l and w are given in meters, so is the
                                    perimeter. ▲

EXAMPLE **3** Find the perimeter of a regular pentagon whose sides measure 7 cm each.

7 cm

Solution $P = 7 + 7 + 7 + 7 + 7$ Since a regular pentagon has five equal
                                              sides, each 7 cm long, the perimeter of this
$P = 5 \cdot 7$ figure is equal to the sum of five 7's, which
$P = 35$ cm   simplifies to the product of $5 \cdot 7$. ▲

The formula for finding the perimeter of *any regular polygon* with n sides, each of
length s, is presented in the following.

The perimeter of any regular polygon is the product of the number of sides, *n*, times
the length of each side, *s*.

$P = ns$ ▲

For example,

$P = 3s$ gives the perimeter of an equilateral triangle.

$P = 4s$ gives the perimeter of a square.

$P = 5s$ gives the perimeter of a regular pentagon.

$P = 6s$ gives the perimeter of a regular hexagon.

$P = 8s$ gives the perimeter of a regular octagon.

EXAMPLE 4 How many yards of trimming are needed to decorate the border of a 54-inch square tablecloth?

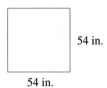

54 in.

54 in.

Solution Find the perimeter of the tablecloth.

$P = 4s$ Use the formula for finding the perimeter of a square.

$P = 4(54)$ Substitute 54 for s.
$P = 216$ in.
Find the number of yards of trimming needed.

$$216 \text{ in.} = 216 \text{ in.} \times \frac{1 \text{ yd}}{36 \text{ in.}}$$ We must convert 216 in. to yd.

$$= \frac{216}{36} \text{ yd} = 6 \text{ yd}$$ ▲

If all the dimensions of a geometric figure are not given in the same units of measurement, we must first convert them to the same unit of measurement before finding the perimeter of the figure.

EXAMPLE 5 The backyard of a suburban home is 40 ft long and 12 yd wide. How much fencing is needed to completely enclose the yard?

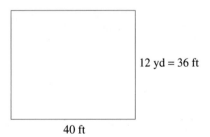

12 yd = 36 ft

40 ft

Solution Find the perimeter.

$$12 \text{ yd} = 12 \text{ yd} \times \frac{3 \text{ ft}}{1 \text{ yd}} = 36 \text{ ft}$$ First convert 12 yd to ft.

$P = 2l + 2w$ Use the formula for finding the perimeter of a rectangle.

$P = 2(40) + 2(36)$ Substitute $l = 40$ and $w = 36$.
$P = 80 + 72$ Do multiplication before addition.
$P = 152$ ft   Since both length and width are in ft, so is the perimeter.   ▲

Quick Quiz | **Find the perimeters of each of the following figures.**

		Answers
1.	Rectangle of length 5 in. and width 8 in.	**1.** 26 in.
2.	Equilateral triangle whose sides measure 9 m.	**2.** 27 m
3.	Regular hexagon whose sides measure 30 cm.	**3.** 180 cm

10.2 Exercises

State which of the following statements are true and which are false. For those that are false, change the italicized expression to make the statement true.

1. The distance around a figure is called its *perimeter.*
2. The perimeter of a rectangle is always *less than* its length.
3. The perimeter of any object is the *product* of the lengths of all of the sides.
4. The perimeter of a regular *octagon* is six times the length of a side.

Find the perimeters of each of the following polygons. All figures are drawn to scale.

5.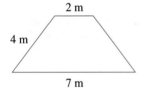

6.

7.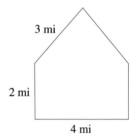

8.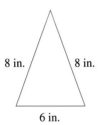

9.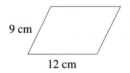

10.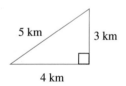

11.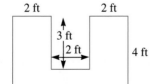

12.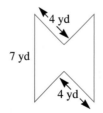

Find the perimeter of each of the following figures.

13. Rectangle of length 9 in. and width 5 in.
14. Rectangle of length 5 yd and width 2 yd.
15. Equilateral triangle whose sides measure 8 km.
16. Equilateral triangle whose sides measure 15 hm.

17. Square whose sides measure 1.2 m.

18. Square whose sides measure 3.1 cm.

19. Rectangle of length $7\frac{1}{2}$ yd and width $1\frac{1}{4}$ yd.

20. Rectangle of length $3\frac{5}{8}$ ft and width $2\frac{1}{8}$ ft.

21. Regular hexagon whose sides measure 3 cm.

22. Regular hexagon whose sides measure 20 ft.

23. Square whose sides measure 15 mm.

24. Square whose sides measure 42 m.

25. Regular pentagon whose sides measure 9 dm.

26. Regular octagon whose sides measure 20 dam.

27. Rectangle of length 2 ft and width 9 in.

28. Rectangle of length 1 yd and width 2 ft.

Solve each of the following word problems.

29. A garden measures 10 ft long and 8 ft wide. How many feet of fencing are needed to enclose the garden?

30. A farmer wishes to fence in a corral that is 80 ft long and 60 ft wide. How much fencing does he need?

31. A television screen that is 42 cm long and 36 cm wide is surrounded by a chrome border. How many centimeters of chrome were used to make the border?

32. A tablecloth that is 72 in. long and 60 in. wide has a decorated border. How many feet of trimming were used to make it?

33. How many feet of caulking are needed to insulate a rectangular window of length 4 ft and width $2\frac{1}{2}$ ft?

34. A shopkeeper wishes to put a string of colored lights along the border of a sign that measures $1\frac{1}{2}$ m long and $\frac{3}{4}$ m wide. How long must the light string be?

35. The front of a house is decorated with a wooden trim in the shape of an equilateral triangle. If one side is 7.28 m long, how many meters of wood were needed to construct the triangle?

36. How many meters of baseboard are needed for a room that is 4.15 m long and 3.45 m wide and has one door 1.2 m wide?

10.3

Area

Another characteristic of geometric figures that has practical applications is **area**.

▶ **DEFINITION** *Area* is the measurement of the surface of a figure. ◀

To measure area we use square units of measurement, because area is a two-dimensional quantity. For example, a square inch (abbreviated as sq in. or in.^{2}) represents the area of a square whose sides are each 1 inch in length. A square centimeter (abbreviated as sq cm or cm^2) represents the area of a square whose sides are each 1 centimeter in length. These are both illustrated in Figure 10.10.

Figure 10.10

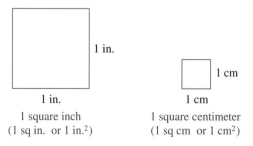

1 in.

1 in.

1 square inch
(1 sq in. or 1 in.^{2})

1 cm

1 cm

1 square centimeter
(1 sq cm or 1 cm^2)

Let us now consider how to find the area of the rectangle shown in Figure 10.11. Here we have a rectangle 7 inches long and 3 inches wide. The problem of finding the area of this rectangle is the same as determining how many square blocks, each measuring 1 inch along an edge, must be used to completely cover the region enclosed by the rectangle. The area of each of these blocks is 1 square inch. Looking at Figure 10.11, we see that 21 of these blocks are needed. Thus, the area of this figure is 21 square inches. Notice that the same result is obtained by multiplying the length of the rectangle by its width.

$$A = (7 \text{ in.})(3 \text{ in.})$$
$$A = 21 \text{ in.}^2$$

Figure 10.11

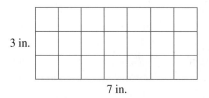

3 in.

7 in.

We therefore have established the formula, stated in the following, for finding the area of a rectangle.

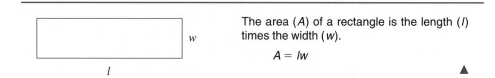

w

The area (A) of a rectangle is the length (l) times the width (w).

$$A = lw$$

l ▲

EXAMPLE 1 Find the area of a rectangle that is 8 m long and 3.2 m wide.

3.2 m

8 m

Solution $A = lw$ 　　　　　　　　 Use the formula for finding the area of a rectangle.

$A = (8)(3.2)$ 　　　　　　　 Substitute $l = 8$, $w = 3.2$

$A = 25.6 \text{ m}^2$ 　　　　　　 Since the length and width are in meters, the area is in square meters. ▲

EXAMPLE 2 Find the area of a rectangle that is $1\frac{1}{2}$ ft long and 10 in. wide.

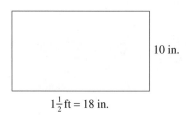

10 in.

$1\frac{1}{2}$ ft = 18 in.

Solution To find the area, we must express the length and width in terms of the same unit of measurement.

$$1\frac{1}{2} \text{ ft} = \frac{3}{2} \text{ ft} \times \frac{12 \text{ in.}}{1 \text{ ft}} = 18 \text{ in.}$$ Convert $1\frac{1}{2}$ ft to in.

$A = lw$ Use the formula for finding the area of a rectangle.

$A = (18)(10)$ Substitute $l = 18$ and $w = 10$.

$A = 180$ sq in. Since length and width are given in inches, the resulting area is in square inches. ▲

To illustrate how equivalent units of area are derived, let us consider a square whose sides are each 1 yard long, as shown in Figure 10.12. By definition, this square has an area of 1 square yard. Since 1 yard = 3 feet, we can also determine this area in feet by multiplying the length, which is 3 feet, by the width, which is also 3 feet.

Figure 10.12

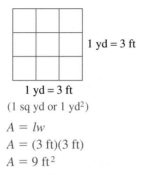

1 yd = 3 ft

1 yd = 3 ft
(1 sq yd or 1 yd²)

$A = lw$
$A = (3 \text{ ft})(3 \text{ ft})$
$A = 9 \text{ ft}^2$

Therefore, an area of 1 square yard is equivalent to 9 square feet.

Notice that since the length and width of a square are the same, to determine its area, we can simply square the measurement of one of its sides (see below). Using this procedure, we can also show that 1 square foot = 144 square inches.

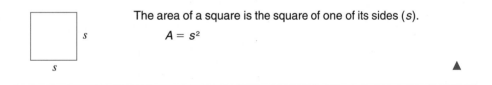

The area of a square is the square of one of its sides (s).

$A = s^2$

▲

Table 10.1 summarizes equivalent units of area in the English system. Square inches, square feet, and square yards are used to measure relatively small areas. Acres and square miles are used to measure large areas of land.

TABLE 10.1
English Units of Area

1 ft²	=	144 in.²
1 yd²	=	9 ft²
1 acre	=	4,840 yd²
1 acre	=	43,560 ft²
1 mi²	=	640 acres

EXAMPLE 3 A biologist owns a piece of land that measures 330 ft × 330 ft (the × is read as "by"). How many acres does she own?

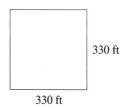

330 ft

330 ft

Solution Find this area first in terms of square feet.

$A = s^2$ 330 ft × 330 ft gives the dimensions of a square. Use the formula for finding the area of a square.

$A = (330)^2$ Substitute $s = 330$.

$A = 108,900$ sq ft   Since s is given in ft, the resulting area is in sq ft.

Now convert 108,900 sq ft to acres.

$$108,900 \text{ sq ft} = 108,900 \text{ sq ft} \times \frac{1 \text{ acre}}{43,560 \text{ sq ft}} = 2.5 \text{ acres}$$

The biologist owns 2.5 acres of land. ▲

In the metric system, each unit of area differs from the next larger unit by a factor of 100. For example, 1 square centimeter is equivalent to 100 square millimeters, as shown in Figure 10.13. The metric units that are used to measure large areas of land are the are (a), which is pronounced "air," and the hectare (ha). Table 10.2 summarizes equivalent metric units of area.

Figure 10.13

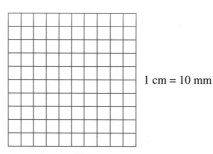

1 cm = 10 mm

1 cm = 10 mm
$1 \text{ cm}^2 = 100 \text{ mm}^2 = 1$ square centimeter

TABLE 10.2
Metric Units of Area

1 cm^2	=	100 mm^2
1 dm^2	=	100 cm^2
1 m^2	=	100 dm^2
1 a	=	100 m^2
1 ha	=	100 a

We will now determine formulas for finding areas of parallelograms, triangles, and trapezoids. To find the area of a parallelogram, we need to know two dimensions—the base (*b*), which is the measurement of one of its sides, and the height (*h*), which is the distance between the base and the side parallel to it.

Figure 10.14 shows a parallelogram with a base of 6 cm and a height of 3 cm. We can determine the area of this figure by cutting off the shaded piece on the left and attaching it to the right of this figure to form a rectangle, as shown in Figure 10.15.

Figure 10.14

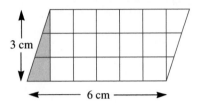

Figure 10.15

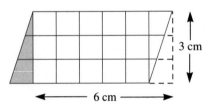

The area of the resulting parallelogram is simply the product of the length (6 cm) and the width (3 cm), which is 18 cm^2:

$$A = (6 \text{ cm})(3 \text{ cm})$$
$$A = 18 \text{ cm}^2$$

Notice that the same result is obtained by finding the product of the base and the height of the original parallelogram. The formula for determining the area of a parallelogram is given here.

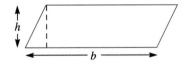

The area of a parallelogram is the base (*b*) times the height (*h*).

$$A = bh$$

EXAMPLE 4 Find the area of a parallelogram with a base of 70 m and a height of 60 m. Express your answer in ares.

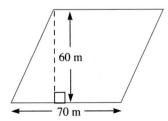

Solution   First find the area in terms of square meters.

$A = bh$ Use the formula for finding the area of a
                                   parallelogram.

$A = (70)(60)$ Substitute $b = 70$ m and $h = 60$ m.
$A = 4,200 \text{ m}^2$
Now convert 4,200 m² to ares.

$$4,200 \text{ m}^2 = 4,200 \text{ m}^2 \times \frac{1 \text{ a}}{100 \text{ m}^2} = 42 \text{ a}$$

The area of the parallelogram is 42 a.   ▲

EXAMPLE 5   The area of a parallelogram with an 80-in. base is 4,800 sq in. What is its height?

Solution $A = bh$ Use the formula for finding the area of a
                                            parallelogram.

$4,800 = 80h$ Substitute $A = 4,800$ and $b = 80$, and
$80h = 4,800$ solve for h.

$\dfrac{1}{80}(80h) = \dfrac{1}{80}(4,800)$

                                            Since the area is given in sq in. and the
$h = 60$ in.   base is in in., the height must also be in in.

The height of the parallelogram is 60 in.   ▲

To find the area of a triangle we also need to know the base (b), which is one of its sides, and the height (h), which is the perpendicular distance from the base to the vertex opposite the base. Figure 10.16 shows the base and height of three different triangles. Figure 10.17 shows that the area of each of these triangles is exactly equal to half of the area of a parallelogram with the same base b and height h. We have thus established the formula, stated on the next page, for calculating the area of a triangle.

Figure 10.16

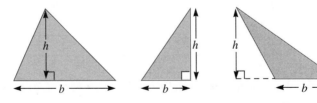

Figure 10.17

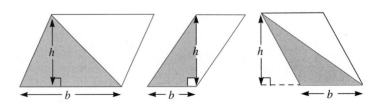

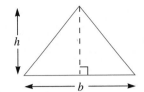

The area of a triangle is one-half the base (b) times the height (h).

$$A = \frac{1}{2}bh$$

▲

EXAMPLE 6   Find the area of the triangle.

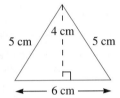

Solution   The dimensions needed are the base, which is 6 cm, and the height, which is 4 cm.

$A = \dfrac{1}{2}bh$ Use the formula for finding the area of a triangle.

$A = \dfrac{1}{2}(6)(4)$ Substitute $b = 6$ and $h = 4$.

$A = \dfrac{1}{2}(24)$

$A = 12 \ \text{cm}^2$ Since the base and height are given in cm, the area is in cm². ▲

To find the area of a trapezoid, we need to know three dimensions—the lengths of the two parallel sides, which are sometimes referred to as the lower base (b_1) and the upper base (b_2), and the height (h), which is the distance between the bases. The area of a trapezoid can be calculated by dividing it into two triangles, as shown in Figure 10.18, and calculating the area of each triangle. The area of a trapezoid is the sum of these two areas.

$$A = A_1 + A_2$$

$$A = \frac{1}{2}b_1h + \frac{1}{2}b_2h$$

$$A = \frac{1}{2}h(b_1 + b_2)$$

The last line in our calculation is the formula for finding the area of a trapezoid. To show that it is equivalent to the previous line, multiply it out, using the distributive property. The formula is summarized on the following page.

Figure 10.18

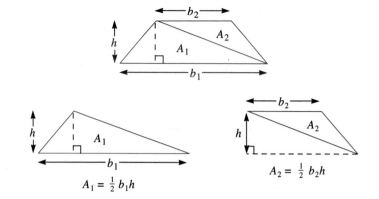

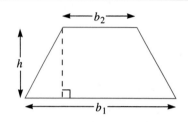

$$A_1 = \frac{1}{2} b_1 h$$

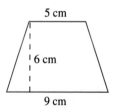

The area of a trapezoid is one-half the height (h) times the sum of the bases (b_1 and b_2).

$$A = \frac{1}{2}h(b_1 + b_2)$$

▲

EXAMPLE 7   Find the area of a trapezoid that has bases of 9 cm and 5 cm and a height of 6 cm.

5 cm

6 cm

9 cm

Solution $A = \dfrac{1}{2}h(b_1 + b_2)$ Use the formula for the area of a trapezoid.

$A = \dfrac{1}{2}(6)(9 + 5)$ Substitute $h = 6$, $b_1 = 9$, $b_2 = 5$

$A = \dfrac{1}{2}(6)(14)$

$A = \dfrac{1}{2}(84)$

$A = 42 \text{ cm}^2$ Since the dimensions are in cm, the area is in cm^2.   ▲

EXAMPLE 8   Find the area of a trapezoid that has bases of 8 yd and 2 yd and a height of 9 ft.

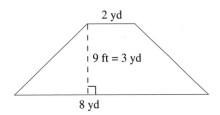

Solution   To find the area, all dimensions must be expressed in the same units.

$$9 \text{ ft} = 9 \, \cancel{ft} \times \frac{1 \text{ yd}}{3 \, \cancel{ft}} = 3 \text{ yd}$$ First convert 9 ft to yd.

$$A = \frac{1}{2}h(b_1 + b_2)$$ Use the formula for finding the area of a trapezoid.

$$A = \frac{1}{2}(3)(8 + 2)$$ Substitute $h = 3$, $b_1 = 8$, $b_2 = 2$.

$$A = \frac{1}{2}(3)(10)$$

$$A = \frac{1}{2}(30)$$

$$A = 15 \text{ yd}^2$$ Since the dimensions were given in yd, the resulting area is in yd². ▲

EXAMPLE 9   A remnant of fabric is shaped as illustrated on the left. What is its area?

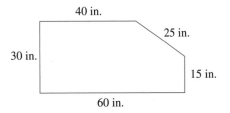 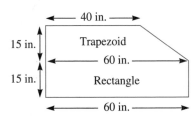

Solution   There are a number of different ways to solve this problem. One way is to view the figure as a trapezoid sitting on top of a rectangle. The dimensions needed to find the area of each are shown in the illustration on the right. Compare this to the original to see that the height and lower base of the trapezoid were obtained by knowing the dimensions of the rectangle. Now find the area of each piece.

Find the area of the rectangle.

$$A_R = lw$$ Use A_R to indicate the area of the rectangle.

$$A_R = (60)(15)$$ Substitute $l = 60$ and $w = 15$.
$$A_R = 900 \text{ in.}^2$$

Find the area of the trapezoid.

$$A_T = \frac{1}{2}h(b_1 + b_2)$$ Use A_T to indicate the area of the trapezoid.

$$A_T = \frac{1}{2}(15)(60 + 40)$$ Substitute $h = 15$, $b_1 = 60$, $b_2 = 40$.

$$A_T = \frac{1}{2}(15)(100)$$

$$A_T = \frac{1}{2}(1,500)$$

$$A_T = 750 \text{ in.}^2$$

To find the total area, add the two pieces together.
$$A = A_R + A_T$$
$$A = 900 + 750$$
$$A = 1,650 \text{ in.}^2$$
The area of the remnant is 1,650 square inches.   ▲

EXAMPLE 10 A living room is shaped like the figure illustrated on the left. How many square feet of carpeting are needed to cover the floor of this living room?

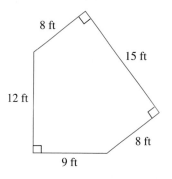 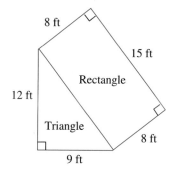

Solution The area of the living room is composed of a triangle and a rectangle, as shown in the figure on the right.

Find the area of the triangle.

$$A_T = \frac{1}{2}bh$$ Use A_T to represent the area of the triangle.

$$A_T = \frac{1}{2}(9)(12)$$ Substitute $b = 9$ ft and $h = 12$ ft.

$$A_T = \frac{1}{2}(108)$$

$$A_T = 54 \text{ ft}^2$$

Find the area of the rectangle.

$$A_R = lw$$

Use A_R to represent the area of the rectangle.

$$A_R = (15)(8)$$
$$A_R = 120 \text{ ft}^2$$

Substitute $l = 15$ ft and $w = 8$ ft.

To find the total area, add the two pieces together.
$$A = A_T + A_R$$
$$A = 54 + 120$$
$$A = 174 \text{ ft}^2$$
174 ft² of carpeting is needed.

▲

Quick Quiz

Find the area of each of the following polygons.   Answers

1. Rectangle of length 8 m and width 7 m.   **1.** 56 m²

2. Square whose sides are 1.3 mi long.   **2.** 1.69 mi²

3. Triangle with base $2\frac{1}{2}$ in. and height 3 in.   **3.** $3\frac{3}{4}$ in.²

4. Trapezoid with bases 5 cm and 7 cm long and a height of 4 cm.   **4.** 24 cm²

10.3 Exercises

Indicate which of the following statements are true and which are false. For those that are false, change the italicized or underlined expression to make the statement true.

1. Area is measured in *linear* units of measurement.

2. The area of a triangle is *one-third* of the area of a parallelogram with the same base and height.

3. An *are* is equivalent to 100 square meters.

4. One square yard is equivalent to <u>3</u> square feet.

Find the area of each of the following polygons.

5. Rectangle of length 12 ft and width 9 ft.

6. Rectangle of length 21 in. and width 16 in.

7. Rectangle of length 8.1 cm and width 7 cm.

8. Rectangle of length 5.6 m and width 3 m.

9. Rectangle of length $4\frac{3}{4}$ mi and width $2\frac{1}{2}$ mi.

10. Rectangle of length $8\frac{1}{2}$ km and width $5\frac{1}{4}$ km.

11. Rectangle of length 4 yd and width 10 ft.

12. Rectangle of length 9 cm and width 30 mm.

13. Square whose sides are 1.3 ft long.

14. Square whose sides are 25 in. long.

15. Square whose sides are 44 cm long.

16. Square whose sides are 1.5 km long.

17. Square whose sides are $4\frac{2}{3}$ yd long.

18. Square whose sides are $7\frac{1}{2}$ m long.

19. Parallelogram with base 3.6 dm and height 2.1 dm.

20. Parallelogram with base 0.4 mi and height 0.3 mi.

21. Parallelogram with base $2\frac{1}{2}$ yd and height 12 ft.

22. Parallelogram with base 70 m and height 9 dam.

23. Triangle with base 2 yd and height 3 yd.

24. Triangle with base 9 dm and height 7 dm.

25. Triangle with base $3\frac{1}{2}$ m and height $2\frac{1}{4}$ m.

26. Triangle with base $4\frac{3}{8}$ ft and height $4\frac{4}{5}$ ft.

27. Triangle with base 2 ft and height 1 yd.

28. Triangle with base 6 m and height 80 dm.

29. Trapezoid with bases 4 ft and 8 ft long and height of $7\frac{1}{2}$ ft.

30. Trapezoid with bases $4\frac{1}{2}$ in. and $9\frac{1}{2}$ in. long and a height of 4 in.

31. Trapezoid with bases 6.2 cm and 4.4 cm long and a height of 3.1 cm.

32. Trapezoid with bases 51 m and 39 m long and a height of 20 m.

33. Trapezoid with bases 8 in. and 1 ft long and a height of 5 in.

34. Trapezoid with bases 60 mm and 4 cm long and a height of 8 cm.

Convert each of the following units of measurement as indicated.

35. $18 \text{ ft}^2 = $ ____ yd^2

36. $3 \text{ acres} = $ ____ yd^2

37. $2 \text{ sq ft} = $ ____ sq in.

38. $320 \text{ acres} = $ ____ sq mi

39. $300 \text{ mm}^2 = $ ____ cm^2

40. $8 \text{ m}^2 = $ ____ dm^2

41. $75 \text{ dm}^2 = $ ____ m^2

42. $42 \text{ cm}^2 = $ ____ mm^2

43. $7 \text{ a} = $ ____ m^2

44. $4 \text{ ha} = $ ____ a

45. $36 \text{ in.}^2 = $ ____ ft^2

46. $1{,}210 \text{ yd}^2 = $ ____ acres

47. $0.75 \text{ acre} = $ ____ ft^2

48. $5\frac{1}{4} \text{ mi}^2 = $ ____ acres

49. $800 \text{ mm}^2 = $ ____ m^2

50. $3{,}500 \text{ cm}^2 = $ ____ a

Solve each of the following word problems.

51. How many square feet of fabric are needed to make a tablecloth that measures 70 in. long and 54 in. wide?

52. How many square feet of shelf paper are needed to cover four shelves, each $3\frac{1}{2}$ ft long and $1\frac{1}{4}$ ft wide?

53. A shawl is made out of a triangular piece of fabric that has a base of 54 in. and a height of 18 in. What is the area of the shawl?

54. The mainsail of a boat is shaped like a right triangle that has a 5-ft base and a 12-ft height. What is the area of the sail?

55. How many square tiles measuring 20 cm along an edge are needed to cover a floor 5 m long and 7 m wide?

56. How large must a piece of plywood be to build a door that is 6 ft 6 in. long and 2 ft 8 in. wide?

57. A rectangular piece of land measures 3 km $\times$ 5 km. What is its area in hectares?

58. A square piece of land measures 8 hm on a side. What is its area in ares?

59. How much would it cost to carpet a room that measures 18 ft $\times$ 12 ft if carpeting sells for $21.95 per square yard?

60. One gallon of paint covers 300 sq ft. How much paint is needed to cover the ceiling of a room that measures 10 ft $\times$ 9 ft?

61. A rectangle with a width of 9 in. has an area of 108 sq in. What is its length?

62. A square has an area of 225 sq ft. What is the length of one of its sides?

63. A parallelogram with a base of $7\frac{1}{2}$ cm has an area of 12 cm^2. What is its height?

64. A rectangle with an area of $9\frac{3}{4}$ m^2 is $6\frac{1}{4}$ m long. What is its width?

65. A triangle whose base measures 2.4 in. has an area of 3.6 in.^{2}. What is its height?

66. A triangle whose area is 84 mm^2 has a height of 16 mm. How long is its base?

67. A trapezoid with bases that measure 3 dm and 7 dm has an area of 45 dm^2. What is its height?

68. A trapezoid with an area of 480 yd^2 has two bases that measure 17 yd and 13 yd. What is its height?

69. How many square feet of wallpaper are needed to cover a wall that measures $9\frac{1}{2}$ ft $\times$ $7\frac{3}{4}$ ft and has a window that measures $4\frac{1}{2}$ ft $\times$ 3 ft? (See the figure below.)

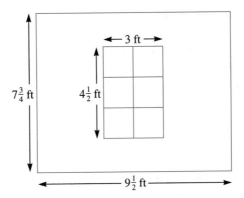

70. How many square feet of Formica are needed to cover the counter shown below?

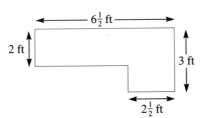

71. How much paneling would it take to panel a wall shaped like the accompanying drawing?

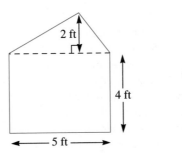

72. How many square feet of carpeting are needed to cover four steps that measure 42 in. long, 11 in. wide, and 9 in. high, as shown in the picture on the next page?

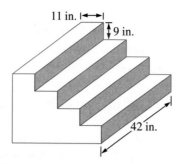

73. Find the area of this figure.

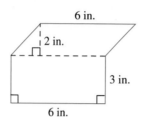

74. Find the area of this figure.

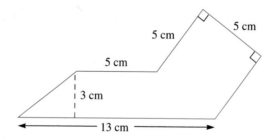

10.4

The Pythagorean Theorem

One of the most famous theorems in mathematics was discovered by the Greek mathematician Pythagoras around 500 B.C. This theorem states a formula that describes the relationship between the sides of any right triangle. The two sides of a right triangle that meet in a right angle are called the **legs**, and the longest side, which joins the two legs, is called the **hypotenuse**. The hypotenuse is the side opposite the right angle.

The Pythagorean Theorem is illustrated in Figure 10.19. Here we have a right triangle with legs 3 units and 4 units in length and a hypotenuse of 5 units. A square has been constructed on each side of the triangle. Notice that the sum of the areas of the two squares drawn adjacent to the two legs is equal to the area of the square drawn adjacent to the hypotenuse.

$$9 + 16 = 25$$

Figure 10.19

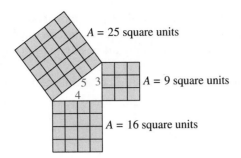

$A = 25$ square units

$A = 9$ square units

$A = 16$ square units

Therefore, the sum of the squares of the two legs is equal to the square of the hypotenuse.

$$(3)^2 + (4)^2 = (5)^2$$

This same relationship holds true for any right triangle. To generalize this result, we let the letters a and b represent the legs of a right triangle and the letter c represent the hypotenuse. Then we write the formula

$$a^2 + b^2 = c^2$$
$$\text{or} \quad c^2 = a^2 + b^2$$

The Pythagorean Theorem is summarized below.

The Pythagorean Theorem states that for any right triangle with legs a and b and hypotenuse c, the sum of the squares of the two legs is equal to the square of the hypotenuse.

$$a^2 + b^2 = c^2$$

▲

EXAMPLE 1   Find the hypotenuse of a right triangle whose legs measure 9 m and 12 m.

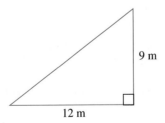

9 m

12 m

Solution $a^2 + b^2 = c^2$ Use the Pythagorean Theorem.
$c^2 = a^2 + b^2$
$c^2 = 9^2 + 12^2$ Substitute $a = 9$ and $b = 12$.
$c^2 = 81 + 144$ Raise to powers first.
$c^2 = 225$ Then do addition.
$c = \sqrt{225}$ Solve for c by taking the square root of both sides.
$c = 15$ m

The hypotenuse is 15 m.   ▲

Generally, when we take the square root of both sides of an equation, we obtain both a positive and a negative result. However, since we are dealing with dimensions of a geometric figure in these examples, only the positive square root makes sense as a solution for c.

EXAMPLE 2   A triangle has sides that measure 7 cm, 8 cm, and 11 cm in length. Is it a right triangle?

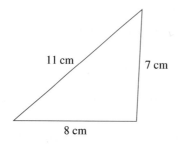

Solution   In order for the figure to be a right triangle, it must satisfy the Pythagorean Theorem.

$a^2 + b^2 \ ? \ c^2$
$7^2 + 8^2 \ ? \ 11^2$ Substitute $a = 7$, $b = 8$, and $c = 11$.
$49 + 64 \ ? \ 121$
$113 \neq 121$ See if both sides of the resulting statement are equal.

The figure is not a right triangle.

Since the two sides are unequal, the Pythagorean Theorem is not satisfied.   ▲

The formula $a^2 + b^2 = c^2$ enables us to find the hypotenuse c, given the measurements of the two legs, a and b. If we are given the measurement of the hypotenuse and one leg, we must rearrange the formula to find the other leg as follows:

$$a^2 + b^2 = c^2$$
$$a^2 + b^2 - b^2 = c^2 - b^2 \qquad \text{Subtract } b^2 \text{ from both sides.}$$
$$a^2 + 0 = c^2 - b^2$$
$$a^2 = c^2 - b^2$$
$$a = \sqrt{c^2 - b^2} \qquad \text{Take the square root of both sides.}$$

Remember that since dimensions of geometric figures are always positive, we are only concerned with the positive square root. Therefore, to find the measure of leg a given hypotenuse c and leg b, use the formula $a = \sqrt{c^2 - b^2}$. Similarly, it can be shown that to find the measure of leg b given the hypotenuse c and leg a, we use the formula $b = \sqrt{c^2 - a^2}$.

EXAMPLE 3   The hypotenuse of a right triangle is 100 ft long, and one leg is 80 ft long. How long is the other leg?

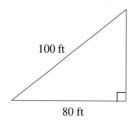

Solution $a = \sqrt{c^2 - b^2}$ | Use the alternate form for the Pythagorean Theorem.

$a = \sqrt{100^2 - 80^2}$ | Substitute $c = 100$ and $b = 40$.

$a = \sqrt{10,000 - 6,400}$ | Simplify expression under the radical. First raise to powers.

$a = \sqrt{3,600}$ | Then do subtraction.

$a = 60$ ft | Finally, take the square root.

The other leg is 60 ft long. ▲

EXAMPLE 4 A kite is flying 15 ft above the person holding the string. How long is the string if the horizontal distance between the kite flyer and the kite is 20 ft?

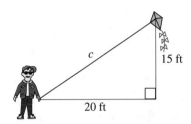

Solution Draw a diagram to illustrate the problem. The length of the string is the hypotenuse of a right triangle. Since it is unknown, assign it the variable c.

Let $c = $ length of the string.

$c^2 = a^2 + b^2$ | Use the Pythagorean Theorem.

$c^2 = 20^2 + 15^2$ | Substitute $a = 20$, $b = 15$.

$c^2 = 400 + 225$

$c^2 = 625$ | First raise to powers. Then do addition.

$c = \sqrt{625}$ | Take the square root of both sides.

$c = 25$ ft

The string is 25 ft long. ▲

Quick Quiz

Given a right triangle with legs a and b and hypotenuse c, find the following dimensions.

Answers

1. Find c if $a = 5$ cm and $b = 12$ cm.
2. Find a if $c = 17$ in. and $b = 15$ in.

1. $c = 13$ cm
2. $a = 8$ in.

10.4 Exercises

For each of the following right triangles with legs a and b and hypotenuse c, use the Pythagorean Theorem to find the unknown side.

1. Find c if $a = 6$ in. and $b = 8$ in.
2. Find c if $a = 8$ cm and $b = 15$ cm.
3. Find c if $a = 12$ m and $b = 16$ m.
4. Find c if $a = 30$ ft and $b = 40$ ft.
5. Find a if $c = 13$ yd and $b = 5$ yd.
6. Find b if $c = 15$ mm and $a = 12$ mm.

7. Find b if $c = 2.5$ cm and $a = 2.0$ cm. 8. Find a if $c = 3.0$ mi and $b = 1.8$ mi.

9. Find c if $a = 900$ mm and $b = 1,200$ mm. 10. Find b if $c = 1,000$ in. and $a = 800$ in.

11. Find c if $a = 12.5$ cm and $b = 17.7$ cm.   Round off to nearest tenth.

12. Find c if $a = 35.8$ ft and $b = 28.4$ ft.   Round off to nearest tenth.

13. Find a if $c = 62.4$ yd and $b = 47.6$ yd.   Round off to nearest tenth.

14. Find b if $c = 8.94$ m and $a = 3.61$ m.   Round off to nearest hundredth.

State whether or not each of the following triangles with sides a, b, and c are right triangles. Assume that the sides of each triangle are given in terms of the same unit of measurement.

15. $a = 6$, $b = 8$, $c = 12$ 16. $a = 5$, $b = 6$, $c = 8$

17. $a = 9$, $b = 40$, $c = 41$ 18. $a = 5$, $b = 13$, $c = 14$

19. $a = 21$, $b = 28$, $c = 35$ 20. $a = 18$, $b = 22$, $c = 27$

21. $a = 90$, $b = 120$, $c = 150$ 22. $a = 50$, $b = 120$, $c = 130$

Solve each of the following word problems.

23. The base of a ladder is 5 ft away from a wall, and its top edge, which rests against the wall, is 12 ft above the ground. How long is the ladder?

24. To straighten a tree, a rope is tied around its trunk at a point 6 ft above the ground and pulled taut. The other end of the rope is tied to a stake in the ground that is 8 ft away from the base of the trunk. What is the length of the rope measured from the tree to the stake?

25. Two airplanes pass each other in the sky, one headed due north at a speed of 600 mph and the other headed due east at a speed of 800 mph. What is the distance between them 1 hr later?

26. A baseball diamond is in the shape of a square whose vertices are the bases and home plate. If the distance between first and second base is 90 ft, what is the distance between second base and home plate? Round off your answer to the nearest foot.

27. A 20-ft-long guy wire that supports a telephone pole is attached to a stake in the ground 12 ft away from the base of the pole. How high above the ground is the other end of the guy wire attached to the pole?

28. The diagonal of a rectangle is 300 cm, and its width is 180 cm. What is the length of the rectangle? (The diagonal of a rectangle is a straight line that connects opposite vertices.)

29. A right triangle has a hypotenuse measuring 7.5 m and a leg measuring 4.5 m. What is its area?

30. A rectangle has a 12.5-in. diagonal and a length of 10 in. What is its area?

31. Find a and b for the triangle shown here. Then find its perimeter and area.

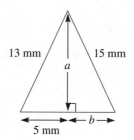

13 mm   15 mm

a

5 mm b

32. Find the value of *h* for the parallelogram shown here. Then find its perimeter and area.

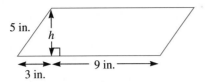

10.5

Circles

In this section, we will learn how to find the perimeter and area of figures called **circles**.

▶ **DEFINITION** A *circle* is a geometric figure that lies in a plane and includes all points that are the same distance away from a fixed point called the *center*. The distance from the center to any point on the circle is called the *radius*. A line that passes through the center and connects two points on a circle is called the *diameter*.

Figure 10.20 illustrates the radius (*r*) and diameter (*d*) of a circle. Notice that the diameter is exactly equal to twice the radius.

Figure 10.20

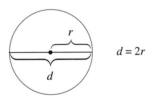

$d = 2r$

For example, a circle with a diameter of 18 in. has a radius of 9 in. A circle with a radius of 21 cm has a diameter of 42 cm.

The perimeter of a circle, or distance around it, is called the *circumference*. Early mathematicians discovered that the circumference of any circle is a little more than three times its diameter. In fact, they found that for any circle, the ratio of the circumference to the diameter is equal to the same fixed number. This number is represented by the symbol π, which is the Greek letter pi. Pi is an irrational number that can be expressed as the following nonrepeating decimal.

$$\pi = 3.14159265\ldots$$

Thus, the ratio of the circumference (*C*) to the diameter (*d*) of any circle is equal to π.

$$\frac{C}{d} = \pi$$

To find a formula for the circumference of a circle, we can multiply both sides of this equation by *d*.

$$d \cdot \frac{C}{d} = \pi \cdot d$$

$$d \cdot \frac{C}{d} = \pi d$$

$$C = \pi d$$

Since the diameter is equal to twice the radius, another formula for the circumference can be obtained by substituting $2r$ for d.

$$C = \pi(2r) \quad \text{or} \quad C = 2\pi r$$

The circumference (perimeter) of a circle is π times the diameter (d) or 2π times the radius (r).

$$C = \pi d$$
$$C = 2\pi r$$

▲

EXAMPLE 1 Find the circumference of a circle whose radius is 8 cm.

8 cm

Solution

$C = 2\pi r$	Use the formula for finding circumference in terms of r.
$C = 2\pi(8)$	Substitute $r = 8$.
$C = 16\pi$ cm	Since the radius is given in cm, so is the circumference.
	Notice that 16π represents the product $16 \cdot \pi$. ▲

In Example 1, we expressed the circumference in terms of π, which is the exact answer. This is not a convenient way to express a measurement for practical applications. Therefore, we frequently use the decimal 3.14 or the fraction $\frac{22}{7}$ to approximate π.

EXAMPLE 2 A circular tablecloth has a diameter of 70 in. How many feet of trimming are needed to decorate its border?

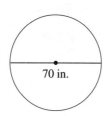

70 in.

Solution   First find the circumference of the table in inches.

$C = \pi d$ — Use the formula for finding circumference in terms of d.

$C = \pi(70)$ — Substitute $d = 70$.

$C = \dfrac{22}{7}(70)$ — Use $\dfrac{22}{7}$ to approximate π.

$C = 220$ in.

Now convert 220 in. to feet.

$$220 \text{ in.} = 220 \text{ in.} \times \frac{1 \text{ ft}}{12 \text{ in.}} = \frac{220}{12} \text{ ft} = \frac{55}{3} \text{ ft} = 18\frac{1}{3} \text{ ft}$$

Thus, $18\frac{1}{3}$ ft of trimming is needed.   ▲

EXAMPLE 3   A record with a diameter of 12 in. is moving at a speed of $33\frac{1}{3}$ revolutions per minute (rpm). After 1 min, how far will a bug that is sitting on the edge of the record have traveled?

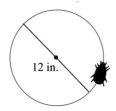

Solution   After one revolution, the bug will have traveled a distance that is equal to the circumference of the circle.

$C = \pi d$
$C = \pi(12)$ — Substitute $d = 12$.
$C = 12\pi$ in.

In 1 min, the record will make $33\frac{1}{3}$ revolutions. Therefore, the distance traveled in 1 min will equal the circumference times $33\frac{1}{3}$.

$$\text{distance} = 12\pi \cdot 33\frac{1}{3}$$

$$= 12\pi \cdot \frac{100}{3}$$

$$= \frac{\overset{4}{\cancel{12}}\pi(100)}{\cancel{3}}$$

$$= 400\pi = 400(3.14) \qquad \text{Use } \pi \cong 3.14.$$

$$\text{distance} = 1{,}256 \text{ in.}$$

After 1 min, the bug will have traveled 1,256 in.   ▲

To find the area of a circle, we can divide it up into a number of wedge-shaped pieces called *sectors* and rearrange them as shown in Figure 10.21. Notice that the resulting figure approximates a parallelogram with bases equal to one-half the circumference, and a height equal to the radius.

Figure 10.21

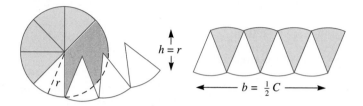

If we divide the circle up into a larger number of sectors, the figure formed in this way more closely resembles a parallelogram, since the curved edges of the sectors that form the bases will approach straight lines. We can find the area of this parallelogram as follows:

$$A = bh$$

$$A = \frac{1}{2}Cr \qquad\qquad \text{Substitute } b = \tfrac{1}{2}C \text{ and } h = r.$$

$$A = \frac{1}{2}(2\pi r)r \qquad\qquad \text{Substitute } C = 2\pi r.$$

$$A = \pi r^2$$

Our result is the formula for finding the area of a circle, which is restated here.

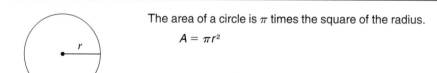

The area of a circle is π times the square of the radius.

$$A = \pi r^2$$

▲

EXAMPLE 4 Find the area of a circle whose radius is 15 ft. Use 3.14 to approximate π.

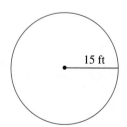

15 ft

Solution $A = \pi r^2$ Use the formula for finding the area of a circle.

$A = \pi(15)^2$ Substitute $r = 15$.

$A = (3.14)(225)$ Use $\pi \cong 3.14$.

$A = 706.5$ sq ft   Since the radius is given in ft, the area is in sq ft.

The area is 706.5 sq ft.   ▲

EXAMPLE 5   The area of a circle is 900π sq yd. What is its diameter?

Solution   Find the radius.

$$A = \pi r^2$$
$$900\pi = \pi r^2$$
Use the formula for finding the area of a circle.

$$\pi r^2 = 900\pi$$
Substitute $A = 900\pi$.

$$\frac{1}{\pi}\pi r^2 = 900\pi\frac{1}{\pi}$$

$$\frac{1}{\pi}\pi r^2 = 900\pi\frac{1}{\pi}$$
Multiply both sides by $\frac{1}{\pi}$ so that the π's cancel.

$$r^2 = 900$$
Take the square root of both sides.

$$r = \sqrt{900}$$
Since area is given in sq yd, the radius is in yd.

$$r = 30 \text{ yd}$$

Find the diameter.

$$d = 2r$$
Multiply the radius by 2.

$$d = 2(30)$$
$$d = 60 \text{ yd}$$

Therefore the diameter is 60 yd.   ▲

EXAMPLE 6   Find the area of the smallest circle that circumscribes a square with a diagonal of 6 cm.

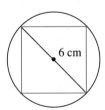

Solution   First draw a sketch. Notice that the vertices of the square touch the circle that circumscribes it. Therefore, the diagonal of the square is also the diameter of the circle. Thus, the radius of the circle is 3 cm.

$A = \pi r^2$
$A = \pi 3^2$
$A = 9\pi \text{ cm}^2$

The area of the circle is $9\pi \text{ cm}^2$.   ▲

Quick Quiz

				Answers
1.	Find the circumference of a circle with a radius of 56 mm. Use $\pi \cong \frac{22}{7}$.		**1.**	352 mm
2.	Find the area of a circle with a radius of 13 ft.		**2.**	169π sq ft
3.	Find the area of a circle with a diameter of 20 cm. Use $\pi \cong 3.14$.		**3.**	314 cm²

10.5 Exercises

Indicate which of the following statements are true and which are false. For those that are false, change the italicized word to make the statement true.

1. The area of a circle is π times the square of its *diameter*.

2. The diameter of a circle is equal to *twice* its radius.

3. For any circle, the ratio of the circumference to the *radius* is equal to π.

4. If the radius of a circle is doubled, its area is *quadrupled*.

Find the circumference of each of the following circles given its diameter d or radius r. Express your answers in terms of π unless otherwise indicated.

5. $d = 31$ in.

6. $d = 8.4$ m

7. $r = 7.2$ cm

8. $r = 18$ ft

9. $d = 63$ in.   Use $\pi \cong \frac{22}{7}$.

10. $d = 2.1$ dm   Use $\pi \cong \frac{22}{7}$.

11. $d = 3\frac{1}{2}$ km   Use $\pi \cong \frac{22}{7}$.

12. $d = 6\frac{1}{8}$ mi   Use $\pi \cong \frac{22}{7}$.

13. $r = 3.5$ ft   Use $\pi \cong 3.14$.

14. $d = 9$ yd   Use $\pi \cong 3.14$.

15. $d = 8\frac{3}{4}$ m   Use $\pi \cong 3.14$.

16. $r = 4\frac{1}{2}$ cm   Use $\pi \cong 3.14$.

17. $r = 0.84$ hm   Use $\pi \cong \frac{22}{7}$.

18. $d = 98$ in.   Use $\pi \cong \frac{22}{7}$.

 19. $d = 542$ ft   Use $\pi \cong 3.14$.

20. $r = 265$ mm   Use $\pi \cong 3.14$.

Find the area of each of the following circles given its diameter d or radius r. Express your answers in terms of π unless otherwise indicated.

21. $r = 19$ dm

22. $r = 150$ in.

23. $d = 5.2$ km

24. $d = 87$ yd

25. $r = 70$ mm   Use $\pi \cong \frac{22}{7}$.

26. $r = 28$ ft   Use $\pi \cong \frac{22}{7}$.

27. $r = 2\frac{5}{8}$ yd   Use $\pi \cong \frac{22}{7}$.

28. $r = 6\frac{2}{9}$ dm   Use $\pi \cong \frac{22}{7}$.

29. $r = 300$ cm   Use $\pi \cong 3.14$.

30. $r = 0.7$ mi   Use $\pi \cong 3.14$.

31. $d = 8\frac{2}{3}$ hm   Use $\pi \cong \frac{22}{7}$.

32. $d = 11\frac{1}{5}$ yd   Use $\pi \cong \frac{22}{7}$.

33. $d = 6$ ft   Use $\pi \cong 3.14$.

34. $d = 40$ cm   Use $\pi \cong 3.14$.

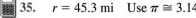 **35.** $r = 45.3$ mi   Use $\pi \cong 3.14$.

 36. $d = 6.2$ m   Use $\pi \cong 3.14$.

Solve each of the following word problems. Leave your answers in terms of π unless otherwise indicated.

37. How much gold braid is needed to decorate the top and bottom edges of a lampshade that has an 11-in. diameter? Use $\pi \cong 3.14$.

38. How much fencing is needed to enclose a circular flower bed if the distance from the center to the edge is 1.4 m? Use $\pi \cong \frac{22}{7}$.

39. What is the area of the top of the stump of a tree with a diameter of $3\frac{1}{2}$ ft? Use $\pi \cong \frac{22}{7}$.

40. How much wood is needed to construct the top of a circular table that has a 70-in. diameter? Use $\pi \cong \frac{22}{7}$.

41. A bicycle wheel has a diameter of 26 in. How far does the air nozzle on the edge of the wheel move when the wheel makes one revolution? Use $\pi \cong 3.14$.

42. A fan is shaped like a quarter of a circle with a radius of 20 cm. What is the area of the fan? Use $\pi \cong 3.14$.

43. What is the radius of a circle that has a circumference of 74π cm?

44. What is the radius of a circle that has an area of 196π mm^2?

45. A skating rink is shaped like a rectangle bounded by two semicircles, as shown in the drawing below. Find its area. Use $\pi \cong 3.14$.

46. Find the area of the largest circle that can be inscribed in a square whose side measures 5 in., as shown in the drawing below.

47. Find the area of the shaded region in this figure. Use $\pi \cong 3.14$.

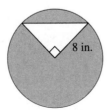

48. Find the area of this figure. Use $\pi \cong 3.14$.

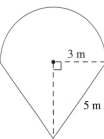
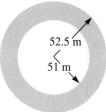

49. Find the area of a sidewalk that surrounds a circular flower bed, as shown in the accompanying drawing. Use $\pi \cong 3.14$.

50. The planet Mercury completes one revolution about the sun in 88 days, moving at a speed of 30 mi/sec. What is the distance between Mercury and the sun? Use $\pi \cong 3.14$. Round off your answer to the nearest million miles.

10.6
Volume

Our discussion of capacity in Section 9.3 involved a characteristic of three-dimensional figures known as **volume**.

▶ **DEFINITION** *Volume* is the measurement of the space enclosed by a solid—that is, the space enclosed by a three-dimensional figure.

To measure volume, we use cubic units of measurement, because volume is a three-dimensional quantity. For example, a cubic inch (abbreviated as cu in. or in.^{3}) represents the volume of a cube whose edges are each 1 inch in length. A cubic centimeter (abbreviated as cu cm or cm^3) represents the volume of a cube whose edges are each 1 centimeter in length. These are both illustrated in Figure 10.22.

Figure 10.22

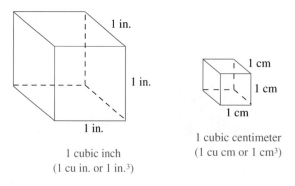

1 cubic inch
(1 cu in. or 1 in.^{3})

1 cubic centimeter
(1 cu cm or 1 cm^3)

Let us now consider how to find the volume of the rectangular box shown in Figure 10.23. Here we have a box that has a length of 5 inches, a width of 3 inches, and a height of 4 inches. The problem of finding the volume of the box is the same as determining how many cubes, each measuring 1 inch along every edge, are needed to completely fill the box. The volume of each of these cubes is one cubic inch. The bottom layer of cubes that forms the base of the box has five cubes along its length and three cubes along its width, making a total of $5 \times 3 = 15$ cubes. The entire box consists of four of these layers of 15 cubes each, making a total of $15 \times 4 = 60$ cubes. Thus, the volume of this box is 60 cubic inches.

Figure 10.23

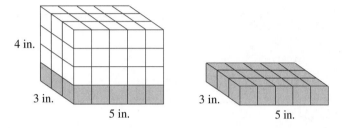

Notice that this same result can be obtained simply by finding the product of the length, width, and height.

$V = (5 \text{ in.})(3 \text{ in.})(4 \text{ in.})$
$V = 60 \text{ in.}^3$

We have thus established the formula for calculating the volume of a rectangular solid (see below).

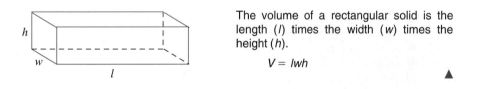

The volume of a rectangular solid is the length (l) times the width (w) times the height (h).

$$V = lwh$$

▲

To illustrate how equivalent units of volume are derived, let us consider a cube whose edges are each 1 yard long, as shown in Figure 10.24. By definition, this cube has a volume

Figure 10.24

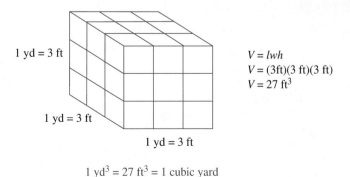

$V = lwh$
$V = (3 \text{ft})(3 \text{ ft})(3 \text{ ft})$
$V = 27 \text{ ft}^3$

1 yd^3 = 27 ft^3 = 1 cubic yard

of 1 cubic yard. Since 1 yard = 3 feet, we can also determine this volume by multiplying the length, which is 3 feet, times the width, which is 3 feet, times the height, which is also 3 feet. Therefore, the volume of 1 cubic yard is equivalent to 27 cubic feet.

Notice that since the length, width, and height of a cube are all the same, to determine its volume, we can cube the measurement of one of its edges (see below). Using the procedure, we can show that $1 \text{ ft}^3 = 1{,}728 \text{ in.}^3$. Table 10.3 summarizes equivalent units of volume in the English system. Notice that the last two relationships can be used for conversion to gallons, introduced as a unit of capacity in Section 9.3.

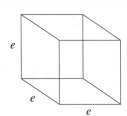

The volume of a cube is the cube of one of its edges (e).

$$V = e^3$$

▲

TABLE 10.3
English Units of Volume

1 ft^3	=	$1{,}728 \text{ in.}^3$
1 yd^3	=	27 ft^3
1 ft^3	=	7.48 gal
1 gal	=	231 in.^3

EXAMPLE 1 An aquarium has a length of 30 in., a width of 15.4 in., and a height of 20 in. How many gallons of water does it hold when full?

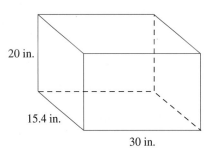

Solution First determine the volume of the aquarium in cubic inches.

$V = lwh$ Use the formula for finding the volume of a rectangular solid.

$V = (30)(15.4)(20)$ Substitute $l = 30$, $w = 15.4$, $h = 20$.

$V = 9{,}240 \text{ in.}^3$

Now convert $9{,}240 \text{ in.}^3$ to gallons.

$$9{,}240 \text{ in.}^3 = 9{,}240 \text{ in.}^3 \times \frac{1 \text{ gal}}{231 \text{ in.}^3} = \frac{9{,}240}{231} \text{ gal} = 40 \text{ gal}$$

Thus, the aquarium holds 40 gal when full.   ▲

In the metric system, each unit of volume differs from the next larger unit by a factor of 1,000. For example, 1 cubic centimeter is equivalent to 1,000 cubic millimeters, as shown in Figure 10.25.

Figure 10.25

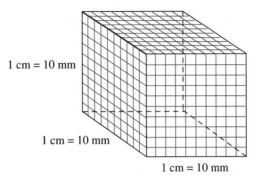

1 cm = 10 mm

1 cm = 10 mm

1 cm = 10 mm

$1 \text{ cm}^3 = 1,000 \text{ mm}^3 = 1$ cubic centimeter

One liter (L) is sometimes defined to be the volume contained in a cube that measures one decimeter along every edge. Thus,

$$1 \text{ L} = 1 \text{ dm}^3$$

Since $\quad 1 \text{ dm}^3 = 1,000 \text{ cm}^3$

$$1 \text{ L} = 1,000 \text{ cm}^3$$

and $\quad \dfrac{1}{1,000} \text{ L} = 1 \text{ mL} = 1 \text{ cm}^3$

These relationships are illustrated in Figure 10.26.

Figure 10.26

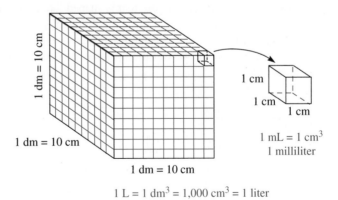

1 dm = 10 cm

1 dm = 10 cm

1 dm = 10 cm

1 cm

1 cm

1 cm

$1 \text{ mL} = 1 \text{ cm}^3$
1 milliliter

$1 \text{ L} = 1 \text{ dm}^3 = 1,000 \text{ cm}^3 = 1$ liter

Similarly, since $\quad 1 \text{ kL} = 1,000 \text{ L}$

$$1 \text{ kL} = 1,000 \text{ dm}^3$$

and $\qquad\qquad\quad 1 \text{ kL} = 1 \text{ m}^3$

Table 10.4 summarizes equivalent metric units of volume.

TABLE 10.4
Metric Units of Volume

1 cm^3	=	1,000 mm^3
1 dm^3	=	1,000 cm^3
1 m^3	=	1,000 dm^3
1 L	=	1 dm^3 = 1,000 cm^3
1 mL	=	1 cm^3
1 kL	=	1,000 dm^3 = 1 m^3

EXAMPLE 2   Convert each of the following metric units of measurement as instructed.

(a)   Convert 500 mL to cubic centimeters.

$$500 \text{ mL} = 500 \text{ m}\cancel{L} \times \frac{1 \text{ cm}^3}{1 \text{ m}\cancel{L}} = 500 \text{ cm}^3$$

(b)   Convert 30 cm^3 to cubic millimeters.

$$30 \text{ cm}^3 = 30 \text{ c}\cancel{\text{m}^3} \times \frac{1,000 \text{ mm}^3}{1 \text{ c}\cancel{\text{m}^3}} = 30,000 \text{ mm}^3$$

(c)   Convert 7 m^3 to cubic centimeters.

$$7 \text{ m}^3 = 7 \text{ }\cancel{\text{m}^3} \times \frac{1,000 \text{ dm}^3}{1 \text{ }\cancel{\text{m}^3}} = \frac{7,000 \text{ }\cancel{\text{dm}^3}}{1} \times \frac{1,000 \text{ cm}^3}{1 \text{ }\cancel{\text{dm}^3}} = 7,000,000 \text{ cm}^3 \quad \blacktriangle$$

We will now present the formulas for finding the volume of a cylinder, a sphere, and a cone. A *cylinder* is a figure that is shaped like a can. It has two circular bases joined by sides that are perpendicular to them. The radius of a cylinder (r) is the radius of the circular base. The height of a cylinder (h) is the distance between the two bases. To find the volume of a cylinder, we can multiply the area of the base, which is πr^2, by the height h (see below).

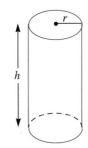

The volume of a cylinder is the area of the base (πr^2) times the height (h).

$$V = \pi r^2 h$$

$\blacktriangle$

EXAMPLE 3   Find the volume of a cylinder with a radius of 3 cm and a height of 80 mm.

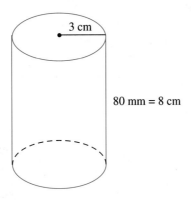

Solution   Both radius and height must be in terms of the same units.

$$80 \text{ mm} = 80 \text{ mm} \times \frac{1 \text{ cm}}{10 \text{ mm}} = 8 \text{ cm} \qquad \text{Convert 80 mm to cm.}$$

$$V = \pi r^2 h$$
$$V = \pi (3)^2 (8) \qquad \text{Substitute } r = 3 \text{ and } h = 8.$$
$$V = \pi (9)(8) \qquad \text{Raise to powers before multiplying.}$$
$$V = 72\pi \text{ cm}^3 \qquad \text{Since the dimensions are in cm, the}$$
$$\text{volume is in cm}^3. \quad \blacktriangle$$

A *sphere* is a figure shaped like a ball. The radius (r) of a sphere is the distance from the center to any point on the sphere. Since the technique needed to derive the formula for the volume of a sphere requires a knowledge of calculus, we simply state the result (see below).

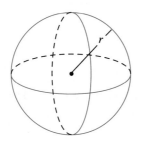

The volume of a sphere is $\frac{4}{3}\pi$ times the cube of the radius (r).

$$V = \frac{4}{3}\pi r^3$$

$\blacktriangle$

EXAMPLE 4   Find the volume of a hemisphere with a 3-ft radius.

Solution Find the volume of the sphere and take one-half of the result.

$$V = \frac{4}{3}\pi r^3$$

Since a hemisphere is one-half of a sphere.

$$V = \frac{4}{3}\pi(3)^3$$

Substitute $r = 3$.

$$V = \frac{4}{3}\pi(27)$$

$$V = 36\pi \text{ ft}^3$$

36π ft^3 is the volume of a sphere of radius 3 ft.

$$\frac{1}{2}(36\pi) = 18\pi$$

Find one-half of the volume of the sphere.

The volume of the hemisphere is 18π ft^3. ▲

A *cone* is a figure that is shaped like a funnel. It has a circular base and sides that come to a point directly above or below the center of the base. The radius (r) of a cone is the radius of the circle that forms the base. The height (h) of a cone is the distance from the center of the base to the point where the sides meet. The radius and height are perpendicular to each other. In order to derive the formula for the volume of a cone, one must know calculus, so we simply state the result (here).

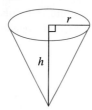

The volume of a cone is $\frac{1}{3}\pi$ times the square of the radius (r) times the height (h).

$$V = \frac{1}{3}\pi r^2 h$$

▲

EXAMPLE 5 Find the volume of a cone that has a height of 6 in. and a base diameter of 4 in.

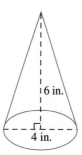

6 in.

4 in.

Solution   If the base has a diameter of 4 in., its radius is 2 in.

$$V = \frac{1}{3}\pi r^2 h$$

$$V = \frac{1}{3}\pi(2)^2(6)$$ Substitute $r = 2$ and $h = 6$.

$$V = \frac{1}{3}\pi(4)(6)$$

$$V = 8\pi \ \text{in.}^3$$ ▲

EXAMPLE 6   Find the volume of the accompanying figure, which is composed of a cylinder of height 10 cm and radius 2 cm, topped by a cone with the same radius and a height of 9 cm. Use $\pi \cong 3.14$.

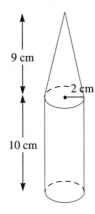

9 cm

2 cm

10 cm

Solution   First find the volume of the cylinder and cone separately.

$$V_{\text{cyl}} = \pi r^2 h$$ Formula for finding the volume of a cylinder.

$$V_{\text{cyl}} = \pi(2)^2(10)$$ Substitute $r = 2$ and $h = 10$.

$$V_{\text{cyl}} = (3.14)(4)(10)$$ Use $\pi \cong 3.14$.

$$V_{\text{cyl}} = 125.6 \ \text{cm}^3$$

$$V_{\text{cone}} = \frac{1}{3}\pi r^2 h$$ Formula for finding the volume of a cone.

$$V_{\text{cone}} = \frac{1}{3}\pi(2)^2(9)$$ Substitute $r = 2$ and $h = 9$.

$$V_{\text{cone}} = \frac{1}{3}(3.14)(4)(9)$$ Use $\pi \cong 3.14$.

$$V_{\text{cone}} = 37.68 \ \text{cm}^3$$

The total volume is equal to the volume of the cylinder plus the volume of the cone.

$$V_{total} = V_{cyl} + V_{cone}$$
$$V_{total} = 125.6 + 37.68$$
$$V_{total} = 163.28 \text{ cm}^3$$

▲

Quick Quiz

Find the volume of each of the following figures.   **Answers**

1. Rectangular solid with length 9 cm, width 5 cm, and height 6 cm.

   1. 270 cm³

2. Cylinder with radius 3 in. and height 7 in.

   2. 63π in.³

3. Sphere with radius of 6 ft.

   3. 288π ft³

4. Cone with radius 7 m and height 12 m. Use $\pi \cong \frac{22}{7}$.

   4. 616 m³

10.6 Exercises

Indicate which of the following statements are true and which are false. For those that are false, change the italicized or underlined expression to make the statement true.

1. Volume is measured in *square* units of measurement.

2. A liter is equivalent to one cubic *meter.*

3. A cubic yard is equivalent to 27 cubic *inches.*

4. A cubic centimeter is equivalent to <u>100</u> cubic millimeters.

Find the volume of each of the following geometric solids. Leave your answers in terms of π unless otherwise indicated.

5. Rectangular solid with length 8 in., width 4 in., and height 3 in.

6. Rectangular solid with length 11 dm, width 2 dm, and height 4 dm.

7. Rectangular solid with length $8\frac{1}{3}$ yd, width $4\frac{1}{5}$ yd, and height $3\frac{1}{3}$ yd.

8. Rectangular solid with length $5\frac{1}{2}$ m, width $2\frac{3}{4}$ m, and height 6 m.

9. Rectangular solid with length 6 ft, width 3 ft, and height 9 in.

10. Rectangular solid with length 9 cm, width 20 mm, and height 7 cm.

11. Cube that measures 4 cm along every edge.

12. Cube that measures 5 ft along every edge.

13. Cube that measures 1.2 in. along every edge.

14. Cube that measures 80 mm along every edge.

15. Cube that measures $2\frac{1}{4}$ dm along every edge.

16. Cube that measures $1\frac{1}{2}$ yd along every edge.

17. Cylinder with radius 2 cm and height 9 cm.

18. Cylinder with radius 3 in. and height 7 in. Use $\pi \cong \frac{22}{7}$.

19. Cylinder with radius 1 ft and height 5 ft. Use $\pi = 3.14$.

20. Cylinder with radius $2\frac{1}{2}$ m and height 40 dm.

21. Cylinder with diameter 8 dm and height 120 cm.

22. Cylinder with diameter 6 ft and height $1\frac{1}{3}$ ft. Use $\pi \cong 3.14$.

23. Sphere with radius 7 ft.

24. Sphere with radius 1.2 m.

25. Sphere with radius 3 cm. Use $\pi \cong 3.14$.

26. Sphere with radius 6 in. Use $\pi \cong 3.14$.

27. Sphere with diameter 15 in.

28. Sphere with diameter 18 dm.

29. Hemisphere with radius 9 ft. Use $\pi \cong 3.14$.

30. Hemisphere with diameter 6 yd.

31. Cone with radius 10 ft and height 30 ft. Use $\pi \cong 3.14$.

32. Cone with radius 1.5 dm and height 2.5 dm.

33. Cone with radius 4 in. and height 12 in. Use $\pi \cong 3.14$.

34. Cone with diameter 14 m and height 9 m. Use $\pi \cong \frac{22}{7}$.

35. Cone with diameter 8 in. and height 1 ft.

36. Cone with radius 3 cm and height 80 mm.

37. Cone with radius 5.7 cm and height 9.2 cm. Use $\pi \cong 3.14$. Round off to the nearest tenth.

38. Sphere with diameter 8.5 in. Use $\pi \cong 3.14$. Round off to the nearest tenth.

Convert each of the following units of measurement as instructed.

39. $3 \text{ ft}^3 = \underline{\hspace{0.5in}} \text{ in.}^3$

40. $54 \text{ ft}^3 = \underline{\hspace{0.5in}} \text{ yd}^3$

41. $\frac{1}{2} \text{ gal} = \underline{\hspace{0.5in}} \text{ in.}^3$

42. $7 \text{ ft}^3 = \underline{\hspace{0.5in}} \text{ gal}$

43. $9 \text{ cm}^3 = \underline{\hspace{0.5in}} \text{ mm}^3$

44. $200 \text{ cm}^3 = \underline{\hspace{0.5in}} \text{ dm}^3$

45. $7,000 \text{ mm}^3 = \underline{\hspace{0.5in}} \text{ cm}^3$

46. $8 \text{ dm}^3 = \underline{\hspace{0.5in}} \text{ cm}^3$

47. $19 \text{ cm}^3 = \underline{\hspace{0.5in}} \text{ mL}$

48. $80 \text{ dm}^3 = \underline{\hspace{0.5in}} \text{ L}$

49. $60 \text{ m}^3 = \underline{\hspace{0.5in}} \text{ kL}$

50. $75 \text{ cm}^3 = \underline{\hspace{0.5in}} \text{ L}$

51. $800 \text{ mm}^3 = \underline{\hspace{0.5in}} \text{ mL}$

52. $7 \text{ dm}^3 = \underline{\hspace{0.5in}} \text{ mL}$

53. $600 \text{ m}^3 = \underline{\hspace{0.5in}} \text{ cm}^3$

54. $70,000 \text{ mm}^3 = \underline{\hspace{0.5in}} \text{ dm}^3$

Solve each of the following word problems.

55. A swimming pool is 200 ft long, 100 ft wide, and $3\frac{1}{2}$ ft deep. How many gallons of water are needed to fill the pool?

56. A refrigerator is 2.5 ft long, 2 ft wide, and 5.5 ft high. What is its capacity in cubic feet?

57. A beach ball has a 21-in. diameter. What is its volume? Use $\pi \cong \frac{22}{7}$.

58. A dome is shaped like a hemisphere that has a 20-ft diameter. What is the volume it encloses? Use $\pi \cong 3.14$. Round off to the nearest cubic foot.

59. A can of soda is 12 cm high and 7 cm in diameter. How many milliliters can it hold? Use $\pi \cong \frac{22}{7}$.

60. A cylindrical gas tank has a radius of 2 m and a height of 5 m. How many liters of gas are in the tank when it is full? Use $\pi \cong 3.14$.

61. A sand pile is shaped like a cone that is 10 ft high and has a radius of 6 ft. How many cubic feet of sand are in the pile? Use $\pi \cong 3.14$.

62. A water tank in the shape of a cone is 7 m high and has a radius of 3 m. How many kiloliters of water can it hold? Use $\pi \cong \frac{22}{7}$.

63. Find the volume of the following figure, which is composed of a cylinder of height 10 in. and radius 6 in., topped by a hemisphere with the same radius. Use $\pi \cong 3.14$.

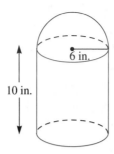

64. Find the volume of the accompanying figure, which is a cone of height 8 m and radius 3 m, topped by a hemisphere with the same radius. Use $\pi \cong 3.14$.

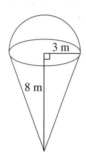

10.7
Summary and Review

Key Terms

(10.1)   A **point** indicates a location or position and is represented by a dot.

A **line** is a set of points extending indefinitely in opposite directions.

A **plane** is a set of points that form a flat surface extending indefinitely in all directions.

A **polygon** is a figure that lies in a plane and has many sides. A **regular polygon** is a polygon in which all sides and all angles are equal.

A **triangle** is a three-sided figure.

A **quadrilateral** is a four-sided figure.

A **pentagon** is a five-sided figure.

A **hexagon** is a six-sided figure.

An **octagon** is an eight-sided figure.

An **angle** is formed whenever two lines meet. The two lines are called the sides of the angle, and the point at which they meet is called the **vertex** of the angle.

A **degree** is a unit of measurement that is used to indicate the size of an angle.

A **right angle** is a 90° angle.

An **acute angle** is an angle greater than 0° and less than 90°.

An **obtuse angle** is an angle greater than 90° and less than 180°.

Perpendicular lines meet in a right angle.

An **equilateral triangle** has three equal sides and three equal angles.

An **isosceles triangle** has two equal sides and two equal angles.

A **right triangle** has a right angle.

Parallel lines are lines in the same plane that never meet.

A **trapezoid** is a quadrilateral that has two parallel sides.

A **parallelogram** is a quadrilateral whose opposite sides are equal and parallel.

A **rectangle** is a parallelogram that has four right angles.

A **square** is a rectangle that has four equal sides.

(10.2) **Perimeter** is the distance around a geometric figure.

(10.3) **Area** is the measurement of the surface of a figure.

(10.4) The **legs** of a right triangle are the two sides that meet in a right angle.

The **hypotenuse** of a right triangle is the longest side, which joins the two legs.

(10.5) A **circle** is a geometric figure that lies in a plane and includes all points that are the same distance away from a fixed point called the *center*.

The **radius** of a circle is the distance from the center to any point on the circle.

The **diameter** of a circle is a line that connects two points on a circle and passes through its center.

The **circumference** of a circle is the perimeter of, or distance around, a circle.

Pi (π) is the ratio of the circumference of a circle to its diameter. π is an irrational number approximately equal to 3.14.

(10.6) **Volume** is the measurement of the space enclosed by a solid, that is, the space enclosed by a three-dimensional figure.

Calculations

(10.2) To find the **perimeter of a rectangle** of length l and width w,

$$P = 2l + 2w$$

To find the **perimeter of a regular polygon** with n sides, each of length s,

$$P = ns$$

(10.3) To find the **area of a rectangle** of length l and width w,

$$A = lw$$

To find the **area of a square** with sides of length s,

$$A = s^2$$

To find the **area of a parallelogram** with base b and height h,

$$A = bh$$

To find the **area of a triangle** with base b and height h,

$$A = \frac{1}{2}bh$$

To find the **area of a trapezoid** with bases b_1 and b_2 and height h,

$$A = \frac{1}{2}h(b_1 + b_2)$$

(10.4)   The **Pythagorean Theorem** states that for any right triangle with legs a and b and hypotenuse c,

$$a^2 + b^2 = c^2$$

(10.5)   To find the **circumference of a circle** of diameter d and radius r,

$$C = \pi d \qquad \text{or} \qquad C = 2\pi r$$

To find the **area of a circle** of radius r,

$$A = \pi r^2$$

(10.6)   To find the **volume of a rectangular solid** of length l, width w, and height h,

$$V = lwh$$

To find the **volume of a cube** with edges of length e,

$$V = e^3$$

To find the **volume of a cylinder** of radius r and height h,

$$V = \pi r^2 h$$

To find the **volume of a sphere** of radius r,

$$V = \frac{4}{3}\pi r^3$$

To find the **volume of a cone** of radius r, and height h,

$$V = \frac{1}{3}\pi r^2 h$$

Chapter **10**
Review Exercises

Indicate which of the following statements are true and which are false. For those that are false, change the italicized word to make the statement true.

1. Parallel lines *sometimes* intersect.
2. Perimeter is measured in *linear* units of measurement.
3. The region enclosed by a two-dimensional figure is called *volume*.
4. The Pythagorean Theorem applies to *equilateral* triangles.
5. The circumference of a circle is another name for its *area*.
6. If the radius of a sphere is doubled, its volume is increased by a factor of *eight*.

(10.3, 10.6)   **Convert each of the following units of measurement as instructed.**

7. $2 \text{ mi}^2 = \underline{\hspace{1cm}}$ acres

8. $72 \text{ in.}^2 = \underline{\hspace{1cm}}$ ft^2

9. $400 \text{ m}^2 = \underline{\hspace{1cm}}$ a

10. $3.5 \text{ dm}^2 = \underline{\hspace{1cm}} \text{ cm}^2$ 11. $36 \text{ ft}^2 = \underline{\hspace{1cm}} \text{ yd}^2$ 12. $36 \text{ in.}^2 = \underline{\hspace{1cm}} \text{ ft}^2$

13. $250 \text{ cm}^2 = \underline{\hspace{1cm}} \text{ m}^2$ 14. $0.8 \text{ ha} = \underline{\hspace{1cm}} \text{ m}^2$ 15. $0.2 \text{ ft}^3 = \underline{\hspace{1cm}} \text{ in.}^3$

16. $693 \text{ in.}^3 = \underline{\hspace{1cm}} \text{ gal}$ 17. $2.7 \text{ ft}^3 = \underline{\hspace{1cm}} \text{ yd}^2$ 18. $748 \text{ gal} = \underline{\hspace{1cm}} \text{ ft}^3$

19. $14 \text{ mL} = \underline{\hspace{1cm}} \text{ cm}^3$ 20. $9,000 \text{ cm}^3 = \underline{\hspace{1cm}} \text{ L}$ 21. $5.4 \text{ kL} = \underline{\hspace{1cm}} \text{ m}^3$

22. $400 \text{ mm}^3 = \underline{\hspace{1cm}} \text{ cm}^3$

(10.2, 10.5) **Find the perimeter (or circumference) of each of the following figures. Leave answers in terms of π where appropriate.**

23. Rectangle of length 8 cm and width 3 cm.

24. Square whose sides are each 4.3 m long.

25. Triangle with sides 7 in., $4\frac{1}{2}$ in., and $6\frac{1}{4}$ in. long.

26. Regular octagon whose sides are each 10.5 ft long.

27. Circle of radius 7 mm.

28. Circle of diameter 5 in.

29. Right triangle whose legs measure 3 ft and 4 ft.

30. Right triangle whose legs measure 5 cm and 12 cm.

(10.3, 10.5) **Find the area of each of the following figures. Leave answers in terms of π where appropriate.**

31. Rectangle of length 7 ft and width 9 ft.

32. Square whose sides are 12 cm long.

33. Parallelogram with base 9.5 m and height 6 m.

34. Triangle with base 10 in. and height 3.2 in.

35. Trapezoid with bases 2 ft and 8 ft long and height 6 ft.

36. Circle of radius 9 mm.

37. Right triangle with hypotenuse 10 m and leg 8 m.

38. Right triangle with hypotenuse 13 ft and leg 12 ft.

39. Circle of diameter 9 in.

40. Right triangle whose legs measure 5 in. and 7 in.

41. Rectangle of length 90 cm and width 3 dm.

42. Parallelogram with base 2 ft and height 8 in.

(10.6) **Find the volume of each of the following solids. Leave answers in terms of π where appropriate.**

43. Cube that measures 8 in. on every edge.

44. Cylinder with radius 4 ft and height 7 ft.

45. Sphere with radius 3 in.

46. Hemisphere with radius 9 cm.

47. Cone with radius 2 m and height 6 m.

48. Sphere with diameter 12 ft.

49. Cylinder with radius 9 cm and height 5 dm.

50. Cone with radius 8 in. and height $1\frac{1}{2}$ ft.

(10.2–10.6) **Solve each of the following word problems.**

51. A door that measures 2.15 m long and 90 cm wide is to be decorated with a string of lights. How long must the light string be in order to reach around the door?

52. A rectangular piece of land that measures 160 yd long × 120 yd wide is bounded on one side by a river that runs parallel to the longest side. How much fencing is needed to enclose the three remaining sides?

53. The top of 13-ft ladder rests against a wall at a point that is 12 ft above the ground. How far away from the wall is the base of the ladder?

54. Two boats leave from the same starting point, one headed due south at a speed of 40 mph, and the other headed due east at a speed of 30 mph. How far apart are they $\frac{1}{2}$ hr later?

55. How many square feet of carpeting are needed to cover the floor of a room shaped like the accompanying figure?

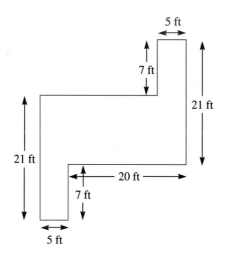

56. How much would it cost to paint the four walls of a room that measures 20 ft long, 18 ft wide, and $8\frac{1}{2}$ ft high if paint sells for $12 per gallon, and 1 gal covers 200 sq ft? Assume that there are no windows in the room and that the door is to be painted.

57. A square has a perimeter of 150 cm. What is the length of one of its sides?

58. An equilateral triangle has a perimeter of 54 in. What is the length of one of its sides?

59. A triangle with a 72-ft base has an area of 2,160 ft^2. What is its height?

60. A rectangle with an area of 69 cm^2 has a length of 10 cm. What is its width?

61. If the side of a square measures 5 in., what is the length of its diagonal? Round off to the nearest tenth of an inch.

62. The diagonal of a rectangle measures 20 cm. If the width of the rectangle is 12 cm, what is its length?

63. What is the diameter of a circle whose circumference measures 62π m?

64. What is the radius of a circle whose area is 81π ft^2?

65. The volume of a cube is 64 cm^3. What is the length of one of its edges?

66. The volume of a cylinder with a 3-in. radius is 72π in.^{3}. What is its height?

67. A window is 60 in. high and 72 in. wide. In order to calculate the amount of fabric needed for draperies, you must add 1 foot to the height of the window and double the width of the window. If the fabric is 36 in. wide, how many yards are needed to make draperies for this window? (Assume that you will need four panels of fabric.)

68. A living room is $11\frac{1}{2}$ ft wide and 25 ft in length. If carpeting comes in rolls that are 12 ft wide, how many yards must be purchased to cover the living room? How much carpeting is wasted?

69. The earth has a diameter of 7,927 mi at the equator. What is the circumference of the earth measured along the equator? Use $\pi \cong 3.14$. Round off your answer to the nearest mile.

70. The moon makes one revolution around the earth in about 27 days while traveling at a speed of 0.63 mile per second. What is the approximate distance between the earth and the moon? Use $\pi \cong 3.14$. Round off your answer to the nearest thousand miles.

 71. A punch bowl is shaped like a hemisphere with a radius of 20 cm. How many liters of punch can it hold when full? Use $\pi \cong 3.14$. Round off to the nearest tenth of a liter.

 72. A barrel is shaped like a cylinder that has a diameter of 85 cm and a height of 120 cm. How many liters can it hold when full? Use $\pi \cong 3.14$. Round off your answer to the nearest liter.

Chapter **10**
Explain It in Words

73. What is the difference between perimeter and area? Give an example of something that would be measured using perimeter and something that would be measured using area.

74. What is the difference between area and volume? Give an example of something that would be measured using area and something that would be measured using volume.

75. Explain how to find the hypotenuse of a right triangle, given the measurement of its two legs. Use an example.

76. Explain how to find the area of a circle, given the measurement of its radius. Use an example.

77. If the measurement of the side of a square is equal to the measurement of the side of an equilateral triangle, which figure has a larger perimeter? Explain why.

78. If the measurement of the base and height of a parallelogram is equal to the measurement of the base and height of a triangle, which figure has the larger area? Explain why.

Chapter **10**
Chapter Test

[10.3, 10.6]   Convert each of the following units of measurement.

1. $4 \, m^2 = $ _____ cm^2

2. $83 \, cm^3 = $ _____ ml

3. $5 \, yd^3 = $ _____ ft^3

4. $30 \, L = $ _____ cm^3

[10.2, 10.5]   Find the perimeter (or circumference) of each of the following figures. Leave your answers in terms of π where appropriate.

5. Rectangle of length 5 m and width 4 m.

6. Regular hexagon whose sides measure 7 in.

7. Circle of radius 9 cm.

8. Right triangle whose legs measure 12 ft and 16 ft.

[10.3, 10.5]   Find the area of each of the following figures. Leave your answers in terms of π where appropriate.

9. Parallelogram with base 6 dm and height 2 dm.

10. Triangle with base 9 ft and height 2 ft.

11. Circle with 7 m diameter.

12. Trapezoid with bases 8 in. and 1 ft and height 6 in.

[10.6] **Find the volume of each of the following figures. Leave your answers in terms of π where appropriate.**

13. Cube that measures 4 in. on every edge.

14. Cylinder with radius 2 cm and height 5 cm.

15. Sphere with radius 3 yd.

16. Cone with radius 3 dm and height 8 dm.

[10.2–10.6] **Solve each of the following word problems.**

17. The top of a 10-ft ladder rests against a wall at a point that is 8 ft above the ground. How far from the wall is the base of the ladder?

18. How many feet of ribbon are needed to decorate the border of a circular tablecloth that has a radius of 4 ft?

19. A rectangle having an area of 28 m^2 has a width of 4 m. What is its length?

20. What is the radius of a circle whose area is 64π cm^2?

Chapter **11**

Introduction to Algebra

Algebra is considered to be a generalization of arithmetic. It is a branch of mathematics that provides us with an extremely powerful tool for solving a wide variety of problems that involve finding an unknown quantity. Before we begin our discussion of algebra, it is important that we understand the vocabulary that is used to explain ideas that pertain to it.

The notation used in algebraic expressions consists of **constants** and **variables**. Constants are numbers or quantities that do not change in value. For example, the number 9 and the symbol π are both constants. Variables are unknown quantities that may have different values and are represented by letters. For example, the letters x, y, and z are often said to represent variables. The **coefficient** of a variable is the number that is multiplied by the variable. For example, the expression $7x$ means 7 multiplied by x. The variable is x and the constant is 7. The coefficient of x is 7.

The **terms** of an algebraic expression are the quantities that are added or subtracted. For example, the expression

$$4x + 9y$$

terms

has two terms, $4x$ and $9y$. The expression

$$8a - 9b + 5c$$

terms

has three terms, $8a$, $9b$, and $5c$. The term $8a$ is composed of two factors, the constant 8 and the variable a. Eight is also the coefficient of a. In the term $-9b$, -9 is the coefficient of the variable b. In the term $5c$, 5 is the coefficient of the variable c.

The procedure for evaluating an algebraic expression is given here.

To evaluate an algebraic expression, substitute numbers for the unknown variables and simplify the result using the order of operations rules. ▲

413

EXAMPLE 1   Evaluate $5x + 7y$ when $x = -2$ and $y = 4$.

Solution
$$
\begin{aligned}
5x + 7y &= 5(-2) + 7(4) && \text{Substitute } -2 \text{ for } x \text{ and } 4 \text{ for } y.\\
&= -10 + 28 && \text{First do multiplication.}\\
&= 18 && \text{Then do addition.} \qquad \blacktriangle
\end{aligned}
$$

EXAMPLE 2   Evaluate $5(2a - b) + 4c$ when $a = 4$, $b = -2$, and $c = -5$.

Solution
$$
\begin{aligned}
5(2a - b) + 4c &= 5[2(4) - (-2)] + 4(-5) && \text{Substitute } a = 4, b = -2, c = -5.\\
&= 5[8 - (-2)] + 4(-5) && \text{Simplify expression in [ ] first by}\\
& && \text{multiplying, then subtracting.}\\
&= 5(10) + 4(-5)\\
&= 50 + (-20) && \text{Do multiplication.}\\
&= 30 && \text{Finally, do addition.} \qquad \blacktriangle
\end{aligned}
$$

Since algebra is an important tool for solving a wide variety of word problems, we will now illustrate how words can be translated into algebraic expressions.

EXAMPLE 3   Translate each of the following word expressions into algebraic expressions.

(a)   Three more than x.
$x + 3$

(b)   The product of eight and n.
$8n$

Notice that the multiplication sign is omitted when expressing the product of a number and a variable.

(c)   Four less than w.
$w - 4$

Note that "four less than w" means that 4 is subtracted from w.

(d)   Five times the sum of seven and y.
$5(7 + y)$

Notice that the parentheses indicate that addition is to be performed before multiplication.

(e)   Two-fifths the quantity eight minus t.

$\dfrac{2}{5}(8 - t)$

Notice that the expression that follows the words *the quantity* is enclosed in parentheses. $\qquad \blacktriangle$

Let us now look at some examples of algebraic expressions that can be translated into words.

EXAMPLE 4   Translate each of the following algebraic expressions into words.

(a) $5 + 3w$
Five plus the product of three and w.

(b) $6(a + b)$
Six times the sum of a and b, or six times the quantity a plus b.

Remember that the words *the quantity* indicate that the expression that follows appears in parentheses. $\qquad \blacktriangle$

Quick Quiz

Answers

1. Evaluate $5x + 7y - 3z$ when $x = 2$, $y = -1$, and $z = -2$.

   1. 9

2. Write an algebraic expression that represents eight subtracted from t.

   2. $t - 8$

3. Translate the expression $(x + 5) \div 2$ into words.

   3. The quantity x plus five, divided by two.

11.1 Exercises

Indicate which of the following statements are true and which are false. For those that are false, change the italicized word to make the statement true.

1. A number that appears in an algebraic expression is called a *constant*.
2. Quantities that are added or *multiplied* are called terms.
3. The *coefficient* of a variable is a number that is multiplied by a variable.
4. Algebra is a generalization of *geometry*.

Indicate how many terms are in each of the following algebraic expressions. For each term, identify the variable and its coefficient.

5. $3w$

6. t

7. $2x + 5y$

8. $8a - 7b$

9. $3p + 9q + r$

10. $x + 6y - 4z$

11. $\frac{1}{2}c - d$

12. $x + \frac{3}{4}y + \frac{1}{8}z - 12w$

Evaluate each of the following algebraic expressions for the values given.

13. $a + 4b$ when $a = -8$, $b = 3$

14. $5x - y$ when $x = 2$, $y = 6$

15. $4s - 7t$ when $s = 5$, $t = -1$

16. $3w + 2z$ when $w = -6$, $z = -5$

17. $\frac{1}{2}x - \frac{2}{3}y$ when $x = 4$, $y = 9$

18. $\frac{3}{4}a + \frac{7}{8}b$ when $a = 12$, $b = -8$

19. $4p - 3q - 5r$ when $p = 3$, $q = 7$, $r = -1$

20. $-7a - 9b - 11c$ when $a = -2$, $b = 3$, $c = 1$

21. $-5x + y - 2z$ when $x = \frac{4}{5}$, $y = 9$, $z = -\frac{1}{2}$

22. $s - 3t - 7u$ when $s = -4$, $t = \frac{2}{3}$, $u = \frac{5}{7}$

23. $-2(a + 5b)$ when $a = -15$, $b = 12$

24. $4(3x - y)$ when $x = 21$, $y = 30$

25. $5(2s - 5t + 3u)$ when $s = 3$, $t = -2$, $u = 6$

26. $-7(3x - 4y - 2z)$ when $x = -5$, $y = -4$, $z = -1$

27. $3(w - 2t) + 5(3w + t)$ when $w = -1$, $t = 6$

28. $9(x - y) - 2(3x - 4y)$ when $x = 4$, $y = -7$

Translate each of the following word expressions into algebraic expressions.

29. x decreased by five.

30. Nine more than y.

31. The product of t and negative two.

32. Thirteen divided by w.

33. Two-thirds of z.

34. One-half of q.

35. Thirty-five less than x.

36. Twenty-four increased by y.

37. Five more than the product of seven and *a*.
38. Two less than the product of negative four and *n*.
39. Negative eight times the sum of *a* and *b*.
40. Seven times the quantity *x* minus *y*.
41. The quantity fourteen minus *x*, divided by two.
42. The quantity *y* plus nine, divided by four.
43. Twice the sum of *a* and five.
44. The quantity nine minus *t*, tripled.
45. Five-eighths the quantity *u* plus seven.
46. Three-fifths the quantity one minus *w*.

Translate each of the following algebraic expressions into words.

47. $x - 4$
48. $y + 9$
49. $6a$
50. $14 \div a$
51. $9 + t$
52. $3 - n$
53. $w \div 11$
54. $-5z$
55. $2y + 8$
56. $7x - 13$
57. $3n - 2$
58. $4p + 19$
59. $5(z - 7)$
60. $6(t + 4)$
61. $(4 - u) \div 3$
62. $(y + 9) \div 2$
63. $\frac{3}{8}(x + 5)$
64. $\frac{1}{3}(w - 8)$

11.2
Simplifying Algebraic Expressions

In this section, we will show how to simplify algebraic expressions using a technique called **combining like terms**.

▶ **DEFINITION** *Like terms*, or *similar terms*, are terms whose variable factors are the same.

For example, the terms $8x$ and $\frac{3}{5}x$ are like terms since they have the same variable factor, *x*. The terms $7xy$ and $-2xz$ are not like terms, because the variable factors of $7xy$ are xy, and the variable factors of $-2xz$ are xz. The procedure for combining like terms is summarized below and illustrated in the following examples.

To simplify an algebraic expression by combining like terms, first rewrite the expression using the distributive property, and then add or subtract the coefficients of the variable. ▲

EXAMPLE 1 Simplify each of the following algebraic expressions by combining like terms.

(a) $8x + 3x$
$$8x + 3x = (8 + 3)x$$ Rewrite using the distributive property.
$$= 11x$$ Add the coefficients of *x*.

(b) $9y - 4y$
$$9y - 4y = (9 - 4)y$$ Rewrite using the distributive property.
$$= 5y$$ Subtract the coefficients of *y*.

(c) $5x + x + 8x$
$$5x + x + 8x = (5 + 1 + 8)x$$ Rewrite using the distributive property.
$$= 14x$$ Add the coefficients of *x*. Keep in mind that the second term, *x*, has a coefficient of 1.

(d) $3x - 7x - 9x$

$3x - 7x - 9x = (3 - 7 - 9)x$ Think, $3 - 7 - 9 = 3 + (-7) + (-9)$.

$\qquad\qquad\quad = -13x$ ▲

Notice that when we simplify algebraic expressions, we usually do not write the operation subtraction as adding the opposite, even though mentally we may be using that technique to combine the coefficients. [See Example 1(d).]

Sometimes, before we can combine like terms, we need to use the commutative property of addition to change the order of the terms when rewriting the problem, as shown in Example 2.

EXAMPLE 2   Simplify each of the following algebraic expressions.

(a) $5x - 8 + 7x + 3$

$5x - 8 + 7x + 3 = 5x + 7x - 8 + 3$	By the commutative property of addition.
$\qquad\qquad\qquad = (5 + 7)x - 8 + 3$	By the distributive property.
$\qquad\qquad\qquad = 12x - 8 + 3$	Add the coefficients of x.
$\qquad\qquad\qquad = 12x - 5$	Combine the constant terms.

(b) $2a - 5 - 8a + 9 - 4a - 3$

$2a - 5 - 8a + 9 - 4a - 3$

$= 2a - 8a - 4a - 5 + 9 - 3$	By the commutative property of addition.
$= (2 - 8 - 4)a - 5 + 9 - 3$	By the distributive property.
$= -10a - 5 + 9 - 3$	Combine the coefficients of a.
$= -10a + 1$	Combine the constant terms. ▲

If an algebraic expression that appears in parentheses cannot be simplified, we multiply each term in parentheses by the factor that precedes the parentheses, using the distributive property. We then simplify the resulting expression by combining like terms. Consider Example 3.

EXAMPLE 3   Simplify each of the following algebraic expressions.

(a) $3(2y - 5) + 8$

$3(2y - 5) + 8 = 3(2y) - 3(5) + 8$	Use the distributive property to remove the parentheses by multiplying each term in parentheses by 3.
$\qquad\qquad = (3 \cdot 2)y - 3(5) + 8$	By the associative property.
$\qquad\qquad = 6y - 15 + 8$	Do multiplication first.
$\qquad\qquad = 6y - 7$	Add constant terms.

(b) $2(7x - 3) + 5(-2x + 5)$

$2(7x - 3) + 5(-2x + 5)$

$= 2(7x) - 2(3) + 5(-2x) + 5(5)$	Use the distributive property to remove parentheses.
$= 14x - 6 + (-10x) + 25$	
$= 14x + (-10x) - 6 + 25$	By the commutative property.
$= (14 - 10)x - 6 + 25$	To add $-10x$ we subtract $10x$.
$= 4x + 19$	

(c) $8 - (4w + 3)$

$8 - (4w + 3) = 8 + (-1)(4w + 3)$ To subtract a quantity in parentheses, we add the opposite of each term in parentheses. This can be shown using the distributive property. Multiply each term in parentheses by -1.

$\qquad\qquad = 8 + (-1)(4w) + (-1)(3)$

$\qquad\qquad = 8 + (-4w) + (-3)$

$\qquad\qquad = -4w + 8 - 3$ By the commutative property.

$\qquad\qquad = -4w + 5$

(d) $-8(x - 9) - 3(x + 5)$

$-8(x - 9) - 3(x + 5)$

$\qquad = -8(x - 9) + (-3)(x + 5)$

$\qquad = -8x - 8(-9) + (-3)x$ Use the distributive property to remove parentheses.
$\qquad\quad + (-3)(5)$

$\qquad = -8x + 72 - 3x - 15$

$\qquad = -8x - 3x + 72 - 15$ By the commutative property.

$\qquad = -11x + 57$ Combine like terms.   ▲

Quick Quiz

Simplify each of the following algebraic expressions by combining like terms.

Answers

1. $-2x + 9x - 4x$ 1. $3x$
2. $5z - 3 - 7z + 8$ 2. $-2z + 5$
3. $4(5t + 2) - (t - 3)$ 3. $19t + 11$

11.2 Exercises

Indicate which of the following statements are true and which are false. For those that are false, change the italicized word to make the statement true.

1. Like terms are terms in which the *constant* factors are the same.
2. When combining like terms, we add or *subtract* the coefficients of the variable.
3. To multiply a quantity in parentheses, we use the *distributive* property.
4. To change the order of terms when simplifying an algebraic expression, we use the *associative* property.

Simplify each of the following algebraic expressions by combining like terms.

5. $8x + 9x$ 6. $7y - 2y$ 7. $-3a + 5a$

8. $-4b - 6b$ 9. $t - 9t$ 10. $5w + w$

11. $\dfrac{3}{4}a - \dfrac{1}{4}a$ 12. $\dfrac{1}{5}z + \dfrac{3}{5}z$ 13. $-35u - 18u$

14. $-61t + 23t$

15. $7.2p - 4.9p$

16. $5.1z - 3.4z$

17. $8y + 4y + 3y$

18. $7x + 5x + 9x$

19. $-2z + 4z + 12z$

20. $8s - 3s - 9s$

21. $23c + c + 19c$

22. $69r - 47r - 51r$

23. $-8.5u + u - 6.7u$

24. $w - 4.8w + 2.5w$

25. $5(y + 2) - 3$

26. $7(q - 6) + 4$

27. $8(5 - n) + 3$

28. $-3(7 + y) - 5$

29. $5 - (9w + 4)$

30. $2 - (5 + 8t)$

31. $-3 - (5p - 8)$

32. $-7 - (4 - d)$

33. $7x - 5(2x - 4)$

34. $-3z + 7(-4z - 3)$

35. $4s - 3(5s - 2)$

36. $-2u - 4(-7u + 5)$

37. $-3(6x + 7) + 9(2x + 1)$

38. $8(3a + 4) - 4(6a + 2)$

39. $6(2z - 4) + 8(7z - 5)$

40. $2(7m - 6) - 4(3m - 2)$

41. $4(-3t + 6) - 2(5t - 8)$

42. $-6(3b - 5) + 7(5b - 3)$

43. $-7(6k + 2) + 5(1 - 4k)$

44. $9(2a - 3) - 8(-a - 1)$

11.3
The Addition Principle of Equality

The term **equation** was defined in Section 1.2 as a mathematical expression that states that two quantities are equal. We have already worked with some simple equations in Section 5.4, which dealt with finding the unknown term in a proportion, and in Section 6.3, which dealt with equations involving percent. In the next three sections, we will learn how to solve a variety of similar equations, called **linear equations**.

▶ **DEFINITION** A *linear equation* in one variable is an equation that can be written in the form $ax + b = c$, where a, b, and c are constants and $a \neq 0$.

Some examples of linear equations in one variable are:

$$3x + 9 = 0 \qquad 6x + 5 = 2x - 8$$
$$x - 7 = 2 \qquad 3(x - 2) = 4$$

Notice that each of these equations has only a single variable, x, and that the exponent of x is 1 since $x = x^1$. For this reason, equations of this form are often called first-degree equations of a single variable.

▶ **DEFINITION** The *solution* to a linear equation in one variable is the value that can be substituted for the unknown variable so that the resulting statement is true.

For example, the number 4 is a solution to the linear equation $3x - 5 = 7$, since the resulting statement is true when we substitute 4 for x.

$$3x - 5 = 7$$
$$3(4) - 5 = 7$$
$$12 - 5 = 7$$
$$7 = 7 \qquad \text{This is a true statement.}$$

EXAMPLE 1 Determine if the number 2 is a solution to the linear equation $7x - 11 = 8 - 3x$.

Solution

$$7x - 11 = 8 - 3x$$
$$7(2) - 11 = 8 - 3(2) \qquad \text{Substitute 2 for } x.$$
$$14 - 11 = 8 - 6$$
$$3 = 2 \qquad \text{This is a false statement.}$$

Therefore, the number 2 is not a solution to the linear equation. ▲

In order to solve a linear equation, we will perform a series of algebraic operations on both sides of the equation that will result in a series of **equivalent equations**.

▶ **DEFINITION** *Equivalent equations* **are equations that have the same solution.**

For example, the equations

$$-5x + 3 = 18 \quad \text{and} \quad x = -3$$

are equivalent equations since they have the same solution, -3. That is, when -3 is substituted for x in both equations, the resulting statements are true.

EXAMPLE 2 Determine whether or not each of the following pairs of equations is equivalent.

(a) $y = -4$ and $3y - 2 = y - 9$

$$3y - 2 = y - 9$$
$$3(-4) - 2 = (-4) - 9 \qquad \text{Substitute } y = -4.$$
$$-12 - 2 = -13$$
$$-14 = -13 \qquad \text{This is a false statement.}$$

Therefore, $y = -4$ and $3y - 2 = y - 9$ are not equivalent equations. ▲

EXAMPLE 3 Determine if $x = -\frac{1}{2}$ and $2(2 - 3x) = 3 - 8x$ are equivalent equations.

Solution

$$2(2 - 3x) = 3 - 8x$$
$$2\left[2 - 3\left(-\frac{1}{2}\right)\right] = 3 - 8\left(-\frac{1}{2}\right) \qquad \begin{array}{l}\text{Substitute } -\frac{1}{2} \text{ for } x \text{ in the second equation}\\ \text{and determine if the resulting statement}\\ \text{is true.}\end{array}$$
$$2\left[2 - \left(-\frac{3}{2}\right)\right] = 3 - (-4)$$
$$2\left(2 + \frac{3}{2}\right) = 3 + 4$$
$$2\left(\frac{7}{2}\right) = 7$$
$$7 = 7 \qquad \text{This is a true statement.}$$

Therefore, $x = -\frac{1}{2}$ and $2(2 - 3x) = 3 - 8x$ are equivalent equations. ▲

The first type of linear equation that we will consider can be solved using the **addition principle of equality**.

According to the *addition principle of equality,* if the same quantity is added to both sides of an equation, the resulting equation is equivalent to the original one. ▲

Recall from Section 5.3 that the goal in solving an equation is to isolate the variable on one side of the equation so that the number that is a solution to the equation appears on the other side. If an equation is of the form $x + b = c$, where b and c are constants, we can isolate the variable x on the left by adding the opposite of b, or $-b$, to both sides of the equation, as follows:

$x + b + (-b) = c + (-b)$	By the addition property of equality.
$x + 0 = c + (-b)$	Since $b + (-b) = 0$.
$x = c - b$	Combine constants on the right.

This procedure is illustrated by Example 4.

EXAMPLE 4   Solve $x + 5 = 3$.

Solution

$x + 5 = 3$	
$x + 5 + (-5) = 3 + (-5)$	Add the opposite of 5, or -5, to both sides.
$x + 0 = -2$	
$x = -2$	▲

EXAMPLE 5   Solve $-6 = a - 2$.

Solution

$-6 + 2 = a - 2 + 2$	Add the opposite of -2, or 2, to both sides.
$-4 = a + 0$	
$-4 = a$	Notice that the statements $-4 = a$ and
$a = -4$	$a = -4$ represent the same solution.

Though it is correct to express the solution with the variable on either the left- or right-hand side of the equal sign, the variable usually appears on the left. ▲

Often, we need to combine like terms on both sides of the equation before applying the addition principle of equality, as shown in Example 6.

EXAMPLE 6   Solve $-8y - 3 + 9y - 7 = 4$.

Solution

$-8y - 3 + 9y - 7 = 4$	
$-8y + 9y - 3 - 7 = 4$	By the commutative property.
$y - 10 = 4$	Combine like terms.
$y - 10 + 10 = 4 + 10$	Add the opposite of -10, or 10, to both sides.
$y + 0 = 14$	
$y = 14$	▲

Whenever variable terms appear on both sides of the equation, we use the addition principle of equality to move all variable terms to one side and all constant terms to the other side, as shown in the following examples.

EXAMPLE 7 Solve $-5x = 8 - 6x$.

Solution
$$-5x = 8 - 6x$$
$$-5x + 6x = 8 - 6x + 6x$$ Add the opposite of $-6x$, or $6x$, to both sides.
$$x = 8 + 0$$
$$x = 8$$ ▲

EXAMPLE 8 Solve $5x - 2 = 9 + 4x$.

Solution
$$5x - 2 = 9 + 4x$$
$$5x - 2 + (-4x) = 9 + 4x + (-4x)$$ Add the opposite of $4x$, or $-4x$, to both sides.
$$5x - 2 + (-4x) = 9 + 0$$
$$5x + (-4x) - 2 = 9$$ By the commutative property.
$$x - 2 = 9$$ Combine like terms.
$$x - 2 + 2 = 9 + 2$$ Add 2 to both sides.
$$x + 0 = 11$$
$$x = 11$$ ▲

Whenever quantities appear in parentheses on either side of the equation, we must first multiply them using the distributive property, and then combine like terms. This procedure is illustrated in the following examples.

EXAMPLE 9 Solve $7(x - 3) = 6x$.

Solution
$$7(x - 3) = 6x$$
$$7x - 21 = 6x$$ By the distributive property.
$$(-6x) + 7x - 21 = 6x + (-6x)$$ Add the opposite of $6x$, or $-6x$, to both sides.
$$x - 21 = 0$$ Combine like terms.
$$x - 21 + 21 = 0 + 21$$ Add the opposite of -21, or 21, to both sides.
$$x + 0 = 21$$
$$x = 21$$ ▲

EXAMPLE 10 Solve $-3(t + 2) + 2(2t - 1) = -5$.

Solution
$$-3(t + 2) + 2(2t - 1) = -5$$
$$-3t - 6 + 4t - 2 = -5$$ By the distributive property.
$$-3t + 4t - 6 - 2 = -5$$ By the commutative property.
$$t - 8 = -5$$ Combine like terms.
$$t - 8 + 8 = -5 + 8$$ Add 8 to both sides.
$$t + 0 = 3$$
$$t = 3$$ ▲

Quick Quiz

Solve each of the following linear equations.

1. $x - 9 = 6$
2. $11 - 7y + 8y = 5$
3. $-4t + 3 = 7 - 5t$
4. $5z = 4(z + 6)$

Answers

1. $x = 15$
2. $y = -6$
3. $t = 4$
4. $z = 24$

11.3 Exercises

Indicate which of the following statements are true and which are false. For those that are false, change the italicized word to make the statement true.

1. Equivalent equations have the same *variables*.

2. Linear equations are sometimes called *second*-degree equations.

3. To move a term to the other side of a linear equation, we add its *opposite* to both sides.

4. The *solution* to a linear equation in one variable is the value that can be substituted for the variable so that the resulting statement is true.

Determine whether or not each of the following numbers is a solution to the linear equation that follows it.

5. 2: $3x + 8 = 14$

6. $\frac{1}{2}$: $2y + 5 = 4$

7. -5: $2z + 1 = z - 3$

8. 3: $4x + 3 = 18 - x$

9. -11: $7(t - 3) = 9t$

10. $-\frac{2}{3}$: $-2(3x + 5) + 7 = 1$

Determine whether or not each of the following pairs of equations is equivalent.

11. $x = -4$ and $3x + 7 = 5$

12. $x = \frac{13}{5}$ and $5x - 9 = 3$

13. $z = \frac{8}{3}$ and $-2z + 7 = 4z - 9$

14. $y = -2$ and $9y = 2y - 14$

15. $a = 1$ and $3(2a - 5) + 2 = -8$

16. $b = -3$ and $2(b - 1) = 9(2b + 5)$

Use the addition principle of equality to solve each of the following linear equations.

17. $x - 2 = 6$

18. $x - 9 = 1$

19. $y + 5 = 2$

20. $y + 3 = 11$

21. $w - 6 = -8$

22. $t + 4 = -7$

23. $5 = x + 7$

24. $8 = x - 3$

25. $-3 = y - 11$

26. $-6 = y + 5$

27. $a + 17 = 64$

28. $b + 61 = 24$

29. $z - 52 = 29$

30. $w - 33 = 47$

31. $x + 3.6 = 8.4$

32. $y - 7.2 = 2.3$

33. $5x + 8 - 4x = 3$

34. $9x + 3 - 8x = 5$

35. $-8y - 6 + 9y = -2$

36. $-3y - 7 + 4y = -6$

37. $7z + 2 - 6z - 1 = 4$

38. $5z - 8 - 4z + 3 = 7$

39. $8 - 2t + 5 + 3t = 9$

40. $-3 + 6t + 8 - 5t = 2$

41. $5 = 4t - 6 - 3t + 7$

42. $7 = 9t + 5 - 8t + 1$

43. $3p - 9p + 7p + 5 = -8$

44. $5p - 8p - 3 + 4p = -5$

45. $-7 = 4w + 9 + 3w - 4 - 6w$

46. $-4 = 8w + 5 - 5w - 7 - 2w$

47. $8 - 5x + 6 - 2x + 8x = 3$

48. $7 - 3x - 4 - 5x + 9x = -6$

49. $9x + 2 = 8x$

50. $5x - 7 = 4x$

51. $5x = 4x - 9$

52. $7x = 6x - 5$

53. $7c + 3 = 6c - 5$

54. $3c - 8 = 2c + 5$

55. $8 - 4d = -5d - 3$

56. $11 - 2d = -3d + 4$

57. $-3x = -4x + 7$

58. $-4x = -5x + 9$

59. $9 - 5w = 11 - 6w$

60. $7 - 2w = 1 - 3w$

61. $8a - 3 - a + 11 = 9 - 3a + 5 + 9a$

62. $5a - 9 + a - 7 = 3 + 9a - 6 - 4a$

63. $9 - 5b + 3 + 8b = 7b + 3 - 5b$

64. $3 - 7b + 6 + 5b = 2b + 8 - 5b$

Solve each of the following linear equations by first using the distributive property to remove parentheses.

65. $5(x - 7) - 4x = 5$

66. $3(x - 8) - 2x = 1$

67. $3(x + 5) - 2x = -9$

68. $9(x + 3) - 8x = -5$

69. $-6(x - 8) + 7x = 4$

70. $-2(k + 6) + 3k = -1$

71. $3(4x - 7) = 11x - 8$

72. $6(3x + 8) = 17x + 9$

73. $19x = 6(3x + 5)$

74. $25x = 8(3x - 1)$

75. $5(2t + 3) - 9t = 7$

76. $7u - 2(3u + 4) = 4$

77. $4(5x - 7) + 6 = 6(2x - 1) + 7x$

78. $3(4x + 5) - 7 = 5(3x + 1) - 4x$

79. $2(8x - 3) - 5x = 5(2x + 1)$

80. $-3(2x + 7) + 11x = 2(2x - 5)$

11.4

The Multiplication Principle of Equality

Notice that in each of the examples presented in Section 11.3, once we simplified the example by combining like terms, the coefficient of the variable was 1. We were then able to solve the equation using the addition principle of equality. To solve linear equations in which the coefficient of the variable is not 1, we must use the **multiplication principle of equality** to isolate the variable.

According to the *multiplication principle of equality,* if both sides of an equation are multiplied by the same nonzero quantity, the resulting equation is equivalent to the original one. ▲

Recall that we used this principle in Sections 5.3 and 6.3 to solve equations of the form $ax = c$, where a and c are constants. To isolate the variable x, we multiplied both sides by the reciprocal of the coefficient of x, which is $\frac{1}{a}$, as shown.

$$\frac{1}{a} \cdot ax = c \cdot \frac{1}{a}$$
By the multiplication principle of equality.

$$1 \cdot x = c \cdot \frac{1}{a}$$
Since $\frac{1}{a} \cdot a = 1$.

$$x = \frac{c}{a}$$

We will use the following examples to review this technique.

EXAMPLE 1 Solve $6x = 108$.

Solution

$$6x = 108$$

$$\left(\frac{1}{6}\right)(6x) = (108)\left(\frac{1}{6}\right)$$
Multiply both sides by the reciprocal of 6, which is $\frac{1}{6}$.

$$1 \cdot x = \frac{108}{6}$$

$$x = 18$$
Reduce to lowest terms. ▲

EXAMPLE 2 Solve $-\frac{x}{4} = \frac{3}{8}$.

Solution

$$-\frac{1}{4}x = \frac{3}{8}$$
Recall that $-\frac{x}{4} = -\frac{1}{4}x$.

$$(-4)\left(-\frac{1}{4}x\right) = \left(\frac{3}{8}\right)(-4)$$
Multiply both sides by the reciprocal of $-\frac{1}{4}$, which is -4.

$$1 \cdot x = -\frac{(3)(4)}{8}$$

$$x = -\frac{3}{2}$$
Reduce to lowest term. ▲

Remember that according to the commutative property of multiplication, we can change the order of the numbers being multiplied without affecting the result. For example,

$$(5)(-9) = (5)(-1)(9) = (-1)(5)(9) = -(5)(9) = -45$$

Similarly, in Example 2,

$$\left(\frac{3}{8}\right)(-4) = \frac{(3)(-4)}{8} = \frac{(3)(-1)(4)}{8} = \frac{(-1)(3)(4)}{8} = \frac{-(3)(4)}{8} = \frac{-12}{8} = \frac{-3}{2}$$

EXAMPLE 3 Solve $\frac{2}{3}z = \frac{5}{9}$.

Solution

$$\frac{2}{3}z = \frac{5}{9}$$

$$\left(\frac{3}{2}\right)\left(\frac{2}{3}z\right) = \left(\frac{5}{9}\right)\left(\frac{3}{2}\right)$$
Multiply both sides by the reciprocal of $\frac{2}{3}$, which is $\frac{3}{2}$.

$$1 \cdot z = \frac{15}{18}$$
Since $(\frac{3}{2})(\frac{2}{3}) = 1$.

$$z = \frac{5}{6}$$
Reduce to lowest terms. ▲

EXAMPLE 4 Solve $-5\frac{1}{4}x = 63$.

Solution
$$-5\frac{1}{4}x = 63$$

$$-\frac{21}{4}x = 63 \qquad\qquad \text{Rewrite } -5\frac{1}{4} \text{ as an improper fraction.}$$

$$\left(-\frac{4}{21}\right)\left(-\frac{21}{4}x\right) = (63)\left(-\frac{4}{21}\right) \qquad\qquad \text{Multiply both sides by the reciprocal of } -\frac{21}{4}, \text{ which is } -\frac{4}{21}.$$

$$x = \frac{-(63)(4)}{21}$$

$$x = -12 \qquad\qquad \text{Reduce to lowest terms.} \qquad \blacktriangle$$

EXAMPLE 5 Solve $3.7x = 18.5$.

Solution
$$3.7x = 18.5$$

$$\left(\frac{1}{3.7}\right)(3.7x) = (18.5)\left(\frac{1}{3.7}\right) \qquad\qquad \text{Multiply both sides by the reciprocal of } 3.7, \text{ which is } \frac{1}{3.7}.$$

$$1 \cdot x = \frac{18.5}{3.7}$$

$$x = 5 \qquad\qquad \text{Reduce to lowest terms.} \qquad \blacktriangle$$

Sometimes it is necessary to combine like terms before applying the multiplication principle, as shown in the following examples.

EXAMPLE 6 Solve $2x - 9x = 63$.

Solution
$$2x - 9x = 63$$
$$(2 - 9)x = 63 \qquad\qquad \text{By the distributive property.}$$
$$-7x = 63 \qquad\qquad \text{Combine like terms.}$$

$$\left(-\frac{1}{7}\right)(-7x) = (63)\left(-\frac{1}{7}\right) \qquad\qquad \text{Multiply both sides by the reciprocal of } -7, \text{ which is } -\frac{1}{7}.$$

$$1 \cdot x = \frac{-63}{7}$$

$$x = -9 \qquad\qquad \text{Reduce to lowest terms.} \qquad \blacktriangle$$

EXAMPLE 7 Solve $2 = 8x - 4x - 7x$.

Solution

$$2 = 8x - 4x - 7x$$
$$2 = (8 - 4 - 7)x \qquad \text{By the distributive property.}$$
$$2 = -3x \qquad \text{Combine like terms.}$$
$$-3x = 2 \qquad \text{Rewrite equation so the variable term is on the left.}$$
$$\left(-\frac{1}{3}\right)(-3x) = (2)\left(-\frac{1}{3}\right) \qquad \text{Multiply both sides by the reciprocal of } -3, \text{ which is } -\frac{1}{3}.$$
$$1 \cdot x = -\frac{2}{3} \qquad \text{Since } \left(-\frac{1}{3}\right)(-3) = 1.$$
$$x = -\frac{2}{3}$$

▲

Notice that in Example 7, the statements $2 = -3x$ and $-3x = 2$ represent the same equation. Therefore, to get the variable term on the left, we performed no algebraic operation, but merely switched the terms to the other side of the equal sign.

Quick Quiz

Solve each of the following linear equations. **Answers**

1. $-9x = 72$ 1. $x = -8$

2. $\dfrac{z}{3} = \dfrac{2}{9}$ 2. $z = \dfrac{2}{3}$

3. $-\dfrac{5}{8}w = 35$ 3. $w = -56$

4. $-3x + 7x - 2x = 12$ 4. $x = 6$

11.4 Exercises

Use the multiplication principle of equality to solve each of the following linear equations.

1. $5x = 35$ 2. $8x = 24$ 3. $-7x = 63$ 4. $-5x = 25$

5. $72 = 9x$ 6. $56 = 7x$ 7. $45 = -5x$ 8. $42 = -6x$

9. $-24z = -120$ 10. $-18w = -72$ 11. $7p = 19$ 12. $11q = 4$

13. $-96 = 16a$ 14. $-132 = 33b$ 15. $8x = 0$ 16. $-3x = 0$

17. $\dfrac{1}{3}w = 5$ 18. $\dfrac{1}{7}y = 2$ 19. $\dfrac{3}{5}x = 9$ 20. $\dfrac{4}{9}x = 16$

21. $-\dfrac{5}{8}x = 45$ 22. $-\dfrac{3}{7}x = 15$ 23. $\dfrac{x}{7} = 4$ 24. $-\dfrac{x}{5} = 4$

25. $\dfrac{t}{6} = -\dfrac{2}{3}$ 26. $\dfrac{w}{8} = \dfrac{3}{4}$ 27. $\dfrac{2}{7}y = \dfrac{5}{9}$ 28. $\dfrac{3}{4}y = \dfrac{2}{5}$

29. $\dfrac{7}{9}z = \dfrac{3}{4}$ **30.** $\dfrac{3}{8}w = \dfrac{5}{6}$ **31.** $6.7m = 53.6$ **32.** $2.8m = 16.8$

33. $-3.8y = 19$ **34.** $6.2z = -43.4$

Solve each of the following linear equations by combining like terms before applying the multiplication principle of equality.

35. $5x + 2x = 42$ **36.** $8x - 3x = 45$ **37.** $5x - 8x = 27$

38. $9w - 2w + 5w = 48$ **39.** $4.5t - 3.8t = 6.3$ **40.** $2.9t + 5.6t = 34$

41. $3w - 6w + 7w = 16$ **42.** $5w - 8w + 6w = 33$ **43.** $-8z + 3z - 6z = 55$

44. $-9z - 2z + 5z = -36$ **45.** $-80 = -5y - 8y + 3y$ **46.** $-64 = 3y - 9y - 2y$

47. $17r + 9r - 35r = 27$ **48.** $52s - 6s - 76s = -6$

11.5

Steps for Solving Linear Equations

We are now prepared to tackle some linear equations that require us to use both the addition principle of equality and the multiplication principle of equality. To solve these equations, we can use the following procedure.

Steps for Solving Linear Equations

1. Simplify both sides of the equation as much as possible by using the distributive property to multiply all quantities in parentheses, and then combining like terms.

2. Use the addition principle of equality to move all variable terms to one side of the equation and all constant terms to the other side. This is done by adding to both sides of the equation the opposite of each term you wish to move, and then combining like terms.

3. Use the multiplication principle of equality to isolate the variable on one side of the equation and the solution on the other side. This is done by multiplying both sides of the equation by the reciprocal of the coefficient of the variable.

4. Check that your solution is correct by substituting it back into the original equation to see if the resulting statement is true. ▲

We will illustrate this procedure in the following examples.

EXAMPLE 1   Solve $9y - 7 = 11$.

Solution
$$9y - 7 = 11$$
$$9y - 7 + 7 = 11 + 7$$ Add 7 to both sides to move the constant term to the right.
$$9y = 18$$
$$\left(\frac{1}{9}\right)(9y) = (18)\left(\frac{1}{9}\right)$$ Multiply both sides by $\frac{1}{9}$.
$$y = 2$$

Check:

$$9y - 7 = 11$$
$$9(2) - 7 \ ? \ 11 \qquad \text{Substitute } y = 2.$$
$$18 - 7 \ ? \ 11$$
$$11 = 11 \quad \checkmark \qquad \text{This is a true statement.}$$

Therefore, $y = 2$ is the correct solution. ▲

Before we begin moving terms using the addition principle when solving more complicated linear equations, we must decide on which side to put the variable terms and on which side to put the constant terms. Otherwise, we might end up moving the same terms back and forth without simplifying the equation.

EXAMPLE 2   Solve $-7z = 3z - 40$.

Solution   Let's collect the variable terms on the left and the constant terms on the right.

$$-7z = 3z - 40$$
$$-3z + (-7z) = -3z + 3z - 40 \qquad \text{Add } -3z \text{ to both sides to move the variable term to the left.}$$
$$-10z = -40$$
$$\left(-\frac{1}{10}\right)(-10z) = (-40)\left(-\frac{1}{10}\right) \qquad \text{Multiply both sides by } -\frac{1}{10}.$$
$$z = 4$$

Check:

$$-7z = 3z - 40$$
$$-7(4) \ ? \ 3(4) - 40 \qquad \text{Substitute } z = 4.$$
$$-28 \ ? \ 12 - 40$$
$$-28 = -28 \quad \checkmark \qquad \text{This is a true statement.}$$

Therefore, $z = 4$ is the correct solution. ▲

EXAMPLE 3   Solve $2x - 3 = -5x + 9$.

Solution   Let's collect the variable terms on the left and the constant terms on the right.

$$2x - 3 = -5x + 9$$
$$5x + 2x - 3 = 5x - 5x + 9 \qquad \text{Add } 5x \text{ to both sides to move the variable term to the left.}$$
$$7x - 3 = 9$$
$$7x - 3 + 3 = 9 + 3 \qquad \text{Add 3 to both sides to move the constant term to the right.}$$
$$7x = 12$$
$$\left(\frac{1}{7}\right)(7x) = (12)\left(\frac{1}{7}\right) \qquad \text{Multiply both sides by } \frac{1}{7}.$$
$$x = \frac{12}{7}$$

Check:

$$2x - 3 = -5x + 9$$

$$2\left(\frac{12}{7}\right) - 3 \ ? \ -5\left(\frac{12}{7}\right) + 9 \qquad \text{Substitute } x = \frac{12}{7}.$$

$$\frac{24}{7} - \frac{21}{7} \ ? \ -\frac{60}{7} + \frac{63}{7}$$

$$\frac{3}{7} = \frac{3}{7} \ \checkmark \qquad \text{This is a true statement.}$$

Therefore, $x = \frac{12}{7}$ is the correct solution. ▲

Let us now look at some examples that first require us to multiply expressions in parentheses using the distributive property.

EXAMPLE 4   Solve $4(x - 2) = 5x + 3$.

Solution

$$4(x - 2) = 5x + 3 \qquad \text{Use the distributive property to multiply the quantity in parentheses.}$$

$$4x - 8 = 5x + 3$$

$$-5x + 4x - 8 = -5x + 5x + 3 \qquad \text{Add } -5x \text{ to both sides.}$$

$$-x - 8 = 3 \qquad \text{Combine like terms.}$$

$$-x - 8 + 8 = 3 + 8 \qquad \text{Add 8 to both sides.}$$

$$-x = 11$$

$$(-1)(-x) = (11)(-1) \qquad \text{Since we need to solve for } x \text{ and not } -x, \text{ we must multiply both sides by } -1.$$

$$x = -11$$

Check:

$$4(x - 2) = 5x + 3$$

$$4(-11 - 2) \ ? \ 5(-11) + 3 \qquad \text{Substitute } x = -11.$$

$$4(-13) \ ? \ -55 + 3$$

$$-52 = -52 \ \checkmark \qquad \text{This is a true statement.}$$

Therefore, $x = -11$ is the correct solution. ▲

EXAMPLE 5   Solve $5(w - 3) + 7 = 7w - 2$.

Solution

$$5(w - 3) + 7 = 7w - 2 \qquad \text{Use the distributive property to multiply the quantity in parentheses.}$$

$$5w - 15 + 7 = 7w - 2$$

$$-7w + 5w - 15 + 7 = 7w + (-7w) - 2 \qquad \text{Add } -7w \text{ to both sides.}$$

$$-2w - 8 = -2 \qquad \text{Combine like terms.}$$

$$-2w - 8 + 8 = -2 + 8 \qquad \text{Add 8 to both sides.}$$

$$-2w = 6$$

$$\left(-\frac{1}{2}\right)(-2w) = (6)\left(-\frac{1}{2}\right) \qquad \text{Multiply both sides by } -\frac{1}{2}.$$

$$w = -3$$

Check: $5(w - 3) + 7 = 7w - 2$

$5(-3 - 3) + 7 \; ? \; 7(-3) - 2$ Substitute $w = -3$.

$5(-6) + 7 \; ? \; -21 - 2$

$-30 + 7 \; ? \; -23$

$-23 = -23 \; \checkmark$ This is a true statement.

Therefore, $w = -3$ is the correct answer.   ▲

EXAMPLE 6   Solve $3(z - 1) = 4(8 - z)$.

Solution $3(z - 1) = 4(8 - z)$ Use the distributive property to multiply the quantity in parentheses.

$3z - 3 = 32 - 4z$

$4z + 3z - 3 = 32 - 4z + 4z$ Add $4z$ to both sides.

$7z - 3 = 32$ Combine like terms.

$7z - 3 + 3 = 32 + 3$ Add 3 to both sides.

$7z = 35$

$\left(\dfrac{1}{7}\right)(7z) = (35)\left(\dfrac{1}{7}\right)$ Multiply both sides by $\frac{1}{7}$.

$z = 5$

Check: $3(z - 1) = 4(8 - z)$

$3(5 - 1) \; ? \; 4(8 - 5)$ Substitute $z = 5$.

$3(4) \; ? \; 4(3)$

$12 = 12 \; \checkmark$ This is a true statement.

Therefore, $z = 5$ is the correct solution.   ▲

EXAMPLE 7   Solve $5(x + 5) = 9 - 4(x - 3)$.

Solution $5(x + 5) = 9 - 4(x - 3)$ Use the distributive property to multiply the quantity in parentheses.

$5x + 25 = 9 - 4x + 12$

$5x + 25 = 21 - 4x$ Combine constants on the right.

$4x + 5x + 25 = 21 - 4x + 4x$ Add $4x$ to both sides.

$9x + 25 = 21$ Combine like terms.

$9x + 25 + (-25) = 21 + (-25)$ Add -25 to both sides.

$9x = -4$ Combine constants.

$\left(\dfrac{1}{9}\right)(9x) = (-4)\left(\dfrac{1}{9}\right)$ Multiply both sides by $\frac{1}{9}$.

$x = -\dfrac{4}{9}$

Check: $5(x + 5) = 9 - 4(x - 3)$

$$5\left(-\frac{4}{9} + 5\right) \ ? \ 9 - 4\left(-\frac{4}{9} - 3\right) \qquad \text{Substitute } x = -\frac{4}{9}.$$

$$5\left(-\frac{4}{9} + \frac{45}{9}\right) \ ? \ 9 - 4\left(-\frac{4}{9} - \frac{27}{9}\right)$$

$$5\left(\frac{41}{9}\right) \ ? \ 9 - 4\left(-\frac{31}{9}\right)$$

$$\frac{205}{9} \ ? \ \frac{81}{9} + \frac{124}{9}$$

$$\frac{205}{9} = \frac{205}{9} \ \checkmark \qquad \text{This is a true statement.}$$

Therefore, $x = -\frac{4}{9}$ is the correct solution. ▲

Quick Quiz

Solve each of the following linear equations.

Answers

1.	$3y + 2 = 12$	**1.**	$y = \dfrac{10}{3}$
2.	$-7x = 2x + 27$	**2.**	$x = -3$
3.	$-2(z + 8) = 10$	**3.**	$z = -13$
4.	$4(2w - 5) + 7 = 3$	**4.**	$w = 2$

11.5 Exercises

Indicate which of the following statements are true and which are false. For those that are false, change the italicized word to make the statement true.

1. When solving linear equations, we use the addition principle of equality *before* the multiplication principle of equality.

2. To isolate the variable when solving a linear equation, we *divide* both sides by the reciprocal of its coefficient.

3. To remove parentheses in a linear equation, we use the *commutative* property.

4. When solving a linear equation, we combine like terms *after* using the multiplication principle of equality.

Use both the addition principle of equality and the multiplication principle of equality to solve each of the following equations.

5. $7y - 8 = 13$		**6.** $8k + 2 = 18$		**7.** $3x + 5 = -22$
8. $4a - 9 = 11$		**9.** $4 - 5z = 54$		**10.** $-2w + 9 = 15$
11. $5 + 3t = 6$		**12.** $8 - 7p = 3$		**13.** $2 = 5b - 7$
14. $12 = 4 - 3r$		**15.** $43n + 11 = 97$		**16.** $27q - 32 = 76$
17. $1.7x + 4.1 = 9.2$		**18.** $-2.1x + 3.7 = -8.9$		**19.** $8x + 3 = 11x$

20. $2x - 7 = -12x$

21. $-5x = 2x - 9$

22. $3x = 8x + 5$

23. $3a + 7 = 5a + 4$

24. $-6b - 5 = 8b - 9$

25. $8 + 7u = 12u + 23$

26. $29 - 3m = 4m - 6$

27. $-2t + 9 = 6 + 5t$

28. $11 - 3t = -7t - 4$

29. $7s + 8 - 3s = 5s$

30. $5 + 9c = 2 - 6c + 8$

31. $21 - 22p = 71 - 47p$

32. $28 - 54q = 35q - 61$

Solve each of the following linear equations by first using the distributive property to remove parentheses.

33. $5(x + 7) = 15$

34. $7(x - 3) = 42$

35. $-3(x - 8) = 36$

36. $-2(x + 4) = 12$

37. $4(3y - 8) = 16$

38. $5(2y - 9) = 45$

39. $-8 = 2(5y + 1)$

40. $-18 = 3(5y + 4)$

41. $3(4t - 8) = 15$

42. $8(4t + 7) = 83$

43. $4(3y - 5) = 8y$

44. $2(7y + 3) = 8y$

45. $-6(6 - y) = 4y$

46. $-4(5 - 2y) = -2y$

47. $9(3p - 4) = 15(6 - p)$

48. $3(6p - 1) = 9(2 + p)$

49. $-4(2 - 5z) = 2(3z + 10)$

50. $-6(2z + 1) = 2(9 - 2z)$

51. $6(2z - 3) + 3z = 22$

52. $4(5z + 2) - 7z = -18$

53. $5z - 2(3z + 1) = 18$

54. $3z - 5(2z - 3) = -6$

55. $5(2t - 7) - 1 = 3(t + 2)$

56. $7(t + 4) - 10 = 4(3t + 2)$

57. $6 + 8(2 - s) = 2(4s - 5)$

58. $5(3 - s) - 1 = 3(2s + 1)$

59. $7(2x - 1) = 9(2x - 3) - 6x$

60. $8x - 3(x - 7) = 4(3x + 2)$

61. $3(x + 2) - 2(x - 1) = 5(x + 2) + 6$

62. $-4(x - 2) + 3(x - 1) = 2(x - 7) - 5$

11.6
Solving Word Problems Using Linear Equations

In this section, we will discuss an important application of linear equations, which is solving word problems. Since it will often be impossible for us to absorb all the information given in a word problem in a single reading, it is important that we approach each problem in a systematic fashion, keeping in mind the particular details we wish to extract from the problem in each reading. The following procedure will help us to organize our thinking when solving the word problems covered in this section.

Five Steps for Solving Word Problems

1. Determine what you are asked to find. It may help to write this down.

2. Define all variables that you will need to solve the problem. Represent one unknown quantity with a letter and any other unknown quantities in terms of that variable.

3. Write an equation that relates the variables. Note the key words in the problem that indicate relationships between the quantities.

4. Solve the equation for the unknown and label your answer.

5. Answer the question asked in the problem. Keep in mind that this may or may not be the same as the solution to your equation. Check to make sure that your answer satisfies the conditions of the problem. ▲

EXAMPLE 1 A mother is 5 years older than twice the age of her daughter. If the mother is 37 years old, how old is her daughter?

Solution Find the daughter's age.

Let $\quad x =$ daughter's age
$\qquad 2x + 5 =$ mother's age
mother's age $= 37$

$$2x + 5 = 37$$
$$2x + 5 - 5 = 37 - 5$$
$$2x = 32$$

$$\frac{1}{2}(2x) = \frac{1}{2}(32)$$

$$x = 16 \text{ yr}$$
The daughter is 16 years old.

1. Determine what are you asked to find.
2. Define all variables. The mother's age is 5 more than twice her daughter's.
3. Since the mother's age is 37, we can write this equation to find x.
4. Solve this equation for x.

$x = 16$ yr represents the daughter's age.
5. Answer the question.

Check: $5 + 2(16) = 37 \ \checkmark$

This the mother's age, so the answer checks. ▲

EXAMPLE 2 Two numbers have a sum of 24. If the second number is three times as large as the first, what are these two numbers?

Solution Find two numbers.

Let $\quad x =$ the first number
$\qquad 3x =$ the second number
first number + second number $= 24$
$$x + 3x = 24$$
$$4x = 24$$

$$\frac{1}{4}(4x) = \frac{1}{4}(24)$$

$$x = 6 \text{ is the first number}$$
$$3x = 3(6) = 18 \text{ is the second number}$$
The two numbers are 6 and 18.

1. Determine what you are asked to find.
2. Define all variables. The second number is 3 times as large as the first.
3. Since the two numbers have a sum of 24, we can write this equation to find them.
4. Now solve this equation for x.

Since $x = 6$ is the first number.
5. Answer the question.

Check: $6 + 18 = 24 \ \checkmark$

These answers check. ▲

Consecutive integers are integers that differ by one. For example, 7 and 8 are consecutive integers. Two unknown consecutive integers can be represented by x and $x + 1$. Both consecutive even integers and consecutive odd integers differ by two. For example, 10 and 12 are consecutive even integers, and 15 and 17 are consecutive odd integers. Two unknown consecutive even integers can be represented by x and $x + 2$. Two unknown consecutive odd integers can also be represented by x and $x + 2$. Example 3 involves consecutive odd integers.

EXAMPLE 3   Find two consecutive odd integers whose sum is 88.

Solution   Find two consecutive odd integers.

Let $x = $ first integer

$x + 2 = $ second integer

first integer + second integer = 88

$$x + x + 2 = 88$$
$$2x + 2 = 88$$
$$2x + 2 + (-2) = 88 + (-2)$$
$$2x = 86$$

$$\left(\frac{1}{2}\right)(2x) = \left(\frac{1}{2}\right)(86)$$

$x = 43$ is the first number

$x + 2 = 45$ is the second number

The two integers are 43 and 45.

Check: $43 + 45 = 88$ $\checkmark$

1. Determine what you are asked to find.

2. Define the variables. Consecutive odd integers differ by 2.

3. The sum of the integers is 88.

4. Solve this equation for x.

5. Answer the question.
The sum of the two integers is 88, so the answer checks.   ▲

EXAMPLE 4   The length of a rectangle is four times the width. What are the dimensions of the rectangle if its perimeter is 55 ft?

Solution   Find the length and width of the rectangle.

Let $w = $ width

$4w = $ length

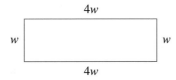

$$w + 4w + w + 4w = 55$$
$$10w = 55$$

$$\frac{1}{10}(10w) = \frac{1}{10}(55)$$

$$w = \frac{55}{10} = 5\frac{1}{2} \text{ ft is the width}$$

$$4w = 4\left(5\frac{1}{2}\right) = 4\left(\frac{11}{2}\right) = 22 \text{ ft is the length}$$

The rectangle has a length of 22 ft and a width of $5\frac{1}{2}$ ft.

Check: $2(22) + 2\left(5\frac{1}{2}\right) = 44 + 11 = 55$ $\checkmark$

1. Determine what you are asked to find.

2. Define all variables. The length is four times the width.

3. Draw a rectangle and label the four sides as shown. Since perimeter is the distance around a figure, we can obtain an equation just by setting the sum of the sides equal to 55.

4. Now solve this equation for w.

Since $w = 5\frac{1}{2}$ ft is the width.

5. Answer the question.

This is the perimeter, so the answers check.   ▲

EXAMPLE 5 A student bought three textbooks for his history, chemistry, and mathematics courses. The chemistry text cost $10 more than the history text. The mathematics text cost twice as much as the chemistry text. If the total bill for these three books was $90, what is the price of each book?

Solution Find the price of each of the three books.   1. What are you to find?

Since the chemistry text cost $10 more than the history text, if we let p represent the price of the history text, $p + 10$ represents the price of the chemistry text. Since the math text costs twice as much as the chemistry text, $2(p + 10)$ represents the price of the math text.

Let $\quad p$ = price of history text $p + 10$ = price of the chemistry text Let $2(p + 10)$ = price of math text	2. Define all variables. The chemistry text cost $10 more than the history text. The math text cost twice as much as the chemistry text.

$$\underbrace{\text{history text}}_{p} + \underbrace{\text{chemistry text}}_{p + 10} + \underbrace{\text{math text}}_{2(p + 10)} = \$90$$

3. Since the total cost of the three books is $90, we can write this equation to find the three prices.

$$p + p + 10 + 2(p + 10) = 90$$

4. Now solve this equation for p.

$$p + p + 10 + 2p + 20 = 90$$
$$4p + 30 = 90$$
$$4p = 60$$
$$p = 15 \text{ dollars for the history text.}$$

$$p + 10 = 15 + 10$$
$$= 25 \text{ dollars for chemistry text.}$$
$$2(p + 10) = 2(15 + 10) = 50 \text{ dollars for math text.}$$

Since $p = 15$ dollars for history text.

Therefore, the history text cost $15, the chemistry text cost $25, and the math text cost $50.

5. Answer the question.

Check: $15 + 25 + 50 = 90$ $\checkmark$

This is the total cost, so the answers check. ▲

EXAMPLE 6 A cellular phone is on sale for $150. If this price represents a 25% discount off the original price, what is the original price of the cellular phone?

Solution Find the original price.   1. What are you asked to find?

Let $\quad p$ = original price $25\% p$ = discount amount	2. Define the variables.

Recall from Section 6.5 that

$$\text{original price} - \text{discount} = \text{sale price}$$
$$p - 25\% p = 150$$

3. Write an equation.

$$p - 0.25p = 150$$

4. Solve for p.

$$1.00p - 0.25p = 150$$
$$0.75p = 150$$
$$p = 200$$

The original price is $200.

5. Answer the question.

Check: $200 - .25(200) = 200 - 50 = 150$ $\checkmark$

This is the sale price. ▲

11.6 Exercises

Solve each of the following word problems.

1. A woman's age is 7 years less than twice the age of her brother. If the woman is 29 years old, how old is her brother?

2. A doctor is two years older than three times the age of her patient. If the doctor is 35 years old, how old is her patient?

3. Two numbers have a sum of 32. If the second is eight more than seven times the first, what are the two numbers?

4. Two numbers have a sum of 20. If the second is five less than four times the first, what are the two numbers?

5. Five more than the product of 8 and a number is 53. What is the number?

6. Seven less than the quotient of a number and 3 is 23. What is the number?

7. Find two consecutive integers whose sum is 55. (*Hint*: Let x be the first integer and $x + 1$ be the second.)

8. Find two consecutive even integers whose sum is 78. (*Hint*: Let x be the first integer and $x + 2$ be the second.)

9. In a statistics class, there are five more men than there are women. If there are a total of 31 students in the class, how many are men?

10. A youngster practiced his piano 15 minutes longer on Sunday than on Saturday. If he practiced a total of 41 minutes that weekend, how long did he practice on Sunday?

11. The distance between New York and Dallas is 54 miles more than twice the distance between Memphis and Pittsburgh. If New York and Dallas are 1,374 miles apart, what is the distance between Memphis and Pittsburgh?

12. A father is 7 inches shorter than his son. If the sum of their heights is 155 inches, how tall is the father?

13. The cost of repairing a VCR was $85. If this included a $50 charge for parts and a $20-per-hour charge for labor, how much time did it take to fix the VCR?

14. A refrigerator repairman charged $195 to repair a refrigerator. This included a $140 charge for parts and a $22-per-hour charge for labor. How long did it take the repairman to fix the refrigerator?

15. The length of a rectangle is 5 feet longer than its width. Find the dimensions of the rectangle if its perimeter is 22 feet.

16. The width of a rectangle is 8 centimeters shorter than its length. If the perimeter of the rectangle is 84 centimeters, what are its dimensions?

17. A house is now selling for $5,000 more than four times what the owners paid for it in 1950. If the selling price today is $95,000, what did the owners pay for it in 1950?

18. A used car that is 3 years old is selling for $150 more than half its original price. If the car now sells for $2,750, what did it cost 3 years ago when it was brand new?

19. A student bought a notebook, a pen, and a candy bar. The pen cost 14¢ more than the candy bar, and the notebook cost twice as much as the pen. If the student was charged $1.82 for all three items, what was the cost of each one?

20. An intern worked 2 more hours on Monday than on Tuesday, and twice as many hours on Wednesday as on Monday. If he worked a total of 42 hours on those 3 days, how many hours did he work each day?

21. An isosceles triangle has a perimeter of 14 inches. If the two equal sides are each three times as long as the third side, what is the length of each side?

22. An isosceles triangle has a perimeter of 370 millimeters. The base of the triangle is 40 millimeters longer than the two equal sides. What is the length of each side?

23. A bag of apples weighs twice as much as a bag of oranges. If the apples are 3 pounds heavier than the oranges, what is the weight of the oranges?

24. A child has eight more blue marbles than red ones. If there are three times as many blue marbles as red ones, how many red marbles does he have?

25. The profits this year for a computer manufacturer are $600,000 higher than last year's profits. If this represents a 20% increase over last year's profits, what were last year's profits? What are this year's profits?

26. As a result of a cost reduction program, the expenses for a financial institution were down $2,100,000 from last year's expenses. If this represents a 35% reduction from last year's expenses, what were last year's expenses? What are this year's expenses?

27. A laser printer is on sale for $850. If this represents a 15% discount off the original price, what is the original price?

28. A FAX machine is selling for $931. If this price includes a 33% profit added to the cost of the FAX machine, what is the cost of the FAX machine?

29. A computer salesman's annual compensation includes a $30,000 base salary plus an 8% commission on his total sales for the year. If the salesman's annual compensation is $94,000, what were his total sales for the year?

30. An art gallery earned $30,000 in show fees plus 12% of the value of all artwork sold. If the gallery earned $270,000 last year, what was the value of the artwork sold?

11.7
Summary and Review

Key Terms

(11.1) **Algebra** is a generalization of arithmetic.

A **constant** is a number or quantity that does not change in value.

A **variable** is an unknown quantity that may have different values and is represented by a letter.

The **coefficient** of a variable is the number that is multiplied by the variable.

The **terms** of an algebraic expression are the quantities that are added or subtracted.

(11.2) **Like terms**, or **similar terms**, are terms whose variable factors are the same.

(11.3) An **equation** is a mathematical expression that states that two quantities are equal.

A **linear equation** is an equation in one variable that can be written in the form $ax + b = c$, where a, b, and c are constants and $a \neq 0$.

The **solution** to a linear equation in one variable is the value that can be substituted for the variable so that the resulting statement is true.

Equivalent equations are equations that have the same solution.

The **addition principle of equality** states that if the same quantity is added to both sides of an equation, the resulting equation is equivalent to the original one.

(11.4) The **multiplication principle of equality** states that if both sides of an equation are multiplied by the same nonzero quantity, the resulting equation is equivalent to the original one.

Calculations

(11.1) **To evaluate an algebraic expression**, we substitute numbers for the variables and simplify the result using the order of operations rules.

(11.2) **To combine like terms**, we rewrite the expression using the distributive property and then add or subtract the coefficients of the variable.

(11.5) **Steps for Solving Linear Equations**

1. Simplify both sides of the equation as much as possible by using the distributive property to multiply all quantities in parentheses, and then combining like terms.

2. Use the addition principle of equality to move all variable terms to one side of the equation and all constant terms to the other side. This is done by adding to both sides of the equation the opposite of each term you wish to move, and then combining like terms.

3. Use the multiplication principle of equality to isolate the variable on one side of the equation and the solution on the other side. This is done by multiplying both sides of the equation by the reciprocal of the coefficient of the variable.

4. Check that your solution is correct by substituting it back into the original equation to see if the resulting statement is true.

(11.6) **Steps for Solving Word Problems**

1. *Determine what you are asked to find.* It may help to write this down.

2. *Define all variables* that you will need to solve the problem. Represent one unknown quantity with a letter and any other unknown quantities in terms of that variable.

3. *Write an equation* that relates the variables. Note the key words in the problem that indicate relationships between the quantities.

4. *Solve the equation* for the unknown and label your answer.

5. *Answer the question asked in the problem.* Keep in mind that this may or may not be the same as the solution of your equation. Check to make sure your solution satisfies the conditions of the problem.

Chapter **11**
Review Exercises

(11.1) **Evaluate each of the following algebraic expressions.**

1. $5x - 2y + z$ when $x = 2$, $y = -3$, $z = -5$

2. $-7a + 4b - 9c$ when $a = -1$, $b = 4$, $c = 3$

3. $8(x - 6y)$ when $x = 5$, $y = 2$

4. $3(4t + 5u)$ when $t = -2$, $u = 4$

5. $\frac{3}{4}p - \frac{5}{8}q$ when $p = 12$, $q = 16$

6. $-\frac{1}{9}r + \frac{2}{3}s$ when $r = 18$, $s = -6$

(11.1) **Translate each of the following word expressions into algebraic expressions.**

7. Seven less than x.

8. The product of three and a.

9. Five times the sum of z and six.

10. Two less than the product of nine and y.

11. The quantity seven plus t, divided by three.

12. Six times the quantity x minus four.

(11.1) **Translate each of the following algebraic expressions into words.**

13. $-5y$ 14. $t + 7$ 15. $3(x - 2)$ 16. $(4 + z) \div 8$ 17. $6w - 2$ 18. $\frac{3}{5}(n + 4)$

(11.2) **Simplify each of the following algebraic expressions by combining like terms.**

19. $4x - 9x$ 20. $a - 7a$ 21. $5y - 2 - 8y - 6$

22. $-3t + 4 - 6t - 5$ 23. $4(2x - 7) + 8$ 24. $3(6y + 2) - 13$

25. $2(5a - 7) - 3(4a + 5)$ 26. $8(4 - t) - 7(3t - 1)$

(11.5) **Solve each of the following linear equations.**

27. $4x - 7 = 25$ 28. $-5y + 2 = 47$ 29. $9u + 1 = 28$

30. $5t - 8 = -43$ 31. $7 - 3z = 31$ 32. $-6 - 5w = 54$

33. $3p + 2 = 5p - 8$ 34. $9q - 4 = 12 - 7q$ 35. $8 - 5a = 7a + 2$

36. $6b + 2 = 9b - 4$ 37. $3(2m + 5) = 5m$ 38. $4(3 - n) = 11n$

39. $8(k - 7) = 2(5k + 2)$ 40. $6(2s - 5) = 3(s + 4)$ 41. $4(8x + 2) + 3x = 17$

42. $9 - 2(4x - 7) = 2x$

(11.6) **Solve each of the following word problems.**

43. Find two consecutive odd integers whose sum is 68.

44. Three numbers have a sum of 39. If the second number is four less than the first, and the third number is equivalent to three times the second, what are the three numbers?

45. It took a student $1\frac{1}{2}$ hours longer to complete her math homework on Tuesday night than it took her on Monday night. If on both nights she spent a total of $5\frac{1}{2}$ hours doing math homework, how much time did she devote to math homework on Tuesday night?

46. Five-eighths of the people employed by a community hospital smoke cigarettes. If 138 hospital employees do not smoke cigarettes, how many people are employed by the hospital?

47. The width of a rectangle is one-third as long as its length. What are the dimensions of the rectangle if its perimeter is 32 centimeters?

48. An isosceles triangle has a perimeter of 43 inches. What is the length of each side if the base is 5 inches shorter than the sum of the two equal sides?

49. The height of the World Trade Center in New York is 123 feet less than twice the height of the Prudential Tower in Boston. If the World Trade Center is 1,377 feet high, how high is the Prudential Tower?

50. The diameter of the planet Jupiter is 1,503 miles larger than 11 times the diameter of the earth. If the diameter of Jupiter is 88,700 miles, what is the diameter of the earth?

Chapter **11**
Explain It in Words

51. What is the difference between a constant and a variable? Give an example of each.

52. Explain what is meant by "like terms." Give some examples of expressions that are like terms and ones that are not like terms.

53. What is meant by the "solution" to a linear equation? Give an example.

54. What is meant by "defining the variable" in solving a word problem?

55. What is meant by "the addition principle of equality"?

56. What is meant by "the multiplication principle of equality"?

Chapter **11**
Chapter Test

[11.1]   Evaluate each of the following algebraic expressions.

1. $-9x + 3y - 4z$ when $x = -1, y = 3, z = 2$

2. $4(3x + y)$ when $x = -2, y = 4$

3. $\frac{5}{8}a + \frac{1}{5}b$ when $a = -8, b = 5$

[11.1]   Translate each of the following into algebraic expressions.

4. 2 less than x

5. 8 more than the product of 3 and t

6. 5 times the quantity w plus 6

[11.1]   Translate each of the following algebraic expressions into words.

7. $y + 8$ 8. $2(x - 7)$

[11.2]   Simplify each of the following by combining like terms.

9. $3x - 9x$ 10. $4a + 7 - 3a + 2$ 11. $3(7t - 8) + 11$

[11.5]   Solve each of the following linear equations.

12. $x + 9 = -3$ 13. $-\frac{2}{9}a = 12$ 14. $6x - 5 = 49$

15. $5(2t - 3) = 25$ 16. $3(y - 6) + 5 = 7 - y$

[11.6]   Solve each of the following word problems.

17. A woman's age is 3 less than 6 times the age of her brother. If the woman is 21 years old, how old is her brother?

18. Two numbers have a sum of 38. If the second is 7 less than 4 times the first, what are the two numbers?

19. In a biology class, there are 7 more men than women. If there are a total of 43 students in the class, how many are men and how many are women?

20. The width of a rectangle is 2 inches shorter than its length. Find the dimensions of the rectangle if its perimeter is 48 inches.

Chapter *12*

Graphing

12.1
Inequalities and Line Graphs

In Chapter 11 we learned a technique for solving first-degree equations containing a single variable, which are also known as linear equations. Recall that the solution set for a linear equation is a real number.

In this section, we will learn how to solve *linear inequalities* that contain a single variable. The expression $x > 5$ is an example of a linear inequality. The solution set for a linear inequality can be represented by an *interval* on a number line, which contains an infinite number of real numbers. The solution set to the inequality $x > 5$ includes all real numbers greater than 5. The following examples illustrate how to graph the solution set to a linear inequality on a number line.

EXAMPLE 1 Graph the solution set represented by each of the following on a number line.

(a) $x \leq 4$

The solution set is all real numbers less than or equal to 4. We graph the solution set by shading the portion of a number line to the left of and including 4.

The closed circle at 4, the endpoint of the interval, indicates that 4 is included in the solution set.

(b) $y < \dfrac{1}{2}$

The solution set is all real numbers less than $\frac{1}{2}$. Shade the region of the number line to the left of $\frac{1}{2}$.

The open circle at $\frac{1}{2}$, the endpoint of the interval, indicates that $\frac{1}{2}$ is not included in the solution set.

(c) $t \geq -5$

The solution set is all real numbers greater than or equal to -5. Shade the region of the number line to the right of and including -5.

Remember that the closed circle at -5, the endpoint of the interval, indicates that -5 is included in the solution set.

(d) $2.3 > w$

Rewrite the inequality with the variable on the left. Remember that the open end of the inequality sign must face the larger quantity, which is 2.3. When you switch the positions of the two quantities being compared, you must reverse the inequality sign.

$w < 2.3$

The solution set is all real numbers less than 2.3. Shade the interval of the number line to the left of 2.3.

The open circle at 2.3, the endpoint of the interval, indicates that 2.3 is not included in the solution set. ▲

The technique we used to solve a linear equation involved adding the same quantity to both sides of the equation until all variable terms were on one side and the constants were on the other side. To solve for the unknown, we then multiplied both sides by the reciprocal of the coefficient of the variable.

We will now use a similar technique to solve linear inequalities of a single variable. Before we can proceed, we must first determine what happens when a quantity is added to both sides of the inequality. For example, let us consider the expression, $2 < 8$, which states that 2 is less than 8. If we add the number 5 to both sides, we obtain the following:

$$2 < 8$$
$$2 + 5 \; ? \; 8 + 5$$
$$7 < 13$$

Thus, when a positive number is added to both sides, the left-hand side of the inequality is still less than the right. If we add the number -9 to both sides of the original inequality, we obtain the following:

$$2 < 8$$
$$2 + (-9) \; ? \; 8 + (-9)$$
$$-7 < -1$$

Thus, when a negative number is added to both sides, the left-hand side of the inequality is still less than the right. This property, called the **addition property of inequalities**, may be stated as follows:

According to the *addition property of inequalities,* if any quantity is added to both sides of an inequality, the inequality sign remains unchanged.   ▲

Using symbols, we can let a, b, and c represent any algebraic expression and summarize this property as follows:

Addition Property of Inequalities

If $a < b,$
then $a + c < b + c$ ▲

Let us now use the addition property to solve some linear inequalities.

EXAMPLE 2   Solve each of the following linear inequalities, and graph the solution set on a number line.

(a) $x + 5 \leq 3$
$x + 5 + (-5) \leq 3 + (-5)$ Add -5 to both sides.
$x \leq -2$

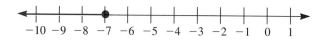

(b) $9t + 4 \leq 8t - 3$
$-8t + 9t + 4 \leq -8t + 8t - 3$ Add $-8t$ to both sides.
$t + 4 \leq -3$
$t + 4 + (-4) \leq -3 + (-4)$ Add -4 to both sides.
$t \leq -7$

(c)

$$3w + 2(w - 4) > 4w$$

$3w + 2w - 8 > 4w$	Remove ( ) using distributive property.
$5w - 8 > 4w$	Combine like terms on left.
$-4w + 5w - 8 > -4w + 4w$	Add $-4w$ to both sides.
$w - 8 > 0$	Combine like terms on left.
$w - 8 + 8 > 0 + 8$	Add 8 to both sides.
$w > 8$	

Let us now examine what happens when both sides of a linear inequality are multiplied by a positive number or a negative number. If we begin with the inequality $2 < 8$ and multiply both sides by 3, we obtain the following:

$$2 < 8$$
$$2(3) \ ? \ 8(3)$$
$$6 < 24$$

Thus, when both sides of the inequality are multiplied by a positive number, the left-hand side remains less than the right-hand side. If we multiply both sides of the original inequality by -1, we obtain the following:

$$2 < 8$$
$$2(-1) \ ? \ 8(-1)$$
$$-2 > -8$$

Thus, when both sides of the inequality are multiplied by a negative number, the inequality *reverses*, and the left-hand side becomes greater than the right-hand side. This property is called the **multiplication property of inequalities** and may be stated as follows:

According to the *multiplication property of inequalities,* if both sides of an inequality are multiplied by a positive number, the inequality sign remains unchanged. If both sides are multiplied by a negative number, the inequality sign reverses. ▲

Using symbols, we can let a, b, and c represent any algebraic expression and summarize this property as follows:

Multiplication Property of Inequalities

If $\quad a < b,$
then $\quad ac < bc,$ when c is positive
and $\quad ac > bc,$ when c is negative $\qquad$ ▲

We will now use the multiplication property to solve some linear inequalities.

EXAMPLE 3 Solve each of the following linear inequalities and graph the solution set on a number line.

(a) $6t < 72$

$$\left(\frac{1}{6}\right)(6t) < \left(\frac{1}{6}\right)(72)$$ Multiply both sides by $\frac{1}{6}$.

$$t < 12$$

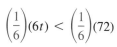

(b) $-7z \le 14$

$$\left(-\frac{1}{7}\right)(-7z) \ge \left(-\frac{1}{7}\right)(14)$$ Multiply both sides by $-\frac{1}{7}$ and reverse the inequality sign.

$$z \ge -2$$

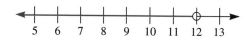

(c) $2 < \frac{1}{3}t$

$$(3)(2) < (3)\left(\frac{1}{3}t\right)$$ Multiply both sides by 3. The inequality sign remains unchanged.

$$6 < t$$ Rewrite inequality with variable on left.

$$t > 6$$ Notice that $6 < t$ and $t > 6$ mean the same thing. In each case, the open end of the inequality sign faces the larger quantity, t.   ▲

Many linear inequalities will require us to see both the addition property and the multiplication property in order to solve them. For these problems, we can use the following procedure:

Procedure for Solving Linear Inequalities

1. Simplify both sides as much as possible, using the distributive property to multiply all quantities in parentheses and combine like terms.

2. Use the addition property of inequalities to move all variable terms to one side of the inequality and all constant terms to the other side.

3. Use the multiplication property of inequalities to isolate the variable on one side of the inequality. Remember to reverse the inequality sign if you multiply both sides by a negative number.

4. Graph the solution set on a number line.   ▲

Let us now look at some examples that illustrate this procedure.

EXAMPLE 4   Solve each of the following linear inequalities and graph the solution set on a number line.

(a) $5x - 7 \leq 2x + 8$

$-2x + 5x - 7 \leq -2x + 2x + 8$ Add $-2x$ to both sides.

$3x - 7 \leq 8$

$3x - 7 + 7 \leq 8 + 7$ Add 7 to both sides.

$3x \leq 15$

$\dfrac{1}{3}(3x) \leq \dfrac{1}{3}(15)$ Multiply both sides by $\dfrac{1}{3}$.

$x \leq 5$

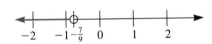

(b) $-4(x + 3) + 7 > 5x + 2$

$-4x - 12 + 7 > 5x + 2$ Remove ( ) using distributive property.

$-4x - 5 > 5x + 2$

$-5x - 4x - 5 > -5x + 5x + 2$ Add $-5x$ to both sides.

$-9x - 5 > 2$

$-9x - 5 + 5 > 2 + 5$ Add 5 to both sides.

$-9x > 7$

$\left(-\dfrac{1}{9}\right)(-9x) < \left(-\dfrac{1}{9}\right)(7)$ Multiply both sides by $-\dfrac{1}{9}$ and reverse the inequality sign.

$x < -\dfrac{7}{9}$

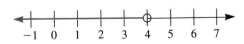

▲

EXAMPLE 5   The sum of three times a number and 8 is greater than five times the number. What numbers satisfy this condition?

Solution   Let $x =$ the number

$3x + 8 > 5x$

$-2x + 8 > 0$ Add $-5x$ to both sides.

$-2x > -8$ Add -8 to both sides.

$x < 4$ Multiply both sides by $-\dfrac{1}{2}$ and reverse the inequality sign.

Any number less than 4 satisfies this condition.   ▲

Quick Quiz

Find the solution set for each of the following linear inequalities and graph the solution on a number line.

1. $x + 3 \le 2$

2. $-\frac{3}{4}y < 9$

3. $5t + 2 < 4 - t$

4. $7(t - 3) \ge 4(1 + 2t)$

Answers

1. $x \le -1$

2. $y > -12$

3. $t < \frac{1}{3}$

4. $t \le -25$

12.1 Exercises

Graph the solution set represented by each of the following on a number line.

1. $x > 7$ **2.** $x < 3$ **3.** $y \le -2$ **4.** $y > -5$ **5.** $t \ge \frac{1}{3}$

6. $z \le 1.5$ **7.** $p < -4.9$ **8.** $w \ge -\frac{3}{4}$ **9.** $9 \ge x$ **10.** $2 < x$

Use the addition property of inequalities to solve each of the following and graph the solution set on a number line.

11. $x + 3 > 9$ **12.** $x + 8 \le 2$ **13.** $x - 7 \le 5$ **14.** $x - 5 < 1$

15. $5y - 2 \ge 4y + 7$ **16.** $8y + 6 > 7y + 3$ **17.** $9 < t + 6$ **18.** $5 \ge t - 4$

19. $2(3t - 5) \le 5t$ **20.** $4(2t - 1) \le 7t$

21. $8t - 3(2t + 1) > 5(t - 2) - 4t$ **22.** $6(t - 3) + 3t < 2(4t + 5) - 7$

Use the multiplication property of inequalities to solve each of the following and graph your solution set on a number line.

23. $3x \le 12$ **24.** $4x < 32$ **25.** $-7t < 21$ **26.** $-5w \ge 35$

27. $\frac{1}{4}y > 8$ **28.** $\frac{3}{5}p < 6$ **29.** $-\frac{2}{3}x \ge 16$ **30.** $\frac{1}{7}y \ge 4$

31. $9 < 3p$ **32.** $6 > -2q$ **33.** $-2z \le -5$ **34.** $-4t \ge -7$

Solve each of the following linear inequalities and graph your solution on a number line.

35. $5x - 3 \ge 7$ **36.** $3x + 5 \le 2$ **37.** $2 - 7x < 16$ **38.** $-4x - 7 > 5$

39. $8y + 5 \ge 2$ **40.** $7y - 8 < -3$ **41.** $4t + 9 < 7t - 2$ **42.** $2t - 5 \ge 6t - 3$

43. $4(t - 5) \le 8$ **44.** $5(t + 3) > 10$ **45.** $3(2y + 8) < 5$ **46.** $4(3y + 7) \ge -3$

47. $-2(x - 7) > 4$ **48.** $-3(x - 5) < 12$

49. $5(y + 3) - 2y < 8y + 9$ **50.** $6(y - 4) + y \ge 5y - 4$

51. $4(7 - t) \le -3(2t + 5)$ **52.** $5(2t + 1) < -2(9 - t)$

53. $2(4w + 5) - 3w \ge -5(7 - 2w)$ **54.** $-3(9 - 2z) + z < 4(3z - 8)$

Solve each of the following word problems.

55. The sum of three times a number and 5 is less than twice the number. What numbers satisfy this condition?

56. The sum of eight times a number and 9 is greater than or equal to five times the number. What numbers satisfy this condition?

57. The product of a number and 7 is greater than the number plus 12. Find all possible solutions.

58. The product of a number and 4 is less than or equal to the number minus 6. Find all possible solutions.

59. A number decreased by 2 is greater than or equal to three times the number minus 16. What numbers satisfy this condition?

60. A number increased by 5 is less than seven times the number plus 41. What numbers satisfy this condition?

12.2

The Cartesian Coordinate System

We have already seen, in Section 8.5, that any real number corresponds to a point on a number line. For example, the points that represent the numbers $3, \frac{1}{2}, -2, 0,$ and $\sqrt{3}$ are shown in the number line illustrated in Figure 12.1.

Figure 12.1

$$-2 \qquad 0 \quad \tfrac{1}{2} \quad 1 \quad \sqrt{3} \qquad 3$$

In this section, we will learn how to graphically represent an *ordered pair*, which consists of two real numbers. Just as a real number is associated with a point on a number line, an ordered pair is associated with a point in a two-dimensional plane called the *Cartesian coordinate system* or *rectangular coordinate system*.

To draw the Cartesian coordinate system, we join two number lines, one vertical and the other horizontal, at their zeros. The horizontal number line is called the *x*-axis, and the vertical number line is called the *y*-axis. The *x*- and *y*-axes are perpendicular; that is, they meet in a right angle. Their point of intersection is called the *origin*. The *x*- and *y*-axes divide the Cartesian plane into four quadrants numbered I, II, III, and IV moving counter-clockwise, as shown in Figure 12.2.

Figure 12.2

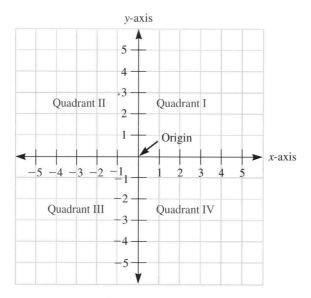

To locate a point in the Cartesian plane (or *x*-*y* plane), we must know its *x*-coordinate, which is the horizontal distance from the *y*-axis, and the *y*-coordinate, which is the vertical distance from the *x*-axis. In general, any point in the Cartesian plane can be represented by the ordered pair (x, y), where x is the *x*-coordinate or *abscissa*, and y is the *y*-coordinate or *ordinate*. Note that the *x*-coordinate always appears before the *y*-coordinate in the ordered pair.

Points having positive *x*-coordinates are located to the right of the *y*-axis, those having negative *x*-coordinates are located to the left of the *y*-axis, and those for which the *x*-coordinate is zero are *on* the *y*-axis. Points having positive *y*-coordinates are located above the *x*-axis, those having negative *y*-coordinates are located below the *x*-axis, and those for which the *y*-coordinate is zero are *on* the *x*-axis. The *origin* has coordinates (0, 0).

For example, in the ordered pair (2, 3), 2 is the *x*-coordinate or abscissa and 3 is the *y*-coordinate or ordinate. To graph this point in the *x*-*y* plane we begin at the origin and move two units to the right of the *y*-axis and three units up from the *x*-axis. To graph the point (-3, 1), we begin at the origin and move 3 units to the left of the *y*-axis and 1 unit up from the *x*-axis. To graph the point (1, -3), we begin at the origin and move 1 unit to the right of the *y*-axis and 3 units down from the *x*-axis. To graph the point (-2, -4), we begin at the origin and move 2 units to the left of the *y*-axis and 4 units down from the *x*-axis. To graph the point (4, 0), we begin at the origin and move 4 units to the right of the *y*-axis. The points associated with the ordered pairs (-3, 1), (1, -3), (-2, -4), and (4, 0) are shown in Figure 12.3.

The order of the coordinates in an ordered pair specifies the location of the point it represents. If the *x*- and *y*-coordinates are reversed, a different point is represented. For example, note that the ordered pairs (-3, 1) and (1, -3) are associated with two different points in Figure 12.3.

This procedure of graphing a point in the *x*-*y* plane is also called **plotting** a point. Let us now consider these examples.

Figure 12.3

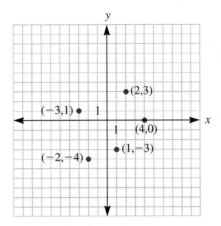

EXAMPLE 1 Plot the points associated with the ordered pairs (5, 2), (−3, 0), (−1, −4), (0, 6), and (2, −2). Be sure to label each point plotted.

Solution

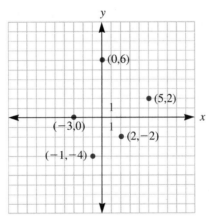

▲

EXAMPLE 2 Give the ordered pairs represented by the points *A*, *B*, *C*, *D*, *E*, and *F* in this diagram.

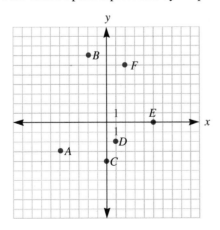

Solution $A(-5, -3)$ $B(-2, 7)$ $C(0, -4)$ $D(1, -2)$ $E(5, 0)$ $F(2, 6)$ ▲

12.2 Exercises

Indicate which of the following statements are true and which are false. For those that are false, change the italicized word to make the statement true.

1. The rectangular coordinate system is also called the *Cartesian* coordinate system.

2. In an ordered pair that represents a point in the x-y plane, the y-coordinate appears *first*.

3. The x-coordinate of an ordered pair is also called the *ordinate*.

4. The Cartesian plane is divided into *four* quadrants by the x- and y-axes.

5. If a point is located in the second quadrant, its abscissa is *greater* than zero.

6. The *ordinate* of every point on the x-axis is zero.

7. The *y-coordinate* of every point on the y-axis is zero.

8. Both the x- and y-coordinates of every point in the *fourth* quadrant are less than zero.

Plot the points associated with each of the following ordered pairs. Be sure to label each point you plot.

9. (6, 1)	**10.** (1, 6)	**11.** (5, 0)	**12.** (0, 8)	**13.** (−3, 4)
14. (7, −2)	**15.** (5, −9)	**16.** (−4, 4)	**17.** (−8, −5)	**18.** (−3, −7)
19. (0, −6)	**20.** (−9, 0)	**21.** (−5, 7)	**22.** (7, −5)	**23.** (6, 9)
24. (8, 1)	**25.** (−4, 0)	**26.** (5, 0)	**27.** (−9, −9)	**28.** (−7, 7)

In Exercises 29–38 give the ordered pairs represented by the points shown in this diagram.

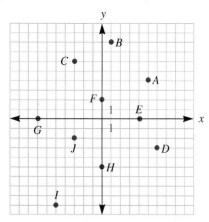

29. A	**30.** B	**31.** C	**32.** D	**33.** E
34. F	**35.** G	**36.** H	**37.** I	**38.** J

12.3
Graphing Linear Equations

In Chapter 11, we found that the solution to a linear equation in one variable is a single real number. For example, the solution to the equation

$$2x + 3 = 5$$

is the number 1. When 1 is substituted for the variable x, the resulting statement is true:

$$2(1) + 3 \ ? \ 5$$
$$2 + 3 \ ? \ 5$$
$$5 = 5 \quad \checkmark$$

The solution set for a linear equation in two variables is a set of ordered pairs. For example, one solution to the linear equation

$$x + 2y = 7$$

is the ordered pair (3, 2). This solution can also be expressed in the form $x = 3$, $y = 2$. Thus, when 3 is substituted for the variable x and 2 is substituted for the variable y, the resulting statement is true:

$$3 + 2(2) \ ? \ 7$$
$$3 + 4 \ ? \ 7$$
$$7 = 7 \quad \checkmark$$

There are an infinite number of ordered pairs that are solutions to a linear equation of two variables.

Other solutions to the linear equation

$$x + 2y = 7$$

are the ordered pairs (7, 0), (11, -2), (-1, 4), and (0, $\frac{7}{2}$). Notice that for each ordered pair, the statements that result from substituting the appropriate values for x and y are true:

(7, 0): $\quad 7 + 2(0) \ ? \ 7$
$\qquad\qquad 7 + 0 \ ? \ 7$
$\qquad\qquad\qquad 7 = 7 \quad \checkmark$

(11, -2): $\quad 11 + 2(-2) \ ? \ 7$
$\qquad\qquad\qquad 11 + -4 \ ? \ 7$
$\qquad\qquad\qquad\qquad 7 = 7 \quad \checkmark$

(-1, 4): $\quad -1 + 2(4) \ ? \ 7$
$\qquad\qquad\quad -1 + 8 \ ? \ 7$
$\qquad\qquad\qquad\quad 7 = 7 \quad \checkmark$

$\left(0, \dfrac{7}{2}\right)$: $\quad 0 + 2\left(\dfrac{7}{2}\right) \ ? \ 7$
$\qquad\qquad\qquad 0 + 7 \ ? \ 7$
$\qquad\qquad\qquad\quad 7 = 7 \quad \checkmark$

To determine whether an ordered pair is a solution to a linear equation in two variables, substitute the first coordinate for x and the second coordinate for y. If the resulting statement is true, the ordered pair is a solution to the equation. ▲

EXAMPLE 1 Determine whether or not each of the following ordered pairs are solutions to the linear equations given.

(a) (1, -2): $\quad 5x - 3y = 8$
$\qquad\qquad 5(1) - 3(-2) \ ? \ 8$
$\qquad\qquad\qquad 5 + 6 \ ? \ 8$
$\qquad\qquad\qquad\qquad 11 \neq 8$

Thus, (1, -2) is not a solution to the equation $5x - 3y = 8$.

(b) $(-1, 9)$: $y = -8x + 1$
9 ? $-8(-1) + 1$
9 ? $8 + 1$
$9 = 9$ $\checkmark$
Thus, $(-1, 9)$ is a solution to the equation $y = -8x + 1$.

(c) $(-5, -3)$: $3x - 7y = 6$
$3(-5) - 7(-3)$? 6
$-15 + 21$? 6
$6 = 6$ $\checkmark$
Thus, $(-5, -3)$ is a solution to the equation $3x - 7y = 6$. ▲

To find a solution to a linear equation in two variables, we can use this procedure:

1. Choose a value for either the x- or y-coordinate of the ordered pair.

2. Substitute into the given equation the value chosen for the first variable, and then solve the equation for the other coordinate of the ordered pair.

Sometimes we are given one coordinate of an ordered pair that is a solution to a linear equation, and we are asked to find the other coordinate, as shown in the following examples.

EXAMPLE 2 Find the missing coordinates in the ordered pairs that are solutions to the following equations.

(a) $3x + y = 7$: $(0, \ )$
$3(0) + y = 7$ Substitute in $x = 0$ and solve for y.
$0 + y = 7$
$y = 7$
The solution is $(0, 7)$.

(b) $y = 9x + 2$: $(\ , -1)$
$-1 = 9x + 2$ Substitute in $y = -1$ and solve for x.
$9x + 2 = -1$
$9x = -3$
$x = -\dfrac{1}{3}$

The solution is $\left(-\dfrac{1}{3}, -1\right)$. ▲

The graph of a linear equation is a straight line which includes the set of points representing the ordered pairs of the solution set. Since a straight line is determined by two points, we need to find at least two ordered pairs that are solutions to the linear equation in order to graph it.

To Graph a Linear Equation	**1.** Find at least two ordered pairs that are solutions to the linear equation.
	2. Plot the points that correspond to those ordered pairs on the *x-y* plane.
	3. Draw a straight line through those points. ▲

EXAMPLE 3 Graph the equation $4x - 3y = 6$.

Solution Find a few ordered pairs of the solution set. In order to keep the arithmetic simple we will choose the values $x = 0$, $x = 3$, and $y = 0$.

Let $x = 0$: $4(0) - 3y = 6$

$-3y = 6$

$y = -2$

$(0, -2)$ is a solution.

Let $x = 3$: $4(3) - 3y = 6$

$12 - 3y = 6$

$-3y = -6$

$y = 2$

$(3, 2)$ is a solution.

Let $y = 0$: $4x - 3(0) = 6$

$4x = 6$

$x = \dfrac{3}{2}$

$(\frac{3}{2}, 0)$ is a solution.

When graphing linear equations, the solutions are often listed in a table as shown here. Plot these three points and draw a straight line through them to obtain the graph.

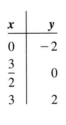

x	y
0	-2
$\dfrac{3}{2}$	0
3	2

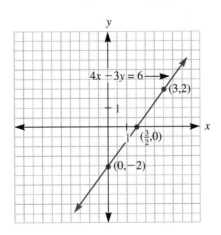

Though it is only necessary to find two points, the third point acts as a check. If the three points did not all lie in a straight line, we made a mistake in calculating one of the coordinates. ▲

The x-coordinate of the point where the graph of an equation crosses the x-axis is called the x-*intercept*, and the y-coordinate of the point where it crosses the y-axis is called the y-*intercept*. In Example 3, the equation $4x - 3y = 6$ crosses the x-axis at the point $(\frac{3}{2}, 0)$, and it crosses the y-axis at the point $(0, -2)$. Thus, the x-intercept is $\frac{3}{2}$ and the y-intercept is -2.

In general, if a is an x-intercept of a graph, then the graph passes through the point $(a, 0)$. If b is a y-intercept of a graph, then the graph passes through the point $(0, b)$. This is illustrated in Figure 12.4.

Figure 12.4

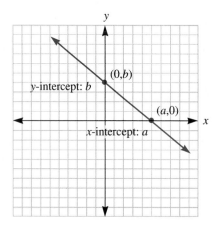

Therefore, to find the x-intercept of an equation, set $y = 0$ and solve for x. To find the y-intercept, set $x = 0$ and solve for y. *The simplest way to graph many linear equations is to find the x- and y-intercepts and connect them with a straight line.*

EXAMPLE 4   Graph the equation $3x + 5y = 15$.

Solution   Find the x-intercept: Find the y-intercept:

Let $y = 0$: $3x + 5(0) = 15$ Let $x = 0$: $3(0) + 5y = 15$
$$3x = 15$$ $$5y = 15$$
$$x = 5$$ $$y = 3$$

Plot $(5, 0)$ and $(0, 3)$ and draw a straight line through them.

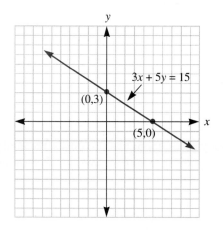

EXAMPLE 5   Graph the equation $y = -5x + 7$.

Solution   Choose values for x and mentally calculate the corresponding y-coordinates. List the solutions using a table format. Plot these points and draw a straight line through them.

x	y
0	7
1	2
2	-3

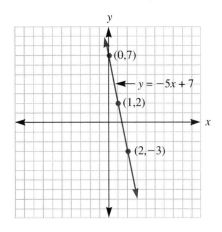

EXAMPLE 6   Graph the equation $x = 3$.

Solution   The x-coordinate of every point on this line must be 3.
The y-coordinate can be anything.
Thus, two points on this line are $(3, 0)$ and $(3, 5)$.

x	y
3	0
3	5

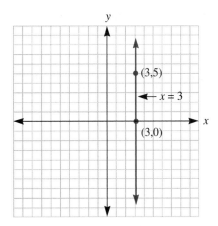

Notice that the graph of the equation $x = 3$ is parallel to the y-axis and will never cross it. The equation $x = 3$ is an example of a **vertical line**. ▲

EXAMPLE 7   Graph the equation $y = -1$.

Solution   The y-coordinate of every point on this line must be -1.
The x-coordinate can be anything.
Thus, two points on this line are $(-2, -1)$ and $(2, -1)$.

x	y
-2	-1
2	-1

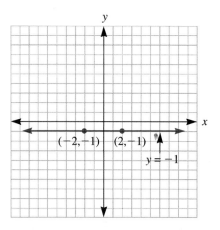

Notice that the graph of the equation $y = -1$ is parallel to the x-axis and will never cross it. The equation $y = -1$ is an example of a **horizontal line**. ▲

Examples 6 and 7 illustrate the following:

The equation of a vertical line is $x = a$, where a is the x-coordinate of every point on the line. The equation of a horizontal line is $y = b$, where b is the y-coordinate of every point on the line. ▲

EXAMPLE 8 Find the equation of the vertical line that passes through the point $(-5, 2)$.

Solution The equation of a vertical line is $x = a$ where a is the x-coordinate of every point on the line. Since one point on this line has an x-coordinate of -5, the equation of this line is $x = -5$.

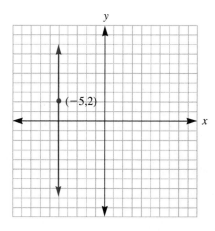

▲

12.3 Exercises

Indicate which of the following statements are true and which are false. For those that are false, change the italicized expression to make the statement true.

1. A solution to a linear equation in two variables is an *ordered pair*.
2. To find the x-intercept of an equation, set the *variable x equal to zero*.
3. A *horizontal* line has the equation $y = b$ where b is the y-coordinate of every point on the line.
4. The *x-intercept* of the line $y = b$ is b.

Determine whether or not each of the following ordered pairs are solutions to the linear equation given.

5. $(1, -8)$: $y = -8x$
6. $(3, 1)$: $y = 3x$
7. $(4, 1)$: $x + y = 4$
8. $(2, -4)$: $x - y = 6$
9. $(-2, 3)$: $x + 4y = 10$
10. $(4, 1)$: $2x + y = 8$
11. $(-3, -8)$: $y = 5x + 7$
12. $(-2, -9)$: $y = 3x - 4$
13. $(3, 4)$: $5x - 2y = 6$
14. $(2, 1)$: $-3x + 7y = 1$

Find the missing coordinates in the ordered pairs which are solutions to the following equations.

15. $y = -\dfrac{1}{3}x$ $(6, \ ), (\ , 0)$
16. $y = \dfrac{2}{5}x$ $(\ , 3), (5, \ )$
17. $x + 2y = 6$ $(\ , 2), (-4, \ )$
18. $x - 3y = 14$ $(-2, \ ), (\ , 3)$
19. $y = 5x$ $(-1, \ ), (\ , -10)$
20. $y = -8x$ $(\ , -8), (0, \ )$
21. $y = 5x + 2$ $(3, \ ), (-2, \ )$
22. $y = 8x - 4$ $(-1, \ ), (3, \ )$
23. $4x + y = 11$ $(\ , 3), (-6, \ )$
24. $-7x + y = 5$ $(\ , -9), (1, \ )$
25. $3x + 6y = 18$ $(\ , 3), (-2, \ )$
26. $2x - 4y = 8$ $(4, \ ), (\ , 3)$
27. $-3x + 2y = -8$ $(\ , -4), (\ , 5)$
28. $-5x - 4y = 3$ $(\ , 3), (\ , -2)$
29. $x = 7$ $(\ , -5), (7, \ )$
30. $y = -3$ $(\ , -3), (9, \ )$

Find the x-intercept and y-intercept of each of the following linear equations.

31. $x - y = 8$
32. $x + y = 3$
33. $y = 5x$
34. $y = -7x$
35. $y = 3x + 2$
36. $y = -5x + 1$
37. $x + 4y = 8$
38. $3x - y = 6$
39. $2x - 5y = 10$
40. $4x + 7y = 28$
41. $3x + 7y = 21$
42. $9x - 2y = 18$

Graph each of the following linear equations.

43. $x + y = 7$
44. $x - y = 4$
45. $y = 3x + 6$
46. $y = -5x + 10$
47. $y = -5x$
48. $y = 3x$
49. $3x + 4y = 12$
50. $3x + 6y = 18$
51. $5x - 3y = 15$
52. $7x - 2y = 14$
53. $y = 5x - 2$
54. $y = 4x + 1$
55. $3x + y = 9$
56. $4x - y = 8$
57. $x - 5y = 10$
58. $x + 2y = 6$
59. $x = 7$
60. $y = 3$
61. $y = -4$
62. $x = -6$

Answer the following questions.

63. What is the equation of the vertical line that passes through the point $(2, -5)$?
64. What is the equation of the horizontal line that passes through the point $(-1, -7)$?

65. What is the equation of the horizontal line that passes through the point $(\frac{1}{2}, \frac{2}{3})$?

66. What is the equation of the vertical line that passes through the point $(-3.5, -6.8)$?

67. Plot three points on the x-axis. What is the y-coordinate of every point on the x-axis? What is the equation of the x-axis?

68. Plot three points on the y-axis. What is the x-coordinate of every point on the y-axis? What is the equation of the y-axis?

69. Graph the lines $y = 2x + 3$ and $y = 2x - 5$ on the same set of coordinate axes. What do you notice about these two lines?

70. Graph the lines $y = -3x - 1$ and $y = -3x + 4$ on the same set of coordinate axes. What do you notice about these two lines?

12.4
Summary and Review

Key Terms

(12.2) The **Cartesian coordinate system** is the two-dimensional plane in which equations in two variables are graphed. It is also called the rectangular coordinate system or x-y plane.

An **ordered pair** represents a point in the x-y plane.

The **abscissa** is the x-coordinate of an ordered pair.

The **ordinate** is the y-coordinate of an ordered pair.

The **x-intercept** of an equation is the x-coordinate of the point where the graph of the equation crosses the x-axis.

The **y-intercept** of an equation is the y-coordinate of the point where the graph of the equation crosses the y-axis.

Calculations

(12.1) According to the **addition property of inequalities**, if any quantity is added to both sides of an inequality the inequality remains unchanged.

According to the **multiplication property of inequalities**, if both sides of an inequality are multiplied by a positive number, the inequality remains unchanged, and if both sides are multiplied by a negative number, the inequality reverses.

(12.3) **To graph a linear equation**, find at least two ordered pairs that are solutions to the equation, plot the points that correspond to those ordered pairs in the x-y plane, and draw a straight line through these points.

To find the x-intercept of an equation, set $y = 0$ and solve for x.

To find the y-intercept of an equation, set $x = 0$ and solve for y.

(12.3) The equation of a **vertical line** is $x = a$, where a is the x-coordinate of every point on the line.

The equation of a **horizontal line** is $y = b$, where b is the y-coordinate of every point on the line.

Chapter **12**
Review Exercises

Indicate which of the following statements are true and which are false. For those that are false, change the italicized expression to make the statement true.

1. If both sides of an inequality are multiplied by a negative number, the inequality *reverses*.
2. The y-coordinate of an ordered pair is called the *abscissa*.
3. For any point located in the second quadrant, the abscissa is *less than* the ordinate.
4. To find the *x-intercept* of an equation, set $x = 0$ and solve for y.
5. The equation of a *vertical* line is $x = a$, where a is the x-coordinate of every point on the line.
6. The x-y plane is also called the *Cartilage* coordinate system.

(12.1) Solve each of the following linear inequalities and graph the solution set on a number line.

7. $2x + 9 < 5$
8. $3x - 7 \geq 2$
9. $3 - 4t \geq 7$
10. $5 - 6t > 7$
11. $4y + 6 > 8y - 3$
12. $2y - 8 < 5y + 4$
13. $7(z - 3) > 6$
14. $9(z + 2) \geq 4$
15. $3(2w - 5) - 4w \leq 5(4 - w)$
16. $-2(4w - 1) \leq 3(3w - 2) + 7w$

(12.3) Graph each of the following linear equations.

17. $x - y = 9$
18. $x + y = 6$
19. $3x + 8y = 24$
20. $5x - 4y = 20$
21. $x - 3y = 9$
22. $6x + y = 12$
23. $y = -8x$
24. $y = 3x$
25. $y = 5x + 2$
26. $y = -2x - 1$
27. $y = 7$
28. $x = -5$

Chapter **12**
Explain It in Words

29. How do you find the x-intercept of an equation? Give an example.
30. How do you find the y-intercept of an equation? Give an example.
31. What is the difference between the abscissa and the ordinate? Give an example.
32. What is the difference between the equation of a horizontal line and the equation of a vertical line? Give examples of each.
33. What is meant by "the multiplication principle of inequalities"? Give an example.
34. What is meant by "the addition principle of inequalities"? Give an example.

Chapter **12**
Chapter Test

[12.1] Graph the solution set represented by each of the following.

1. $x \leq 3$
2. $y > -6$
3. $t \geq 8.3$
4. $w < -\dfrac{2}{3}$

[12.2] Solve each of the following linear inequalities and graph the solution set on a number line.

5. $x - 7 \leq 3$

6. $y + 4 > 9$

7. $5x - 7 < 8$

8. $3x + 1 \geq 13$

9. $9w - 7 > 6w - 11$

10. $2(z - 6) \leq 4z$

11. $3x > 5(x - 2)$

12. $4(t + 5) \geq 3(t + 8)$

[12.3] Find the x-intercept and y-intercept of each of the following equations.

13. $x - 5y = 10$

14. $4x + 6y = 24$

15. $y = -7x + 2$

[12.4] Graph each of the following linear equations.

16. $x + y = 7$

17. $4x - 8y = 32$

18. $6x + 3y = 18$

19. $x = -5$

20. $y = -3x$

Appendix

How to Use a Calculator

There are basically two kinds of calculators—those that use algebraic logic, and those that use reverse Polish notation (RPN). Generally, a calculator with an equal sign ⟨=⟩ will use algebraic logic, and one with an enter key ⟨ENTER⟩ will use RPN. The best way to learn how to use your calculator properly is to read its instruction manual.

We will briefly discuss how a few arithmetic operations are performed on a calculator that uses algebraic logic. Some important keys you will use are the ten numerical keys, the four basic operations keys ⟨+⟩⟨−⟩⟨×⟩⟨÷⟩, the decimal point key ⟨·⟩, the square root key ⟨√⟩, and the clear key ⟨C⟩, which should always be pressed before beginning a new calculation in order to clear the memory. The examples in Table A.1 illustrate the use of these keys.

TABLE A.1

Operation	Example	Keys to Press	Answers in Display Window
Addition	$53 + 84$	⟨5⟩⟨3⟩⟨+⟩⟨8⟩⟨4⟩⟨=⟩	137.
Subtraction	$95 - 27$	⟨9⟩⟨5⟩⟨−⟩⟨2⟩⟨7⟩⟨=⟩	68.
Multiplication	5.7×8	⟨5⟩⟨·⟩⟨7⟩⟨×⟩⟨8⟩⟨=⟩	45.6
Division	$-265 \div 5$	⟨2⟩⟨6⟩⟨5⟩⟨+/−⟩⟨÷⟩⟨5⟩⟨=⟩	−53.
Square Root	$\sqrt{361}$	⟨3⟩⟨6⟩⟨1⟩⟨√⟩	19.

When more than one operation is performed in sequence, some calculators will automatically do multiplication and division before addition and subtraction, according to the order of operations rules. Others will simply perform the operations in the order in which they are entered into the calculator. Thus, pressing the sequence of keys

will yield one of two possible answers, depending on your calculator. If the answer that appears in the display window is 17, your calculator automatically performed multiplication before addition.

$$5 + (3 \times 4) = 17$$

If the answer that appears in the display window is 32, your calculator performed the addition first.

$$(5 + 3) \times 4 = 32$$

Regardless of the kind of calculator you have, you can dictate the order in which you want the operations to be performed by entering the operation you want performed first and pressing the equal sign $\boxed{=}$ before proceeding on to the next operation. For example, to calculate $(34 - 8) \div 2$, press this series of buttons:

$$\boxed{3}\boxed{4}\boxed{-}\boxed{8}\boxed{=}\boxed{\div}\boxed{2}\boxed{=}$$

The result in the display window should be 13.

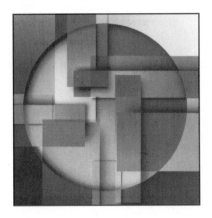

Answers to Odd-Numbered Exercises and Chapter Tests

1.1 Exercises

1. false; does 3. false; 7,008 5. tens; fifty-eight
7. hundreds; eight hundred six
9. thousands; four thousand thirty
11. ten thousands; twenty-one thousand, nine
13. hundred thousands; five hundred thousand, one
15. hundred thousands; one million, four hundred ninety-six thousand, six hundred twenty-eight
17. millions; thirty-eight million, two thousand, five hundred forty
19. hundred millions; seven hundred fifteen million, one hundred fifty-one thousand, five hundred fifteen
21. billions; eighty-nine billion, six hundred forty-one million, three hundred seventy-two thousand, fifty-five
23. trillions; seven trillion, four hundred eighty-six thousand, five hundred thirty-two
25. 47 27. 602 29. 1,066 31. 8,008 33. 50,600
35. 944,026 37. 7,000,008 39. 11,064,218
41. 2,053,103,481 43. 50,009,300,500
45. 5,705,006,474,302 47. $(3 \times 10) + (3 \times 1)$
49. $(6 \times 100) + (5 \times 1)$ 51. $(4 \times 1,000) + (2 \times 1)$
53. $(1 \times 10,000) + (4 \times 10)$
55. $(6 \times 100,000) + (3 \times 10,000) + (8 \times 1,000)$ $+ (5 \times 100) + (1 \times 10) + (8 \times 1)$
57. $(3 \times 10,000,000) + (4 \times 10,000) + (1 \times 1,000)$ $+ (3 \times 1)$
59. $(9 \times 100,000,000) + (9 \times 10,000,000)$ $+ (9 \times 1,000,000) + (9 \times 100,000) + (9 \times 10,000)$ $+ (9 \times 1,000) + (9 \times 100) + (9 \times 10) + (9 \times 1)$
61. $(5 \times 1,000,000,000) + (3 \times 10,000,000)$ $+ (8 \times 1,000,000) + (6 \times 100,000) + (2 \times 1,000)$ $+ (1 \times 100) + (2 \times 10) + (4 \times 1)$
63. $(3 \times 100,000,000,000) + (2 \times 1,000,000)$

$+ (4 \times 100,000) + (4 \times 1,000) + (1 \times 100)$
65. $(4 \times 10,000,000,000,000) + (4 \times 1,000,000,000,000)$ $+ (2 \times 100,000,000,000) + (9 \times 10,000,000,000)$ $+ (5 \times 10,000,000) + (3 \times 1,000,000) + (7 \times 100,000)$ $+ (7 \times 10,000) + (7 \times 1,000) + (1 \times 100) + (1 \times 10)$ $+ (9 \times 1)$
67. Multiply 7 and 3. 69. Divide 18 by 2.
71. a is equal to 10. 73. w is greater than t. 75. $<$
77. $<$ 79. $>$ 81. $=$ 83. $\neq; <$ 85. $=$ 87. $=$
89. $\neq; <$ 91. $=$ 93. 64,000,000
95. five trillion, eight hundred eighty billion
97. hundred millions
99. fourteen million, six hundred forty-eight thousand

1.2 Exercises

1. true 3. false; rightmost 5. false; less 7. 4 9. 14
11. 6 13. 13 15. 17 17. 15 19. 17 21. 16
23. 18 25. 19
27. commutative property of addition
29. associative property of addition
31. additive identity
33. commutative property of addition
35. commutative and associative properties of addition
37. 70; 72 39. 600; 648 41. 130; 134 43. 900; 901
45. 1,400; 1,426 47. 49 49. 133 51. 1,371 53. 785
55. 146 57. 725 59. 13,682 61. 492 63. 98,588
65. 9,361 67. 7,200,158 69. 44,037 71. $4 + 9 = 13$
73. $3 + 8$ 75. $5 + x = 9$ 77. $x + y$ 79. 1,656 books
81. 1,037 miles 83. $10,150 85. 437 tickets
87. 9,075,129 cars

465

1.3 Exercises

1. true **3.** true **5.** 9 **7.** 8 **9.** 7 **11.** 8 **13.** 5
15. 9 **17.** 3 **19.** 9 **21.** 10 **23.** 500; 536
25. 500; 523 **27.** 4,000; 4,224 **29.** 19,000; 18,732
31. 45 **33.** 13 **35.** 46 **37.** 645 **39.** 117 **41.** 226
43. 4,619 **45.** 1,884 **47.** 4,166 **49.** 27,728
51. 5,607,775 **53.** 31,677,314 **55.** $12 - 5$
57. $15 - 7 = 8$ **59.** $10 - 3 = 7$ **61.** $x - y$ **63.** $361
65. 181,618,000 **67.** $3,700 **69.** 3,032,100 returns

1.4 Exercises

1. 16 ft 1 in. **3.** 6 lb 4 oz **5.** 12 wk 1 da **7.** 7 hr 1 min
9. 3 ft 8 in. **11.** 5 lb 11 oz **13.** 4 da
15. 5 hr 52 min 33 sec **17.** 2 hr 25 min **19.** 5 lb 2 oz
21. 5 in. **23.** 2 lb 9 oz

1.5 Exercises

1. false; addition **3.** true **5.** 25 **7.** 0 **9.** 42 **11.** 49
13. 5 **15.** 30 **17.** 45 **19.** 63 **21.** 81 **23.** 21
25. 56 **27.** 42 **29.** 72 **31.** 0 **33.** 63
35. associative property of multiplication
37. distributive property of multiplication over addition
39. distributive property of multiplication over subtraction
41. multiplicative identity
43. commutative and associative properties of multiplication
45. $3(6) + 3(2) = 18 + 6 = 24$ **47.** $8(7 + 3) = 8(10) = 80$
49. $5(7) - 5(2) = 35 - 10 = 25$
51. $9(3) + 8(3) = 27 + 24 = 51$
53. $6(12 + 8) = 6(20) = 120$ **55.** $7(9 - 4) = 7(5) = 35$
57. $8(40) + 8(5) = 320 + 40 = 360$
59. $4(38 - 8) = 4(30) = 120$ **61.** $5(73 + 7) = 5(80) = 400$
63. $3(94 - 4) = 3(90) = 270$ **65.** 318 **67.** 567 **69.** 294
71. 1,312 **73.** 2,904 **75.** 5,696 **77.** 10,011
79. 58,926 **81.** 144,666 **83.** 612,306 **85.** 2,608,320
87. 29,925,372 **89.** 26,171,653

1.6 Exercises

1. 270 **3.** 1,500 **5.** 30,000 **7.** 640,000 **9.** 4,800,000
11. 140,000,000 **13.** 74,958 **15.** 434,654
17. 4,602,070 **19.** 4,605,484 **21.** 6,852,920
23. 32,888,706 **25.** 9,542,019 **27.** 1,500; 1,377
29. 12,000; 11,603 **31.** 560,000; 577,163
33. 1,500,000; 1,531,242 **35.** $5 \cdot 4 = 20$ **37.** $3a$
39. 504 miles **41.** $3,910 **43.** 7,179,396 books
45. 525,600 minutes

1.7 Exercises

1. false; dividend **3.** false; subtraction **5.** false; less
7. 9 **9.** 0 **11.** 8 **13.** 3 **15.** 7 **17.** 8 **19.** 4
21. 5 **23.** 3 **25.** 1 **27.** 7 R3 **29.** 5 R4 **31.** 6 R1
33. 18 **35.** 52 **37.** 41 **39.** 37 R2 **41.** 153 R2
43. 1,460 R1 **45.** 1,276 R2 **47.** 402 **49.** 6,351 R5
51. 16,434 R1 **53.** 2,156,831 R3 **55.** 1,333,333 R1
57. 4 R2 **59.** 1 R24 **61.** 11 R11 **63.** 5 R57
65. 185 R8 **67.** 590 R7 **69.** 25 R346 **71.** 25 R279
73. 73 R197 **75.** 92 R570 **77.** 35 **79.** 239
81. $36 \div 9$ **83.** $x \div y$ **85.** $860 **87.** 17 cards; 1 card
89. 4 pounds **91.** $645 **93.** 14 seats **95.** 13 hours

Review Exercises for Chapter 1

1. false; right **3.** false; division **5.** false; multiplication
7. five hundred forty-three; $(5 \times 100) + (4 \times 10) + (3 \times 1)$
9. four thousand, eight hundred ninety-two; $(4 \times 1,000)$
  $+ (8 \times 100) + (9 \times 10) + (2 \times 1)$
11. one hundred fifty-three thousand, eight hundred seventy-two; $(1 \times 100,000) + (5 \times 10,000) + (3 \times 1,000)$
  $+ (8 \times 100) + (7 \times 10) + (2 \times 1)$
13. four million, seventy-nine thousand, five hundred eighty-one; $(4 \times 1,000,000) + (7 \times 10,000) + (9 \times 1,000)$
  $+ (5 \times 100) + (8 \times 10) + (1 \times 1)$
15. eight billion, four hundred ninety-one million, seven hundred thirty-seven thousand, four hundred ninety-five;
  $(8 \times 1,000,000,000) + (4 \times 100,000,000)$
  $+ (9 \times 10,000,000) + (1 \times 1,000,000) + (7 \times 100,000)$
  $+ (3 \times 10,000) + (7 \times 1,000) + (4 \times 100) + (9 \times 10)$
  $+ (5 \times 1)$
17. $>$ **19.** $=$ **21.** $>$ **23.** $>$
25. commutative property of multiplication
27. distributive property
29. commutative property of multiplication
31. multiplicative identity
33. commutative property of multiplication
35. commutative property of multiplication
37. commutative property of addition
39. 1,259 **41.** 2,914 **43.** 33 R4 **45.** 247 **47.** 351
49. 157 R7 **51.** 95,238 **53.** 6,058 **55.** 905,148
57. 526,183 **59.** 24,187,113 **61.** 47,163,214 **63.** $5 - 2$
65. $33 \div 3$ **67.** $8a$ **69.** $2 \cdot 8 = 16$ **71.** 349 mi
73. 132,497 immigrants **75.** 2 hours 15 minutes **77.** $135
79. 15,580 miles **81.** 11,793,000 people **83.** 32 points
85. $23
87. The associative property of addition means that you can change the grouping of the numbers being added without affecting the result. For example, $(3 + 7) + 5 = 3 + (7 + 5)$. If we simplify the left-hand side of this expression, we add

Review Exercises for Chapter 1 (cont.)

3 and 7 first to get 10, and then add 5 to that result to get 15. If we simplify the right-hand side of the expression, we add 7 and 5 first to get 12, and then add 3 to that result to get 15. It does not matter if we add 3 and 7 first or if we add 7 and 5 first. Both methods result in the same answer, 15.

89. The identity element of addition is 0, since we can add 0 to any number and the result is the same as that number. For example, $0 + 5 = 5, 37 + 0 = 37$.

91. Our number system is called the base ten system because it includes ten digits: 0, 1, 2, 3, 4, 5, 6, 7, 8, and 9.

Chapter 1 Test

1. seven thousand, ninety-three
2. $(6 \times 1,000) + (4 \times 100) + (8 \times 10) + (5 \times 1)$
3. distributive property
4. associative property of multiplication
5. multiplicative identity property
6. commutative property of addition
7. 157   **8.** 822   **9.** 7,436   **10.** 18   **11.** 656   **12.** 8,541
13. 2,692   **14.** 3,648   **15.** 4,967,352   **16.** 60,800,000
17. 94   **18.** 379   **19.** 871   **20.** $72 \div 9$ **21.** $8 + 3 = 11$
22. $15 - 4 = 11$ **23.** $6 \times 5 = 30$ **24.** \$440   **25.** 1:23 P.M.

2.1 Exercises

1. false; base   **3.** true   **5.** $3^3 \cdot 8^2$ **7.** $2^6 \cdot 6^1 \cdot 7^2$
9. $5^3 \cdot 8^2 + 5^2 \cdot 7^3$ **11.** 2^2 **13.** 5^2 **15.** 2^5 **17.** 3^4
19. 5^4 **21.** 100   **23.** 9   **25.** 216   **27.** 720   **29.** 49,000
31. 160,000   **33.** 33,872,256
35. $(5 \times 10^2) + (3 \times 10^1) + (8 \times 10^0)$
37. $(6 \times 10^2) + (2 \times 10^0)$
39. $(7 \times 10^3) + (4 \times 10^2) + (3 \times 10^1) + (6 \times 10^0)$
41. $(4 \times 10^4) + (1 \times 10^3) + (7 \times 10^2) + (8 \times 10^1)$
$+ (3 \times 10^0)$
43. $(7 \times 10^4) + (4 \times 10^3) + (6 \times 10^1) + (2 \times 10^0)$
45. $(3 \times 10^5) + (4 \times 10^4) + (5 \times 10^0)$
47. $(6 \times 10^6) + (9 \times 10^5) + (4 \times 10^4) + (7 \times 10^3)$
$+ (8 \times 10^1) + (9 \times 10^0)$
49. $(5 \times 10^9) + (2 \times 10^8) + (1 \times 10^6) + (8 \times 10^5)$
$+ (4 \times 10^4) + (6 \times 10^2) + (9 \times 10^1) + (3 \times 10^0)$
51. $(7 \times 10^1) \times (4 \times 10^2) = (7 \times 4) \times (10^1 \times 10^2)$
$= 28 \times 10^3$
53. $(6 \times 10^2) \times (8 \times 10^3) = (6 \times 8) \times (10^2 \times 10^3)$
$= 48 \times 10^5$

2.1 Exercises (cont.)

55. $(11 \times 10^2) \times (7 \times 10^3) = (11 \times 7) \times (10^2 \times 10^3)$
$= 77 \times 10^5$
57. $(4 \times 10^4) \times (5 \times 10^4) = (4 \times 5) \times (10^4 \times 10^4)$
$= 20 \times 10^8$
59. $(7 \times 10^5) \times (3 \times 10^5) = (7 \times 3) \times (10^5 \times 10^5)$
$= 21 \times 10^{10}$
61. $25 \times 10^4 \times 2 \times 10^5 = 50 \times 10^9 = 5 \times 10^{10}$
63. 3^9 **65.** 4^6 **67.** 2^{14} **69.** 5^{11} **71.** 3^{12} **73.** 4^{10}
75. 7^{10} **77.** 5^{11} **79.** $8^6 \cdot 6^7$ **81.** $2^9 \cdot 10^8$ **83.** $7^{13} \cdot 8^8$
85. 6^{20} **87.** 4^{38}

2.2 Exercises

1. false; first   **3.** true   **5.** 17   **7.** 100   **9.** 13   **11.** 16
13. 77   **15.** 16   **17.** 40   **19.** 35   **21.** 9   **23.** 2
25. 135   **27.** 5   **29.** 3,850   **31.** 8,888   **33.** 17   **35.** 3
37. 23   **39.** 2   **41.** 32   **43.** $7^5 + 7^7$ **45.** $5^{10} + 5^6$
47. 2,003   **49.** 116,447   **51.** multiply first, then add: 14
53. divide first, then subtract: 5
55. raise to power first, then multiply: 18
57. subtract first, then add: 4
59. add numbers in parentheses first, then raise to power: 81
61. subtract numbers in parentheses first, then multiply: 6
63. $2(6 + 2) = 16$ **65.** $5 \cdot 4 - 2 = 18$ **67.** $2^3 + 6 = 14$
69. $(3 \cdot 4) + (10 \div 2) = 17$ **71.** $(2 + 5)(6 + 3) = 63$

2.3 Exercises

1. true   **3.** false; less than or equal to   **5.** 46   **7.** 344
9. 41,576   **11.** 72°F   **13.** 61 beats per minute
15. 29 minutes   **17.** 62 words per minute   **19.** 214 miles
21. 8   **23.** \$1,000   **25.** 140 deaths per day
27. 142,845 square miles

2.4 Exercises

1. true   **3.** false; 3   **5.** 109   **7.** 131 R2   **9.** 672
11. 464 R7   **13.** 24,613 R2   **15.** 7,449   **17.** 143,350 R6
19. 2   **21.** 2, 3, 4, 5, 6, 8, 10   **23.** 2, 3, 4, 6, 8   **25.** 3
27. 3, 5, 9   **29.** 2, 3, 4, 6, 9   **31.** 3, 9   **33.** 2, 4, 8   **35.** 2
37. 2, 4   **39.** 2, 3, 4, 6, 8   **41.** 2, 3, 4, 6, 8   **43.** 3, 5, 9
45. 2, 3, 6   **47.** 2, 4, 5, 8, 10   **49.** 0, 2, 4, 6, 8   **51.** 0, 4, 8
53. 4   **55.** 1   **57.** 0, 2, 4, 6, 8   **59.** 2, 6   **61.** 4   **63.** 1
65. yes   **67.** yes   **69.** no   **71.** no   **73.** no   **75.** no
77. no   **79.** yes   **81.** yes   **83.** no   **85.** no   **87.** yes

2.5 Exercises

1. false; prime 3. true 5. composite 7. prime
9. prime 11. composite 13. prime 15. composite
17. composite 19. composite 21. prime 23. prime
25. composite 27. prime 29. composite 31. composite
33. composite 35. 0, 4, 8, 12, 16 37. 0, 7, 14, 21, 28
39. 0, 10, 20, 30, 40 41. 0, 15, 30, 45, 60
43. 0, 20, 40, 60, 80 45. 2^4 47. $2 \cdot 3^3$ 49. $2 \cdot 3 \cdot 7$
51. 2^5 53. $2 \cdot 3 \cdot 13$ 55. $2^3 \cdot 7$ 57. 2^6 59. $2^2 \cdot 3 \cdot 5$
61. $2^2 \cdot 3^3$ 63. $3^2 \cdot 5^2$ 65. $2 \cdot 3 \cdot 5^2$ 67. $2 \cdot 89$
69. $2^3 \cdot 3^2 \cdot 5$ 71. $2 \cdot 3 \cdot 5 \cdot 23$ 73. $2^2 \cdot 5^2 \cdot 7$
75. $2 \cdot 5^2 \cdot 19$ 77. $2^2 \cdot 5^4$ 79. $2 \cdot 5^5$ 81. $2^4 \cdot 5^4$

2.6 Exercises

1. false; product 3. false; less than 5. 6 7. 8 9. 12
11. 5 13. 12 15. 2 17. 8 19. 13 21. 12 23. 14
25. 8 27. 15 29. 12 31. 30 33. 40 35. 108
37. 96 39. 168 41. 24 43. 200 45. 126 47. 720
49. 2,940 51. 288 53. GCF, 3; LCM, 252
55. GCF, 6; LCM, 270 57. GCF, 7; LCM, 294
59. GCF, 12; LCM, 1,512 61. GCF, 4; LCM, 1,008

Review Exercises for Chapter 2

1. false; is 3. false; 6 5. true 7. 5^9 9. 2^8
11. 7^{20} 13. 2^{27} 15. $4^5 + 4^7$ 17. 44 19. 648
21. 36 23. 16 25. 15 27. 3 29. 5,326 31. 12
33. 24 35. 23 37. 3 39. 135 41. 59,031
43. 515,363 45. 3, 9 47. 5 49. none
51. 2, 3, 4, 5, 6, 8, 10 53. 3, 5 55. prime
57. composite 59. prime 61. composite
63. composite 65. $2^2 \cdot 17$ 67. $2^3 \cdot 3 \cdot 5$ 69. $5 \cdot 7^2$
71. $2^2 \cdot 3 \cdot 5^3$ 73. GCF, 8; LCM, 224
75. GCF, 18; LCM, 432 77. GCF, 4; LCM, 96
79. GCF, 3; LCM, 1,656 81. GCF, 7; LCM, 168
83. $6 \cdot 9 - 8 = 46$ 85. $6(4 + 9) = 78$
87. $(17 - 3)^2 = 196$ 89. $63 \div 7 - 3 = 6$ 91. 72
93. 82 people per square mile 95. 131,075 passengers
97. A prime number is a whole number that is greater than 1
and only divisible by 1 and itself. A composite number is
a whole number that is greater than 1 and is divisible by
numbers other than 1 and itself. For example, 5 is a prime
number because it is only divisible by 1 and 5. The
number 6 is a composite number because it is not only
divisible by 1 and 6, but it is also divisible by 2 and 3.
99. A number is divisible by 2 if its ones digit is 0, 2, 4, 6, or
8. For example, the number 64 is divisible by 2 because

Review Exercises for Chapter 2 (cont.)

its ones digit is 4. The number 65 is not divisible by 2
because its ones digit is 5.
101. The greatest common factor of a group of numbers is the
largest number by which every number in the group is
divisible. For example, the greatest common factor of 8,
12, and 16 is 4 because 4 is the largest number by which
8, 12, and 16 are divisible.

Chapter 2 Test

1. 7^8 2. 4^8 3. 2^{16} 4. 5^{17} 5. 30 6. 3 7. 28
8. 95 9. 3 10. 2, 3, 4, 5, 6, 9, 10 11. 5 12. prime
13. $2 \cdot 7^2$ 14. 27 15. 7 16. 160 17. 144
18. $4^3 - 8$; 56 19. $3(7 + 4)$; 33
20. $(64 + 83 + 76 + 78 + 89) \div 5$; 78

3.1 Exercises

1. true 3. true 5. false; improper 7. false; $\dfrac{20}{5}$
9. true 11. $\dfrac{2}{5}$ 13. $\dfrac{39}{8}$ 15. $\dfrac{1}{11}$ 17. $\dfrac{6}{1}$ 19. $\dfrac{13}{52}$
21. improper 23. improper 25. proper 27. improper
29. improper 31. $54 \div 6 = 9$ 33. $0 \div 6 = 0$
35. $21 \div 7 = 3$ 37. $\dfrac{2}{3}$ 39. $1\dfrac{2}{3} = \dfrac{5}{3}$ 41. $1\dfrac{3}{4}$ 43. $6\dfrac{1}{3}$
45. $8\dfrac{1}{5}$ 47. $27\dfrac{1}{3}$ 49. $14\dfrac{3}{7}$ 51. $35\dfrac{5}{9}$ 53. $77\dfrac{4}{7}$ 55. $9\dfrac{7}{12}$
57. $40\dfrac{13}{21}$ 59. $\dfrac{16}{5}$ 61. $\dfrac{74}{9}$ 63. $\dfrac{31}{7}$ 65. $\dfrac{85}{8}$ 67. $\dfrac{99}{8}$
69. $\dfrac{131}{6}$ 71. $\dfrac{151}{20}$ 73. $\dfrac{359}{10}$ 75. $\dfrac{680}{13}$ 77. $\dfrac{201}{101}$
79. $\dfrac{6,559}{94}$ 81. $\dfrac{3,168}{23}$ 83. $\dfrac{3}{32}$ 85. $\dfrac{13}{52}$ 87. $\dfrac{7}{12}$
89. $\dfrac{7}{11}$ (men); $\dfrac{4}{11}$ (women) 91. $7\dfrac{21}{60} = \dfrac{441}{60}$ minutes
93. $1\dfrac{5}{16} = \dfrac{21}{16}$ pounds

3.2 Exercises

1. false; dividing 3. false; division 5. $\dfrac{18}{42}$ 7. $\dfrac{3}{7}$ 9. $\dfrac{16}{26}$
11. $\dfrac{5}{9}$ 13. $\dfrac{35}{28}$ 15. $\dfrac{6}{7}$ 17. $\dfrac{1}{2}$ 19. $\dfrac{21}{4}$ 21. $\dfrac{11}{14}$ 23. $\dfrac{6}{5}$

3.2 Exercises (cont.)

25. $\frac{15}{8}$ **27.** $\frac{7,497}{10,682}$ **29.** $\frac{13}{81}$ **31.** 1 **33.** 40 **35.** 15

37. 12 **39.** 48 **41.** 4 **43.** 3 **45.** 1 **47.** 14 **49.** 5

51. 108 **53.** 1,122 **55.** 2 **57.** 5 **59.** 150 **61.** 9

63. 3,355 **65.** 3 **67.** $\frac{2}{5}$ **69.** $\frac{1}{3}$ **71.** $\frac{3}{4}$ **73.** $\frac{3}{4}$ **75.** 6

77. $\frac{4}{7}$ **79.** $\frac{1}{6}$ **81.** $\frac{5}{2}$ **83.** $\frac{1}{2}$ **85.** $\frac{2}{7}$ **87.** $\frac{7}{9}$ **89.** $\frac{21}{44}$

91. $\frac{11}{18}$ **93.** $\frac{1}{6}$ **95.** $\frac{4}{9}$ **97.** $\frac{25}{7}$ **99.** $\frac{11}{12}$ **101.** $\frac{3}{16}$

103. $\frac{64}{3}$ **105.** $\frac{1}{6}$ **107.** $\frac{11}{25}$ **109.** $\frac{8}{9}$ **111.** $\frac{1}{21}$ week

3.3 Exercises

1. true **3.** true **5.** $\frac{4}{7}$ **7.** 1 **9.** $\frac{4}{5}$ **11.** $\frac{15}{17}$ **13.** $\frac{23}{20}$

15. $\frac{13}{12}$ **17.** $\frac{5}{4}$ **19.** $\frac{25}{33}$ **21.** $\frac{41}{50}$ **23.** $\frac{5}{28}$ **25.** $\frac{123}{35}$

27. $\frac{127}{100}$ **29.** $\frac{35}{144}$ **31.** $\frac{49}{150}$ **33.** $\frac{23}{54}$ **35.** $\frac{17}{63}$ **37.** $\frac{91}{132}$

39. $\frac{183}{20}$ **41.** $\frac{93}{10}$ **43.** $\frac{85}{98}$ **45.** $\frac{31}{30}$ **47.** $\frac{1,091}{1,000}$ **49.** $\frac{34}{75}$

51. $\frac{25}{24}$ **53.** $\frac{117}{112}$ **55.** $\frac{104}{213}$ **57.** $\frac{1,460}{1,421}$ **59.** $\frac{16}{7}$ **61.** $\frac{52}{17}$

63. $\frac{111}{8}$ **65.** $1\frac{13}{30}$ hours **67.** $\frac{59}{60}$. **69.** $1\frac{1}{8}$ pounds

3.4 Exercises

1. true **3.** true **5.** < **7.** < **9.** < **11.** < **13.** <

15. > **17.** < **19.** > **21.** > **23.** < **25.** $\frac{4}{7}$ **27.** $\frac{1}{9}$

29. $\frac{11}{72}$ **31.** $\frac{13}{18}$ **33.** $\frac{31}{90}$ **35.** $\frac{7}{72}$ **37.** $\frac{22}{39}$ **39.** $\frac{23}{48}$

41. $\frac{84}{275}$ **43.** $\frac{25}{168}$ **45.** $\frac{25}{6}$ **47.** $\frac{79}{12}$ **49.** $\frac{3}{28}$ **51.** $\frac{138}{2,183}$

53. $\frac{37}{72}$ **55.** $\frac{5}{6}$ **57.** $\frac{29}{42}$ **59.** $\frac{13}{60}$ **61.** $\frac{79}{144}$ **63.** $\frac{1}{12}$ cup

65. The teammate swam $\frac{1}{40}$ mile farther. **67.** Democrats

3.5 Exercises

1. $7\frac{3}{4}$ **3.** $7\frac{5}{8}$ **5.** $10\frac{5}{8}$ **7.** $135\frac{7}{8}$ **9.** $47\frac{1}{16}$ **11.** $16\frac{7}{12}$

13. $17\frac{17}{24}$ **15.** $16\frac{51}{80}$ **17.** $26\frac{29}{36}$ **19.** $19\frac{5}{56}$

21. $140\frac{19}{25}$ **23.** $10\frac{7}{180}$ **25.** $17\frac{43}{144}$ **27.** $84\frac{13}{27}$ **29.** $14\frac{3}{5}$

31. $46\frac{1}{3}$ **33.** $15\frac{31}{54}$ **35.** $9\frac{11}{12}$ **37.** $11\frac{1}{48}$ **39.** $18\frac{13}{16}$

41. $252\frac{358}{817}$ **43.** $993\frac{187}{910}$ **45.** $6\frac{1}{2}$ **47.** $5\frac{5}{8}$ **49.** $2\frac{5}{12}$

51. $9\frac{17}{27}$ **53.** $27\frac{7}{15}$ **55.** $3\frac{2}{3}$ **57.** $1\frac{1}{3}$ **59.** $3\frac{1}{2}$

61. $15\frac{16}{27}$ **63.** $13\frac{1}{18}$ **65.** $\frac{13}{24}$ **67.** $2\frac{4}{7}$ **69.** $7\frac{13}{18}$

71. $24\frac{3}{8}$ **73.** $3\frac{16}{21}$ **75.** $6\frac{27}{56}$ **77.** $2\frac{19}{144}$ **79.** $2\frac{5}{12}$

81. $7\frac{43}{60}$ **83.** $7\frac{111}{160}$ **85.** $9\frac{26}{75}$ **87.** $8\frac{269}{288}$ **89.** $5\frac{173}{378}$

91. 8 **93.** $15\frac{1}{20}$ **95.** $47\frac{1}{6}$ **97.** $13\frac{361}{1,000}$ **99.** $5\frac{5}{96}$

101. $4\frac{46}{55}$ miles **103.** $16\frac{19}{60}$ gallons **105.** $2\frac{3}{8}$ inches

107. Saturday: $2\frac{11}{60}$ hours; Sunday: $1\frac{59}{60}$ hours; Total: $4\frac{1}{6}$ hours

109. $6\frac{5}{48}$ ft.

3.6 Exercises

1. false; multiplied by **3.** false; $\frac{15}{8}$ **5.** $\frac{2}{21}$ **7.** $\frac{5}{32}$

9. $\frac{12}{77}$ **11.** $\frac{10}{27}$ **13.** $\frac{2}{3}$ **15.** $\frac{1}{56}$ **17.** $\frac{25}{4}$ **19.** $\frac{5}{66}$

21. $\frac{4}{21}$ **23.** $\frac{5}{24}$ **25.** $\frac{15}{224}$ **27.** $\frac{14}{9}$ **29.** $\frac{2}{121}$ **31.** 7

33. $1\frac{5}{7}$ **35.** $34\frac{2}{5}$ **37.** $1\frac{1}{16}$ **39.** $52\frac{1}{2}$ **41.** $7\frac{7}{12}$ **43.** $32\frac{1}{2}$

45. $33\frac{1}{3}$ **47.** $38\frac{1}{2}$ **49.** $16\frac{13}{36}$ **51.** $10\frac{5}{7}$ **53.** $73\frac{1}{3}$

55. $\frac{185}{324}$ **57.** $1,615\frac{59}{899}$ **59.** $\frac{5}{24}$ **61.** $\frac{71}{80}$ **63.** $\frac{7}{20}$

65. $\frac{1}{108}$ **67.** $\frac{1,311}{1,220}$ **69.** 31 **71.** $\frac{9}{5}$ **73.** $\frac{5}{21}$ **75.** $\frac{11}{10}$

3.6 Exercises (cont.)

77. $1\frac{2}{3}$ cups **79.** $\frac{8}{35}$ **81.** $133\frac{1}{3}$ yards **83.** $21\frac{7}{8}$ feet

3.7 Exercises

1. true **3.** false; 9 **5.** $\frac{8}{3}$ **7.** 7 **9.** $\frac{1}{11}$ **11.** 1 **13.** $\frac{4}{31}$

15. $\frac{2}{27}$ **17.** $\frac{13}{28}$ **19.** 6 **21.** $\frac{14}{15}$ **23.** $\frac{10}{21}$ **25.** $\frac{6}{5}$ **27.** $\frac{5}{14}$

29. $\frac{6}{5}$ **31.** $\frac{4}{3}$ **33.** $\frac{8}{21}$ **35.** $\frac{35}{4}$ **37.** 14 **39.** $\frac{1}{12}$ **41.** 9

43. $\frac{2}{5}$ **45.** $6\frac{2}{3}$ **47.** $\frac{2}{5}$ **49.** $1\frac{47}{88}$ **51.** $2\frac{14}{15}$ **53.** $\frac{5}{9}$

55. $3\frac{1}{2}$ **57.** $\frac{110}{279}$ **59.** $\frac{1,961}{1,944}$ **61.** $2\frac{6,412}{20,925}$ **63.** $\frac{2}{9}$

65. $\frac{2}{7}$ **67.** $\frac{31}{7}$ **69.** $\frac{13}{6}$ **71.** $\frac{73,358}{19,095}$ **73.** $5\frac{1}{3}$ laps

75. $\frac{7}{20}$ oz **77.** $\frac{5}{16}$ pound **79.** $4\frac{1}{3}$ boxes **81.** $\frac{3}{4}$ pt

83. $4\frac{1}{2}$ times

3.8 Exercises

1. $\frac{9}{20}$ **3.** $\frac{16}{5}$ **5.** $\frac{39}{20}$ **7.** 12 **9.** $\frac{19}{21}$ **11.** $\frac{7}{5}$ **13.** $\frac{26}{63}$

15. $\frac{9}{43}$ **17.** $\frac{9}{38}$ **19.** $21\frac{3}{4}$ **21.** $\frac{9}{200}$ **23.** $6\frac{33}{40}$ **25.** $1\frac{54}{83}$

27. $\frac{37}{249}$ **29.** $\frac{139}{143}$ **31.** $\frac{74}{129}$

Review Exercises for Chapter 3

1. false; sum **3.** true **5.** false; equivalent

7. incorrect; $\frac{15}{8}$ **9.** incorrect; 0 **11.** correct

13. incorrect; greater **15.** $\frac{5}{8}$ **17.** $\frac{7}{24}$ **19.** $\frac{40}{33}$ **21.** $\frac{35}{4}$

23. $\frac{19}{54}$ **25.** $1\frac{1}{2}$ **27.** 21 **29.** $\frac{31}{84}$ **31.** $5\frac{1}{3}$ **33.** $11\frac{19}{24}$

35. $4\frac{23}{24}$ **37.** $\frac{1}{16}$ **39.** $26\frac{2}{5}$ **41.** $2\frac{28}{33}$ **43.** $\frac{2}{9}$ **45.** $9\frac{35}{54}$

47. $\frac{19}{6}$ **49.** $\frac{8}{7}$ **51.** $\frac{19}{36}$ **53.** $\frac{13}{32}$ **55.** $\frac{15}{7}$ **57.** $\frac{7}{16}$

Review Exercises for Chapter 3 (cont.)

59. $\frac{58}{33}$ **61.** $\frac{9}{10}$ **63.** 1 **65.** $7\frac{7}{33}$ **67.** $\frac{41}{85}$ **69.** $3\frac{1}{8}$

71. $6\frac{1}{8}$ mi **73.** $1\frac{7}{8}$ points **75.** $5\frac{1}{3}$ cups sugar; 7 cups flour

77. Republicans; $\frac{5}{36}$ **79.** 18 bottles **81.** 104 people

83. To convert an improper fraction to a mixed number, you divide the numerator by the denominator. The quotient is the whole number portion of the mixed number, and the fractional portion of the mixed number is the remainder divided by the divisor. For example, to convert $\frac{37}{5}$ to an improper fraction you divide 37 by 5. The quotient is 7 with a remainder of 2. Therefore, $\frac{37}{5}$ is equivalent to the mixed number $7\frac{2}{5}$.

85. To multiply two fractions, you multiply the numerators and place the result over the product of the denominators. For example, to multiply $\frac{3}{5} \cdot \frac{2}{7}$ you multiply the numerators to obtain $3 \cdot 2 = 6$. You multiply the denominators to obtain $5 \cdot 7 = 35$. The answer is $\frac{6}{35}$.

87. To reduce a fraction to lowest terms, you divide both the numerator and the denominator by their greatest common factor. For example, to reduce the fraction $\frac{12}{16}$ to lowest terms you divide both 12 and 16 by their greatest common factor, which is 4. Since $12 \div 4 = 3$ and $16 \div 4 = 4$, the fraction $\frac{12}{16}$ reduces to $\frac{3}{4}$.

Chapter 3 Test

1. $10\frac{1}{4}$ **2.** $\frac{23}{8}$ **3.** $\frac{1}{3}$ **4.** $\frac{4}{19}$ **5.** $\frac{5}{9}$ **6.** $\frac{11}{20}$ **7.** $\frac{3}{10}$

8. $\frac{1}{12}$ **9.** $9\frac{1}{12}$ **10.** $4\frac{7}{9}$ **11.** $5\frac{11}{14}$ **12.** $2\frac{23}{24}$ **13.** $\frac{8}{45}$

14. $\frac{7}{15}$ **15.** 51 **16.** $12\frac{3}{5}$ **17.** $\frac{3}{4}$ **18.** $5\frac{1}{7}$ **19.** $\frac{13}{25}$

20. $1\frac{19}{33}$ **21.** 1 **22.** $\frac{4}{15}$ **23.** $2\frac{1}{2}$ **24.** 56 miles per hour

25. $4\frac{1}{8}$ miles

4.1 Exercises

1. false; hundredths **3.** false; denominator **5.** 0.3
7. 0.08 **9.** 0.005 **11.** 0.529 **13.** 0.0021 **15.** 0.00003
17. 0.00127 **19.** 0.006589 **21.** 7.3 **23.** 5.19
25. 2.837 **27.** 7.0803 **29.** tenths; three tenths

4.1 Exercises (cont.)

31. hundredths; fifty-three hundredths
33. tenths; five and nine tenths
35. tenths; forty-seven and eight tenths
37. thousandths; thirty-eight and nine hundred seventy-four thousandths
39. hundred-thousandths; eight hundred and sixty-five thousand, three hundred twenty-seven hundred-thousandths
41. tenths; three thousand, six hundred three and forty-six hundredths
43. ten-thousandths; five hundred and four thousand, six hundred eighty-nine ten-thousandths
45. ten-thousandths; eight thousand, sixty-seven and four ten-thousandths
47. millionths; three hundred sixty-two thousand, six and two hundred thousand, eight millionths
49. 0.4   **51.** 0.006   **53.** 0.0004   **55.** 0.0386
57. 0.000006   **59.** 0.002059   **61.** 2.6   **63.** 308.12

65. 11,000.00058   **67.** $(5 \times 10^0) + \left(2 \times \dfrac{1}{10^1}\right)$

69. $(3 \times 10^1) + (6 \times 10^0) + \left(4 \times \dfrac{1}{10^1}\right)$
$+ \left(6 \times \dfrac{1}{10^2}\right)$

71. $(2 \times 10^2) + (3 \times 10^1) + \left(5 \times \dfrac{1}{10^3}\right)$

73. $(3 \times 10^4) + (5 \times 10^0) + \left(6 \times \dfrac{1}{10^1}\right) + \left(2 \times \dfrac{1}{10^2}\right)$
$+ \left(7 \times \dfrac{1}{10^3}\right) + \left(8 \times \dfrac{1}{10^4}\right)$

75. $(5 \times 10^7) + (8 \times 10^4) + (3 \times 10^3) + (7 \times 10^1)$
$+ \left(4 \times \dfrac{1}{10^2}\right) + \left(6 \times \dfrac{1}{10^4}\right)$

77. $0.16; zero and 16/100 dollars
79. $3.74; three and 74/100 dollars
81. $46.48; forty-six and 48/100 dollars
83. $805.67; eight hundred five and 67/100 dollars
85. $7.14; seven and 14/100 dollars   **87.** thousandths
89. 27.32

4.2 Exercises

1. true   **3.** false; thousandth   **5.** 3.3; 3.28; 3.279
7. 14.0; 13.96; 13.963   **9.** 7.0; 7.01; 7.006
11. 0.5; 0.50; 0.502   **13.** 62.8; 62.83; 62.830
15. 38.1; 38.07; 38.071   **17.** 4.8; 4.80; 4.798
19. 68.0; 68.00; 68.000   **21.** 6,800; 6,850; 6,847

4.2 Exercises (cont.)

23. 7,000; 7,050; 7,046   **25.** 700; 710; 709
27. 700; 700; 696   **29.** 7,100; 7,080; 7,080
31. 500; 500; 500   **33.** 1,000; 990; 986
35. 60,700; 60,710; 60,710   **37.** 330,000   **39.** 6.0504
41. 2.587410   **43.** 70.0   **45.** 10,000.0   **47.** $368
49. $32.85   **51.** $88.6   **53.** $5.031   **55.** $6.8340
57. 51 miles   **59.** 53.7 seconds   **61.** $500

4.3 Exercises

1. false; vertically   **3.** false; right to left   **5.** 12.1
7. 46.95   **9.** 16.564   **11.** $46.75   **13.** 154.2783
15. 26.187   **17.** 1,308.9149   **19.** $863.58   **21.** 41.15147
23. 4.6   **25.** $8.12   **27.** 45.668   **29.** $159.06
31. 5.4376   **33.** 2.7226   **35.** 410.066   **37.** 631.8205
39. 6.4   **41.** 424.519   **43.** $76.45   **45.** 8,644.6516
47. $3,151.87   **49.** 22.172315   **51.** 170; 170.14
53. 350; 349.2   **55.** 820; 820.9   **57.** 5,400; 5,392.4
59. $3.21   **61.** 20.8 miles   **63.** $15.14   **65.** $5.70
67. $185.99

4.4 Exercises

1. true   **3.** false; product   **5.** 0.014   **7.** 0.00224
9. 3.078   **11.** 1.4418   **13.** $46.472   **15.** 0.0002698
17. 45.39084   **19.** 0.3023172   **21.** 512.144448
23. $35,999.88   **25.** 1.371816   **27.** 185,353.38
29. 80,000   **31.** 0.31   **33.** 3,602.005   **35.** 570,000,000
37. 280,300   **39.** 14,400   **41.** 0.0282   **43.** 22,740
45. 270   **47.** 140   **49.** 4,400,000   **51.** 0.089   **53.** 0.064
55. 1,750   **57.** 91   **59.** 0.00002   **61.** 4.8   **63.** 7.68
65. 289   **67.** 30   **69.** 0.081   **71.** $7.80
73. 1,062.99 inches   **75.** 5,151,600   **77.** $648.31
79. $9,205   **81.** 25,200,000 square miles   **83.** $2,280.61

4.5 Exercises

1. false; left   **3.** true   **5.** 9   **7.** 0.007   **9.** 13,000
11. 0.013   **13.** 0.14   **15.** 23,000   **17.** 0.52   **19.** 56
21. 470,000   **23.** 4.6667   **25.** 186.7   **27.** 693.333
29. $2.98   **31.** 2,551.85   **33.** 3,052   **35.** $14.2
37. 24,831   **39.** 0.06818   **41.** 0.000783   **43.** 0.0006005
45. 0.90009   **47.** 0.0002817   **49.** 0.005386   **51.** 5,000
53. 3.2   **55.** 40   **57.** 0.00288   **59.** 0.0000128   **61.** 3
63. 6.12   **65.** 20   **67.** 0.0816   **69.** 2,000   **71.** $711.54
73. 6.3 pounds   **75.** 5.793 persons per square mile   **77.** 2.4
79. 160.9 miles per hour

4.6 Exercises

1. false; repeating   **3.** true   **5.** $\frac{2}{25}$ **7.** $\frac{4}{25}$ **9.** $\frac{3}{125}$

11. $\frac{3}{400}$ **13.** $\frac{5}{16}$ **15.** $\frac{5}{2}$ **17.** $\frac{64}{5}$ **19.** $\frac{309}{20}$ **21.** $\frac{803}{20}$

23. $\frac{323}{80}$ **25.** 0.25   **27.** 0.4   **29.** $0.\overline{45}$ **31.** 0.15625

33. $0.2\overline{6}$ **35.** $0.\overline{4}$ **37.** $0.3\overline{8}$ **39.** 0.75   **41.** $0.\overline{2}$ **43.** $2.\overline{6}$

45. $1.\overline{4}$ **47.** 1.1875   **49.** $30.58\overline{3}$ **51.** $242.\overline{3}$

53. $0.\overline{095238}$ **55.** $0.6\overline{78}$ **57.** 0.158   **59.** 0.23

61. 0.8667   **63.** 0.54545   **65.** 0.286   **67.** 2.57

69. 40.89   **71.** 0.763

Review Exercises for Chapter 4

1. false; right   **3.** false; division   **5.** false; subtracting

7. eight and seven tenths; $(8 \times 10^0) + \left(7 \times \frac{1}{10^1}\right)$

9. five and eighty-seven hundredths;

$$(5 \times 10^0) + \left(8 \times \frac{1}{10^1}\right) + \left(7 \times \frac{1}{10^2}\right)$$

11. forty-three and six hundred twenty-eight thousandths;

$$(4 \times 10^1) + (3 \times 10^0) + \left(6 \times \frac{1}{10^1}\right) + \left(2 \times \frac{1}{10^2}\right)$$
$$+ \left(8 \times \frac{1}{10^3}\right)$$

13. eight hundred six and five thousand, six hundred fourteen

ten-thousandths; $(8 \times 10^2) + (6 \times 10^0) + \left(5 \times \frac{1}{10^1}\right)$
$$+ \left(6 \times \frac{1}{10^2}\right) + \left(1 \times \frac{1}{10^3}\right) + \left(4 \times \frac{1}{10^4}\right)$$

15. 4.5   **17.** 18   **19.** 60   **21.** 1,000   **23.** $5.048

25. 628.432   **27.** 737.47   **29.** 0.1926   **31.** 2,380

33. 9.3607   **35.** 3.24292   **37.** 8.37   **39.** $3,661.49

41. $2,519.72   **43.** 88.0589   **45.** 1.48304   **47.** 7

49. 2,880,000   **51.** 190   **53.** 0.296   **55.** 0.000021

57. 3.6   **59.** 12   **61.** 64   **63.** 627.8428   **65.** 1.4567886

67. 5.3   **69.** 0.000000078   **71.** 16.894   **73.** 9,209,000

75. $\frac{3}{25}$ **77.** $\frac{1}{40}$ **79.** $\frac{1}{6,250}$ **81.** $\frac{9}{4,000}$ **83.** $\frac{22}{5}$ **85.** $\frac{409}{20}$

87. $0.\overline{6}$ **89.** 0.4375   **91.** $0.\overline{72}$ **93.** 0.714   **95.** 2.4

97. $4.\overline{72}$ **99.** 0.246   **101.** first day; 983.3 miles

103. 2.5   **105.** $18.25   **107.** $1.98

109. A terminating decimal is the exact representation of a fraction. It is the quotient that is obtained when you

Review Exercises for Chapter 4 (cont.)

divide the numerator by the denominator until you obtain a remainder of zero. For example, 0.25 is a terminating decimal since it is equivalent to the fraction $\frac{1}{4}$, and when you calculate $1 \div 4$, you obtain a quotient of 0.25 with a remainder of zero.

111. To convert a fraction to a decimal, you divide the numerator by the denominator. For example, to convert $\frac{2}{5}$ to a decimal, you calculate $2 \div 5$ and obtain a result of 0.4.

113. To round off a decimal, examine the digit to the right of the place value to which you are rounding off. If it is 5 or greater, add 1 to the digit in the place value to which you are rounding, and if not, that digit stays the same. For example, to round off 4.725 to the nearest hundredth, since 2 is in the hundredths place, we look at the digit to the right of 2, which is 5. Since $5 = 5$, we add 1 to the 2, and the rounded number is 4.73.

Chapter 4 Test

1. three and ninety-seven hundredths;

$$(3 \times 1) + (9 \times \frac{1}{10^1}) + (7 \times \frac{1}{10^2})$$

2. seventy-three and four hundred seventy-five thousandths;

$$(7 \times 10^1) + (3 \times 1) + (4 \times \frac{1}{10^1}) + (7 \times \frac{1}{10^2})$$
$$+ (5 \times \frac{1}{10^3})$$

3. 8.48   **4.** 63   **5.** $4.783   **6.** 375.79   **7.** 658.4042

8. 89,067.67   **9.** 23.71   **10.** 554.191   **11.** 115.272

12. 0.8334   **13.** 345.6004   **14.** 2300   **15.** 57.9

16. 894.5   **17.** 5400   **18.** 93   **19.** 0.082617   **20.** 15.5

21. 0.38   **22.** 0.2727 . . .   **23.** $10.75   **24.** $418.25

25. 2.9 inches per month

5.1 Exercises

1. false; numerator   **3.** true   **5.** $\frac{3}{5}$ **7.** $\frac{4}{1}$ **9.** $\frac{5}{8}$ **11.** $\frac{22}{1}$

13. $\frac{1}{14}$ **15.** $\frac{1}{12}$ **17.** $\frac{15}{16}$ **19.** $\frac{14}{33}$ **21.** $\frac{3}{1}$ **23.** 12 feet

25. 4 quarters   **27.** 10 pints   **29.** $1\frac{1}{2}$ ft   **31.** 10 ounces

33. $\frac{7}{6}$ **35.** $\frac{19}{25}$ **37.** $\frac{4}{3}$ **39.** $\frac{7}{8}$ **41.** $\frac{8}{5}$ **43.** $\frac{7}{6}$ **45.** $\frac{3}{8}$

47. $\frac{3}{8}$ **49.** $\frac{3}{2}$ **51.** $\frac{7}{25}$ **53.** $\frac{17}{15}$ **55.** $\frac{4}{3}$ **57.** $\frac{6}{1}$ **59.** $\frac{2}{5}$

5.1 Exercises (cont.)

61. a. $2:7$; **b.** $7:1$; **c.** $\dfrac{1}{5}$; **d.** $\dfrac{3}{10}$; **e.** $\dfrac{9}{10}$ **63. a.** $1:9$;

b. $9:1$; **c.** $9:1,000$; **d.** $\dfrac{100}{501}$; **e.** $\dfrac{400}{501}$ **65.** $\dfrac{1}{4}$

67. the sedan **69.** 5.5¢ per ounce **71.** $1:19$ **73.** $19:50$
75. 2.375¢ per fluid ounce; 2.75¢ per fluid ounce; the frozen orange juice
77. $\dfrac{9}{8}$; Monday **79.** 62 miles per hour
81. 125 persons per square mile
83. 8.75¢ per egg (dozen); 8.5¢ per egg (eight); 8-egg carton
85. $\dfrac{8}{9}$; red ribbon **87.** 6.19 persons per square mile
89. 5.47 miles per hour

5.2 Exercises

1. true **3.** false; less
5. extremes, 6 and 3; means, 9 and 2; 6 is to 9 as 2 is to 3
7. extremes, 5 and 8; means, 2 and 20; 5 is to 2 as 20 is to 8
9. extremes, $\dfrac{1}{2}$ and 22; means, 11 and 1; $\dfrac{1}{2}$ is to 11 as 1 is to 22
11. extremes, $2\dfrac{1}{3}$ and 24; means, 8 and 7; $2\dfrac{1}{3}$ is to 8 as 7 is to 24
13. false **15.** true **17.** false **19.** false **21.** false
23. true **25.** true **27.** false **29.** true **31.** true
33. false **35.** false **37.** $<$ **39.** $>$ **41.** $<$ **43.** $<$
45. $<$ **47.** $>$ **49.** $>$ **51.** $>$ **53.** $>$
55. the White Sox player **57.** deodorant soap
59. U.S. skater **61.** 18.5-ounce bottle **63.** Alabama

5.3 Exercises

1. true **3.** false; cross multiply
5. variable, x; coefficient, 5 **7.** variable, t; coefficient, 17
9. variable, u; coefficient, $\dfrac{5}{8}$ **11.** variable, a; coefficient, 1.4
13. variable, x; coefficient, 18 **15.** $x = 15$ **17.** $y = 14$
19. $x = \dfrac{40}{13}$ **21.** $y = 14$ **23.** $m = 18$ **25.** $q = 78$
27. $y = \dfrac{23}{6}$ **29.** $x = \dfrac{147}{5}$ **31.** $y = \dfrac{24}{7}$ **33.** $z = 243$
35. $y = \dfrac{192}{5}$ **37.** $q = 90$ **39.** $x = 2$ **41.** $s = \dfrac{18}{55}$
43. $x = \dfrac{13}{30}$ **45.** $w = \dfrac{3}{4}$ **47.** $v = 2\dfrac{5}{6}$ **49.** $x = 0.08$
51. $y = 0.0025$ **53.** $w = 0.18$ **55.** $t = 0.025$ **57.** $y = 6$
59. $z = 0.0000005$ **61.** $t = 0.06$ **63.** $x = 144$
65. $w = 672$ **67.** $x = 0.190$ **69.** $w = 22$

5.4 Exercises

1. true **3.** true
5. $\dfrac{7}{2} = \dfrac{x}{12}$; $\dfrac{7}{x} = \dfrac{2}{12}$; $\dfrac{2}{7} = \dfrac{12}{x}$; $\dfrac{x}{7} = \dfrac{12}{2}$; $x = 42$ passes
7. $\dfrac{20}{4} = \dfrac{45}{x}$; $\dfrac{20}{45} = \dfrac{4}{x}$; $\dfrac{4}{20} = \dfrac{x}{45}$; $\dfrac{45}{20} = \dfrac{x}{4}$; $x = 9$ pages
9. $\dfrac{10}{4} = \dfrac{6}{x}$; $\dfrac{10}{6} = \dfrac{4}{x}$; $\dfrac{4}{10} = \dfrac{x}{6}$; $\dfrac{6}{10} = \dfrac{x}{4}$; $x = 2.4$ hours
11. 8 feet **13.** 15 hits **15.** 240 dentists **17.** 14 car sales
19. $1.70 **21.** 2,463 students **23.** 435 balls
25. 77 people **27.** 2 feet **29.** 42 students **31.** 1 gram
33. 50 people **35.** 315 miles **37.** 165 calories **39.** $1.05
41. 1 cup **43.** $8.73 **45.** 14 minutes
47. a. 20 nickels; **b.** 30 quarters; **c.** 62 coins
49. a. 24 juniors; **b.** 18 sophomores; **c.** 42 seniors
51. 11.2 mg **53.** 6,146 women **55.** 144,000 square miles

Review Exercises for Chapter 5

1. true **3.** false; multiplied by **5.** true **7.** $\dfrac{2}{3}$ **9.** $\dfrac{8}{1}$
11. $\dfrac{5}{6}$ **13.** $\dfrac{6}{5}$ **15.** $\dfrac{28}{9}$ **17.** $\dfrac{22}{5}$ **19.** $>$ **21.** $<$ **23.** $>$
25. $>$ **27.** $<$ **29.** $<$ **31.** $<$
33. variable, x; coefficient, 8
35. variable, z; coefficient, 1
37. variable, t; coefficient, $\dfrac{3}{5}$
39. variable, b; coefficient, 15 **41.** $x = 24$ **43.** $x = 14$
45. $t = \dfrac{9}{2}$ **47.** $n = 24$ **49.** $x = \dfrac{1}{14}$ **51.** $y = 3\dfrac{3}{4}$
53. $t = 6$ **55.** $n = 24$ **57.** $x = 0.0432$ **59.** $\dfrac{1}{13}$
61. $0.258 per pound **63.** $\dfrac{21}{20}$; math **65.** the shortstop
67. $246 **69.** 681 mothers **71.** 8 teaspoons **73.** $12,950
75. 4.77 persons per square mile **77.** $123.1 billion
79. A ratio is a fraction that expresses a relationship between two quantities. For example, if there are 7 men and 4 women in a meeting, the ratio of men to women is $\dfrac{7}{4}$, or $7:4$. The ratio of women to men is $\dfrac{4}{7}$, or $4:7$.
81. The fundamental property of proportions states that in any proportion, the product of the extremes is equal to the product of the means. For example, in the proportion

$$\dfrac{2}{5} = \dfrac{6}{15}$$

the product of the extremes is $2 \cdot 15 = 30$, the product of the means is $5 \cdot 6 = 30$, and these products are equal.
83. The kinds of word problems that can be solved using a proportion include those that express a relationship

Review Exercises for Chapter 5 (cont.)

between two quantities in which a ratio between two known quantities can be set equal to a ratio in which one quantity is known and the other is unknown. An example of such a word problem is as follows: If a computer consultant earns $300 for working 8 hours how much would she earn for working 10 hours? This problem could be solved by finding the unknown in the proportion

$$\frac{\$300}{8 \text{ hr}} = \frac{x}{10 \text{ hr}}$$

The first ratio includes the known quantities $300 and 8 hours of work, and the second ratio includes an unknown dollar amount for 10 hours of work.

Chapter 5 Test

1. $\frac{9}{7}$ **2.** $\frac{3}{4}$ **3.** $\frac{3}{2}$ **4.** $\frac{6}{7}$ **5.** $\frac{8}{5}$ **6.** $<$ **7.** $<$ **8.** $>$

9. $<$ **10.** $<$ **11.** $x = 16$ **12.** $x = 45$ **13.** $w = 10\frac{5}{6}$

14. $x = 1\frac{9}{10}$ **15.** $z = 2.4$ **16.** $3.29 **17.** compact

18. 52 **19.** $6.95 **20.** 7 ft 6 in.

6.1 Exercises

1. true **3.** false; less **5.** $\frac{3}{4}$ **7.** $\frac{1}{50}$ **9.** $\frac{57}{100}$ **11.** $\frac{16}{25}$

13. $\frac{6}{5}$ **15.** 5 **17.** $\frac{11}{250}$ **19.** $\frac{9}{2,000}$ **21.** $\frac{307}{10,000}$

23. $\frac{173}{1,000}$ **25.** $\frac{3}{16}$ **27.** $\frac{15}{16}$ **29.** $\frac{1}{200}$ **31.** $\frac{5}{16}$ **33.** $\frac{351}{500}$

35. $\frac{2}{3}$ **37.** 13% **39.** 1% **41.** 3.2% **43.** $5\frac{1}{3}\%$

45. 75% **47.** 40% **49.** 28% **51.** 65% **53.** 80%

55. 2.7% **57.** 3.18% **59.** 17.5% **61.** 62.5%

63. $11\frac{1}{9}\%$ **65.** $46\frac{2}{3}\%$ **67.** 60% **69.** $\frac{7}{20}$

6.2 Exercises

1. false; left **3.** true **5.** 0.71 **7.** 0.05 **9.** 0.248

11. 0.0409 **13.** 0.0028 **15.** 2.25 **17.** 36.84

19. 0.05529 **21.** 0.1803 **23.** 0.085 **25.** 0.456

27. 1.30125 **29.** 0.0214 **31.** 37% **33.** 4% **35.** 66%

6.2 Exercises (cont.)

37. 61.3% **39.** 0.9% **41.** 60% **43.** 6.75% **45.** 450%

47. 300% **49.** 1,320% **51.** 81.3% **53.** 507%

55. 6.05% **57.** 50% **59.** 87.5% **61.** $11\frac{1}{9}\%$

63. 18.75% **65.** 67.5% **67.** $41\frac{2}{3}\%$ **69.** $13\frac{1}{3}\%$

71. $28\frac{4}{7}\%$ **73.** $38\frac{6}{13}\%$ **75.** 360% **77.** 550%

79. 275% **81.** 150% **83.** 13.0% **85.** 60.4%

87. 82.5% **89.** 0.05; 5% **91.** 0.1875; 18.75%

93. $\frac{1}{3}$; $33\frac{1}{3}\%$ **95.** $\frac{2}{3}$; 0.66 . . . **97.** 0.833 . . . ; $83\frac{1}{3}\%$

99. 37.5% **101.** 48%

6.3 Exercises

1. true **3.** false; base **5.** 52 **7.** 75 **9.** 25%

11. 23.75 **13.** 0.34% **15.** 8.75% **17.** 100 **19.** 15.5

21. 340% **23.** 81.25 **25.** 3,000 **27.** 1.875% **29.** 21

31. 250% **33.** 62.5 **35.** $18\frac{3}{4}$ **37.** $77\frac{7}{9}\%$ **39.** 18

41. 0.8619204 **43.** 397.15 **45.** 18 students

47. 480 students **49.** 600 people **51.** 144 students

53. 16.25% **55.** 350 people **57.** 75% **59.** 2%

61. 31.25% **63.** 1,105 workers **65.** 60 grams

67. 7 people **69.** 2.1% **71.** 62.1%

6.4 Exercises

1. true **3.** false; 15% **5.** $3,255 **7.** $18,979.45

9. $20,980.35 **11.** $90,737.14 **13.** $2.25 **15.** $0.60

17. $9.00 **19.** $26.25 **21.** $8,078.50 **23.** $305.50

25. $36,980 **27.** $1.11 **29.** $405 **31.** 7% **33.** $1,200

35. $17,000 salary plus 8% commission **37.** $12,504.50

39. 5.5% **41.** $636 **43.** $20,000 **45.** $3.30 **47.** $160

49. $18.30 **51.** $68.43 **53.** $2,547.88

6.5 Exercises

1. false; sum of **3.** true **5.** 30% **7.** $6,296.50

9. $6,839.10 **11.** $307.34 **13.** 40% **15.** 5,950

17. $18.45 **19.** $424.15 **21.** 62.5% **23.** 140 patients

25. $4.92 **27.** $11,243.75 **29.** $241.50 **31.** $166.140

33. 5% **35.** $30 **37.** 638 students **39.** $53.57

41. $13,123,000 **43.** $9,359.35 **45.** $3,433.83

47. 4.94%

6.6 Exercises

1. a. third grade; 200 students; **b.** 800 students; **c.** 12.5%
3. a. Sault Ste. Marie, MI; 115 inches; **b.** Cheyenne, WY;
c. 65 inches
5. a. June; **b.** 0.6%; **c.** 12.1%
7. a. 14–24: 553,401; 25–44: 1,605,948; 45–64: 3,624,234;
65 and over: 5,067,417; **b.** 1,443,183;
c. 85.2%; 9,245,052 people
9. a. 1970 to 1975; **b.** 5.5%; **c.** 2,546,000;
d. 1.75 times greater
11. a. Arizona and Louisiana; 1,000,000 bales;
b. 1,300,000 bales; **c.** Arkansas; **d.** 1,200,000 bales
13. a. April; 89¢ per gallon; **b.** April to June; 8¢ per gallon;
c. 5%; **d.** 98¢ per gallon
15. a. transportation: $250; miscellaneous expenses: $350;
books and supplies: $400; room and board: $3,500;
tuition: $8,000; **b.** $4,500; **c.** 8%; $1,000

Review Exercises for Chapter 6

1. true **3.** false; salesperson **5.** false; denominator
7. $\dfrac{57}{100}$ **9.** $\dfrac{41}{500}$ **11.** $\dfrac{9}{5}$ **13.** $\dfrac{3}{400}$ **15.** $\dfrac{5}{6}$ **17.** 9%

19. 80% **21.** 250% **23.** 43.75% **25.** $38\dfrac{6}{13}\%$

27. 18.6% **29.** 0.18 **31.** 0.02 **33.** 1.23 **35.** 0.0009
37. 0.182 **39.** 28% **41.** 31.9% **43.** 80% **45.** 0.43%
47. 720% **49.** 92.4 **51.** 112.5 **53.** 16% **55.** 15.5
57. 3,000% **59.** 1.35% **61.** 60% **63.** 110 people

65. $39.69 **67.** 396 students **69.** $44\dfrac{4}{9}\%$ **71.** $611.50

73. $31.96 **75.** $825 **77.** 10% **79.** 548 people
81. 14.35%
83. a. 15,018,450 families; **b.** 458,190 families
85. To convert a percent to a decimal, remove the percent sign
and move the decimal point two places to the left. For
example, 37% = 0.37. This technique works because
percent means "per one hundred," so 37% means 37
hundredths or $37 \times \dfrac{1}{100}$, which is equal to $\dfrac{37}{100}$, or 0.37.
87. To convert a fraction to a percent, change the fraction to a
decimal, move the decimal point two places to the right,
and attach a percent sign. For example, $\dfrac{5}{8} = 0.625$
$= 62.5\%$. This technique works because converting a
fraction to a decimal makes it possible to express the
number as a fraction with a denominator of 100. Since
percent means "per one hundred," the numerator of a
fraction with a denominator of 100 is the value of the
percent. For example,

$$0.625 = \frac{0.625 \times 100}{1 \times 100} = \frac{62.5}{100} = 62.5\%$$

Review Exercises for Chapter 6 (cont.)

89. Use a pie chart to display data when the data can be represented as parts of a whole, and the sum of the pieces is
100%. For example, a pie chart could be used to represent
a family budget. Each piece of the pie would represent the
percentage spent on items such as food, clothing, housing,
transportation, and miscellaneous items. Use a bar graph to
illustrate how one quantity compares to another type of
data. For example, a bar graph could be used to show the
annual family budget for each year over a 5-year period.
91. Calculate the interest on an amount of money borrowed
by multiplying the annual interest rate by the amount
borrowed by the number of years needed to pay back
the loan. For example, if someone borrows $10,000 for
2 years at an interest rate of 8%, the total interest is
$10,000 \times 2 \times .08 = $1,600.

Chapter 6 Test

1. $\dfrac{13}{25}$ **2.** $\dfrac{1}{250}$ **3.** $\dfrac{33}{400}$ **4.** 38% **5.** 137.5%

6. 7.5% **7.** 0.64 **8.** 0.0275 **9.** 54% **10.** 0.28%
11. 57 **12.** 600 **13.** 20% **14.** 28 **15.** $8,100
16. $490 **17.** $10 **18.** 5% **19.** 9.5% **20.** $28
21. $105 **22.** 30% **23.** 216 calories **24.** 198 calories
25. 1,170 calories

7.1 Exercises

1. true **3.** false; greater than **5.** < **7.** = **9.** >
11. < **13.** 7; positive **15.** 51; positive
17. -5.8; negative **19.** $-\dfrac{1}{5}$; negative **21.** $3\dfrac{1}{2}$; positive
23. -0.23; negative **25.** -6; negative **27.** $-a$; positive

29. 3 **31.** 51 **33.** $\dfrac{3}{5}$ **35.** 3.9 **37.** -62 **39.** $-\dfrac{5}{9}$

41. 82 **43.** -52 **45.** $-8\dfrac{7}{9}$ **47.** $-9\dfrac{3}{5}$ **49.** -5.4

51. -1.85 **53.** -5.48 **55.** -28 **57.** = **59.** >
61. > **63.** = **65.** > **67.** < **69.** > **71.** = **73.** >

7.2 Exercises

1. false; negative **3.** true **5.** 2 **7.** -11 **9.** 1
11. -14 **13.** 11 **15.** -41 **17.** 21 **19.** -25 **21.** 44
23. -82 **25.** -2.5 **27.** 2.9 **29.** 6.6 **31.** -0.12
33. 222 **35.** $\dfrac{2}{9}$ **37.** $\dfrac{1}{8}$ **39.** $-\dfrac{1}{4}$ **41.** $1\dfrac{1}{4}$ **43.** -5

7.2 Exercises (cont.)

45. 1   **47.** 6   **49.** -535 **51.** $\dfrac{2}{3}$ **53.** 34,335,564

55. -1.71271 **57.** -28 **59.** $\dfrac{3}{5}$ **61.** -0.675

63. 534,987   **65.** $-7°\,\text{F}$ **67.** 14,494 feet
69. 15-yard line

7.3 Exercises

1. true   **3.** false; greater than   **5.** -2 **7.** -6 **9.** -7
11. 14   **13.** -1 **15.** -23 **17.** 36   **19.** -1.8
21. 14.1   **23.** -751 **25.** -0.334 **27.** -0.525

29. $-2,266$ **31.** $-\dfrac{1}{4}$ **33.** $-\dfrac{3}{2}$ **35.** $-8\dfrac{5}{8}$

37. $-2,637,949$ **39.** -12.16497 **41.** 2   **43.** 0

45. -5 **47.** 7   **49.** -3.9 **51.** -50 **53.** $\dfrac{1}{8}$ **55.** $4\dfrac{1}{2}$

57. 2,164,054   **59.** $11°\,\text{F}$ **61.** $44,920

7.4 Exercises

1. true   **3.** true   **5.** -21 **7.** 30   **9.** -0.21

11. -24.6 **13.** -1.68 **15.** -0.66 **17.** $\dfrac{3}{10}$ **19.** $\dfrac{1}{10}$

21. 3   **23.** -9 **25.** 45   **27.** 0   **29.** -54 **31.** -24

33. 48   **35.** $\dfrac{1}{12}$ **37.** $\dfrac{1}{12}$ **39.** -420 **41.** 240

43. -144 **45.** 76.93443   **47.** 146,730.24   **49.** 81
51. 1   **53.** 81   **55.** -64 **57.** 10,000   **59.** -20
61. -35 **63.** -300 **65.** 37   **67.** -30 **69.** -62
71. 73   **73.** 34   **75.** $-2,409$ **77.** $-307,695$

7.5 Exercises

1. false; positive   **3.** true   **5.** -8 **7.** 8   **9.** -5

11. 12   **13.** 3   **15.** -0.09 **17.** -0.8 **19.** $-\dfrac{1}{6}$

21. $-\dfrac{1}{10}$ **23.** $-1\dfrac{41}{55}$ **25.** 12   **27.** $\dfrac{3}{8}$ **29.** $-\dfrac{3}{4}$

31. $-\dfrac{1}{3}$ **33.** $\dfrac{10}{3}$ **35.** 1   **37.** $-\dfrac{3}{4}$ **39.** $-\dfrac{9}{7}$ **41.** $-\dfrac{1}{4}$

43. -5 **45.** 1   **47.** 9   **49.** 4   **51.** -29 **53.** -3

55. -7 **57.** $-\dfrac{2}{3}$ **59.** -8 **61.** $\dfrac{6}{7}$ **63.** $\dfrac{2}{3}$

Review Exercises for Chapter 7

1. incorrect; 2   **3.** incorrect; 63   **5.** correct
7. correct   **9.** incorrect; -4 **11.** $<$ **13.** $=$ **15.** $<$
17. $>$ **19.** $=$ **21.** $<$ **23.** $<$ **25.** 2   **27.** -5
29. 42   **31.** -3 **33.** -5 **35.** -69 **37.** -0.56

39. -40 **41.** $\dfrac{1}{6}$ **43.** $\dfrac{13}{30}$ **45.** $-\dfrac{1}{6}$ **47.** $\dfrac{1}{6}$ **49.** -1

51. -10 **53.** 42   **55.** -180 **57.** $\dfrac{8}{3}$ **59.** $-\dfrac{9}{7}$

61. -21 **63.** 26   **65.** -80 **67.** $-4,700$ **69.** -28

71. -9 **73.** $\dfrac{4}{3}$ **75.** $-84,260$ **77.** $-10,610,873$

79. The "$-$" sign may be used to indicate the operation subtraction, as in the equation $9 - 2 = 7$. The "$-$" sign may also be used to indicate a negative number, such as in the number -5, which is located five units to the left of zero on a number line. The "$-$" sign may also be used to indicate the opposite of a number, such as in the first "$-$" sign in the expression $-(-3)$, which is read as "the opposite of negative 3," which is equal to 3.

81. You have to be more careful when adding positive and negative numbers than when just adding positive numbers because when you add positive numbers you just need to perform the operation addition, and the sign of the answer is always positive. When you add positive and negative numbers, however, you need to find the difference between their absolute values and then determine the sign of the answer, which may be positive or negative depending upon which number has the larger absolute value. For example, to add $5 + 9$ the answer is simply the sum of 5 and 9, which is 14. To add $5 + (-9)$, you need to find the difference between the absolute values of 5 and -9, which is equal to the difference between 9 and 5, which is 4. Since the absolute value of -9 is greater than the absolute value of 5, the answer is negative: $5 + (-9) = -4$.

83. To subtract a negative number from a positive number, you add the opposite of the negative number (which is a positive number) to the positive number. For example, to subtract -3 from 8 you add a positive 3 to 8, which is 11: $8 - (-3) = 8 + (+3) = 11$.

Chapter 7 Test

1. $<$ **2.** $=$ **3.** $<$ **4.** 5   **5.** -2.3 **6.** $-\dfrac{7}{8}$ **7.** 2

8. -11.6 **9.** $1\dfrac{4}{15}$ **10.** 48   **11.** -1.26 **12.** $\dfrac{4}{21}$

13. -7 **14.** 8   **15.** $-\dfrac{3}{2}$ **16.** 6   **17.** 70   **18.** $\dfrac{1}{6}$

19. -24 **20.** -18

8.1 Exercises

1. false; perfect square   3. false; positive   5. 1   7. 7
9. -9 11. ± 8 13. 12   15. -3 17. 18   19. 15
21. 17   23. 13   25. 30   27. 1.41   29. 2.24   31. 3.87
33. 9.22   35. 7.87   37. 7.94   39. 24   41. 27   43. 32
45. 64   47. 71   49. 62   51. 95   53. 2.32   55. 7.84
57. 13.71   59. 27.60   61. 23.54   63. 25.27   65. 73.13
67. 168.60   69. 0.66   71. 0.18   73. 0.05   75. 2,713.89

8.2 Exercises

1. false; product   3. true   5. 20   7. 50   9. 90
11. 110   13. 800   15. $\dfrac{1}{5}$ 17. $\dfrac{2}{9}$ 19. $\dfrac{6}{11}$ 21. $\dfrac{8}{13}$
23. 0.5   25. 0.7   27. 1.1   29. 0.3   31. 0.08   33. 0.03
35. $4\sqrt{3}$ 37. $2\sqrt{5}$ 39. $5\sqrt{2}$ 41. $4\sqrt{5}$ 43. $7\sqrt{2}$
45. $8\sqrt{2}$ 47. $6\sqrt{10}$ 49. $12\sqrt{2}$

8.3 Exercises

1. 5.2   3. 6.7   5. 31.4   7. 36.1   9. 357   11. 794
13. 520   15. 0.72   17. 0.29   19. 0.694   21. 0.068
23. 1,200   25. 2.4   27. 4.2   29. 8.24   31. 30.4
33. 19.30   35. 65.7   37. 242   39. 0.85   41. 0.52
43. 0.114   45. 0.21   47. 0.07   49. 2,596

8.4 Exercises

1. 3   3. 2   5. 4   7. 5   9. 10   11. $\dfrac{1}{2}$ 13. $\dfrac{1}{10}$
15. 0.2   17. 0.4   19. -2 21. -8 23. $-\dfrac{1}{4}$ 25. $2\sqrt[3]{2}$
27. $2\sqrt[3]{3}$ 29. $5\sqrt[3]{2}$ 31. $3\sqrt[3]{10}$ 33. $10\sqrt[4]{2}$ 35. $10\sqrt[5]{5}$

8.5 Exercises

1. true   3. true   5. false; a rational number
7. false; real
9. integers, rational numbers, real numbers
11. irrational numbers, real numbers
13. rational numbers, real numbers
15. irrational numbers, real numbers
17. rational numbers, real numbers
19. rational numbers, real numbers
21. counting numbers, whole numbers, integers, rational
   numbers, real numbers

8.5 Exercises (cont.)

23. rational numbers, real numbers
25. irrational numbers, real numbers
27. whole numbers, integers, rational numbers, real numbers
29. $<$ 31. $=$ 33. $>$ 35. $=$ 37. $<$ 39. $<$ 41. $<$
43. $>$ 45. $>$ 47. $=$

Review Exercises for Chapter 8

1. 7   3. $\pm\dfrac{3}{5}$ 5. -0.6 7. -70 9. -0.03 11. 100
13. 3.7   15. 18   17. 127   19. 0.28   21. 6,580
23. 0.365   25. 2.43   27. 25.2   29. 197   31. 0.530
33. 7,331   35. 0.797   37. $10\sqrt{2}$ 39. $2\sqrt{2}$ 41. $2\sqrt{6}$
43. 2   45. 3   47. -1
49. irrational numbers, real numbers
51. rational numbers, real numbers
53. rational numbers, real numbers
55. rational numbers, real numbers
57. counting numbers, whole numbers, integers, rational
   numbers, real numbers
59. irrational numbers, real numbers   61. $>$ 63. $=$
65. $=$ 67. $>$ 69. $>$ 71. $>$
73. The perfect square of a number is a number whose square
   root is a whole number. For example, 49 is a perfect square
   because the square root of 49 is 7, which is a whole
   number. $\sqrt{49} = 7$ because $7 \cdot 7 = 49$.
75. All integers are rational numbers because any integer can
   be expressed as a fraction in the form of $\frac{a}{b}$, where a and b
   are integers and $b \neq 0$. Every integer can be written as a
   fraction with a numerator equal to the integer and a denominator equal to 1. For example, 7 is both an integer and
   a rational number because 7 can be written as $\frac{7}{1}$. -53 is
   both an integer and rational number because -53 can be
   written as $-\frac{53}{1}$.
77. Rational numbers can be written in the form $\frac{a}{b}$, where a
   and b are integers and $b \neq 0$. Irrational numbers cannot be
   expressed in this form. For example, $\frac{3}{8}$ is a rational number
   since both 3 and 8 are integers. $\sqrt{5}$ is an irrational number
   because it cannot be expressed in the form $\frac{a}{b}$, where a and
   b are integers and $b \neq 0$.

Chapter 8 Test

1. 5   2. -0.9 3. $\dfrac{3}{8}$ 4. ± 70 5. 0.06   6. $6\sqrt{2}$
7. $4\sqrt{3}$ 8. 2   9. -3 10. $\dfrac{1}{10}$
11. counting numbers, whole numbers, integers, rational
   numbers, real numbers

Chapter 8 Test (cont.)

12. integers, rational numbers, real numbers
13. rational numbers, real numbers
14. real numbers
15. rational numbers, real numbers
16. = **17.** > **18.** < **19.** < **20.** =

9.1 Exercises

1. true **3.** false; centimeters **5.** 96 **7.** $2\frac{1}{3}$ **9.** 17

11. 86 **13.** $7\frac{1}{3}$ **15.** 3,168 **17.** 282 **19.** 9 **21.** 880

23. 9.5 **25.** 15 **27.** 500 **29.** 19,000 **31.** 3,500
33. 38 **35.** 5.3 **37.** 700 **39.** 5.9 **41.** 20,000
43. 85.7 **45.** 72 **47.** 81 **49.** 700,000 **51.** 9 cm
53. 91 m **55.** 68.5 cm **57.** 3.2 m **59.** 15 cm
61. 12,755 km **63.** 14; 13.78 **65.** 150; 152.4
67. 2; 1.829 **69.** 720; 724.1 **71.** 450; 457.2
73. Minneapolis **75.** 26 tiles **77.** 66.6 ft **79.** $15.94

9.2 Exercises

1. false; 16 **3.** false; pounds **5.** 2 **7.** 1,000 **9.** 124
11. 137 **13.** 106 **15.** 86 **17.** 0.125 **19.** 5 **21.** 6,800
23. 2.4 **25.** 5,000 **27.** 3.5 **29.** 5,000 **31.** 0.0067
33. 0.75 **35.** 347,000 **37.** 4,180 **39.** 3,000,000
41. 0.088 **43.** 0.0095 **45.** 625 **47.** 70,000 **49.** 454 g
51. 300 g **53.** 5 g **55.** 1 t **57.** 15; 13.61 **59.** 8.8; 8.82
61. 7.2; 6.532 **63.** 21,000; 19,845 **65.** 96; 105.8
67. 50 g **69.** chocolate bar without nuts **71.** 4.5 kg
73. 420-g package

9.3 Exercises

1. false; more than **3.** false; 100 **5.** 144 **7.** $2\frac{1}{2}$

9. 1.75 **11.** 21 **13.** 12 **15.** 11 **17.** 74 **19.** $1\frac{5}{8}$

21. $1\frac{1}{8}$ **23.** 3 **25.** 0.85 **27.** 0.062 **29.** 57,600

31. 8.8 **33.** 897 **35.** 7,500 **37.** 5,750 **39.** 3.62
41. 69 **43.** 0.92 **45.** 2,100 mL **47.** 3.8 L **49.** 70 L
51. 0.4 kL **53.** 9; 8.517 **55.** 5; 5.073 **57.** 32; 30.43
59. 20; 20.29 **61.** 100; 101.5 **63.** milk **65.** 24 doses
67. 1.893 L **69.** $0.50

9.4 Exercises

1. false; C **3.** true **5.** 38°C **7.** 190°C **9.** 30°C
11. 20 **13.** −5 **15.** 95 **17.** 239 **19.** −20 **21.** 23
23. 153 **25.** 75 **27.** 11 **29.** 54 **31.** 28 **33.** 22; 25
35. 142; 149 **37.** 14; 15 **39.** 192; 194 **41.** −36; −40

43. −2; $-2\frac{2}{9}$ **45.** 294; $285\frac{4}{5}$ **47.** 40°C **49.** 200°C

Review Exercises for Chapter 9

1. false; hundredth **3.** true **5.** 23.8 m **7.** 232 g

9. 2 L **11.** 6 cm **13.** 37°C **15.** $\frac{2}{3}$ **17.** 2.25 **19.** 3

21. 0.3 **23.** 8.2 **25.** 53,000 **27.** 66 **29.** 10,500

31. $1\frac{2}{3}$ **33.** 920 **35.** 230 **37.** 860 **39.** 10 **41.** 185

43. 75 **45.** 18,000 **47.** 0.063 **49.** $-6\frac{1}{9}$ **51.** 60; 59.06

53. 20; 19.96 **55.** 12; 11.36 **57.** 25; $22\frac{2}{9}$

59. 16 sandwiches **61.** 8,914 km **63.** 237 mL bottle
65. 102.2°F
67. To convert a measurement from inches to feet, multiply the measurement by the unit fraction $\frac{1 \text{ ft}}{12 \text{ in.}}$ so that inches cancel. For example, to convert 18 inches to feet, multiply 18 inches by the unit fraction $\frac{1 \text{ ft}}{12 \text{ in.}}$ as follows:

$$18 \text{ in.} = 18 \text{ in.} \times \frac{1 \text{ ft}}{12 \text{ in.}} = \frac{18}{12} \text{ ft} = 1\frac{1}{2} \text{ ft.}$$

69. To convert a measurement from gallons to quarts, multiply the measurement by the unit fraction $\frac{4 \text{ qt}}{1 \text{ gal}}$ so that gallons cancel. For example, to convert 5 gallons to quarts, multiply 5 gallons by the unit fraction $\frac{4 \text{ qt}}{1 \text{ gal}}$ as follows:

$$5 \text{ gal} = 5 \text{ gal} \times \frac{4 \text{ qt}}{1 \text{ gal}} = 20 \text{ qt.}$$

71. You would measure the size of a can of soda in milliliters rather than in hectoliters since a can of soda is about equal to a third of a liter, which is much less than 1 hectoliter, or 100 liters. For example, a typical can of soda measures 355 milliliters. This is equal to 0.00355 hectoliters, so the measurement in milliliters is easier to understand.

Chapter 9 Test

1. $6\frac{2}{3}$ **2.** 96 **3.** 10 **4.** 2,640 **5.** 48 **6.** 60

7. 70,000 **8.** 0.4 **9.** 3.2 **10.** 520,000 **11.** 8.8
12. 30 **13.** 18 **14.** 90 **15.** 150 **16.** 5 **17.** 86
18. 23 **19.** Omaha **20.** 75

10.1 Exercises

1. false; perpendicular **3.** false; two **5.** false; 90°
7. false; five **9.** false; 90° **11.** true **13.** 45°
15. 180° **17.** 30° **19.** parallelogram **21.** right triangle
23. quadrilateral **25.** pentagon

10.2 Exercises

1. true **3.** false; sum **5.** 17 m **7.** 14 mi **9.** 33 mm
11. 26 ft **13.** 28 in. **15.** 24 km **17.** 4.8 m **19.** $17\frac{1}{2}$ yd
21. 18 cm **23.** 60 mm **25.** 45 dm **27.** 66 in.
29. 36 ft **31.** 156 cm **33.** 13 ft **35.** 21.84 m

10.3 Exercises

1. false; square **3.** true **5.** 108 ft^2 **7.** 56.7 cm^2
9. $11\frac{7}{8}$ mi^2 **11.** 120 ft^2 **13.** 1.69 ft^2 **15.** 1,936 cm^2
17. $21\frac{7}{9}$ yd^2 **19.** 7.56 dm^2 **21.** 10 yd^2 = 90 ft^2
23. 3 yd^2 **25.** $3\frac{15}{16}$ m^2 **27.** 3 ft^2 **29.** 45 ft^2
31. 16.43 cm^2 **33.** 50 in.^{2} **35.** 2 **37.** 288 **39.** 3
41. $0.75 = \dfrac{3}{4}$ **43.** 700 **45.** $0.25 = \dfrac{1}{4}$ **47.** 32,670
49. 0.0008 **51.** 26.25 ft^2 = $26\frac{1}{4}$ ft^2 **53.** 486 in.^{2}
55. 875 tiles **57.** 1,500 ha **59.** $526.80 **61.** 12 in.
63. $1\frac{3}{5}$ cm **65.** 3 in. **67.** 9 dm **69.** $60\frac{1}{8}$ ft^2 **71.** 25 ft^2
73. 30 in.^{2}

10.4 Exercises

1. 10 in. **3.** 20 m **5.** 12 yd **7.** 1.5 cm **9.** 1,500 mm
11. 21.7 cm **13.** 40.3 yd **15.** no **17.** yes **19.** yes
21. yes **23.** 13 ft **25.** 1,000 mi **27.** 16 ft **29.** 13.5 m^2
31. $a = 12$ mm; $b = 9$ mm; perimeter = 42 mm;
area = 84 mm^2

10.5 Exercises

1. false; radius **3.** false; diameter **5.** 31π in.
7. 14.4π cm **9.** 198 in. **11.** 11 km **13.** 21.98 ft
15. $27\frac{19}{40}$ m **17.** 5.28 hm **19.** 1,701.88 ft **21.** 361π dm^2
23. 6.76π km^2 **25.** 15,400 mm^2 **27.** $21\frac{21}{32}$ yd^2
29. 282,600 cm^2 **31.** $55\frac{11}{25}$ hm^2 **33.** 28.26 ft^2
35. 6,443.5626 mi^2 **37.** 69.08 in. **39.** $9\frac{5}{8}$ ft^2
41. 81.64 in. **43.** 37 cm **45.** 14,826 ft^2 **47.** 168.96 in.^{2}
49. 487.485 m^2

10.6 Exercises

1. false; cubic **3.** false; feet **5.** 96 in.^{3} **7.** $116\frac{2}{3}$ yd^3
9. $13\frac{1}{2}$ ft^3 **11.** 64 cm^3 **13.** 1.728 in.^{3} **15.** $11\frac{25}{64}$ dm^3
17. 36π cm^3 **19.** 15.7 ft^3 **21.** 192π dm^3 = 192,000 cm^3
23. $457\frac{1}{3}\pi$ ft^3 **25.** 113.04 cm^3 **27.** 562.5π in.^{3}
29. 1,526.04 ft^3 **31.** 3,140 ft^3 **33.** 200.96 in.^{3}
35. $\frac{1}{27}\pi$ ft^3 = 64π in.^{3} **37.** 312.9 cm^3 **39.** 5,184
41. $115\frac{1}{2}$ **43.** 9,000 **45.** 7 **47.** 19 **49.** 60 **51.** 0.8
53. 600,000,000 **55.** 523,600 gal **57.** 4,851 in.^{3}
59. 462 mL **61.** 376.8 ft^3 **63.** 1,582.56 in.^{3}

Review Exercises for Chapter 10

1. false; never **3.** false; area **5.** false; perimeter
7. 1,280 **9.** 4 **11.** 4 **13.** 0.025 **15.** 345.6 **17.** 0.1
19. 14 **21.** 5.4 **23.** 22 cm **25.** $17\frac{3}{4}$ in. **27.** 14π mm
29. 12 ft **31.** 63 ft^2 **33.** 57 m^2 **35.** 30 ft^2 **37.** 24 m^2
39. 20.25π in. **41.** 27 dm^2 = 2,700 cm^2 **43.** 512 in.^{3}
45. 36π in.^{3} **47.** 8π m^3 **49.** 4,050 cm^3 **51.** 6.1 m
53. 5 ft **55.** 420 ft^2 **57.** 37.5 cm **59.** 60 ft **61.** 7.1 in.
63. 62 m **65.** 4 cm **67.** 8 yd **69.** 24,891 mi
71. 16.7 L
73. Perimeter is the distance around a figure and area is a measure of the flat surface covered by a figure. For example, measuring the distance around a backyard in yards to determine how much fencing is needed to enclose it is an application of perimeter. Measuring the region covered by a living room floor in square feet to determine how much carpeting is needed to cover it is an application of area.
75. To find the hypotenuse of a right triangle, given the measurement of its two legs, you find the square of each leg, add those two numbers together, and take the square root of that result. For example, to find the hypotenuse of a right triangle with legs measuring 6 ft and 8 ft, find 6^2, which is 36, and 8^2, which is 64. Then add 36 and 64 to obtain 100, and take the square root of 100, which is 10. The hypotenuse is therefore 10 ft.
77. If the side of a square is the same as the side of an equilateral triangle, the square would have the larger perimeter. This is because the perimeter is the sum of the measurement of all the sides of a figure, and a square has four sides while an equilateral triangle has three sides. For example, a square whose sides measure 5 inches has a perimeter of $4 \cdot 5 = 20$ inches, and an equilateral triangle whose sides measure 5 inches has a perimeter of $3 \cdot 5 = 15$ inches.

Chapter 10 Test

1. 40,000   **2.** 83   **3.** 135   **4.** 30,000   **5.** 18 m
6. 42 in.   **7.** 18π cm   **8.** 48 ft   **9.** 12 dm^2 **10.** 9 ft^2
11. 12.25π m^2 **12.** 60 in.^{2} **13.** 64 in.^{3} **14.** 20π cm^3
15. 36π yd^3 **16.** 24π dm^3 **17.** 6 ft   **18.** 8π ft   **19.** 7 m
20. 8 cm

11.1 Exercises

1. true   **3.** true   **5.** terms, 1; variable, w; coefficient, 3
7. terms, 2; variable, x; coefficient, 2; variable, y;
  coefficient, 5
9. terms, 3; variable, p; coefficient, 3; variable, q; coefficient,
  9; variable, r; coefficient, 1
11. terms, 2; variable, c; coefficient, $\frac{1}{2}$; variable, d;
  coefficient, -1
13. 4   **15.** 27   **17.** -4 **19.** -4 **21.** 6   **23.** -90
25. 170   **27.** -24 **29.** $x - 5$ **31.** $-2t$ **33.** $\frac{2}{3}z$
35. $x - 35$ **37.** $7a + 5$ **39.** $-8(a + b)$
41. $(14 - x) \div 2$ **43.** $2(a + 5)$ **45.** $\frac{5}{8}(u + 7)$
47. four less than x or x decreased by 4
49. the product of six and a **51.** nine more than t
53. w divided by eleven
55. eight more than the product of two and y
57. two less than the product of three and n
59. five times the quantity z minus seven
61. the quantity four minus u, divided by three
63. three-eighths the quantity x plus five

11.2 Exercises

1. false; variable   **3.** true   **5.** $17x$ **7.** $2a$ **9.** $-8t$
11. $\frac{1}{2}a$ **13.** $-53u$ **15.** $2.3p$ **17.** $15y$ **19.** $14z$
21. $43c$ **23.** $-14.2u$ **25.** $5y + 7$ **27.** $43 - 8n$
29. $1 - 9w$ **31.** $5 - 5p$ **33.** $-3x + 20$ **35.** $-11s + 6$
37. -12 **39.** $68z - 64$ **41.** $-22t + 40$ **43.** $-62k - 9$

11.3 Exercises

1. false; solution   **3.** true   **5.** yes   **7.** no   **9.** no
11. no   **13.** yes   **15.** no   **17.** $x = 8$ **19.** $y = -3$
21. $w = -2$ **23.** $x = -2$ **25.** $y = 8$ **27.** $a = 45$
29. $z = 81$ **31.** $x = 4.8$ **33.** $x = -5$ **35.** $y = 4$

11.3 Exercises (cont.)

37. $z = 3$ **39.** $t = -4$ **41.** $t = 4$ **43.** $p = -13$
45. $w = -12$ **47.** $x = -11$ **49.** $x = -2$ **51.** $x = -9$
53. $c = -8$ **55.** $d = -11$ **57.** $x = 7$ **59.** $w = 2$
61. $a = 6$ **63.** $b = -9$ **65.** $x = 40$ **67.** $x = -24$
69. $x = -44$ **71.** $x = 13$ **73.** $x = 30$ **75.** $t = -8$
77. $x = 16$ **79.** $x = 11$

11.4 Exercises

1. $x = 7$ **3.** $x = -9$ **5.** $x = 8$ **7.** $x = -9$ **9.** $z = 5$
11. $p = \frac{19}{7}$ **13.** $a = -6$ **15.** $x = 0$ **17.** $w = 15$
19. $x = 15$ **21.** $x = -72$ **23.** $x = 28$ **25.** $t = -4$
27. $y = \frac{35}{18}$ **29.** $z = \frac{27}{28}$ **31.** $m = 8$ **33.** $y = -5$
35. $x = 6$ **37.** $x = -9$ **39.** $t = 9$ **41.** $w = 4$
43. $z = -5$ **45.** $y = 8$ **47.** $r = -3$

11.5 Exercises

1. true   **3.** false; distributive   **5.** $y = 3$ **7.** $x = -9$
9. $z = -10$ **11.** $t = \frac{1}{3}$ **13.** $b = \frac{9}{5}$ **15.** $n = 2$
17. $x = 3$ **19.** $x = 1$ **21.** $x = \frac{9}{7}$ **23.** $a = \frac{3}{2}$
25. $u = -3$ **27.** $t = \frac{3}{7}$ **29.** $s = 8$ **31.** $p = 2$
33. $x = -4$ **35.** $x = -4$ **37.** $y = 4$ **39.** $y = -1$
41. $t = \frac{13}{4}$ **43.** $y = 5$ **45.** $y = 18$ **47.** $p = 3$
49. $z = 2$ **51.** $z = \frac{8}{3}$ **53.** $z = -20$ **55.** $t = 6$
57. $s = 2$ **59.** $x = -10$ **61.** $x = -2$

11.6 Exercises

1. 18 years old   **3.** 3 and 29   **5.** 6   **7.** 27 and 28
9. 18 men   **11.** 660 miles   **13.** 1.75 hr
15. length, 8 ft; width, 3 ft   **17.** $22,500
19. candy bar, 35¢; pen, 49¢; notebook, 98¢
21. 2 in., 6 in., 6 in.   **23.** 3 lb   **25.** $3,000,000
27. $1,000   **29.** $800,000

Review Exercises for Chapter 11

1. 11   **3.** -56 **5.** -1 **7.** $x - 7$ **9.** $5(z + 6)$
11. $(7 + t) \div 3$ **13.** the product of negative five and y
15. three times the quantity x minus two
17. two less than the product of six and w **19.** $-5x$
21. $-3y - 8$ **23.** $8x - 20$ **25.** $-2a - 29$ **27.** $x = 8$

29. $u = 3$ **31.** $z = -8$ **33.** $p = 5$ **35.** $a = \dfrac{1}{2}$

37. $m = -15$ **39.** $k = -30$ **41.** $x = \dfrac{9}{35}$ **43.** 33 and 35

45. $3\frac{1}{2}$ hr   **47.** length, 12 cm; width, 4 cm   **49.** 750 ft
51. A constant is a number or quantity that does not change in
value. A variable is an unknown quantity that may have
different values and is usually represented by a letter. The
number 9 is a constant and the letter x is often used to represent a variable.
53. The solution to a linear equation is the value that can be
substituted for the unknown variable so that the resulting
statement is true. For example, the number 5 is a solution
to the equation $2x - 7 = 3$ since when 5 is substituted for
x, the resulting statement, $3 = 3$, is true.
55. The addition principle of equality states that if the same
quantity is added to both sides of an equation, the resulting
equation is equivalent to the original one. For example, to
solve the equation $x - 9 = 17$, we can add 9 to both sides
of the equation to obtain the equivalent equation $x = 26$.

Chapter 11 Test

1. 10   **2.** -8 **3.** -4 **4.** $x - 2$ **5.** $3t + 8$
6. $5(w + 6)$ **7.** 8 more than y
8. 2 times the quantity x minus 7   **9.** $-6x$ **10.** $a + 9$
11. $21t - 13$ **12.** $x = -12$ **13.** $a = -54$ **14.** $x = 9$
15. $t = 4$ **16.** $y = 5$ **17.** 4 years old   **18.** 9 and 29
19. 25 men, 18 women   **20.** width, 11 in.; length, 13 in.

12.1 Exercises

1. $x > 7$

3. $y \le -2$

12.1 Exercises (cont.)

5. $t \ge \dfrac{1}{3}$

7. $p < -4.9$

9. $x \le 9$

11. $x > 6$

13. $x \le 12$

15. $y \ge 9$

17. $t > 3$

19. $t \le 10$

21. $t > -7$

23. $x \le 4$

25. $t > -3$

27. $y > 32$

12.1 Exercises (cont.)

29. $x \le -24$

31. $p > 3$

33. $z \ge \dfrac{5}{2}$

35. $x \ge 2$

37. $x > -2$

39. $y \ge -\dfrac{3}{8}$

41. $t > \dfrac{11}{3}$

43. $t \le 7$

45. $y \le -\dfrac{19}{6}$

47. $x < 5$

49. $y > \dfrac{6}{5}$

12.1 Exercises (cont.)

51. $t \le -\dfrac{43}{2}$

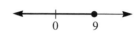

53. $w \le 9$

55. $x < -5$

57. $x > 2$

59. $x \le 7$

12.2 Exercises

1. true   **3.** false; abscissa   **5.** false; less
7. false; x-coordinate
9–27.

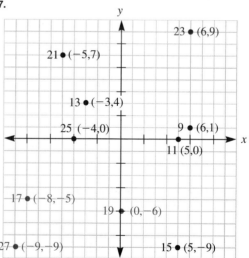

29. $(5, 4)$ **31.** $(-3, 6)$ **33.** $(4, 0)$ **35.** $(-7, 0)$
37. $(-5, -9)$

12.3 Exercises

1. true   **3.** true   **5.** yes   **7.** no   **9.** yes   **11.** yes
13. no   **15.** $-2; 0$ **17.** $2; 5$ **19.** $-5; -2$ **21.** $17; -8$
23. $2; 35$ **25.** $0; 4$ **27.** $0; 6$ **29.** 7; any real number

31. $(8, 0); (0, -8)$ **33.** $(0, 0)$ **35.** $\left(-\dfrac{2}{3}, 0\right); (0, 2)$

37. $(8, 0); (0, 2)$ **39.** $(5, 0); (0, -2)$ **41.** $(7, 0); (0, 3)$

43.

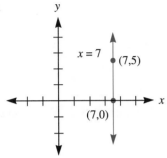

45.

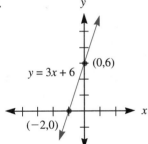

47.

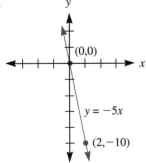

12.3 Exercises (cont.)

49.

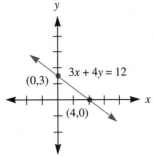

51.

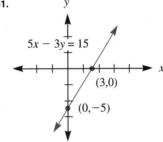

53.

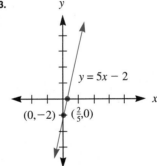

55.

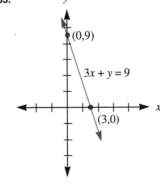

12.3 Exercises (cont.)

57.

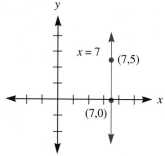

59.

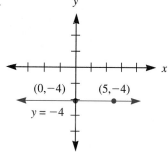

61.

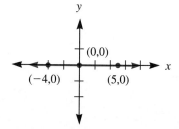

63. $x = 2$ **65.** $y = \dfrac{2}{3}$ **67.** $0; y = 0$

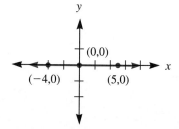

12.3 Exercises (cont.)

69. The lines are parallel.

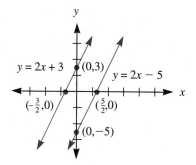

Review Exercises for Chapter 12

1. true   **3.** true   **5.** true

7. $x < -2$

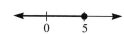

9. $t \le -1$

11. $y < \dfrac{9}{4}$

13. $z > \dfrac{27}{7}$

15. $w \le 5$

17.

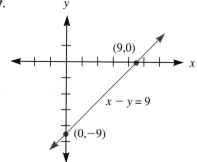

Review Exercises for Chapter 12 (cont.)

19.

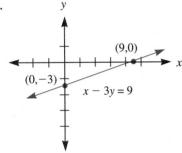

$3x + 8y = 24$
$(0,3)$
$(8,0)$

21.

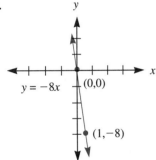

$(9,0)$
$(0,-3)$
$x - 3y = 9$

23.

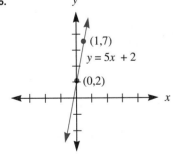

$y = -8x$
$(0,0)$
$(1,-8)$

25.

$(1,7)$
$y = 5x + 2$
$(0,2)$

Review Exercises for Chapter 12 (cont.)

27.

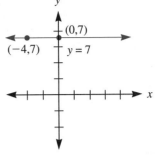

$(0,7)$
$(-4,7)$
$y = 7$

29. To find the x-intercept of an equation, set y equal to zero and solve for x. For example, to find the x-intercept of the equation $3x - y = 18$, set y equal to zero and solve for x, to obtain $x = 6$. The x-intercept is the point $(6, 0)$.

31. The abscissa is the x-coordinate, or the first number, in an ordered pair. The ordinate is the y-coordinate, or the second number, in an ordered pair. For example, in the ordered pair $(-5, 8)$ the abscissa is -5 and the ordinate is 8.

33. The multiplication principle of inequalities states that if both sides of an inequality are multiplied by the same positive number, the inequality remains unchanged. If both sides of an inequality are multiplied by the same negative number, the inequality reverses. For example, if both sides of the inequality $5x < -30$ are multiplied by $\frac{1}{5}$, the resulting inequality is $x < -6$. If both sides of the inequality $-2x < -14$ are multiplied by $-\frac{1}{2}$, the resulting inequality is $x > 7$.

Chapter 12 Test

1.

$x \leq 3$

0   3

2.

$y > -6$

-6 0

3.

$t \geq 8.3$

0   8.3

4.

$w < -\frac{2}{3}$

$-\frac{2}{3}$ 0

5.

$x \leq 10$

0   10

Chapter 12 Test (cont.)

6. $y > 5$

7. $x < 3$

8. $x \geq 4$

9. $w > -\dfrac{4}{3}$

10. $z \geq -6$

11. $x < 5$

12. $t \geq 4$

13. $(10, 0); (0, -2)$ **14.** $(6, 0); (0, 4)$ **15.** $\left(-\dfrac{2}{7}, 0\right); (0, 2)$

16.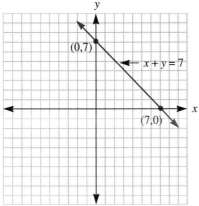

17.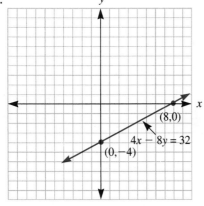

Chapter 12 Test (cont.)

18.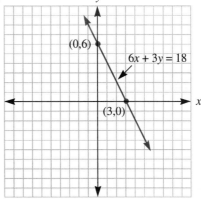

19.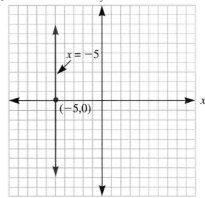

20.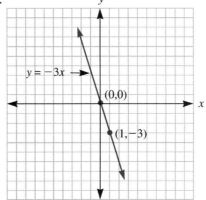

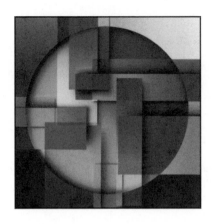

Index